Tissue Expansion

Tissue Expansion

ROLF E. A. NORDSTRÖM, M.D., Ph.D.

Professor of Plastic and Reconstructive Surgery,
University of Tromsø, Norway;
Chairman, Department of Plastic and Reconstructive Surgery,
University Hospital of Tromsø, Norway;
Chief, The Nordström Hospital for Plastic and Reconstructive Surgery,
Helsinki, Finland

BUTTERWORTH–HEINEMANN

Boston • Oxford • Johannesburg • Melbourne • New Dehli • Singapore

Every effort has been made to ensure that the drug dosage schedules within this text are accurate and conform to standards accepted at time of publication. However, as treatment recommendations vary in the light of continuing research and clinical experience, the reader is advised to verify drug dosage schedules herein with information found on product information sheets. This is especially true in cases of new or infrequently used drugs.

Recognizing the importance of preserving what has been written, Butterworth-Heinemann prints its books on acid-free paper whenever possible.

Library of Congress Cataloging-in-Publication Data

Tissue expansion / [edited by] Rolf E. A. Nordstrom.
p. cm.
Includes bibliographical references and index.
ISBN 0-7506-9711-3
1. Tissue expansion. I. Nordstrom, Rolf E. A.
[DNLM: 1. Tissue Expansion—methods. WO 600 T616 1996]
RD119.5.T57T57 1996
617.9'5—dc20
DNLM/DLC
for Library of Congress 96-6179
CIP

British Library Cataloguing-in-Publication Data
A catalogue record for this book is available from the British Library.

The publisher offers discounts on bulk orders of this book.
For information, please write:
Manager of Special Sales
Butterworth-Heinemann
313 Washington Street
Newton, MA 02158-1626
Tel: 617-928-2500
Fax: 617-928-2620

For information on all B–H medical publications available, contact our World Wide Web home page at:
http://www.bh.com/bh/

10 9 8 7 6 5 4 3 2 1

Printed in the United States of America

To my loving wife
Riitta
and my three sons
Richard, Roger, and Robert
without whose patience, support, and understanding
this book would not have seen the world

To all my colleagues and friends who have inspired and encouraged me
throughout the thorny way of my career

Contents

Contributing Authors

Jean-Pierre Aubert, M.D.
Head Registrar, Department of Plastic and Reconstructive Surgery (Head: Prof. Guy Magalon), Hospital de la Conception, Marseilles, France
6. External Valves

Eric David Austad, M.D.
Clinical Assistant Professor of Plastic Surgery, University of Michigan Medical School; Chairman, Department of Plastic Surgery, Catherine McAuley Health Center, Ann Arbor, MI
1. A History of Early Tissue Expansion; 2. Benefits, Indications, and Contraindications; 9. Face and Neck

Jaime Anger, M.D.
Chief, Pediatric Plastic Surgery Section, Hospital Infantil Darcy Vargas, Sâo Paulo, Brazil; Plastic Surgeon, Pediatric Social Section, Hospital Israelita Albert Einstein, Sâo Paulo, Brazil.
23. Burn Scar Expansion; 24. Videoscopy in Controlled Tissue Expansion

Andrew G. Batchelor, M.D., FRCS
Consultant Plastic Surgeon, Department of Plastic, Reconstructive and Hand Surgery, St. James's and S eacroft University Hospital, Leeds, England
16. Genitalia

Bruce S. Bauer, M.D., FACS
Associate Professor of Surgery, Northwestern University Medical School; Head, Division of Plastic Surgery, The Children's Memorial Hospital, Chicago, IL
22. Expanded Full-Thickness Skin Grafts

Hilton Becker, M.D., FRCS
Private Practice, Boca Raton, FL
17. Breast Expansion Augmentation; 19. The Expander Mammary Implant for Breast Reconstruction

Jincai Fan, M.D., Ph.D.
Assistant Professor of Plastic Surgery, Plastic Surgery Hospital, Chinese Academy of Medical Sciences, Beijing, China; Researcher, Dept. of Plastic and Reconstructive Surgery, University of Tromsø, Tromsø, Norway; Researcher, The Nordström Hospital for Plastic and Reconstructive Surgery, Helsinki, Finland
3. Geometric Planning; 7. Filling Policy; 25. Alternative Devices and Techniques

Ken-ichi Homma, M.D.
Chief, Section of Plastic and Reconstructive Surgery, Obihiro Kousei Hospital, Obihiro, Hokkaido, Japan
27. Muscle Vascularized Pedicle Flap (MVP Flap) Using Tissue Expanders: An Experimental Study in Rats

Guy Magalon, M.D.
Professor of Plastic and Reconstructive Surgery, University Hospital of Marseilles; Head, Department of Plastic and Reconstructive Surgery, Hospital de la Conception, Marseilles, France
6. External Valves; 14. Trunk; 15. Extremities; 21. Soft Tissue Expansion in Children

Stephen J. Mathes, M.D.
Professor of Plastic and Reconstructive Surgery, University of California, San Francisco, School of Medicine; Head, Division of Plastic and Reconstructive Surgery, University of California Hospitals and Clinics, San Francisco, CA
18. Breast Reconstruction

Rolf Münker, M.D.
Chief of Aesthetic-Plastic Surgery, Clinic Dr. Münker, Stuttgart; Chief, Aesthetic-Plastic and Hand Surgery, Euro-Med-Clinic, Fürth, Germany
10. Nose

Rolf E. A. Nordström, M.D., Ph.D.
Professor of Plastic and Reconstructive Surgery, University of Tromsø, Norway; Chairman, Department of Plastic and Reconstructive Surgery, University Hospital of Tromsø, Norway; Chief, The Nordström Hospital for Plastic and Reconstructive Surgery, Helsinki, Finland
3. Geometric Planning; 7. Filling Policy; 11. Auricle; 13. Scalp; 25. Alternative Devices and Techniques; 26. Topical Agents and Drugs to Enhance Tissue Expansion

Takehiko Ohura, M.D., Ph.D.
Professor Emeritus, Hokkaido University; President and Medical Director, Kei-Jin-Kai Medical Corporation; Honorary President, Memorial Hospital of Plastic Surgery, Sapporo, Japan
27. Muscle Vascularized Pedicle Flap (MVP Flap) Using Tissue Expanders: An Experimental Study in Rats

Edoardo Raposio, M.D., FICS
Ph.D. Student, Department of Plastic and Reconstructive Surgery, University of Tromsø, Norway; Consultant, Department of Plastic and Reconstructive Surgery, University Hospital, Tromsø, Norway; Researcher, The Nordström Hospital for Plastic and Reconstructive Surgery, Helsinki, Finland; Senior Resident, Department of Plastic and Reconstructive Surgery, National Institute for Cancer Research, University of Genova, Genova, Italy
3. Geometric Planning; 7. Filling Policy; 25. Alternative Devices and Techniques; 26. Topical Agents and Drugs to Enhance Tissue Expansion

Julien H. A. van Rappard, M.D., Ph.D.
Plastic Surgeon, Department of Plastic, Reconstructive and Hand Surgery, Catharina Hospital, Eindhoven, Holland
4. Histologic Changes; 5. Available Devices; 8. Complications: How to Avoid and Treat Them

Armand D. Versaci, M.D.
Clinical Professor of Plastic Surgery, Brown University School of Medicine; Staff Physician, Department of Plastic Surgery, Rhode Island Hospital, Providence, RI
20. Correction of Anomalies of the Breast

Frank A. Vicari, M.D., FACS, FAAP
Assistant Professor of Surgery, Northwestern University Medical School; Attending Plastic Surgeon, Department of Surgery, The Children's Memorial Hospital, Chicago, IL
22. Expanded Full-Thickness Skin Grafts

Albert R.M. Wittkampf, M.D., D.M.D.
Oral and Maxillofacial Surgeon, Department of Oral and Maxillofacial Surgery, Prosthodontics and Special Dental Care, Utrecht University and University Hospital Utrecht, Utrecht, The Netherlands
12. Alveolar Ridge

Tetsunori Yoshida, M.D., Ph.D.
Visiting Instructor of Plastic and Reconstructive Surgery, School of Medicine, Hokkaido University; Chairman, Department of Plastic and Reconstructive Surgery, Sapporo City General Hospital, Sapporo, Hokkaido, Japan
27. Muscle Vascularized Pedicle Flap (MVP Flap) Using Tissue Expanders: An Experimental Study in Rats

Preface

Tissue expansion has a very important place in reconstructive surgery. It has given us a tool for reconstructions that earlier were not possible.

We are familiar with phenomena such as the expansion of the abdominal wall due to the expanding uterus in the pregnant woman and the expanded lips in members of certain tribes in Africa. Despite this, it was not until 1957 that Dr. Neumann published a case report in *Plastic and Reconstructive Surgery* in which tissue expansion techniques were helpful in the correction of a partially avulsed auricle. However, then the idea fell into oblivion and was not reinvented until the mid 1970s.

My interest in tissue expansion started in February 1983 when I met Dr. Radovan at Dr. Ralph Millard's regular Wednesday afternoon meeting, where he was presenting tissue expansion, especially for breast reconstruction. He gave two expanders to Dr. John Devine, Jr., who forwarded them to me. As I had quite a lot of interest in scalp surgery at that time, I immediately wanted to try them for scalp expansion, which was not published and presented yet at that time. I had a case of scalp avulsion due to a dog bite, and we used one of the expanders successfully for correction of this patient. We did measurements on scalp kinetics and the results were published in *Plastic and Reconstructive Surgery* two years later after some delay in the journal. From this time on we had quite an interest in tissue expansion.

This book is intended as a manual both for the beginner and the experienced surgeon. In it, the reconstructive surgeon will find answers to his or her questions and advice that should lead to successful results. The authors of this book have a lot of experience with tissue expansion and many of them have contributed considerably to the knowledge in this field. Personal opinions have been left unchanged and without comment despite the fact that authors of different chapters may have differing opinions. This is only a sign of the fact that everything has not yet been studied or agreed upon and that several methods can lead to a successful result.

It is difficult, if not impossible, to predict the future, including the future for tissue expansion. However, I am convinced that tissue expansion still has several new contributions to make in reconstructive surgery. But only time will show what these will be.

An extensive bibliography has been added to the book for those needing more specific information.

REAN

Acknowledgments

I want to thank all contributing authors for giving their precious time to write their chapters and share their experiences:

Jaime Anger, M.D., (Brazil)
Eric David Austad, M.D., (U.S.A.)
Jean-Pierre Aubert, M.D. (France)
Andrew G. Batchelor, M.D., FRCS (U.K.)
Bruce S. Bauer, M.D., FACS (U.S.A.)
Hilton Becker, M.D., FRCS (U.S.A.)
Jincai Fan, M.D., Ph.D. (China)
Ken-ichi Homma, M.D. (Japan)
Guy Magalon, M.D. (France)
Stephen J. Mathes, M.D. (U.S.A.)
Rolf Münker, M.D. (Germany)
Takehiko Ohura, M.D., Ph.D. (Japan)
Edoardo Raposio, M.D., FICS (Italy)
Julien H. A. van Rappard, M.D., Ph.D. (the Netherlands)
Armand D. Versaci, M.D. (U.S.A.)
Frank A. Vicari, M.D., FACS, FAAP (U.S.A.)
Albert R. M. Wittkampf, M.D., D.M.D. (the Netherlands)
Tetsunori Yoshida, M.D., Ph.D. (Japan)

I also want to thank Maria Pulkkinen, R.N., O.R.N., N.A., for her valuable contribution in getting this book to the publisher, and Dr. Edoardo Raposio for scrutinizing the bibliography and helping with many details necessary to get the book ready.

REAN

1

A History of Early Tissue Expansion

ERIC DAVID AUSTAD

On 25 October 1956, a landmark paper was presented at the 25th annual meeting of the American Society of Plastic and Reconstructive Surgeons (ASPRS) in Miami Beach, Florida. Entitled "The Expansion of an Area of Skin by Progressive Extension of a Subcutaneous Balloon," it described the use of an expanding implant in the reconstruction of the upper two-thirds of a traumatically avulsed ear. The author, Charles G. Neumann, M.D., was a member of the Department of Surgery of New York University Post Graduate Medical School and worked at Bellevue Hospital in New York City. As with many innovations in medicine and other fields, Neumann's work was greeted with disinterest and viewed as eccentric. Although the published text of his work clearly reflects his insight regarding the value of "Be[ing] able to make available a new local supply of skin and subcutaneous tissue . . . " for reconstructive procedures, this was his only published article pertaining to the technique.[1] Neumann apparently became preoccupied with other matters, and his work was essentially forgotten. Yet, 25 years later (11 June 1982) the *Journal of the American Medical Association* headlined its *Medical News* section with the announcement: "Tissue Expansion: 'Future of Plastic Surgery for Next 20 Years'." To understand the evolution and ultimate acceptance of tissue expansion as a reconstructive technique, it is of interest to gain some understanding of the personalities involved, as well as their work.

Charles G. Neumann, M.D. was born 14 October 1913 in New York City. He died at his home in New Rochelle, New York on 19 October 1972 at the age of 59. His obituary, which appeared in the *New Rochelle Standard Star* on 20 October 1972, is brief in spite of the fact that Dr. Neumann had been a resident of New Rochelle for 50 years. More detail is found in the obituary that appeared in *Plastic and Reconstructive Surgery* in December of 1973.[2] Dr. Neumann received his A.B. degree from Dartmouth College in 1934 and was a 1936 graduate of Cornell University Medical School. His plastic surgery residency occurred at New York Hospital from 1947 through 1949, under the direction of Dr. Herbert Conway. Dr. Neumann was board certified in plastic surgery in 1953. He maintained a private practice in suburban New York City while holding an assistant clinical professorship in plastic surgery at New York University. Dr. Neumann's clinical work with tissue expansion was apparently performed entirely at Bellevue Hospital; at least, this was the origin of his one published case. His busy trauma service, the Fourth Surgical Division, was an obvious source of unusual reconstructive problems. However, the source of his inspired concept "to increase the area of a patch of skin and subcutaneous tissue for use in a reconstructive effort by the gradual distension of a rubber balloon placed beneath the skin . . . " remains unclear. From the limited biographical material available, one might speculate that the concept was derived in some way from his work with Dr. Robert Moran, while "doing research on the relief of pressure in hydrocephalic cases. . . . "[2] This would most probably have involved the use of various types of tubing and valves, and it may not have been entirely coincidental that Dr. Neumann's single reported case of tissue expansion involved scalp expansion for reconstruction of a

FIGURE 1-1
Charles G. Neumann, M.D., at graduation from Cornell University Medical School, 1936. Photo courtesy of the Medical Archives, New York Hospital, Cornell Medical Center.

traumatized ear and used an external filling point exiting the scalp from a low occipital stab wound. While current ventricular shunts for treating hydrocephalus are entirely internal and obviously involve intracranial components, this illustration could easily be mistaken by most neurosurgeons as an example of early attempts at ventricular decompression.

Dr. Neumann's clinical experience with tissue expansion was recalled clearly by J. M. Lawson, M.D., who was a senior resident on Dr. Neumann's service at Bellevue Hospital (Lawson, 1989, personal communication). Dr. Lawson described Dr. Neumann as "dry and laconic," always impeccably dressed and somewhat aloof from his colleagues. The only retrievable photograph of Dr. Neumann (Figure 1-1) is taken from his medical school graduation in 1936 and is a recent discovery that has not previously been presented or published. Dr. Lawson, whose assistance was acknowledged in Dr. Neumann's article dealing with expansion, found that his responsibilities as a chief resident included the fabrication, in the Bellevue machine shop, of the first tissue expanders used clinically. Although the article describes the implant as consisting of "a collapsed rubber balloon (5″ long and 1″ in diameter)" fastened to a "flexible polyethylene tube, (⅛″ in diameter)," Dr. Lawson states that these balloons were actually condoms. He found them most difficult to affix securely to intravenous tubing, and implant deflation was a constant source of concern. It is equally interesting that, while only one case report using tissue expansion was published, Dr. Neumann apparently used this technique in treating at least five to six patients, according to Dr. Lawson. A nasal reconstruction was included among those cases. No follow-up information regarding these other patients has been identified.

Dr. Lawson was in the audience when Dr. Neumann presented his pioneering work with tissue expansion at the ASPRS meeting in 1956 and recalled the case clearly. The patient, a 52-year-old man who had sustained traumatic amputation of the upper two-thirds of his right ear, had undergone placement of this first tissue expander via a postauricular incision. After 3 weeks of healing, intermittent inflation of the implant with air began via an external tube and stopcock arrangement. Inflations were performed "several times a week . . . based upon the appearance of the skin and subcutaneous tissue overlying the balloon." Two months after the initial procedure, Neumann reported that the skin overlying the implant "had expanded approximately 50% from its initial area of 12 square inches." At a second operation, the implant was removed and a costal cartilage graft was covered with the expanded flap. The flap survived nicely, but the aesthetic appearance of the reconstructed ear was not spectacular, and the case presentation was not well received, according to Lawson. Neumann was farsighted in preparing his work for publication, however, leaving a permanent and quite elegant record of his revolutionary concept. He even provided us with the term *expansion,* used in the title of his article and now universally recognized as a major reconstructive modality. However, there is no information to date indicating that he continued working with expansion after the late 1950s.

If Neumann had survived another 5 years, he might have attended the annual meeting of the ASPRS in Boston in the fall of 1976. At that meeting, an 8-month experience with the use of tissue expansion for breast reconstruction was presented by Chadomir Radovan, M.D., who had completed his residency in Plastic Surgery at Georgetown University earlier that year. Dr. Radovan (Figure 1-2) was born in Yugoslavia in

FIGURE 1-2
Chadomir Radovan, M.D., pioneer in tissue expansion clinical applications. Photo courtesy of Plastic and Reconstructive Surgery.

1932 and emigrated to the United States in his 20s, fleeing the communist regime in power at that time. He enjoyed recounting, with colorful language and byzantine complexity, the nonlinear route that eventually led to his graduation from Georgetown Medical School in 1967. Dr. Radovan was an accomplished accordion player and used this skill as a means of touring most of the United States, playing primarily at weddings and in small lounges. In doing so, he met his future wife, Lindy. In a move that was typical of his determination and brashness, Radovan introduced himself to Lindy and her father while they were dining, and he announced his wish to marry her. He did so; and Lindy became his wife, his muse, and his office manager. I talked with Radovan often about various aspects of tissue expanders and expansion. It was quite clear that his wife—and her pregnancy, specifically—led him to consider the concept of expansion. Sometime during his training as a plastic surgery resident, first under Dr. Frank Ashley at U.C.L.A. and subsequently at Georgetown under Dr. Al Fleury, Radovan's wife became pregnant. The process of physiological tissue expansion was not lost upon him, and he began designing what would ultimately become the Radovan tissue expander, the first implant of its type to become generally available. Although Dr. Radovan never made reference to this specific source of the concept in his formal presentations, to my knowledge, he would frequently begin them with an illustration of a gravid abdomen.

Radovan was fond of recounting the fortuitous manner by which his implant was first produced. On a snowy day in late 1975, the Georgetown plastic surgery program was visited briefly by Rudy Schulte, president of a large implant manufacturing company. Radovan volunteered to drive Mr. Schulte to the airport for his return flight, and during the entire trip assailed him with sketches and diagrams of the concept and the implant he required. Radovan was unsure of Schulte's interest and was well aware that implant manufacturers are constantly besieged by surgeons with various innovative ideas. However, within a month he received his first custom-manufactured implants from Schulte and performed his first tissue expander placement on 26 January 1976 in a Veteran's Administration hospital in Washington.[3] To my knowledge, no animal studies preceded this work.

In the fall of 1975, I was a resident on the service of William C. Grabb, M.D., Chief of Plastic Surgery at the University of Michigan Medical Center (Figure 1-3). Because of an active endocrinology service at our hospital, patients were referred to us who had sustained massive weight loss in the range of 90 to 136 kg (200 to 300 lbs.). These patients were referred to us for "plastys" of various sorts to remove festoons of excess skin from virtually every part of their anatomies. These patients were interspersed among the more common types of reconstructive surgical problems, namely, those requiring additional soft tissue for closure of various types of traumatic, surgical, and congenital deformities. To me, it was an irony that our patients seemed to polarize into two groups, the "haves" and the "have nots," and it was logical to find some means of using this physiologic capability of skin to expand. Like Neumann, I was quite conversant with the various types of tubes, valves, and reservoirs used in the treatment of hydrocephalus because I had previously spent 2 years with the neurosurgeons at the University of Michigan. I discussed this idea of "implant-induced soft tissue

FIGURE 1-3
William C. Grabb, M.D., scholar, clinician, and early champion of tissue expansion. Photo courtesy of Plastic and Reconstructive Surgery.

augmentation" (I didn't know what else to call it then) with Dr. Grabb on many occasions. He was interested and very supportive of my work, and later he became an early champion of the entire concept. With his encouragement, I visited the research and production facilities of Dow Corning Medical Products on 16 January 1976. Interestingly, this was only 10 days before Radovan's first tissue expander placement in Washington. I described the concept to Silas Braley, director of Dow Corning's research activities, and to his staff. The implant I described to them could be either of two types: a manually inflated valve-containing model or a self-inflating implant that would increase in volume by osmotically attracting extracellular water. The former, I was told by Dow Corning, was not feasible due to the inadequacy of currently available valves. However, the simplicity of a self-inflating implant made it my favored approach in any case, and I was very pleased when Dow Corning agreed to pursue the idea and provide me with sealed silicone shells containing dry sodium chloride. To my surprise, their researchers were unable to speculate regarding the ability of an osmotic gradient to cause transport of extracellular water across their membranes, and we simply agreed to test the concept in the laboratory.

In the summer of 1976, the first self-inflating implants arrived from Dow Corning. They consisted of 18-mL silicone testicular prosthesis shells, each containing two to three salt tablets. Apparently, these tablets were the most readily available source of sodium chloride, and no one could tell me what binding agents and other superfluous materials they might contain in addition to sodium chloride, but residents are not usually in a position to quibble about such things. I proceeded to study the inflation characteristics of these implants in saline baths, which simulated the extracellular space, and by early fall of 1976 a small trial study in animals had begun. In October of that year, Dr. Grabb returned from the annual ASPRS meeting in Boston with the exciting news that "tissue expansion" was being used clinically by a California surgeon, Dr. Chadomir Radovan. Our department subsequently became one of the first sites for approved clinical investigation of the Radovan implant, which we used extensively in the late 1970s and early 1980s. Graduates of the University of Michigan plastic surgery residency during those years have remained prominent in the literature and in clinical practice of tissue expansion worldwide: Louis C. Argenta, M.D., Ernest K. Manders, M.D., Malcolm W. Marks, M.D., Rodney J. Rohrich, M.D., Craig VanderKolk, M.D., Richard D. Anderson, M.D., Paul Izenberg, M.D., and James Thornton, M.D. All have made significant contributions. The ongoing plastic surgery research efforts during that period were assisted immensely by the expertise of George Cherry, Ph.D. and Krystyna Pasyk, M.D., Ph.D. Both provided theoretical and practical support for ongoing resident and junior faculty research projects, while concomitantly pursuing independent and highly important work of their own design.

On 6–7 May 1981, the University of Michigan hosted a two-part plastic surgery seminar, which dealt with hemangiomas on the first day and expanded skin on the second. To my knowledge, this was the first seminar that formally addressed tissue expansion, although the topic did not yet warrant extensive treatment. Guests included Dr. Chadomir Radovan and Dr. Gordon Sasaki. Radovan's presentation was entitled "Experience with the Radovan Tissue Expander," and he participated in the subsequent discussions with great interest. We presented our early experience with the self-inflating tissue expander and described "light histology changes in expanded skin." This was the first presentation of any histologic data relating to tissue expansion. It is probably the most significant component of the subsequently published work.[4,5] Sasaki's involvement in the seminar appeared primarily related to his ongoing work with hemangiomas, but he also participated in the tissue expansion discussions by pre-

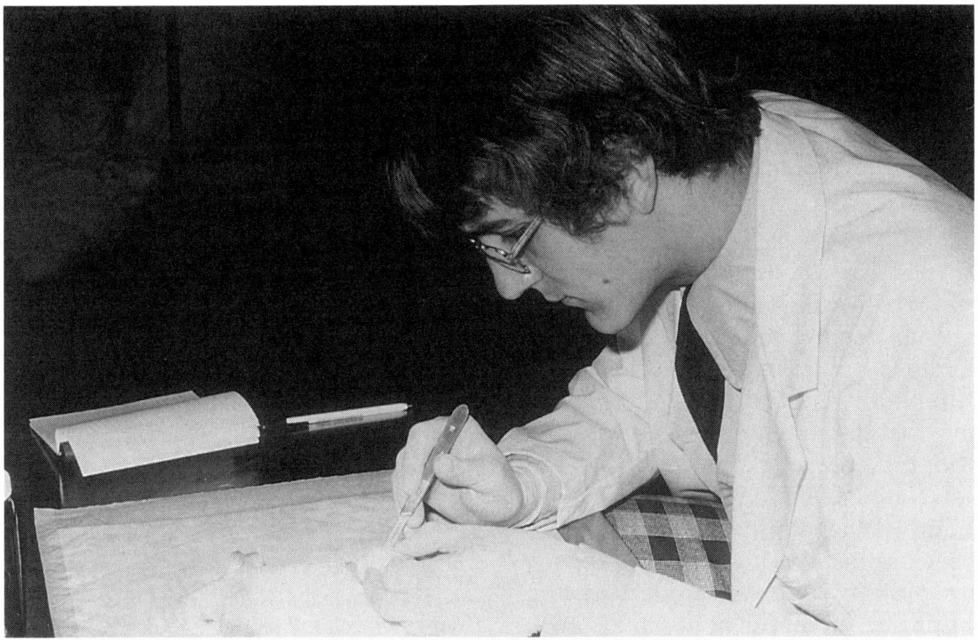

FIGURE 1-4
Gregory L. Rose, early tissue expansion researcher.

senting his work with "the role of the myofibroblast in skin expansion and contraction." He remains an innovative and prolific researcher and clinician, and his contributions to the understanding and practice of tissue expansion are outstanding.

In November 1982, with the encouragement and support of Dr. Grabb, I organized the first national symposium on tissue expansion, sponsored by the Educational Foundation of the ASPRS. This honor was probably due, to a large extent, to work with the self-inflating expander and its subsequent histologic description, which I had submitted for presentation at the 1979 meeting of the ASPRS in Toronto. My co-author for this work, Gregory L. Rose (Figure 1-4), was a bright and dedicated medical student who made many contributions to the ongoing project; his tragic accidental death in 1978 was a great loss. That research was submitted to the 1979 program committee as an abstract, but it was declined. Tissue expansion was still considered by most to be a novelty of unproven value. However, the same data were also submitted as a complete paper to the research competition committee of that meeting, and the paper was ultimately presented there because it was awarded first prize in the Clinical Research category. In addition to the personal satisfaction this afforded, I believe this award had some significance in the ongoing progression of tissue expansion. It lent credibility to this first presentation of any scientific aspect of tissue expansion, and its categorization as a prize for clinical research, rather than basic science, seemed to acknowledge its clinical relevancy.

The 1982 tissue expansion symposium in Ann Arbor, Michigan, was enriched by the contributions of Drs. Radovan, Argenta, Sasaki, Cherry, Rohrich, Manders, Versaci, and Gibney. This group proved to be a nucleus around which almost every subsequent national or international tissue expansion symposium evolved over the subsequent 10 years. This first symposium was short, lasting only one and a half days; there really wasn't much more to say about expansion in 1982. However, a great deal of knowledge and enthusiasm was exchanged, and similar symposia have been held virtually every year subsequently. Greatly missed at this first symposium was the presence of Dr. Grabb, whose untimely death on 21 March 1982 marked the loss of tissue expansion's most prominent proponent. In addition, we lost a valued colleague, mentor, and friend. Dr. Grabb wrote the Invited Discussion to Radovan's first article regarding tissue expansion; it was published in February 1982, only a month before his death.[6,7] In it, his enthusiasm for the technique was quite clear: "Although many innovations are advanced, only a few have had the impact of microsurgery techniques and of muscle flaps. I predict that over the next five years, the technique of tissue expansion will be of equal importance."[5] Although it is difficult to objectively quantify, I believe that Dr. Grabb's prediction has proven to be true.

Tissue expansion suffered another major loss on 23 September 1984, with the accidental death of Dr. Radovan. He drowned while swimming in the ocean off the Florida coast. He is remembered by his colleague and friend, Dr. George T. Craig (1986), for his many contributions to tissue expansion and for the superb care and rapport he offered his patients. Dr. Craig acknowledges that Dr. Radovan was perceived by some as unusual, "not like our own," and noted the "admiration and respect that he had for some of those who did not always respond to him in kind."[8] While Radovan will remain a controversial figure among those who knew him both personally and professionally, it is undeniable that his pioneering work with tissue expansion and his energetic promotion of the technique were key components to its subsequent development. Although neither Radovan nor I became aware of Neumann's original work with tissue expansion until our own projects were well underway, the world literature of plastic surgery is now replete with articles related to various aspects of expansion. Perhaps the greatest testimony to the utility of the technique lies in its widespread application and in the ease with which surgeons-in-training incorporate it.

The future of tissue expansion will be unequivocally altered, at least in the United States, by the decision of several major implant manufacturers to cease production of all types of medically implanted materials. This unforeseen tragedy, related to an ongoing controversy regarding the safety and efficacy of breast implants, holds similar ramifications for many other aspects of surgical research as well. Both the development of new implants and the understanding of underlying physiological mechanisms they involve will be compromised. However, the history of tissue expansion shows it to be a valid technique that is highly beneficial to many patients and that holds a strong position in the mainstream of reconstructive surgery worldwide. Regarding tissue expansion, Dr. Grabb succinctly observed that "the bandwagon is rolling."[7] Its history will continue to be colorful, its goals innovative, and its practitioners ever-increasing.

REFERENCES

1. Neumann CG. The expansion of an area of skin by progressive distention of a subcutaneous balloon. *Plast Reconstr Surg*, 1957;19:124–130.
2. Obituary: Charles G. Neumann, M.D. *Plast Reconstr Surg* 1973;52:691.
3. Radovan C. Tissue expansion in soft tissue reconstruction. *Plast Reconstr Surg* 1984;74:482–490.
4. Austad ED, Pasyk K, McClatchey KD, et al. Histomorphologic evaluation of guinea pig skin and soft tissue after controlled tissue expansion. *Plast Reconstr Surg* 1982;70:704–710.
5. Austad ED, Rose GL. A self-inflating tissue expander. *Plast Reconstr Surg* 1982;70:588–593.
6. Radovan C. Reconstruction of the breast after mastectomy using a temporary expander. *Plast Reconstr Surg* 1982;69:195–206.
7. Grabb WC. Invited discussion: Reconstruction of the breast after mastectomy using a temporary expander. *Plast Reconstr Surg* 1982;69:207–208.
8. Craig C. Obituary: Chadomir Radovan, M.D. *Plast Reconstr Surg* 1986;77:356–357.

2

Benefits, Indications, and Contraindications

Eric David Austad

In his remarkably prescient work with tissue expansion from 1954 to 1958, Neumann[1] articulated many of its indications, benefits, and contraindications. While it is now possible to elaborate upon indications for the use of this technique, the first sentence of Neumann's article begins, "When the need for additional skin arises in the course of reconstructive surgery . . . " This summarizes the prime indication quite well. He then proceeds to describe alternatives that are in the mainstream of reconstructive surgery today, consisting of " . . . the advancement of contiguous tissue . . . the use of free flaps or pedicle flaps." Neumann also described naturally occurring phenomena as illustrations of "biological" tissue expansion: "As breasts develop at puberty or enlarge thereafter, the skin manages to keep pace over the increasing amount of mammary glandular tissue and fat, and as superficial tumors or enlarging herniae develop, the skin and subcutaneous tissue again manage to provide a cover." In addition, he cited the "platter lipped women of Chad," with their self-induced lip expansion, and noted a similar practice among the "Padung women of Burma [who] increase the lengths of their necks by encircling them with an increasing number of wire coils." At the first national symposium on tissue expansion, held in Ann Arbor, Michigan in 1982, almost every presenter had incorporated some variant of this same theme of physiological or self-induced tissue expansion, ranging from illustrations of the gravid abdomen near term (Radovan), to the redundant skin folds of patients who had experienced massive weight loss (Austad). None of us were aware, in 1982, of Neumann's work and his 1957 paper, yet the concept and its benefits were clear to all. As Neumann clearly stated, " . . . gradual tension can lead to an expansion of skin and subcutaneous tissue . . . " leading to " . . . the securing of skin which would be high in quality of match of color and texture with nearby skin, the avoidance of the need for a donor area in the usual sense of the term, and the possible reduction in the number of stages for a particular reconstructive effort."

BENEFITS OF TISSUE EXPANSION

Additional Tissue Is Created with Expansion

Controversy exists regarding the relative contributions of new cell production[2] and undermining[3] to the end result of expansion. In all likelihood, both mitosis and transient recruitment from adjacent, nonexpanded tissue play a role in the ongoing process of expansion. However, Pasyk et al.[4] have reported that epidermis, dermis, and underlying subcutaneous tissue all resumed normal thickness in a patient studied over a 2-year period postexpansion. The methodology of this study is difficult to apply on a large scale in clinical practice because new biopsy scars are generally undesirable. However, if it is true that expanded tissue ultimately becomes indiscernible from its unexpanded counterpart, the conclusion that new tissue has been created is inescapable.

Optimal Matching of Donor Tissue to Recipient Site

The qualities of skin and mucosa that are of prime concern for both aesthetic and functional purposes include color, texture, hair-bearing characteristics, durability, and innervation. Split-thickness grafts, a highly efficient means of obtaining a closed wound, are often deficient in most, if not all, of these criteria. Full-thickness grafts, particularly when obtained from carefully selected donor sites, frequently provide excellent aesthetic and functional restoration. However, they are limited to relatively small defects because of donor site limitations. Skin flaps, including free tissue transfers, quite reliably meet functional requirements. However, because they often are not derived from immediately adjacent tissue, they frequently do not provide optimal color, texture, and hair-bearing characteristics at the recipient site. In most clinical situations, tissue expansion essentially allows primary wound closure with extra surface area elicited from immediately adjacent tissue, which is optimally matched to it.

Retained Sensory Innervation
The specific mechanism by which sensory nerves and sensory end-organs respond to tissue expansion has not been clearly defined, but retention of at least some sensibility clearly occurs with expansion. The working hypothesis for these phenomena involves a preservation of myelinated nerve fibers and a simple redistribution of existing end-organs over a wider surface area. There are no experimental studies indicating an increase in myelinated nerve fibers, as seen with vascular structures during expansion; conversely, there are no reports to date of significant sensory loss associated with expansion. A number of published reports confirm at least grossly retained sensibility after expansion; many of these describe the advancement of sensate expanded flaps in the treatment of pressure ulcers. Esposito et al.[5] describe a series of 11 patients with severe pressure sores using expanded flaps, and the authors are highly optimistic regarding the benefits of moving sensate expanded skin to ulcerated areas. Cole et al.[6] report normal sensibility in expanded flaps used in the lower limbs of five children and one young adult. Masser[7] takes this one step further by using expanded radial free flaps to obtain sensate closure of ankle/heel defects in two patients.

Increased Vascularity
In the early days of tissue expansion, the late 1970s and early 1980s, it was frequently suggested that expansion may cause tissue to outstrip its vascular supply, resulting in extra tissue that would be of little value in reconstruction. Virtually all studies and clinical observations regarding vascularity after expansion show quite the opposite to be true. Radovan was probably the first to note specifically that expanded tissue displays a very robust blood supply, similar to that seen with classic delay. This was studied more specifically by Cherry et al.,[8] and by Sasaki and Pang.[9] While these studies differed in some aspects, both concluded that vascularity is prominently increased with expansion, to a level comparable with delayed flaps. Figure 2-1 illustrates barium angiography of normal, unexpanded pig skin. Figure 2-2 is a comparable barium angiogram of expanded pig skin. The dramatic increase in vascularity, both in the immediate area of expansion and along the perimeter of the implant, is quite obvious.

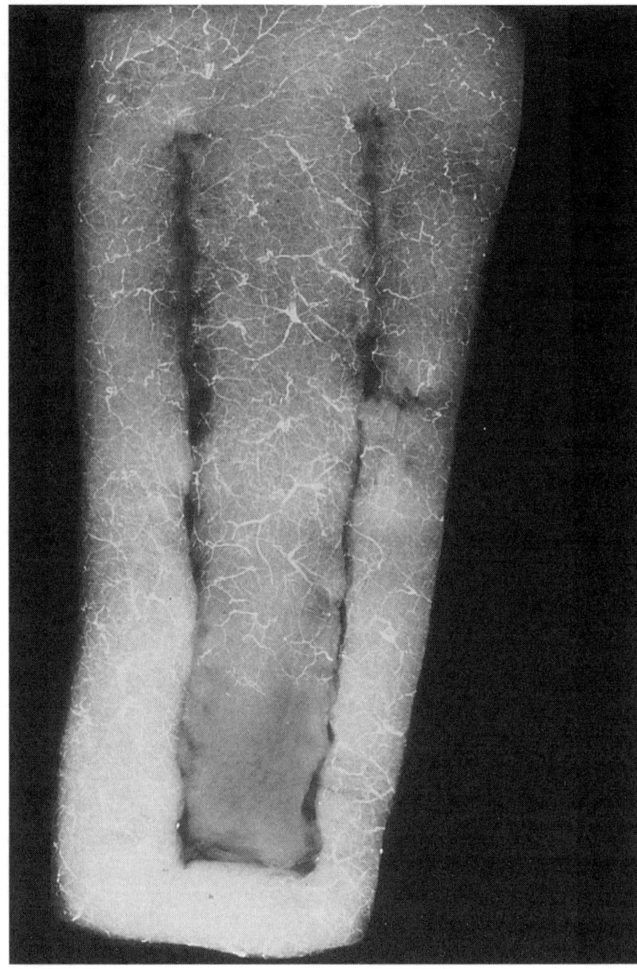

FIGURE 2-1
Barium angiogram of normal unexpanded pig skin.

Expansion Involves Autogenous Tissue
In these days of paranoia and increasing litigation surrounding both "permanent implants" and virally infected nonautogenous tissue (blood, corneal transplants, kidney transplants, and so forth), it is of some comfort to both surgeon and patient to view expansion as a transient procedure that lasts only a few

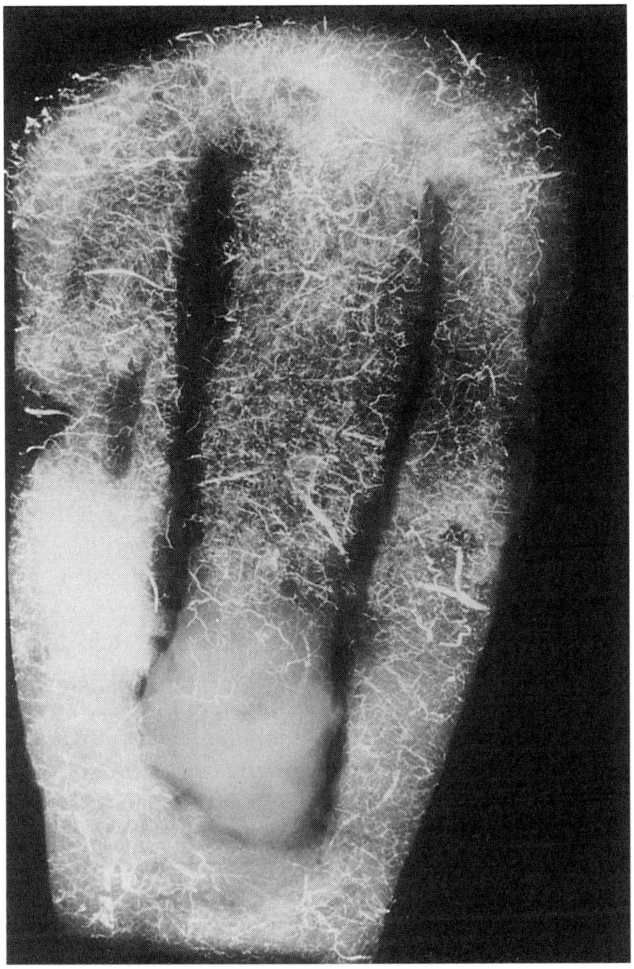

FIGURE 2-2
Barium angiogram of expanded pig skin. Note prominent vascularity.

weeks to a few months and is free of biological cross-contamination. Breast reconstruction with expansion might be viewed as an exception to this observation, because it usually involves a "permanent" implant of some type to provide breast volume, but in almost all other expansion procedures the patient is ultimately implant-free and unexposed to allograft materials.

Tissue Expansion Is Almost Always an Elective Procedure
This has obvious benefit for both patient and surgeon, because surgical schedules can be arranged for the convenience of both. Any surgeon who has been thrust into an emergent, complex reconstruction appreciates the value of more extensive preoperative planning and of establishing patient rapport on a more elective basis. Conversely, patients usually appreciate the opportunity to discuss and plan for the proposed reconstruction with expansion, allowing them latitude in arranging their schedules and addressing their concerns.

Tissue Expansion Is Easily Comprehended by Most Patients
In an age of increasing technological complexity, tissue expansion is a straightforward and easily comprehended alternative for most patients, including children. Analogies to the changes in abdominal skin surface area during pregnancy or to skin changes with weight gain/loss are familiar ones, and most patients are quite enthused by the prospect of "growing their own skin." Complications such as infection and implant erosion or deflation are also readily understood, and patients embarking upon a reconstruction using expansion usually feel quite well prepared for the experience.

Tissue Expansion Is Adaptable to Most Anatomical Areas
Extensive worldwide experience with tissue expansion has been successful or promising in virtually every anatomical region covered by skin or mucosa. While some regions—such as the scalp, forehead, and plantar surfaces—appear to require more protracted and gradual expansion, tissue expansion is a legitimate option for reconstruction of virtually any anatomical site.

Complications of Tissue Expansion Are Virtually Never Disastrous
Worldwide clinical experience with tissue expansion is impossible to accurately quantify, but estimates indicate that somewhere between 200 000 and 500 000 patients have undergone expansion during the period from 1976 to 1992. To date, no tissue expansion-related deaths have been reported in the literature. Unexpected complications such as rib erosion[10] and skull erosion[11] have been reported, but none of these have resulted in significant patient morbidity. Moreover, if skin erosion or infection occurs and the procedure cannot be salvaged, the expansion can be repeated at some later date, or other options such as myocutaneous flaps or free flaps may be used. In general, one does not preclude other means of tissue transfer by proceeding first with expansion.

Tissue Expansion Is Usually Cost-Effective
No prospective studies of the cost-effectiveness of tissue expansion have been published as of mid-1992. However, Gibney has reported cost-related data at several tissue expansion symposia (J. L. Gibney, "Breast Reconstruction Using Tissue Expansion Techniques," Pasadena, CA, November 9, 1984 and Hershey, PA, September 24, 1986), indicating a 30% to 40% reduction

in hospital-related costs when comparing breast reconstructions using expansion with those using myocutaneous flaps. Although expansion always involves at least two surgical procedures, both are usually brief and can often be performed in an outpatient setting. Moreover, there is little doubt that the costs associated with removing a deflated or infected tissue expander are far less than those involved in the treatment of a failed myocutaneous flap.

INDICATIONS FOR THE USE OF TISSUE EXPANSION

To paraphrase Neumann, tissue expansion is a logical alternative when one is confronted with a requirement for additional tissue of specific characteristics, which cannot be acquired by a more efficient or reliable means. The potential benefits of tissue expansion have been previously reviewed, but these may not be pertinent to specific clinical situations. A tenet of reconstructive plastic surgery has been to first consider simple solutions before moving to the complex. For example, one should consider a graft before a local flap, and a local flap before a regional flap or free tissue transfer. Where in this algorithm should tissue expansion be placed? I do not believe it has a tidy niche in this scheme; rather, I believe that it has potential benefit at each level of wound complexity. A small wound, which would otherwise require a graft for closure, might benefit from intraoperative expansion and yield to primary closure. If a graft will clearly be required, an expanded full thickness graft might afford significant aesthetic or functional benefit. Local or regional flaps might become more effective with the increase in surface area and vascularity provided by pretransfer expansion. Preexpansion of free flaps is a highly acceptable alternative when surface area must be maximized and donor site scarring must be minimal.

CONTRAINDICATIONS TO TISSUE EXPANSION

Patient Related Factors

Patients who are not highly motivated to obtain optimal aesthetic and/or functional reconstruction are not, in general, good candidates for tissue expansion. While intraoperative expansion adds little or no morbidity or complexity to postoperative recovery, chronic expansion is virtually always intrusive and occasionally painful. Just as every mastectomy patient does not view herself as a candidate for reconstruction, some patients do not place a high enough priority upon their specific reconstructive problems to warrant consideration of tissue expansion. The financial costs associated with tissue expansion are also viewed by some patients as a contraindication. Because chronic expansion involves at least two surgical procedures and at least one implant, it may not be a feasible means of treating "discretionary" problems such as tattoos or scars. Finally, the patient must have the ability to psychologically and occupationally tolerate a period of weeks during which the expander(s) may be painful, awkward, or grotesque in appearance.

Donor/Recipient Site Factors

It has become a maxim in tissue expansion surgery that high-quality expanded tissue cannot be derived from a marginal donor site. Scar does not expand well or predictably, and expansion in previously injured areas should be avoided when possible. Previous radiation injury at the donor site, seen most frequently in the postmastectomy patient, may be more difficult to assess. Previous radiation is not an absolute contraindication to expansion, but Kuske et al.[12] and others have experienced elevated complication rates when expanding previously irradiated areas. Diabetes mellitus has not yet been identified as a complicating factor in the use of tissue expansion. In fact, one of Radovan's first expansion patients underwent successful transfer of an expanded cross-leg flap for treatment of a diabetic foot ulcer.[13] Osteogenesis imperfecta is the only systemic disease, to date, reported to complicate tissue expansion.[14] A single patient with this disease was unable to undergo expansion; thrombosis of vessels overlying the expander precluded its inflation. As operative experience with tissue expansion increases, other systemic illnesses will probably be associated with increased complication rates.

Societal Factors

Cost may be another issue when considering tissue expansion, in spite of high patient motivation and appropriateness. Even when surgical fees and hospital-related expenses are negligible, implant costs alone can make expansion procedures unattainable for many patients. This is particularly true in developing countries, where imported implants are taxed heavily, sometimes inflating prices five- to tenfold. Manders has urged the careful salvage of intact used expanders, and his plastic surgery department at Hershey Medical Center, Hershey, PA, continues to serve admirably as a clearinghouse for providing used expanders on a cost-free basis to surgeons in developing countries. Anger has also written eloquently regarding this unique opportunity to recycle an important resource.[15] The reuse of tissue expanders in more-developed countries is not straightforward. In a staged series of expansions, many surgeons will remove, inspect, deflate, and replace the same implant for additional expansion if its geometry remains appropriate. While one might speculate that deflation or other mechanical problems may be seen more frequently with such reuse, no clinical data have been gathered to support this. Moreover, one could argue that an implant that has proven reliable in previous use may be more dependable than a new implant that has undergone standard quality control testing but has not withstood the rigors of clinical use. In the United States, no clear "standard of care" has been established regarding the reuse of tissue expanders in the treatment of an individual patient. However, the reuse of these implants in treating two or more separate patients is generally discouraged in countries with litigious climates. No cases of viral or bacterial cross-contamination between patients have been related to the reuse of tissue expanders, although this claim is not difficult to envision. In current practice, some types of silicone implants are reused after sterilization: breast implant "sizers" are probably the most frequent example of this. Resterilization procedures recommended by implant manufacturers are believed to preclude cross-infection with reuse. However, in any economy in which litigation is a major determinant of policy, the reuse of tissue expanders previously exposed to the flora of other patients does not appear judicious.

Surgeon-Related Factors

The protracted nature of the expansion process, with multiple visits at frequent intervals, is labor- and time-intensive for the surgeon, and patients undergoing expansion frequently require high levels of emotional support. Manders and others have addressed this in part by delegating routine expander inflations to nurse-specialists or to referring physicians when the patient resides at some significant distance. In unusually supportive family situations, parents or spouses have been trained to perform routine implant inflations at home. These creative solutions to the inconveniences and logistical problems associated with protracted implant inflation will be evaluated with regard to complication rates; however, until those data are available, many surgeons have refrained from delegating implant inflation to others. In my view, these issues make the use of tissue expansion quite analogous to the use of microsurgical techniques—while both are an integral part of virtually every plastic surgery training program in the world, each surgeon ultimately decides on a personal level the specific array of procedures he or she will provide to patients.

As with microsurgery, tissue expansion techniques do not lend themselves well to occasional use in most clinical practices. While not as technically complex as most microsurgical procedures, the decision to continue using tissue expansion on a regular basis is frequently made in the same all-or-none manner. Though technically competent to employ expansion techniques and cognizant of their value, some reconstructive surgeons make the decision, consciously or not, to avoid embarking on the long and sometimes stressful process of expansion. I believe this is a valid decision. On a regular basis, surgeons identify and attempt to perfect techniques that are best suited to their interests, goals, and talents. Because patient interests are our ultimate concern, however, it is of prime importance that all reconstructive surgeons be aware of the potential benefits of expansion techniques and, when appropriate, provide some means of patient access to the procedure.

REFERENCES

1. Neumann, CG. The expansion of an area of skin by progressive distention of a subcutaneous balloon, *Plast Reconstr Surg* 1957;19:124–130.
2. Austad ED, Thomas SV, Pasyk KA. Tissue expansion: Dividend or loan? *Plast Reconstr Surg* 1986;78:63–67.
3. MacKay DR, Saggers GC, Kotwal N, Manders EK. Stretching skin: Undermining is more important than intraoperative expansion. *Plast Reconstr Surg* 1990;86:722–730.
4. Pasyk K, Argenta LC, Austad ED. Histopathology of human expanded tissue. *Clin Plast Surg* 1987;14:435–445.
5. Esposito G, Di Caprio G, Ziccardi P, Scuderi N. Tissue expansion in the treatment of pressure ulcers. *Plast Reconstr Surg* 1991;87:501–508.
6. Cole WG, Bennett CS, Perks AG, McManamny DS, Barnett, JS. Tissue expansion in the lower limbs of children and young adults. *J Bone Joint Surg* 1990;72:578–580.
7. Masser MR, The preexpanded radial free flap. *Plast Reconstr Surg* 1990;86:295–301.
8. Cherry G, Austad E, Pasyk K, McClatchey K, Rohrich RJ. Increased survival and vascularity of random pattern skin flaps elevated in controlled, expanded skin. *Plast Reconstr Surg* 1983;72:680–685.
9. Sasaki G, Pang C. Pathophysiology of skin flaps raised on expanded pig skin. *Plast Reconstr Surg* 1984;74:59–65.
10. Sinow JD, Halvorsen RA Jr., Matts JP, Schubert W, Letourneau JG, Cunningham BL. Chest-wall deformity after tissue expansion for breast reconstruction. *Plast Reconstr Surg* 1991;88:998–1004.
11. Paletta CE, Bass J, Shehadi SI. Outer table skull erosion causing rupture of scalp expander. *Ann Plast Surg* 1989; 23:538–542.
12. Kuske RR, Schuster R, Klein E, Young L, Perez CA, Fineberg B. Radiotherapy and breast reconstruction: Clinical results and dosimetry. *Int J Rad Onc Bio Phys* 1991; 21:339–346.
13. Radovan C. Tissue expansion in soft-tissue reconstruction. *Plast Reconstr Surg* 1984;74:482–492.
14. Stark GB, Jaeger K. Failed tissue expansion in a patient with osteogenesis imperfecta. *Ann Plast Surg* 1989;22: 156–159.
15. Anger J. The reuse of tissue expanders in developing countries [letter]. *Plast Reconstr Surg* 1991;88:1114–1115.

3

Geometric Planning

ROLF E. A. NORDSTRÖM, EDOARDO RAPOSIO, AND JINCAI FAN

It is seldom sufficient for a good reconstruction to produce just a bulging bulb of tissue in the neighborhood of a defect. A variety of details regarding geometric planning of the tissue expansion are of the utmost importance for a gratifying result.

PLACE OF EXPANDER

The expander is usually placed close to the defect to be reconstructed. This enables one to move the expanded tissue most efficiently to the area to be reconstructed. By leaving more tissue between the expander and the defect, one easily ends up having an expanded tissue that is not moving the distance one has anticipated (that is, a bulge of tissue that is not moving where it should move). This is especially true if one intends to use advancement flaps. On the other hand, one will probably stretch the area to be reconstructed when placing the expander next to it. However, except in particular circumstances, it is advisable to insert the expander(s) close and parallel to the longest side(s) of the lesion to be removed. It is also usually wise to place the expander, if one has a choice, in the most nonmobile area available. When the expander is placed in a joint area with movement, the risk of extrusion is considerably increased.

SHAPE OF EXPANDER

As mentioned by Shively et al.[1] the exact shape that a skin expander assumes is quite variable depending on the initial size and shape of the implant, the thickness and tension of the overlying tissue, and the shape of the underlying base. This shape is constantly changing as the volume of the prosthesis is increased. However, the choice of the initial shape of the expander depends very much on the shape and size of the defect and the flaps one plans to use. Usually one wants to end up with the shortest possible scar in a good location and suitable direction. A hemispheric expander on a flat surface produces tissue that can be maximally utilized only by relaxing incisions close to the base surface as rotation flaps. A rectangular expander produces tissues to be moved usually by relaxing incisions as an advancement flap. When one has a convex surface, a hemispheric expander produces tissue to be advanced directly as long as the base of the expander is sufficiently large in comparison to the radius of the convex area. With a convex area, a spindle-shaped expander will also give advancement without extra relaxing incisions. A round lesion can often be corrected with a crescent-shaped expander without further relaxing incisions, avoiding extra scars. Advancement is achieved with a final linear scar. However, it is worth remembering again that the amount of area gain relies mostly on the projection of the expander.

Van Rappard et al.[2,3] have used a mathematical approach to study the theoretical increase in surface area in relation to the shapes of the expanders. Although the mathematical data was confirmed to be reliable in an *in vitro* model, measurements performed *in vivo* (on pigs) showed that only about 35 percent of the mathematically expected increase in surface area actually takes place, with a clear difference between the different shapes of tissue expanders. The authors found that for expanders with a round base the measured gains are 25 percent of the average of the calculated gains, while for the rectangular and crescentic expanders these percentages are 38 and 32 percent, respectively.

SIZE OF EXPANDER

Morgan and Edgerton[4] and Radovan[5] recommended that the expander base have the same size as the area to be reconstructed, while Gibney[6] and van Rappard et al.[2,3] advised that the same base should be 2.5 to 3 times the size of the defect to be covered. This discrepancy is probably due to the fact that tissue expanders, as noted by Shively et al.[1] are three-dimensional objects and that volume (and not area) is what the clinician measures as the implant is expanded. Thus, the surgeon needs to convert volume to surface area to select the proper size expander for specific needs. This estimation may be attained by resolving specific mathematical functions or, as more often happens, following one's own personal clinical experience. By the latter system, obviously, no objective evaluations may be extrapolated, and this is the main reason there is a lack of agreement about this matter in the literature. But, as reported by Hallock in 1987,[7] also utilizing computer programs or complicated mathematical formulas, the preoperative selection of an appropriate size of implant frequently is erroneous because the size is underestimated. For these reasons, we think it is usually wise to use the largest possible expander, the size being limited only by the anatomy of the donor area.

Since tissue expansion is partly based on mechanical creep (Hirshowitz et al.[8] and Gibson[9]), the phenomenon of stretch-back, which occurs when the expander is removed[10], may noticeably diminish the surface area gained. However, a slight underestimation of the nominal size of the expander will never be a serious clinical problem because it has been proved that overexpansion of tissue expander envelopes, up to 15 times the manufacturer's stated nominal capacity, should be perfectly safe.[7] Furthermore, if one falls short, the expander can be deflated and left in place under the advanced soft tissue and subsequently reexpanded. For this reason, it is advisable not to excise the area to be reconstructed before trial advance of the expanded tissue is performed.

One always has to be aware of the possibility of considerable discrepancy between the injected saline volume and the actual volume of the expander (when it is removed after completed expansion) due to possible injection dome leakage. Testing 22 tissue expanders, we have found that the minimum pressure point of leakage in the injection dome was, on average, 32 mmHg (with a range of 8 to 110 mmHg). Pressures of up to 93 mmHg were shown in the expander when the patient was lying on the implant, leading to considerable leakage of saline through the injection dome.[11] This leakage may have clinical significance in certain situations and may reach disturbing proportions. One has to realize that it is seldom a problem, but everyone working with expanders should be aware of it.

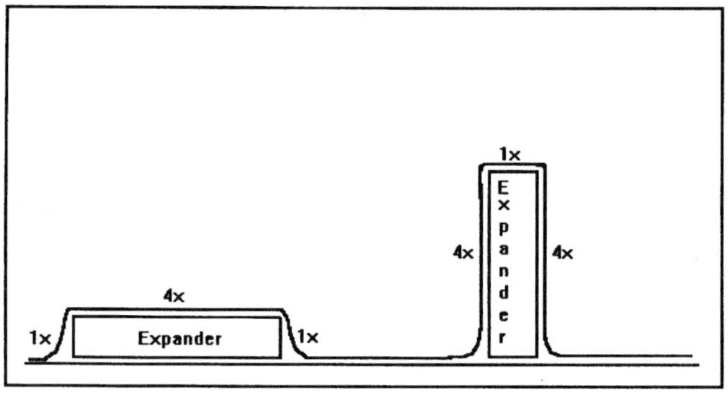

FIGURE 3-1
Outward projection is the most important direction for yielding new tissue by expansion. The area achieved with the expander standing upright is four times greater than the area gained with the implant lying down. The expanders have the same shape and volume.

In addition to size, keep in mind that projection is the most important direction of expansion to gain the maximum amount of tissue. This can easily be seen, for example, using the same amount of fluid in equal-volume expanders, but displacing the expanders in such a way to obtain different projection rates (Figure 3-1); although the expander and fill are the same, the area achieved with the higher projection is four times greater than the lower one. Flat expansion alone produces a minimum of new tissue for the reconstruction!

NUMBER OF EXPANDERS

In 1985, Brobmann and Huber demonstrated that when a defect is irregular, a larger custom-fitted implant is more efficient than several smaller expanders.[12] They have reported that oval implants gain 5 percent more skin in 50 percent less time and with almost 20 percent less intraluminal pressure, while U-shaped implants yield 2.2 percent more skin in 10 percent less time with 35 to 50 percent less pressure when compared with several smaller implants covering the same surface area. These data show that less time, pressure, and volume are needed in a larger expander to gain the same surface area that would be attained from several smaller implants (thus sparing the patient more discomfort, trauma, and risk of infection by reducing the number of filling domes and hence the number of fillings).

On the other hand, the location of the defect to be reconstructed, unfortunately, does not often allow the placing of one single large expander. Thus, in many cases, it is necessary to use two or more expanders simultaneously to achieve the desired goal more efficaciously. In this way, the defect to be reconstructed has to be geometrically split into various figures according to the size and shape of the expanders inserted. In these situations, it is often wise not to have communicating pockets for the expanders so that extrusion or infection of one implant does not affect the others.

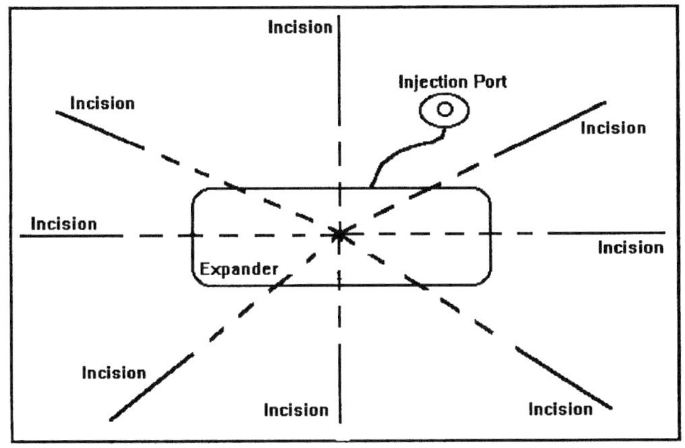

FIGURE 3-2
Examples of optional radial incisions for insertion of the expander.

INJECTION PORT

One may choose to use expanders with integrated, remote internal, or external filling domes. Indications and contraindications of each type are further described in other chapters. However, when using an internal remote injection port, avoid placing it above a joint or a bone prominence to reduce the risk of exposure of the injection dome. If filling port extrusion or leakage occurs, it does not necessarily mean that the expander has to be removed.[11] The exposed dome often can be used as an external port and the expansion can be continued. External filling ports are safe (Jackson)[13] and may also act as an adjunctive draining system.

The connecting tube between the injection dome and the inflatable sack should be placed so it does not migrate on the inflatable sack as the sack grows during expansion. This could cause an erosion of the skin overlying the tube.

INCISIONS FOR INSERTION OF EXPANDERS

The planning of the incisions is of considerable importance because the start of the expansion very much depends on their locations. Furthermore, the locations of the incisions also influences the rate of complications to a great extent. Argenta and Austad[14] advise careful planning before selecting an incision for placement of the prosthesis so that possible alternative flaps are not compromised. The authors state that the incisions should be incorporated into the projected incision ultimately left by implant removal and flap rotation or advancement. Furthermore, the incisions should be placed where they will not interfere with later advancement or compromise the blood supply to the expanded tissue (Manders et al.).[15]

In our opinion, the incisions for insertion of the expander should be "radial," meaning that they should lie in a line drawn from approximately the midpoint of the implant (Figure 3-2). In this way, they are not prone to being forced open when the expansion is started. They should also be as far as possible away from the expander. All this allows one to start the expansion much earlier. It also seems to reduce the rate of complications that would occur later as a result of dehiscence of the scar since the migration or recruitment of tissues during the expansion causes the tissues to move toward the expander. Furthermore, when possible, one should design the incisions within the area to be removed so no additional scars are created in the donor area.

PLANNING OF FLAPS

When one plans the reconstruction procedure, one simultaneously plans the flaps to be used in the reconstruction. Very often an advancement flap is used because it does not necessarily create additional scars as, for example, a transposition flap does. If a straight line closure along one edge or in the middle of a round lesion is to be achieved, one or two crescent-shaped expanders can be used with success. To achieve effective advancement using a rectangular expander, two parallell incisions should be made in the direction of the advancement. The higher up on the expander one places the incisions the more advancement one will achieve, but the narrower the flaps will be. Then one will have two additional lateral expanded areas that may be of no use and may have to be excised. The same problem occurs with a hemispheric expander and a rotation flap. To achieve rotation one has to incise somewhere. When the incision is made along the bottom of the expander, the rotation is minimal. With a higher incision, there is more rotation but a smaller flap. Also, here there is an area of expanded tissue not to be included in the flap, as with the advancement flap mentioned above, but this area can be used to some extent to fill in the area of a back-cut if one is used.

Zide and Karp,[16] who described single and double back-cut flaps, have noted that creation of flaps by cutting around the advancing edge of expanded tissue does not take full advantage of the expansion. Moreover, since the expanded tissue must cover the area where the expander was situated, the width of the implant base helps determine the position of the back-cut (usually the middle of the expander). In cases of triangular lesions or when transposition is required, the authors suggest that the expander be placed parallel to the longest side of the lesion; after expander inflation, a triangular type of flap may be designed to the exact size of the defect to be covered by using a back-cut on one side. In this way, the tissue proximal to the back-cut covers the area where the expander was, while a small dog-ear usually needs to be excised.

HOW TO JUDGE THE AMOUNT OF YIELD CREATED BY EXPANSION

It is important to measure the amount of yield achieved to know when sufficient expansion is performed to allow for reconstruction of the defect. It is also important to be aware that all expanded yield will not be available for the planned reconstruction because it will not all move as desired, and although dog-ears are a yield of the expansion, they are not effectively used for the reconstruction. Rather big differences arise from two further facts: first, there is considerable pressure into the expander and the flap seldom is inserted in its new place with the same amount of tension. Due to its elasticity and contractile forces, the flap will then have a smaller area in the recipient site. Second, when one measures the yield of expansion on the skin, one measures the outer surface of a convex structure covering the expander. On the other hand, it is the inner surface of this same structure that covers the defect. This inner surface is smaller than the surface area measured and thus gives a negative difference. The only case in which surface area loss of this type is not a factor is when there is a convex recipient area and a similar convexity of the expander as, for example, in scalp reconstruction.

How does one know when there is sufficient yield? In 1986, Shively described a computer program that calculates, according to the diameter of the implant base and the desired surface-area gain, the volume needed to expand skin a predetermined amount by means of a spherical expander.[17] In this study, the volume was arbitrarily made equal to the desired area gain. In the same year, Patel developed an analytic solution, which did not require the use of a computer, to solve the equation.[18] He stated, as a conveniently useful rule of thumb, that the final projection (height) of the expander has to be approximately 0.6 times the square root of the area to be reconstructed. More recently, Duits et al.[19] have reported another mathematical model designed to convert tissue-expander volume to surface area.

In our opinion, complex mathematical formulas of any sort will not be a great help because there is a three-dimensional situation with a lot of disturbing local factors. Thus, we routinely use other simple systems that we will describe. When the expander is close to the defect, an advancement flap is going to be used, and the recipient surface as well as the donor area are reasonably flat (this even applies, for example, to the use of small expanders in the scalp), the distance from one end of the base to the other (measured with a caliper) in the direction of the expansion is subtracted from the distance between the same points but measured with a tape measure along the surface of the skin (Figure 3-3). The resulting measurement is then reduced by about one third to compensate for the factors that reduce the effectiveness of the yield to cover the defect. This calculation has served us well during many years and for hundreds of tissue expansions. It is also good to know that in a few cases there has been superfluous tissue available, but many times one would have wished there to be more. So it is better to err on the safe side, especially in cases in which even a small amount of tension will be disastrous (as in the case of tissue pulling the lower lid). If one has too much tissue, it is very easy to trim some away.

It is also worth knowing some tricks to gain a few more millimeters. One is to remove the capsule if this has not already been done. This might somewhat decrease the blood flow in the flap and should be reconsidered in those flaps having a critical circulation. Another trick is to maximally inflate the expander just prior to removing it; this can be done when the patient is under local or general anesthesia. The patient will not be disturbed by pain, and in most cases it is possible to fill until the skin blanches (due to no capillary refill) and even above this point. The capsule around the expander often breaks during this procedure. It is difficult to do studies to evaluate how much such a procedure increases the yield, but our estimate is that it usually gives at least some millimeters and is very well worth a few extra minutes added to the time in the operating room.

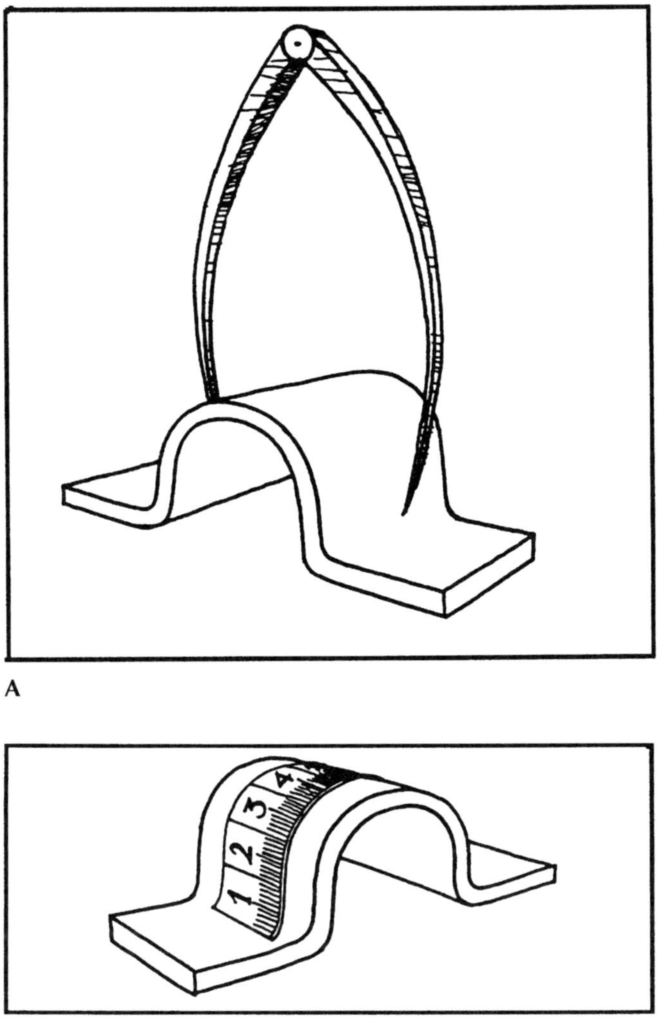

A

B

FIGURE 3-3
(A) Measurement taken with a caliper. (B) Measurement taken with a tape measure.

If there is too little tissue, in most cases an expander can be reinserted; often, the previous expander is suitable. It is obviously better if this possibility has been discussed with the patient before the operation and he or she has agreed to this because it is impossible for the surgeon to know, for example, if the social situation of the patient allows for a continuation of the expansion procedure. Also, in the hair-bearing scalp region it is often advisable to perform multiple expansions, excising the lesion and replacing the expander in a series of operations. An expansion to about 500 milliliters is not very disturbing socially in most patients, and, for this reason, they are able to continue normal life despite the reconstruction. Up to five repeated expansions with excisions have been carried out by us in this way with some patients.

REFERENCES

1. Shively RE, Bermant MA, Bucholz RD. Separation of craniopagus twins utilizing tissue expanders. *Plast Reconstr Surg* 1985;76:765–772.
2. Van Rappard JHA, Sonneveld GJ, Borghouts JMHM. Geometric planning and the shape of the expander. *Facial Plast Surg* 1988a;5:287–290.
3. Van Rappard JHA, Molenaar J, Van Doorn K. et al. Surface-area increase in tissue expansion. *Plast Reconstr Surg* 1988b; 82:833–837.
4. Morgan RF, Edgerton MT. Tissue expansion in reconstructive hand surgery. *J Hand Surg* 1985;10:754–759.
5. Radovan C. Tissue expansion in soft-tissue reconstruction. *Plast Reconstr Surg* 1984;74:482–490.
6. Gibney J. Tissue expansion in reconstructive surgery. Presented at ASPRS Annual Scientific Meeting, Las Vegas, Nevada 1984.
7. Hallock GG. Maximum overinflation of tissue expanders. *Plast Reconstr Surg* 1987;80:567–569.
8. Hirshowitz B, Kaufman T, Ulman J. Reconstruction of the tip of the nose and ala by load cycling of the nasal skin and harnessing of extra skin. *Plast Reconstr Surg* 1986;77: 316–320.
9. Gibson T. The physical properties of the skin. In McCarthy JG (ed.), *Plastic surgery*. Philadelphia, PA: W. B. Saunders Company, 1990, pp. 207–220.
10. Nordström REA. "Stretch-back" in scalp reductions for male pattern baldness. *Plast Reconstr Surg* 1984;73: 422–426.
11. Nordström REA, Pietilä JP, Voutilainen PEJ, et al. Tissue expander injection dome leakage. *Plast Reconstr Surg* 1988;81:26–29.
12. Brobmann GF, Huber J. Effects of different-shaped tissue expanders on transluminal pressure, oxygen tension, histopathologic changes, and skin expansion in pigs. *Plast Reconstr Surg* 1985;76:731–736.
13. Jackson I. Update on tissue expansion using external ports. *Perspect Plast Surg* 1990;4:129–131.
14. Argenta LC, Austad ED. Principles and techniques of tissue expansion. In McCarthy JG (ed.), *Plastic surgery*. Philadelphia, PA: W. B. Saunders Company, 1990, pp. 475–507.
15. Manders EK, Schenden MJ, Furrey JA, et al. Soft-tissue expansion: Concepts and complications. *Plast Reconstr Surg* 1984;74:493–507.
16. Zide BM, Karp NS. Maximizing gain from rectangular tissue expanders. *Plast Reconstr Surg* 1992;90:500–504.
17. Shively RE. Skin-expander volume estimator. *Plast Reconstr Surg* 1986;77:482–483.
18. Patel PK. Estimating the tissue-expander volume: A poor man's recipe. *Plast Reconstr Surg* 1986;78:426.
19. Duits EHAm, Molenaar J, van Rappard JHA. The modeling of skin expanders. *Plast Reconstr Surg* 1989;83: 362–366.

4

Histologic Changes

JULIEN H. A. VAN RAPPARD

The increasing acceptance of tissue expansion led to a growing interest in the underlying physiological and histological mechanisms. Austad et al.[1] and Pasyk et al.[2] performed many experiments in an attempt to answer a recurring question; namely, what changes take place in expanded tissue? Unfortunately, they used self-inflating expanders in their experiments and observed a very gradual expansion. Moreover, they described the cytohistological changes in guinea pigs—animals that are less suitable for experiments on (skin) tissue expansion, particularly because they have very loose skin that can move freely over the deeper tissue.[3] Pig skin is suitable for tissue expansion experiments and resembles human skin more closely than that of most other laboratory animals—in some ways even more than the skin of baboons or monkeys.[4] To obtain more information regarding changes at the cytohistological level during expansion of soft tissue, we performed our tissue expansion experiments in minipigs (Göttinger Minischwein).[5] Recently, these findings have been complemented by us by morphohistological evaluation of tissue samples obtained after tissue expansion procedures in humans. A total number of 722 tissue samples have been evaluated by us, using light and electron microscopy.[6]

BIOMECHANICAL PROPERTIES OF SKIN

A notable physical property that characterizes skin is its ability to adapt to continuous changes. Although skin is an organ composed of several layers and tissues, the overall mechanical properties of the skin are largely determined by the dermis. The amount and distribution of collagen and elastin play a prominent role in mechanical properties. The collagen fibers in relaxed human skin are arranged so that the polypeptides are twined into a rod shape. When stretched, these fibers become oriented in a direction parallel to the stretching force, enabling them to resist further extension. Elastin fibers are abundantly present between the collagen fibers and are linked to one another by end-to-side junctions. Their function is to reestablish the configuration in which the collagen is most relaxed. As collagen fibers have no retracting ability, permanent deformation might take place when these fibres are stressed, exceeding the limits of the elastic fibers.

In experiments using guinea pigs, Schneider et al.[7] showed a significant reduction of tensile strength and elasticity in skin as a result of tissue expansion. Comparable experiments in rabbits by Timmenga et al.[8] demonstrated a decreased stiffness and a significantly larger deformation in expanded skin than in sham-operated and control skin. Maximum stiffness was reached at a shorter extensibility, indicating that the tissue tore sooner and more gradually upon loading. These findings indicate a loss rather than a gain of elasticity. According to Mustoe et al.,[9] changes in the biomechanical properties do not depend on the rate of the expansion.

MORPHOHISTOLOGY OF EXPANDED SOFT TISSUE

Epidermis

In comparison to nonexpanded skin both in animals and humans, epidermal thickness increased under the influence of expansion. Even in rapidly expanded tissues in minipigs, normal or increased thickness was seen after only 7 days. The intercellular spaces did not increase. Occasionally, the polyhedral cells appeared to be somewhat flattened. Microscopically, no increase in mitotic activity could be recorded.[6]

With the normal, clinically applied protocol of tissue expansion, no difference was seen in the histologically distinguishable layers (stratum lucidum is absent in pig skin) or in the different types of cells. By contrast, with extremely acute expansion, spontaneous erosion of the epithelium sometimes occurred. The nuclei became pyknotic and, at the level of the rete ridge, cleavage from the basal membrane was seen. This resulted in slight edema and dermal capillary injection. A marked flattening of the rete ridge was observed, which was most pronounced at locations where expansion was fastest. After more than 12 months, the normal undulating complementary pattern had not returned (Figure 4-1). Olenius (1987, personal communication) described a remarkable increase in the amount of melanin pigment granules in the basal cell layers of human tissue, but this was not confirmed by our findings in humans. Although no increase in mitotic activity was seen in their study, Austad et al.[1] concluded that the constant or increased thickness of the epidermis in guinea pig skin, in combination with no increase of the intercellular spaces, supported the hypothesis that tissue expansion results in a net gain of donor tissue. In 1986, they confirmed this in an animal study in which they used tritiated thymidine to determine that mitotic activity increased under the influence of tissue expansion.[10] Four weeks after expansion, a return to a normal, or perhaps a decreased, situation was seen. Unfortunately, the authors did not evaluate the duration of this gain or *dividend*. In 1987 and 1988, we restudied the epidermopoiesis in controlled tissue expansion.[11] Unlike Austad, we did not use guinea pigs, but for the reasons described, we used Göttinger minipigs. Using flow-cytometric techniques, we were able to analyze extremely large numbers of nuclei at high speed, and this allows for a very objective evaluation of the epidermopoiesis.[12] Also, with this technique, quantitative assessment of stretching is possible. The increase in mitotic activity described by Austad, et al.[10] was confirmed by our study, but it was also shown that with time, the number of epidermal cells per surface unit stayed at the same level in stretched skin as in unstretched skin. This proved that the increase in epidermopoiesis is a temporary activity with a lasting dividend.

Dermis

In our study, the dermis appeared to be more sensitive to expansion than the epidermis. The thickness of the dermis decreased in comparison to that of control and sham-operated animals. This dermal thinning was significantly dependent on the rate of expansion.

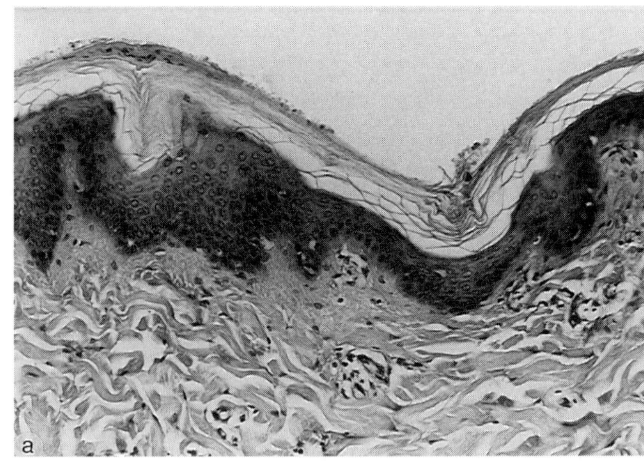

A

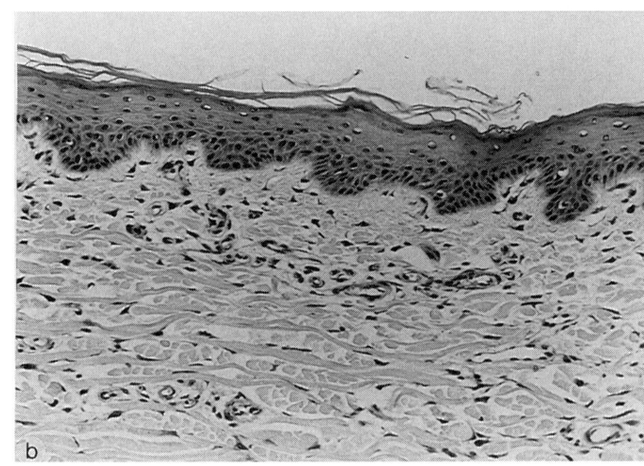

B

FIGURE 4-1
Epidermal-dermal junction. (A) Control skin. Peaked and dome-shaped dermal papillae impact to a wave-like appearance to the border line. Several papillae contribute to one ridge (PAS and hematoxylin × 200). (B) Stretched pig skin. Loss of ridges and reduction of papillar size result in a flattened border line and epidermal surface (PAS and hematoxylin × 200).

Pasyk et al.[13] confirmed our findings of a decrease in thickness of about 20 percent. We did not see a significant total recuperation involving an increase in thickness with time. However, neither the distribution nor the density of fibroblasts was significantly different from that of unexpanded skin. After several weeks of expansion, myofibroblasts appear within the dermis. Myofibroblasts have been found in multiple animal and human granulation tissues. In addition, they have been found in many of the pathological contractile processes of man, such as fibrosis. The presence of the myofibroblasts in the dermis during expansion suggests that the dermis may have taken on a contractile function, possibly in response to progressive expansion. There are also subtle changes in the dermal collagen fibers associated with expansion. The usual three-dimensional texture of collagen fibers is characterized by a cross-linkage of collagen bundles of about 70 degrees from the long axis. With increased expansion, the angle of this cross-linkage decreases. The alignment of collagen fibers starts in the stratum reticulare, and in later stages of expansion, the stratum papillare is also involved. Eventually, all collagen fibers run almost parallel to the surface. After 6 months, the threedimensional image still showed angles less than 70 degrees (Figure 4-2). The collagen fibers showed a tendency to hyalinization. Using a monoclonal procollagen antibody for immunohistochemical staining, Mustoe et al.[9] demonstrated an increase in collagen synthesis by dermal fibroblasts. Notwithstanding these findings, we did not see an increase in the amount of dermal collagen, nor did we see any sign of the production of secondary collagen by electron microscopy (Figure 4-3).

Dermal Appendages

Hair. The papillar epithelium at the base of the follicle of hair roots, although sensitive to anoxia, showed no change in the expanded skin. The distance between the hair follicles in expanded skin became greater, but the number of hair follicles did not increase. During the expansion, considerable hair growth took place. In the animal experiments, temporary and only partial loss of hair was seen in very fast expansion procedures. The increase in distance between the hair follicles is only noticeable when the hair density has been reduced by more than 50 percent (that is, when the area is stretched over 100 percent).

Sebaceous Glands. In our studies, sebaceous glands were relatively insensitive to stretching. Individual

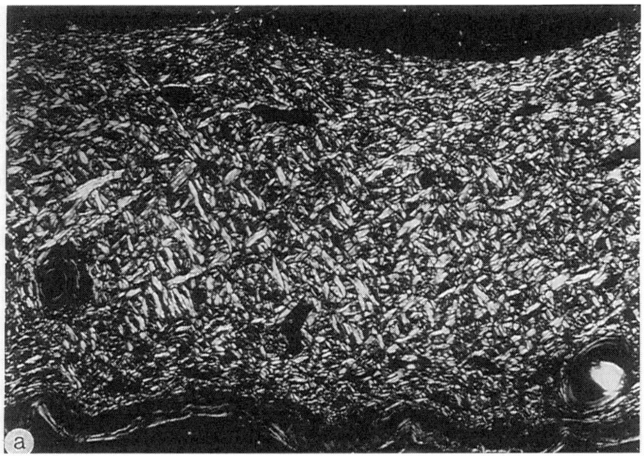

A

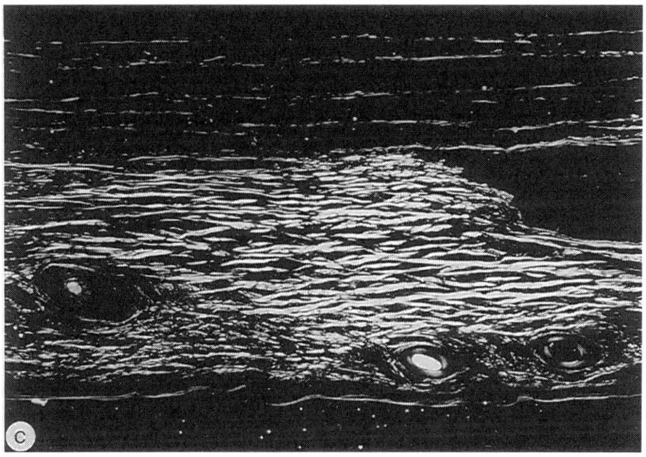

C

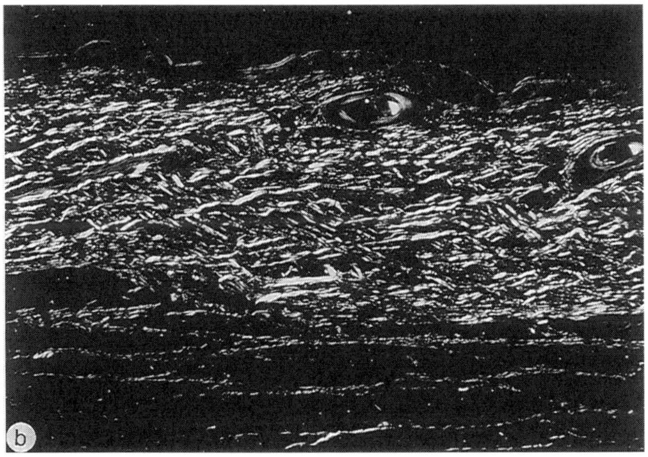

B

FIGURE 4-2
Texture of dermal collagen fibers in minipigs. (A) Control skin. Bundles of collagen fibers cross at a mean angle of about 70 degrees (polarized light × 30). (B) Moderately stretched skin. Crossing of collagen fiber bundles in an acute angle (polarized light × 30). (C) Extremely stretched skin. Collagen fibers run almost parallel to the surface (polarized light × 30). Note the reduction in dermal thickness in B and C.

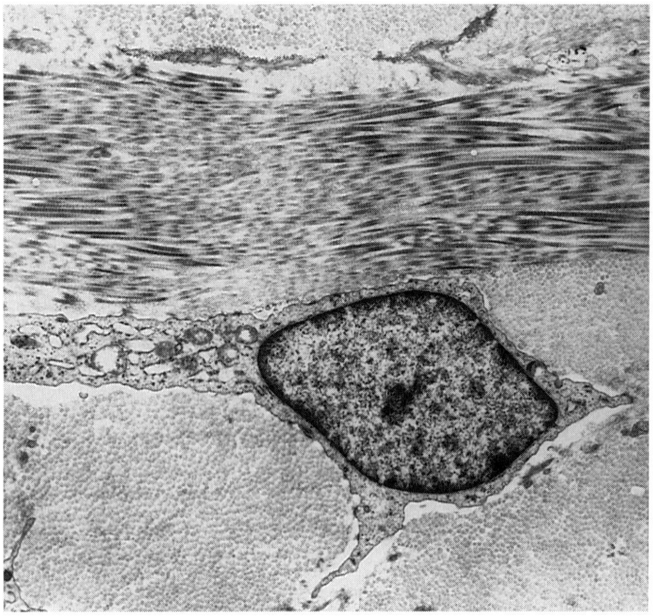

FIGURE 4-3
Fibrocyte from stretched skin with normal appearance. No indications of production of secondary collagen (TEM × 6500).

signs of blockage, either as a result of inactivity or by compression of the excretory duct, were noticed. If the arrectores pilorum muscles showed signs of atrophy, the sebaceous glands were likely to have secretion problems.

Sweat Glands. Sweat glands do not exist in pig skin. In humans, no clear changes in sweat glands are apparent with skin expansion, with the exception of some signs of blockage, which disappear totally a short time after the second operation.

Arrectores Pilorum Muscles. The arrectores pilorum muscles appeared rather insensitive to stretching. Only in situations of extreme expansion did the muscle cells undergo atrophy and even necrosis. In these cases, the effects appeared to be permanent, and significantly more blockage in the ducts of sebaceous glands was seen.

Subcutaneous Fat

The layer of subcutaneous fat is very sensitive to expansion. Its thickness decreased during standard expansion by up to 60 percent, and a significant decrease in the number of fat cells was seen. In the first months after expansion, the thickness was only partly reestablished. This is in contrast to the findings of Pasyk et al.[13] who described, in a series of three human subjects, a total recovery in the subcutaneous fat layer to the normal thickness and even beyond this in the 2 years after removal of the expander. By contrast, our studies show that during faster expansions, necrosis of fat cells frequently occurs, while recovery of thickness in the fat layer was absent even more than 12 months after expansion.

Muscle

The muscle layer in pig skin is a noncontinuous coat called panniculus carnosus. It does not exist in humans, with the exception of the platysma muscle in the subcutaneous tissue of the neck. This layer is sensitive to stretching and pressure. After a threshold of expansion is exceeded, human muscle fibers generally atrophied. The striation pattern decreased or was lost, hyalinization occurred, and the tissue became necrotic. Formation of giant cells and calcification of the damaged tissue was the final step in these catabolic events (Figure 4-4). Although these observations show a sensitivity of the muscular layer to expansion, no functional loss in excursion or strength of the expanded muscle has been demonstrated in our clinical cases. In humans, striated muscle appeared to be more tolerant to expansion than the arrectores pilorum muscles in pigs. Biopsy specimens obtained from the floor of the expanded pockets showed some hyalinization of muscle cells.

Capsule

Usually, a capsule develops around tissue expanders, as it does around all other foreign bodies. In our study, primarily granulation tissue developed consisting of macrophages, fibroblasts, and lymphocytes. Within about 7 days, a double-layered capsule, which consisted of a cell-rich inner layer of macrophages and an outer layer mainly of fibroblasts and some lymphocytes, was present. In time, the outer coat gradually became more rich in collagenous fibers. During the first phase, this capsule and the bordering subcutis were abundantly vascularized. The border between capsule and subcutis became clearer, secondary to the fading of the sterile inflammatory reaction. Later, the capsule close to the tissue expander was covered by a continuous cellular coat consisting of macrophages. Adjacent to the macrophages, a capsule of collagen fibers developed, which condensed progressively. However, the histomorphology of the capsule surrounding the same tissue expander could vary considerably. Differences were also noticed during the course of the study, ranging from only granulation tissue after three weeks to a chronic progressive inflammatory reaction after ten weeks. An increased number of eosinophilic granulocytes was present. No signs of an acute or chronic bacterial infection were seen in any of the histological sections. Giant cells were occasionally associated with hemorrhages. Now and then these multinuclear cells

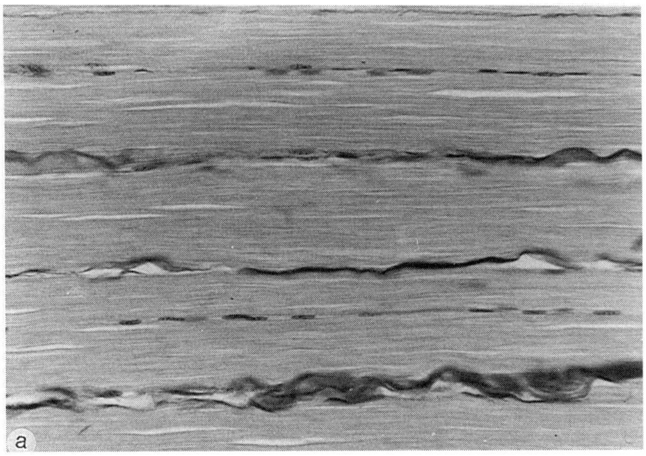

A

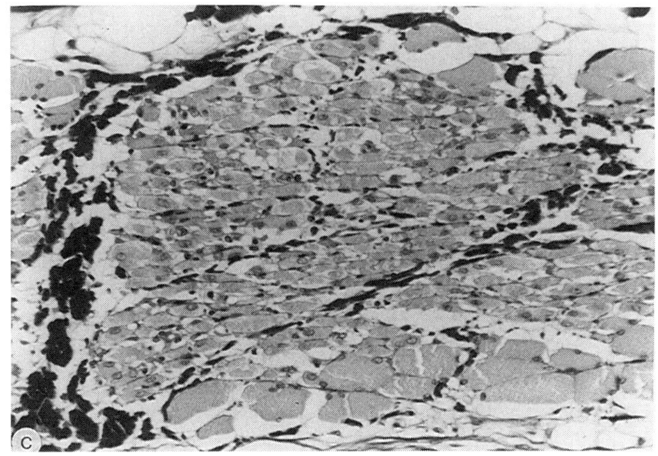

C

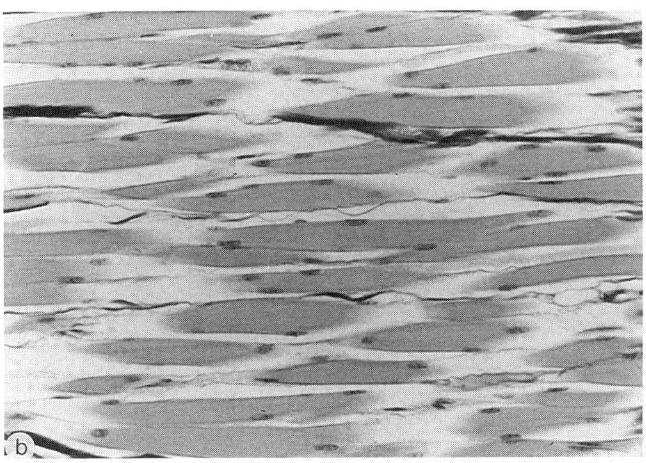

B

FIGURE 4-4
Panniculus carnosus. (A) Control skin. Close arrangement of striated muscle fibers (longitudinal section, van Gieson × 300). (B) Moderately stretched skin. Opening of intercellular space due to shrinkage by atrophy of individual muscle cells (oblique section, van Gieson × 300). (C) Extremely stretched skin. Area of muscle cell necrosis. Note the thickening by condensation of perimysial connective tissue (cross section, van Gieson × 300).

contained a birefrigent material, possibly silicone particles originating from the tissue expander. There was no statistically relevant correlation between the histomorphology of the capsule and the rate of inflation of the expander. The reactivity of the capsule granulation tissue is probably more dependent on local factors such as hematoma, mechanical factors, possible allergic reactions to talcum powder, suture material or gauze, local immunologic dysregulation, and necrosis of overlying fat or muscular tissue (Figures 4-5 and 4-6).

Capsules around tissue expanders are thicker than, for example, those usually found around ordinary breast implants, and this phenomenon presumably is explained by the recurrent tearing forces on the capsule during inflation.[14] The mean thickness of capsules in our animals and in our series in humans was about 3 to 4 mm.[15] Still, 1 year after the removal of the expanders, capsule remnants were found in our experimental animals.

Vascularization

Contrary to the data from Cherry et al.,[16] Saxby,[17] and Leighton et al.,[18] no increase in vascularization or size of the capillary bed of the expanded skin could be demonstrated in our morphohistological study. In our vision, it is the surgical procedure that induces a delay-like phenomenon, which leads to an increase of survival of expanded random pattern flaps. Occasionally, the expansion resulted in extravasation. Rupture of small vessels occurred first in the subcutis (especially in the fat layer following massive expansion) and also in the dermis. This extravasation resulted in a sterile inflammatory reaction characterized by the appearance of macrophages, lymphocytes, and fibroblasts. Although inflammatory reaction can lead to vascular calcification, this was not seen in acute stretching, except in regions of distinct muscular necrosis. Neovascularization was not seen.

Stark et al.[19] describe a rapid elongation of expanded arteries and veins in rats following tissue expansion.

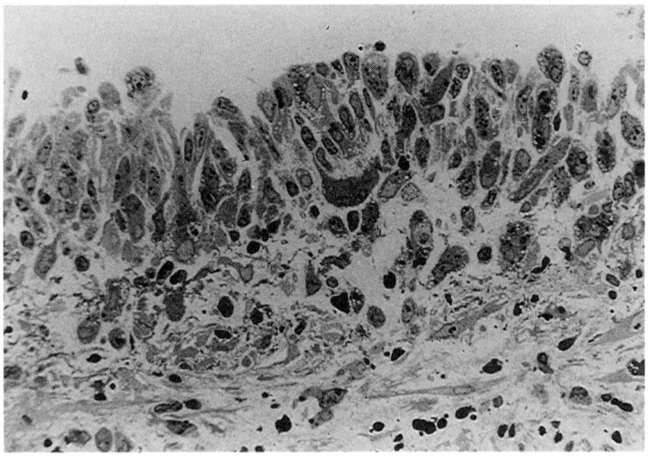

A

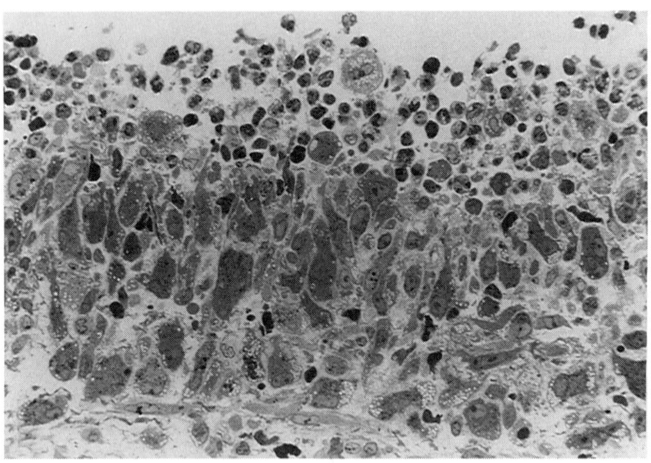

B

FIGURE 4-5
Innermost coat of capsule in minipigs. (A) Consisting of three to four layers of macrophages only with underlying granulation tissue. (B) With neutrophils and cellular debris interspersed between macrophages and expander. A and B are from the same specimen (semithin section for TEM, toluidin blue × 400).

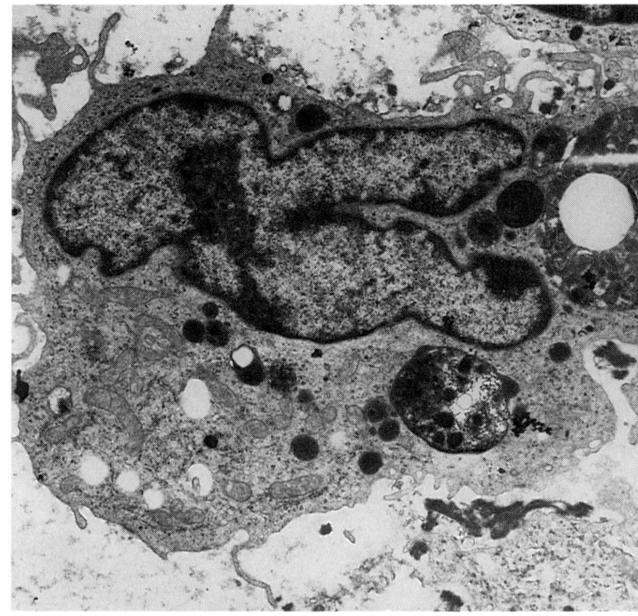

FIGURE 4-6
Macrophage from free border of innermost capsule, exhibiting primary lysosomes and phagolysosomes (TEM, × 5400) in minipigs.

The mean successful elongation of the vessels was 84 ± 47 percent (standard deviation) in 6 weeks. The morphohistology showed no reduction in vessel wall diameter or loss of intimal integrity. No difference in expansion tolerance between arteries and veins was seen.

Nerve Tissue
Nerve tissue appears to be tolerant of expansion. In the cutaneous nerves of expanded skin, no clear difference was seen between the nonexpanded sites and the expanded sites. An occasional condensation of the perineurium in the expanded tissue might indicate a slight atrophy due to pressure. In the extremely overexpanded skin in our animal experiments (25 percent of the original volume per week during 2 weeks, leading to 200 percent volume or 500 ml in 250-ml rectangular expanders in a regular pocket 10 cm × 6 cm), only slight demyelinization of the nerve tissue was seen. There were no signs of necrosis; moreover, reactive sprouting did not appear at a follow-up at 8 months.

Experiments on peripheral nerves of rabbits by van der Wey et al.[20] confirm the findings of Manders et al.[21] in dogs and Milner[22] in guinea pigs. When nerves are expanded gradually, only a moderate reduction in nerve function is seen as reflected by a fall in conduction velocity. Morphohistological changes are restricted to sporadic segmental demyelinization and a slight thinning of the axons, but no axon degeneration is seen. In our material no more than a second degree ischaemic elongation injury (neuropraxia and axontmesis) was seen in the gradually expanded nerves. Only in the extremely rapid expansion (30 percent elongation in 5 days) was a higher degree of ischaemic elongation injury seen, leading to loss of function.[20] It is concluded that nerve tissue is astonishingly tolerant to expansion.

CONCLUSION

Although soft tissue, when adequately expanded, has an apparently normal appearance from the macroscopic point of view, several changes are noticeable at the microscopic level. A review of the literature and our own findings reveal different tolerances toward expansion in the various tissue layers. All layers, with the exception of the epidermal layer, seem to be very sensitive to the expansion process. A limited increase in fibroblasts in the dermis is probably a reaction to the inflammatory process that results from the expansion. There was no apparent increase in the number of dermal collagen bundles. Fat cells and muscle cells usually decreased in number and size, and fast expansion caused necrosis followed by fibrosis of damaged tissues. In our experiments, the sterile inflammation was more pronounced than that described by Austad et al., and, in contrast, the described increase in vascularization was absent. The harm caused by a relatively fast expansion demonstrates that, with the exception of the epidermal layer, tissue expansion results mostly from a passive stretching of tissue. According to Gibson,[23,24] there are three possible ways in which skin can be stretched:

(1) inherent extensibility—the phenomenon that allows small skin defects to be closed directly
(2) mechanical creep—the gradual stretching of skin beyond the limit of its inherent extensibility
(3) biological creep—the gradual stretching of skin lying over a slowly expanding subcutaneous mass; such as the pregnant uterus.

Biological creep is not stretching as it is commonly understood because the skin is not thinned while it extends. In other words, it is effected by metabolic activity, creating new epithelium, collagen and elastical fibers, blood vessels, nerves, and lymphatics, and it is probably stimulated by hormonal factors.

Since striae in the abdominal skin usually remain after pregnancy, it has been suggested that the expansion of the pregnant uterus occurs too fast to give the skin enough time to grow.[24] Our present data indicate that tissue expansion is anything but biological creep. The explanation may be found in the fact that the skin, as an organ, is too specialized to allow regeneration. The same holds true for other highly specialized organ systems such as the central and peripheral nervous systems, the lung, and the kidney. The reaction of these organs to an injury is only healing of the original structure by fibrosis. Only in very primitive animals (for example, worms and reptiles) is full regeneration of parts of the body seen. By contrast, in the epidermis of higher species, a pool of undifferentiated stem cells ensures a constant supply of cells that can undergo specific differentiation (keratinization).[25] This explains the relative tolerance of the epidermis to expansion and the absence of thinning of the layer, and it suggests the formation of new tissue. The experiments described using flow cytometric techniques in expanded skin demonstrated an increase in mitotic activity.[11] Indeed, as demonstrated by Francis and Marks,[26] stretched epidermis always responds with an increase in mitotic activity. The epithelial cells of the epidermis react temporarily to various stresses. For example, exposure to mechanical stress produces horny skin on the hands of mineworkers. When mechanical stress is no longer exerted, this horny coat exfoliates. Exposure to ultraviolet radiation, such as sunlight, not only brings about a tan but also induces an epidermal thickening. Vacationers must notice that after a holiday in the sun, the tan fades and skin regains its smoothness. Given these examples, one can imagine that secondary to exposure to shearing forces, the epidermal cells will react immediately until the stress is gone. Finally, I conclude that in almost all layers, with exception of the epidermis, tissue expansion is really stretching of skin more than formation of new tissue. The widening of scars that frequently occurs after clinical application of the tissue expansion techniques orginates presumably from the process called stretch-back by Nordström[27] and supports these findings.

REFERENCES

1. Austad ED, Pasyk KA, McClatchey KD, Cherry GW. Histomorphologic evaluation of guinea pig skin and soft tissue after controlled tissue-expansion. *Plast Reconstr Surg* 1982;70:704–710.
2. Pasyk KA, Austad ED, McClatchey KD, Cherry GW. Electron microscopic evaluation of guinea pig skin and soft tissues expanded with a self-inflating silicone implant. *Plast Reconstr Surg,* 1982;70:37–45.
3. Bartell TH, Mustoe TA. Animal models of human tissue expansion. *Plast Reconstr Surg* 1989;83:681–686.
4. Ordman LJ, Gillman T. Studies in healing of cutaneous wounds. *Arch Surg* 1966;93:857–882.
5. Van Rappard JHA, Grubben MJAL, Jerusalem C, Borghouts JMHM. Histological changes in expanded soft tissue. In Sasaki G (ed.), *Proceedings of the first international symposium on tissue-expansion.* San Francisco: PSEF, 1987, pp. 143–170.
6. Van Rappard JHA, Grubben MJAL, Jerusalem C, Borghouts JMHM. Histologic changes in expanded soft tissue. In van Rappard JHA (ed.), *Controlled tissue-expansion in reconstructive surgery.* Nijmegen, The Netherlands: SSN Press, 1988, pp. 25–40.
7. Schneider MS, Borkow JE, Cruz IT, Marangoni RD, Shaffer J, Grove J. The tensiometric properties of expanded guinea pig skin. *Plast Reconstr Surg* 1988;81:398–403.
8. Timmenga EJF, Schoorl R, Klopper PJ. Biomechanical and histomorphological changes in expanded rabbit skin. *Br J Plast Surg* 1990;43:101–106.
9. Mustoe TA, Bartell TH, Garner WL. Physical, biomechanical, histologic and biochemical effects of rapid versus conventional tissue expansion. *Plast Reconstr Surg* 1989;83: 687–691.

10. Austad ED, Thomas SB, Pasyk K. Tissue-expansion: dividend or loan? *Plast Reconstr Surg* 1986;78:63–67.
11. Van Rappard JHA, Bauer FW, Grubben MJAL, van Erp P, Borghouts JMHM. Epidermopoiesis in controlled tissue-expansion. In van Rappard JHA (ed.), *Controlled tissue-expansion in reconstructive surgery.* Nijmegen, The Netherlands: SSN Press, 1988, p. 15.
12. Boezeman JBM, Bauer FW, de Grood RM. Flowcytometric analysis of the recruitment of G30 cells in human epidermis in vivo following tape stripping. *Cell Tiss Kinet* 1987;20:99–107.
13. Pasyk KA, Argenta LC, Hassett C. Quantitative analysis of the thickness of human skin and subcutaneous tissue following controlled expansion with a silicone implant. *Plast Reconstr Surg* 1988;81:516–523.
14. Van Rappard JHA. Capsule formation in controlled tissue-expansion. In van Rappard JHA (ed.), *Controlled tissue-expansion in reconstructive surgery.* Nijmegen, The Netherlands: SSN Press, 1988, pp. 51–88.
15. Van Rappard JHA, Jerusalem C, Sonneveld GJ, Borghouts JMHM. Histologic changes in soft tissue due to tissue expansion. *Fac Plast Surg* 1988;5:280–286.
16. Cherry GW, Austad ED, Pasyk KA, McClatchey KD, Rohrig RJ. Increased survival and vascularity of random pattern skin flaps elevated in controlled expanded skin. *Plast Reconstr Surg* 1983;72:580–685.
17. Saxby PJ. Survival of island flaps after tissue-expansion: A pig model. *Plast Reconstr Surg* 1988;81:30–34.
18. Leighton WD, Russel RC, Marcus DE, Eriksson E, Suchy H, Zook EG. Experimental pretransfer expansion of free-flap donor sites: II, Physiology, histology and clinical correlation. *Plast Reconstr Surg* 1988;82:76–84.
19. Stark GB, Hong C, Futrell CW. Rapid elongation of arteries and veins in rats with a tissue-expander. *Plast Reconstr Surg* 1987;80:570–581.
20. Van der Wey LP, Gabreels-Festen AA, Merks MH, Polder TW, Stegeman DF, Spauwen PH, Gabreels FJ. Peripheral nerve elongation by laser Doppler flowmetry controlled expansion: Morphological aspects. *Acta Neuropathol Berl* 1995;89:166–72.
21. Manders EK, Saggers GC, Diaz-Alonso P, Finn L, Sipio JC, Glumac T, et al. Elongation of peripheral nerve and viscera containing smooth muscle. *Clin Plast Surg* 1987; 14:551–562.
22. Milner RH. The effect of tissue expansion on peripheral nerves. *Br J Plast Surg* 1989;42:414–421.
23. Gibson T. The physical properties of skin. In Converse JM (ed.), *Reconstructive plastic surgery.* Philadelphia: Saunders, 1977, pp. 70–77.
24. Gibson T. Reconstruction of the tip of the nose and ala by load cycling of the nasal skin and harnessing of extra skin (discussion). *Plast Reconstr Surg* 1986;70: 320–321.
25. Letterer E. Erwachstum. In *Algemeine pathologie.* Stuttgart: Georg Thieme Verlag, 1959, pp. 470–545.
26. Francis AJ, Marks R. Skin stretching and epidermopoiesis. *Br J Exp Pathol* 1973;58:35–47.
27. Nordström REA. "Stretch-back" in scalp reductions for male pattern baldness. *Plast Reconstr Surg* 1984;73: 422–426.

5

Available Devices

JULIEN H. A. VAN RAPPARD

Tissue expanders have turned out to be an important advance in plastic and reconstructive surgery because they are used to create optimal matching and to avoid unacceptable donor sites. As Muenker[1] observed, the evolution of the tissue expansion technique is closely connected with the technical development of expander devices. The first clinical application of tissue expansion by means of an implant was performed by Neumann in 1957.[2] He used a subcutaneously placed rubber balloon with an external filling mechanism to reconstruct an ear. Apart from the fact that ear reconstruction is one of the more difficult applications of tissue expansion, Neumann was confronted with device problems. The rubber out of which the balloon was made was not sufficiently inert to avoid serious tissue reaction. It was not until 1976 that redevelopment of the technique took place. At the annual meeting of the American Society of Plastic and Reconstructive Surgeons in Boston, Radovan presented his tissue expansion technique. Independently, at the same time, Austad and Rose[3] developed their own tissue expansion technique. The Radovan expander was provided with a remote fill-valve, a small subcutaneously placed compartment, which was connected to the inflatable sac by a small tube. The expanders developed by Austad and Rose utilized an osmotic gradient that drew extracellular water through the wall of the expander causing it to fill very gradually. The inflation period of these prostheses depended on volume and shape, and it was relatively long. Moreover, they contained hypertonic sodium chloride, which could cause necrosis in case of rupture. For this reason these devices did not find accepted clinical application.

At first, tissue expansion was restricted to breast reconstructions. In time, after more experience was gained, the number of indications increased, leading to differentiation of shape and size of expanders. Most of them have been based on the Radovan principle (Figure 5-1).

DIFFERENT DEVICES IN TISSUE EXPANSION

The different devices in tissue expansion can be categorized as standard expanders, custom-made expanders, special devices, and inflation devices. The different manufacturers of these expanders are listed below.

Cox-Uphoff International
1160 Mark Avenue
Carpinteria, California 93013
U.S.A.

Laboratoire Euro-Silicone
21, Rue Francis Combe
95000 CERGY-FRANCE

Laboratoire Sebbin
BP 12-95301 Pontoise
CEDEX-FRANCE

LPI Perouse Implants (Laboratoire)
Zone d'activite d'cu treville
60540 Bornet France

McGhan Medical Corporation
700 Ward Drive
Santa Barbara, California 93111
U.S.A.

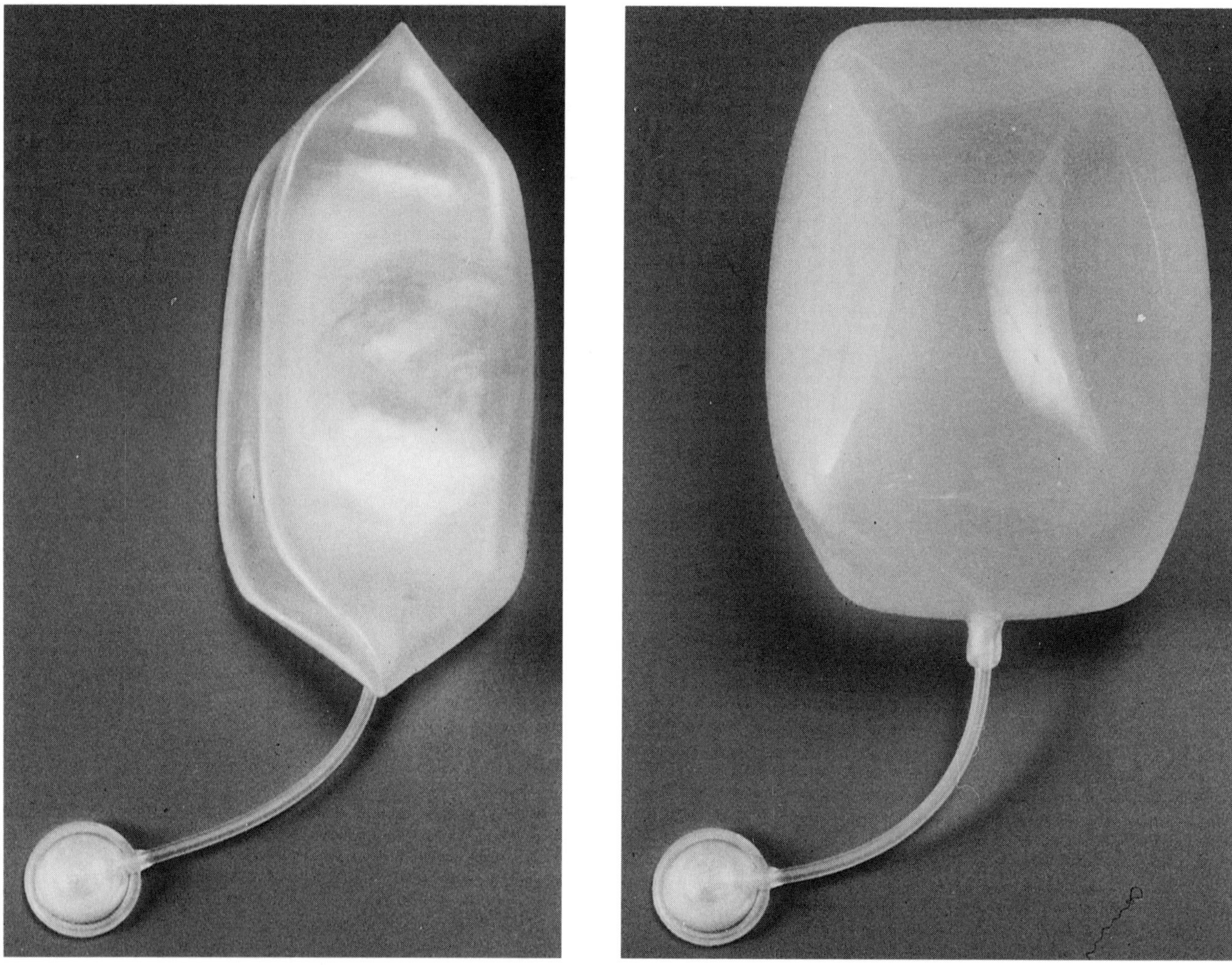

A B

FIGURE 5-1
A basic rectangular (Radovan-like) tissue expander of 250 milliliters with connecting tube and remote fill-valve (A) before and (B) after inflation.

Mentor Corporation
600 Pine Avenue
Goleta, California 93117
U.S.A.

Nagor Limited
P.O. Box 21
Douglas, Isle of Man
British Isles

PIP Poly Implant Prostheses
Ze jean Monet Les Playes
83500 La seyne sur mer France

PMT Corporation
P.O. Box 464
Hopkins, Minnesota 55343
U.S.A.

Silimed/Polytech
Rua Aires Saldhanka, 92 Slj 202 Copacabana
Rio de Janeiro CEP Brazil

Comment by Dr. Eric Austad:

"While expanders remain readily available, regulatory issues and the silicone controversy have resulted in roughly a two-fold increase in the price of these devices. Moreover, while they probably will continue to be classified as Class II devices by the FDA and because they are short-term and do not contain silicone gel, it is unlikely that large commercial suppliers of silicone will return to the market unless legislative protection is afforded them. In the United States, only three sources of silicone are currently available, and this is apparently a worldwide phenomenon. Currently, tissue expanders are available in the United States from four manufacturers: Mentor, PMT, Cox-Uphoff, and McGhan. In spite of the controversy regarding the safety of silicone in its various forms, the manufacturers generally agree that the use of tissue expanders in the past five years has been either "flat" or slightly increasing. None of the remaining manufacturers express pessimism regarding the technique itself, but the continued availability of medical-grade silicon for fabrication is a major concern to most. Other materials are being studied. No new restrictions on the use of expanders have been initiated in this country. With the exception of a significant change in the number of suppliers (Dow Corning and Surgitek no longer produce implants), tissue expanders are readily available and can be used without new protocols or limitations.

"I am aware of a few isolated instances of expander-related lawsuits, claiming either product liability or physician malpractice. These have been related to fairly specific issues such as implant deflation or flap necrosis, however, and have not involved claims of connective tissue disease, hair loss, fatigue, etc. Overall, I believe it is safe to say that tissue expansion and the expanders they rely upon are both alive and well in the United States, but further research will proceed in conjunction with the smaller and privately held companies now producing these devices."

Standard Expanders and Custom-Made Expanders

The tissue expander can best be described as a pre shaped silicone reservoir that can be filled with liquid. This filling is performed by transcutaneous puncture in a small compartment that is connected to the reservoir. This smaller compartment (dome) has a self-sealing wall, which prevents leakage. In the back of the dome, a round stainless steel or hard plastic needle-stop, helps prevent inadvertent perforation of the silicone envelope through the valve (Figure 5-2). Domes can be incorporated into the anterior wall of the prosthesis (self-contained) or can be placed at some distance from the prosthesis and attached by a connecting tube (remote valve). The former, the so-called self-contained valve, has an elevated rim encircling the valve that allows for palpation through the skin, thus facilitating the injection (Figure 5-3). Because the elevated rim was thought to increase the risk of local flap necrosis during expansion, a useful magnetic location system was developed to identify the injection site (Magna Site Locating System, McGhan Medical Corporation) (Figure 5-4). The remote valve can be low profiled or high profiled and small or large cylindrical valves; simpler valves can be supplied for external valves and intraoperative rapid expansion. One of the advantages of incorporated valves is a shorter and easier operation. It is not necessary to create an extra pocket for the connection tube

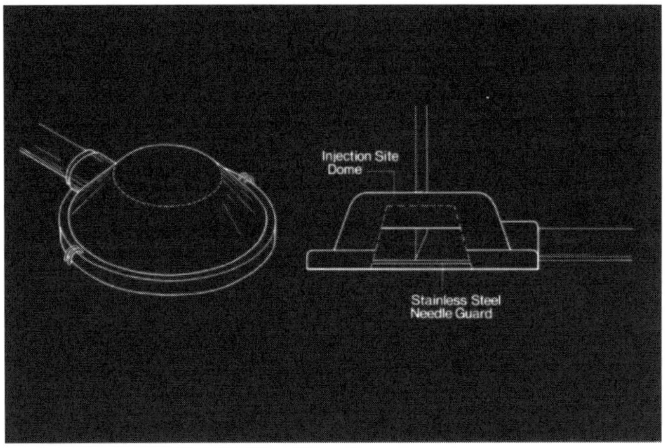

FIGURE 5-2
Schematic drawing of a self-sealing fill-valve.

and dome, and it is much easier to remove the expander. The most obvious disadvantage is the increased risk of damage to the expander. Nevertheless, we have often used incorporated valve expanders that never had to be replaced due to leakage. The industry supplies tissue expanders in various standard shapes: round, tear drop, rectangular, crescent, and longitudinally curved; the volumes vary from 5 to 1000 mL (Figure 5-5).

Custom-made expanders for any location in the body can be ordered. There are almost no limits in volumes and shapes (Figure 5-6). Some designs (Dow Corning Wright) allow the envelope to be sutured to the underlying tissue by means of silicone lips. The problem of wrinkling in the uninflated expander has been avoided by the development of flat expanders. In these *flat-span* expanders, the silicone envelope expands instead of unfolding (Figure 5-7). The flat-span expanders have not been very successful because the base of these expanders does not remain in a horizontal position, and the downward pressure is therefore concentrated on a smaller surface area.

Special Expander Devices

In addition to standard tissue expanders—the most popular of which is the remote filling valve system based on Radovan's prototype—special expander devices have been developed.

Permanent Tissue Expanders for Breast Reconstruction. In 1984, Becker[4–6] was the first to present the technique using permanent tissue expanders for breast reconstruction. In 1985, Gibney[7] started using a similar system. Becker developed his expander with the Mentor Corporation and Gibney developed his with Cox-Uphoff International. The prosthesis is a reverse double-lumen mammary implant that has the silicone gel as the outer implant and the expander device as the

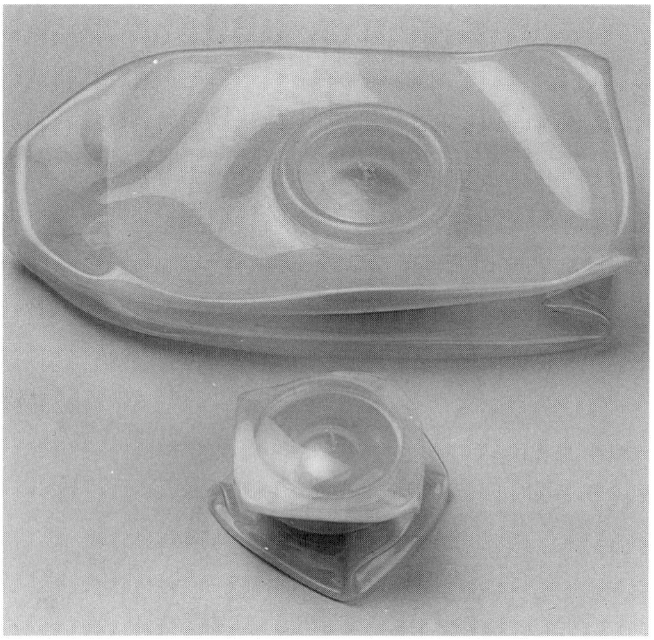

A

B

FIGURE 5-3
Two different tissue expanders with self-contained valves (A) before and (B) after inflation.

FIGURE 5-4
Magna Site Locating System.

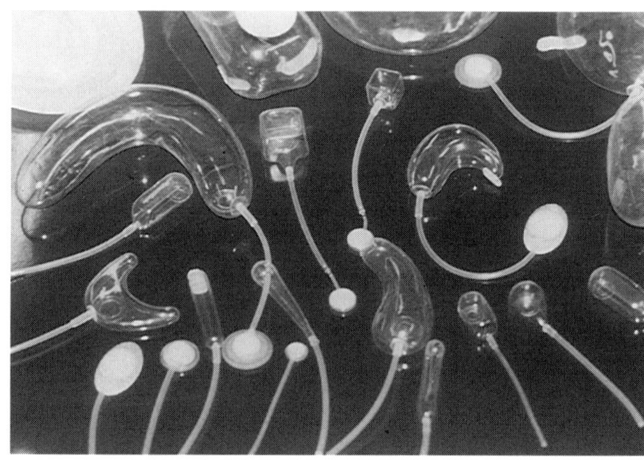

FIGURE 5-5
Tissue expanders can be supplied in many shapes and volumes.

center. The fill system enters the centrally placed expander through multiple self-sealing leaf valves and allows for pulling it out without jeopardizing the integrity of the outer shell or permitting saline leakage. The distal fill-port is low profile and has a stainless steel backing. The long fill-tube enters deeply into the expandable prosthetic device, thereby eliminating the possibility of inadvertent disruptions and subsequent leakage. The basic difference between the Becker and Gibney expanders is that the latter has a larger volume of silicone in the outer implant (Figure 5-8). When a prosthesis with a detachable reservoir is used, the second-stage operation can be reduced to a minor intervention under local anesthesia.

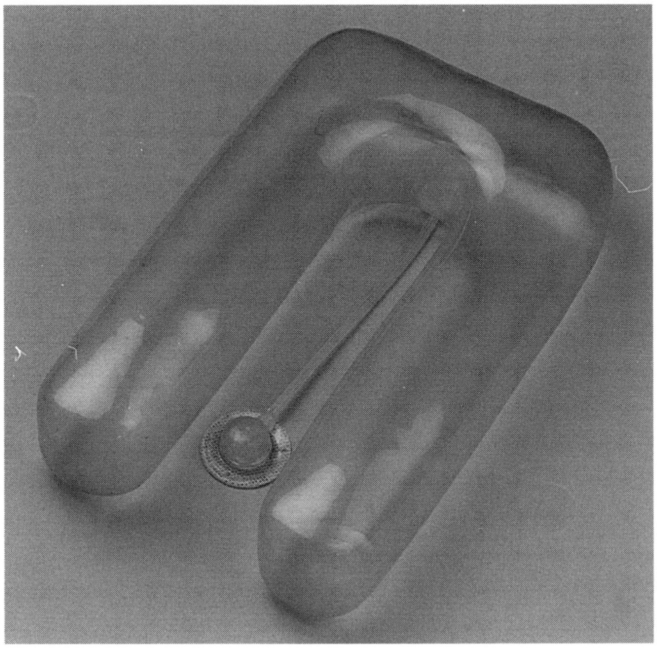

FIGURE 5-6
Custom-made horseshoe-like tissue expander of 1750 milliliters for male pattern baldness.

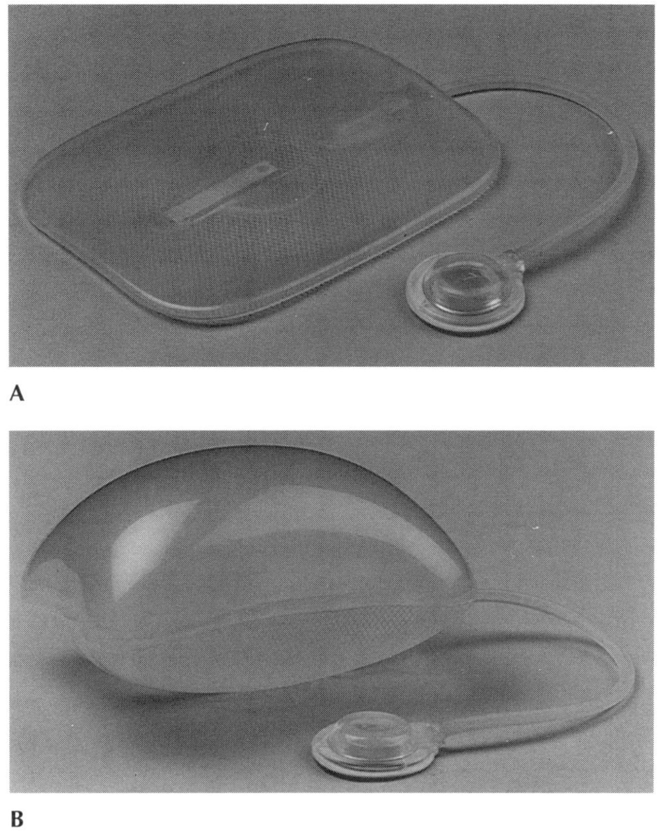

FIGURE 5-7
Flat-span expander, which expands instead of unfolding (A) before and (B) after inflation.

We did not observe any increased risk of leakage in these expanders. Even in situations in which the filling valve and tubing were left in place for volume adjustment, as can be necessary, for example, in patients with Poland syndrome, no problems of leakage occurred over a time period of several years. Experimental studies[8] show that these types of expanders produce a softer final result compared to results seen with the conventional tissue expansion technique.

Differential Expanders. Several implants have been developed that can differentially expand the surface area over different parts of the same area. First, different shapes were used to try to obtain this differential expansion. Later, stacked expanders were developed (Figure 5-9). This type of implant consists of two expanders stacked on each other with each part having its own remote valve. In this way it is possible to expand more in one direction than another. The next step was the development of an expander with only one lumen but with different physicomechanical properties within one envelope. These differential expanders were first used for breast reconstruction (Figure 5-10). The Mentor Corporation developed the Mentor Directa Span expander. Later, several other applications were found, such as scalp reconstruction.

The Pick-a-back Spacer. The pick-a-back spacer (PABS) technique, which was developed by Audretch (Marseille,

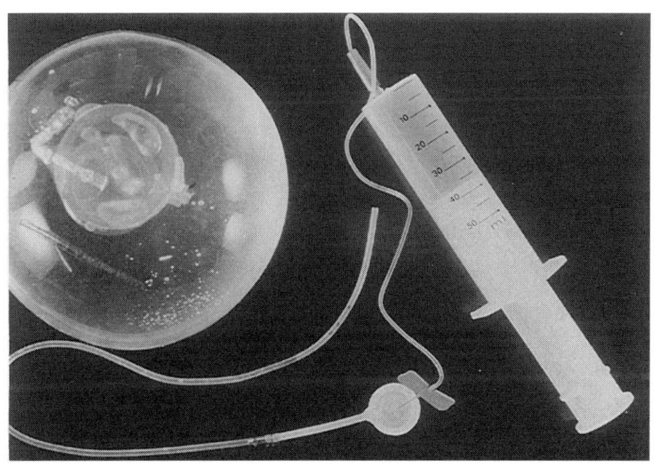

FIGURE 5-8
Permanent tissue expanders for breast reconstruction. The fill system enters the centrally placed expander through multiple self-sealing leaf valves and allows for separation of the fill system without jeopardizing the integrity of the other shell or permitting saline leakage.

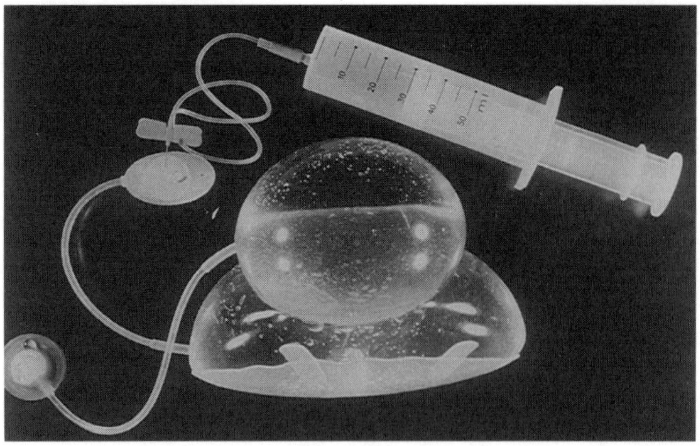

FIGURE 5-9
Stacked expanders, the first step in differential expansion. The two fixed expanders have their own remote valves.

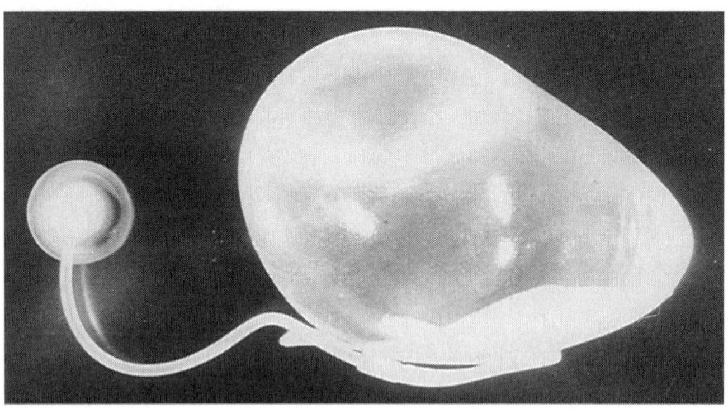

FIGURE 5-10
Differential expanders. Due to varying physicomechanical properties, the caudal part of the expander expands most.

1989, personal communication), addresses the problems of immediate implant insertion after breast excision (Figure 5-11). The device consists of a disk-shaped spacer with an expander in a teardrop position. PABS has a one-point snap fixation between bag and plate—no additional fixation is necessary.

Application is similar to a shoehorn effect. PABS is the only expander system that allows successful delayed inflation of the bag. This helps solve the problem of postoperative excess volume. Especially in immediate breast reconstruction and in irradiated tissues, considerable time must elapse before inflation can be started. The advantages of PABS include having a backing plate to keep the area at the level of the dissection open without bag inflation and being able to adjust the direction of inflation of the expander and to mold the surrounding tissue into a teardrop shape. PABS is also

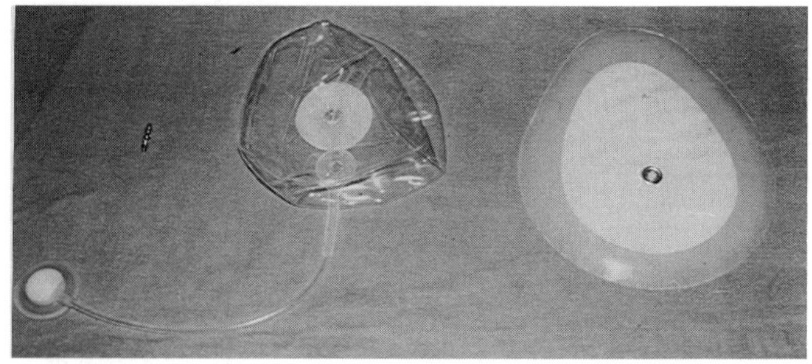

FIGURE 5-11
Pick-a-back spacer (PABS). The spacer (left, disconnected from the expander) helps keep the whole dissection open without having to inflate the expander.

a backup in case of accidental deflation because the spacer prevents shrinking of the pocket.

Intraoral Devices. Special expanders have been developed to create adequate subperiosteal room to reconstruct involuted alveolar ridges. In Chapter 12, Dr. Wittkampf discusses this technique. These intraoral expanders can be supplied with incorporated or remote valves and have a volume of 1 to 5 mL.

Another intraoral application of tissue expansion is used in cleft palate patients. Using very small, usually round, expanders of 1 to 5 mL the palateal soft tissue is enlarged in surface area. The technique is currently more or less experimental. Data from studies by Van Damme of the Nijmegen University in The Netherlands shed light on this subject.[9,10]

Nerve and Vessel Expanders. In 1987, Stark et al.[11] presented a paper about experimental elongation of arteries and veins. Manders et al.[12] and Milner[13] presented papers about nerve expansion. All expansions used prototype expanders, which were round or spherical, had volumes ranging from 10 to 100 mL, and had remote filling valves. Recently, clinical application of these devices has begun.

Textured Expanders. The most recent improvement in tissue expansion devices deserves special attention. With the conventional tissue expansion technique for breast reconstruction, the surgeon can be confronted with several problems. In particular, the formation of the often firm capsule in situations in which rapid expansion cannot be performed (for example, in irradiated tissue) can be a problem. The introduction of the Biocell textured surface silicone expander (Biospan, manufactured by McGhan Medical Corporation) and other textured implants has decreased this problem. As described by Maxwell,[14] the textured surface of this expander also makes it immobile. When the expander is in the desired position, it will not dislocate because of

FIGURE 5-12
Biodimensional system. A textured expander and a matching anatomical textured mammary implant.

pectoral muscle forces. Tissue adherence occurs during expansion, and the capsule contraction is far less than that occurring with conventional smooth surface silicone expanders. The resulting decrease in chest wall compression leads to a more comfortable inflation period. The expanded soft tissue with the thin capsule provides better coverage of tissue for the permanent breast implant. When the conventional tissue expanders have not reached natural ptosis, it can be achieved by the differential expansion of the caudal part of the expander.

As a kind of final touch, McGhan introduced the Biodimensional System (Figure 5-12), which incorporates an anatomical textured tissue expander (Biospan) with a matching anatomical, textured mammary implant (Biocell) for breast reconstruction. After expansion with this textured expander, textured mammary implant (with a volume that is 10 percent less than the expanded volume) is implanted. This texturing activates rapid attachment of the textured surface to the adjacent tissue, imparting mirror image texturization in the developing capsule (Figure 5-13). The Biocell anatomical reconstructive mammary implant is a broad-based gel-filled implant with an additional gel-filled bladder located within the outer shell in the lower pole area for increased projection. The broad base leads to lateral fullness while the anatomical shape results in a more natural upper breast slope, eliminating undesirable upper pole fullness and inadequate projection as seen with round implants. The internal bladder helps eliminate the potential for compression by maintaining fullness inferiorly and avoiding displacement to the superior portion of the chest.

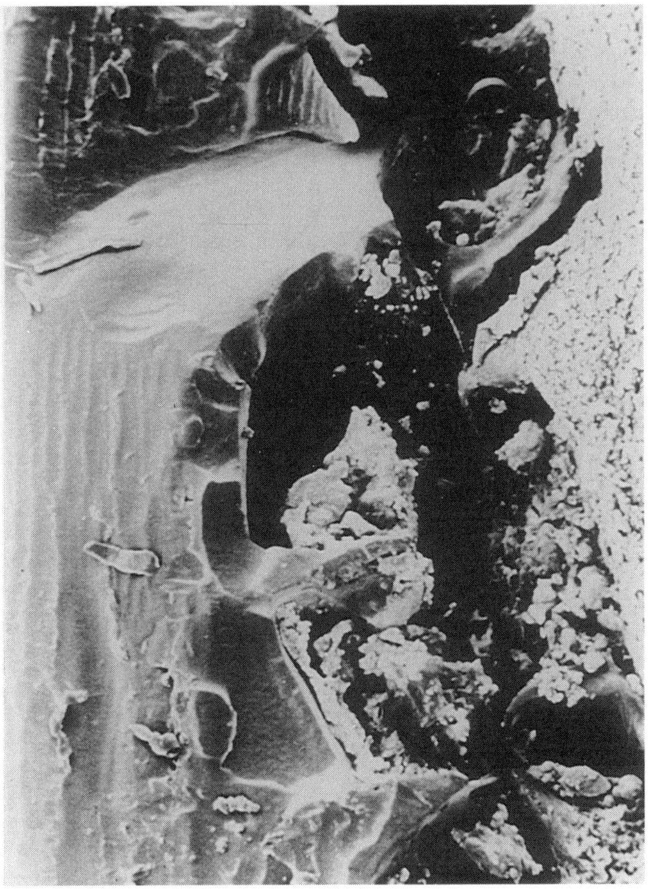

FIGURE 5-13
Textured tissue expanders. The implant surface (left) and the capsule surface (right), show the mirror image development of a textured surface on the capsule following implantation of the textured surface expander (EM scanning).

Inflation Devices and Additions

As described by Nordström et al.,[15] injection-dome leakage can play a role in the tissue expansion technique. In our experience, special care must be taken during the injection phase. The needle size should not be bigger in diameter than gauge 22. We use a 23-gauge scalp vein butterfly-needle, which is attached to the syringe by means of connection tubing (Figure 5-8). In this way, the assistant can inject while the surgeon has both hands free to manipulate the needle at the reservoir site.

Huber developed a needle in which the open site is on the lateral side of the distal part rather than on the end of the needle. This needle minimizes the risk of leaving a small punched out hole in the silicon self-closing layer of the fill-valve.[16]

To minimize the number of injections, the needle should remain in place in the injection site while refilling the syringe. The McGhan Tissue-Expander Fill Kit provides a closed, sterile, single-puncture system to prevent multiple punctures and thus avoids damage of the device and lowers the risk of infection. When multiple expanders are used in one patient, a new needle should be used for each expander to avoid the possibility of cross-contamination if infection is present.

Ditmars[17] described an occupational hazard of the plastic surgeon, "tissue-expander thumb." With the help of a special device, the caulking gun syringe, the expander can be more easily inflated, relieving stress on the thumb.

Cole et al.[18] described the use of a syringe pump for a gradual but faster inflation. The authors reported that this technique avoided pain and discomfort for the patient and was less damaging to the overlying tissue.

Dickson and Batchelor[19] described a simple technique for ensuring accurate puncture of the tissue expansion fill-valve. By instilling 1 mL methylene blue BPC into the expander, the contents of the expander will be permanently stained. The appearance of blue-green fluid in the syringe or extension catheter upon withdrawal after puncture of the dome indicates correct placement of the needle.

In addition to methylene blue, other substances have been used both to relieve pain and to facilitate the expansion. Cohen and Roberts[20] added lidocaine to the inflated liquid to minimize discomfort, and Matt et al.[21] described an improved method to confirm that anticontractile agents provide significantly more rapid expansion.

The different devices for tissue expansion and possible additions or modifications in technique that are described here are available with high technical standards. In almost all situations in which a surgeon uses tissue expansion to perform a reconstruction, he or she can rely on manufacturer support and recommendations from other surgeons who have experience with these techniques. With insight gained from experience in the field, the reconstructive surgeon will achieve the desired results.

ACKNOWLEDGMENTS

Ph.A. van Damme, M.D., D.M.D., member of the scientific library committee of OCITES, is gratefully acknowledged for the literature survey.

REFERENCES

1. Münker R. Various devices available for tissue expansion and clinical experience. *Fac Plast Surg* 1988;5:291–300.
2. Neumann CG. The expansion of an area of skin by progressive distension of a subcutaneous balloon. *Plast Reconstr Surg* 1957;19:124–130.
3. Austad ED, Rose GL. A self-inflating tissue-expander. *Plast Reconstr Surg* 1982;70:588–593.
4. Becker H. Breast reconstruction using an inflatable breast implant with detachable reservoir. *Plast Reconstr Surg* 1984;73:678–683.
5. Becker H. Breast augmentation using the expander mammary prosthesis. *Plast Reconstr Surg* 1987;79:192–199.
6. Becker H. The permanent tissue-expander. *Clin Plast Surg* 1987;14:519–524.
7. Gibney J. Use of a permanent tissue-expander for breast reconstruction. *Plast Reconstr Surg* 1989;84:607–617.
8. Van Rappard JHA. Capsule formation in controlled tissue-expansion. In *Controlled tissue-expansion in reconstructive surgery.* Thesis ScD, University of Groningen, The Netherlands, 1988.
9. Van Damme PA, Freihofer HP, Van't-Hof MA, Kuijpers-Jagtman AM, Maltha JC, Spijkers JM. Radiologic analysis of the effects of subperiosteal palatal soft-tissue expansion in growing cats. *Int J Oral Maxillofac Surg* 1994;23:393–399.
10. Van Damme PA. Soft-tissue expansion in cleft palate surgery. *Plast Reconstr Surg* 1994;93:1307–1312.
11. Stark B, Hong C, Futrell JW. Rapid elongation of arteries and veins in rats with a tissue-expander. *Plast Reconstr Surg* 1987;80:570–581.
12. Manders EK, Saggers GC, Diaz-Alonso P, Finn L, Sipio JC, Glumac T, Au VK, Wong RKM, Mottaleb M. Elongation of peripheral nerve and viscera containing smooth muscle. *Clin Plast Surg* 1987;14:551–562.
13. Milner RH. The effect of tissue-expansion on peripheral nerves. *Brit J Plast Surg* 1989;42:414–421.
14. Maxwell GP. Eighty-four consecutive breast reconstructions using a textured silicone tissue-expander. *Plast Reconstr Surg* (in press).
15. Nordström RE, Pietilä JP, Voutilainen PE, Lilius GP, Virkkunen PJ, Rintala AE. Tissue-expander injection dome leakage. *Plast Reconstr Surg* 1988;81:26–29.
16. Lawrence TW. Huber needles for tissue-expander inflation. *Plast Reconstr Surg* 1986;78:541–542.
17. Ditmars DM. The caulking gun: An aid to inflate tissue-expander prosthesis. *Plast Reconstr Surg* 1983;72:401–402.
18. Cole RP, Gault DT, Mayon BJ, Davis PKB. Pain and forehead expansion. *Brit J Plast Surg* 1991;44:41–43.
19. Dickson MG, Batchelor AG. A technical aid to tissue-expansion. *Br J Plast Surg* 1987;40:305.
20. Cohen IK, Roberts C. Lidocaine to relieve pain with tissue-expansion of the breast. *Plast Reconstr Surg* 1987;79:489.
21. Matt BH, Squier CA, Kelly KM, Bardach J. Enhancement of expansion of guinea pig skin by local delivery of an anticontractable agent using a new bilumen expander. *Ann Plast Surg* 1990;24:335–341.

6

External Valves

JEAN-PIERRE AUBERT AND GUY MAGALON

Three types of valves are used for tissue expansion.

- remote internal valves
- self-contained valves
- external valves

Conventional remote internal-filling valves consist of a silicone dome reinforced with a Dacron mesh and a silicone-covered steel base. Self-contained internal valves are built in the top of the expander. External valves are attached to the expander by a transcutaneous tube.

Valve-related problems account for most complications during the filling phase of tissue expansion. Several problems are associated with remote internal valves. Placement and removal require further undermining, and it is often difficult to locate the valve during filling. With self-contained valves, there is a risk of inadvertently puncturing the prosthesis. Because of these problems and based on the encouraging results of Jackson,[1] who used external valves in a series of 34 cases and did not report a higher complication rate than with conventional valves, we investigated the use of external valves.

The external valve that we tested is made of hard plastic, measures 0.5 centimeter in diameter and 2 centimeters in length, and weighs 2 grams. It has a one-way flow system that prevents reflux and leaking even at high intraprosthetic pressure (Figure 6-1).

SPECIAL TECHNIQUES

The techniques used for external valves differ from those used for conventional remote internal valves in several ways. The length of the placement incision can be considerably shorter for external valves. In some cases an incision of only 1.5 centimeters is sufficient.

We use a drain probe to make the tunnel for the tube. The skin opening should be approximately 8 centimeters from the prosthesis. The tunnel should be as straight as possible because bends and curves can be the source of problems (Figure 6-2).

A metallic adapter is used to connect the valve and filling tube to the expander. After injecting saline to be sure that there are no leaks, the filling tube is secured to the adapter with a braided suture thread (Figure 6-3).

An occlusive dressing is always necessary not only to protect the skin opening from bacterial contamination

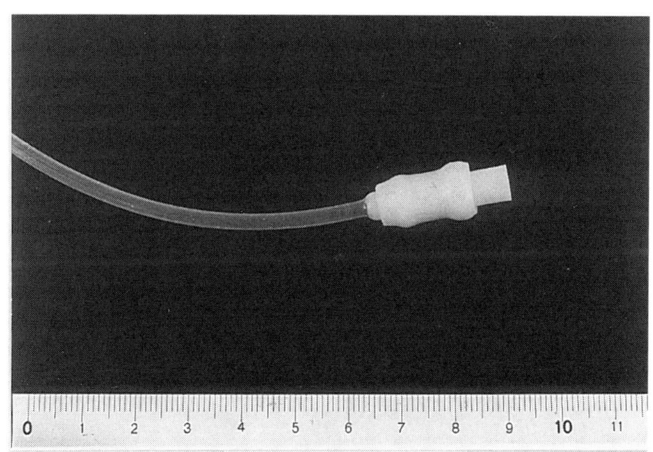

FIGURE 6-1
External filling valve.

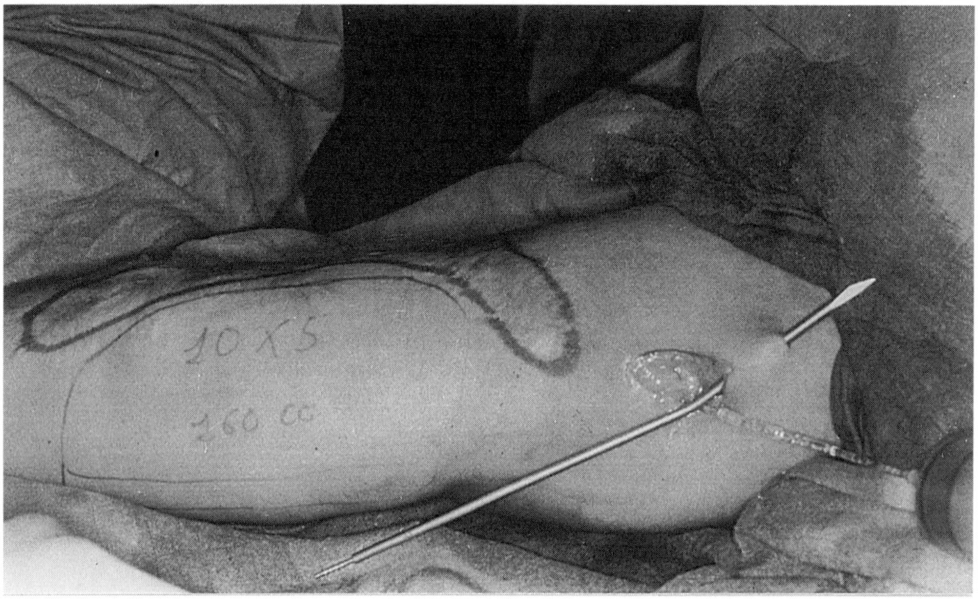

FIGURE 6-2
Insertion of filling valve. Passage of the filling tube from the balloon through the skin via a Redon awl.

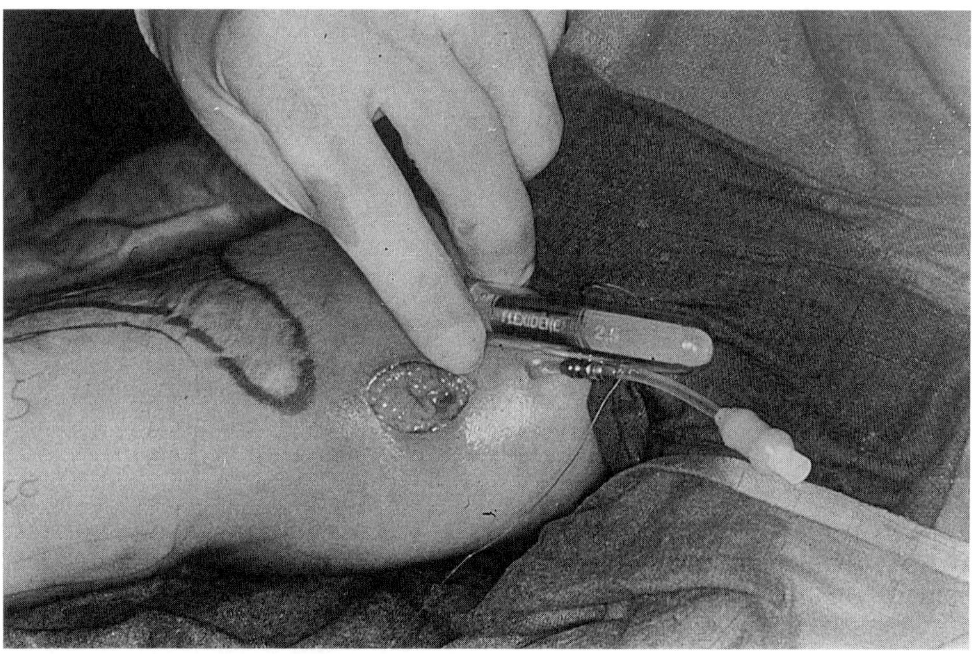

FIGURE 6-3
Insertion of filling valve. Fixation of the tube into the connections by means of braided suture materials.

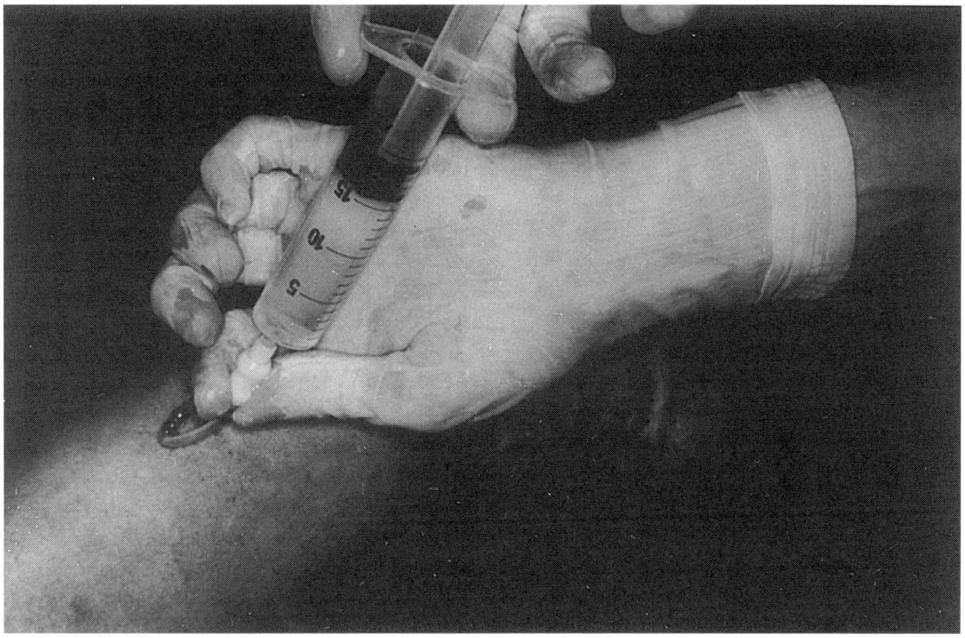

FIGURE 6-4
Filling of the external valve.

but also to prevent tube movement from enlarging the opening. On the extremities, a cast can be applied to immobilize the joints, which will also reduce pain and edema.

It is generally recommended to wait until the stitches are removed before beginning to fill the expander, that is, about 2 weeks after placement. However, our experience demonstrates that if the prosthesis is under healthy skin, filling can begin much earlier. In many cases, we start filling after removal of the drain 48 hours after placement.

Before each filling session, the external valve must be carefully disinfected with an appropriate antiseptic agent. The valve tube should be taped to the skin to prevent enlargement of the skin opening. The operator should work under sterile conditions (sterile gloves and drapes) (Figures 6-4 and 6-5).

The use of external valves simplifies reconstruction in two ways. First, unlike remote internal valves, external valves do not have to be extracted from a periprosthetic capsule. They can be simply pulled out without further undermining. Second, the skin opening can be used to introduce the suction drain without necessitating an additional incision.

INDICATIONS

Pediatric cases are the best indications for external valves for two reasons. The first is that external valves eliminate painful transcutaneous injection, and the second is that they require much less space.[2]

Obese patients are also good indications for external valves because it can be very difficult to locate an internal valve under a thick fat layer. The scalp and the extremities are often good areas for external valves because suitable areas for internal remote domes can be limited (Figure 6-6 through 6-14).

External valves are also indicated when repeating the expansion on the same skin since the skin tends to be thinner and more fragile.

Conversely, external valves must not be used if a permanent prosthesis is to be implanted or if there is any doubt as to the patient's ability to cooperate and maintain hygiene.

COMPLICATIONS

Use of external valves eliminates all complications associated with internal or self-contained valves, namely, loss, capsizing, migration under the expander, exposure, damage or necrosis of overlying skin, and leaks. The only major complication that has been reported with external valves is infection due to contamination of the skin opening. As previously mentioned, preventive treatment consists of keeping the filling tube as straight as possible, placing the adapter away

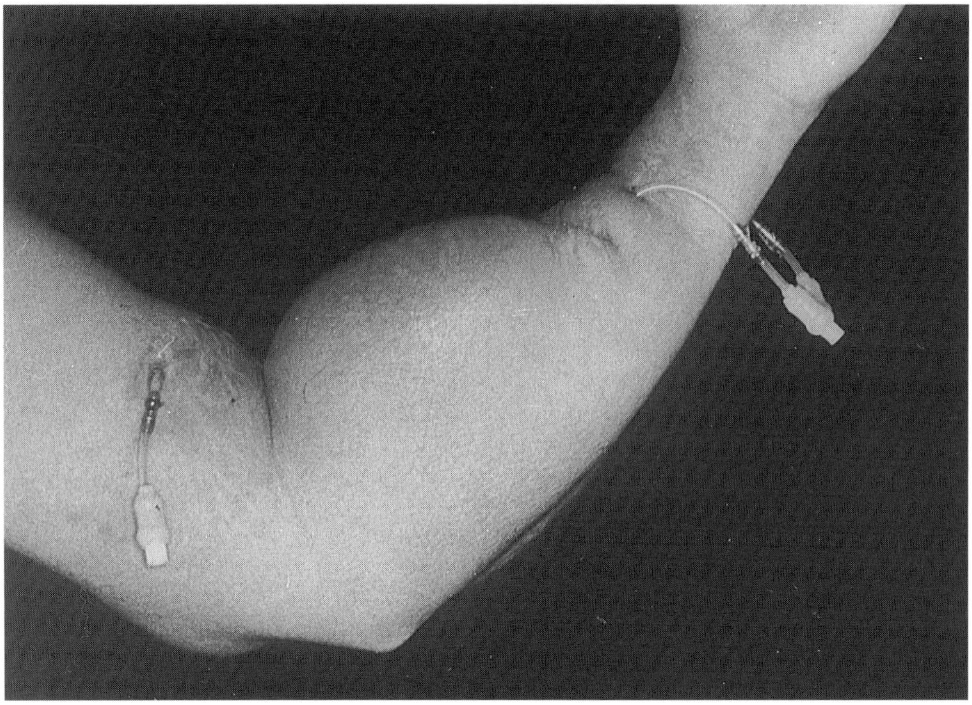

FIGURE 6-5
End of expansion.

from the skin, immobilizing the filling system to prevent enlargement of the skin opening, and maintaining sterility during filling. Using these precautions, the complication rate with external and conventional valves is about the same.[3]

CONCLUSION

If used properly, external valves can eliminate most complications due to valves that occur during the filling phase, thus allowing safer and more efficient treatment.

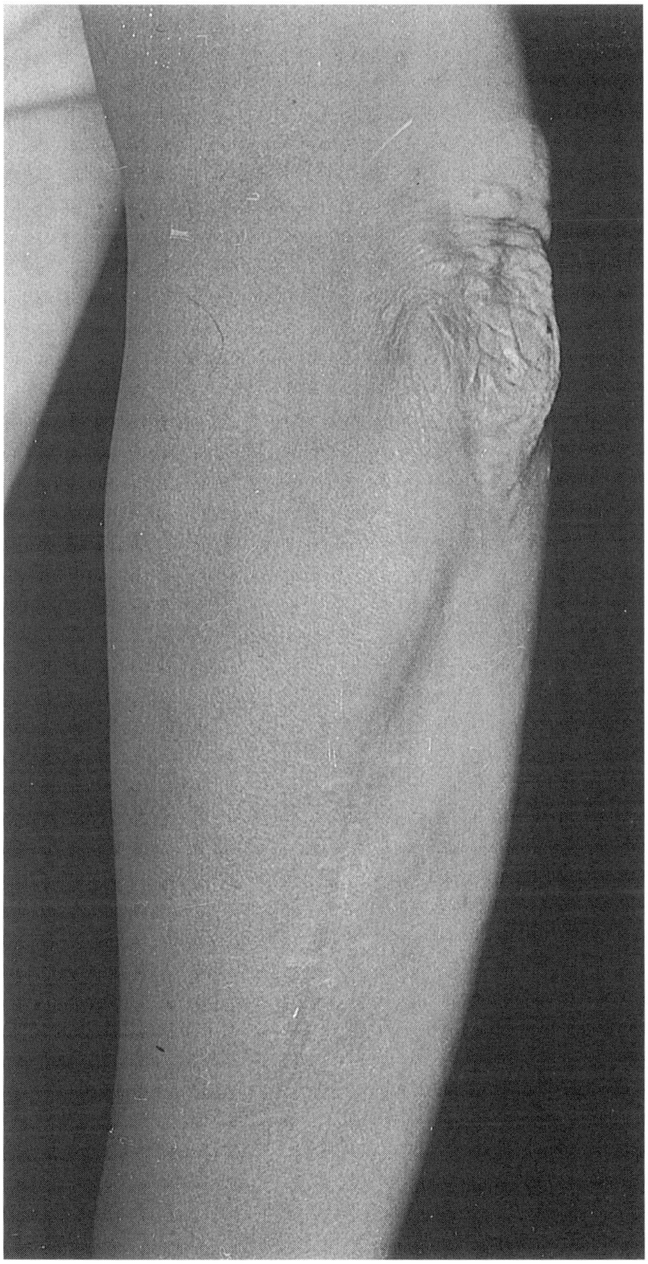

FIGURE 6-6
Unstable scar on the elbow.

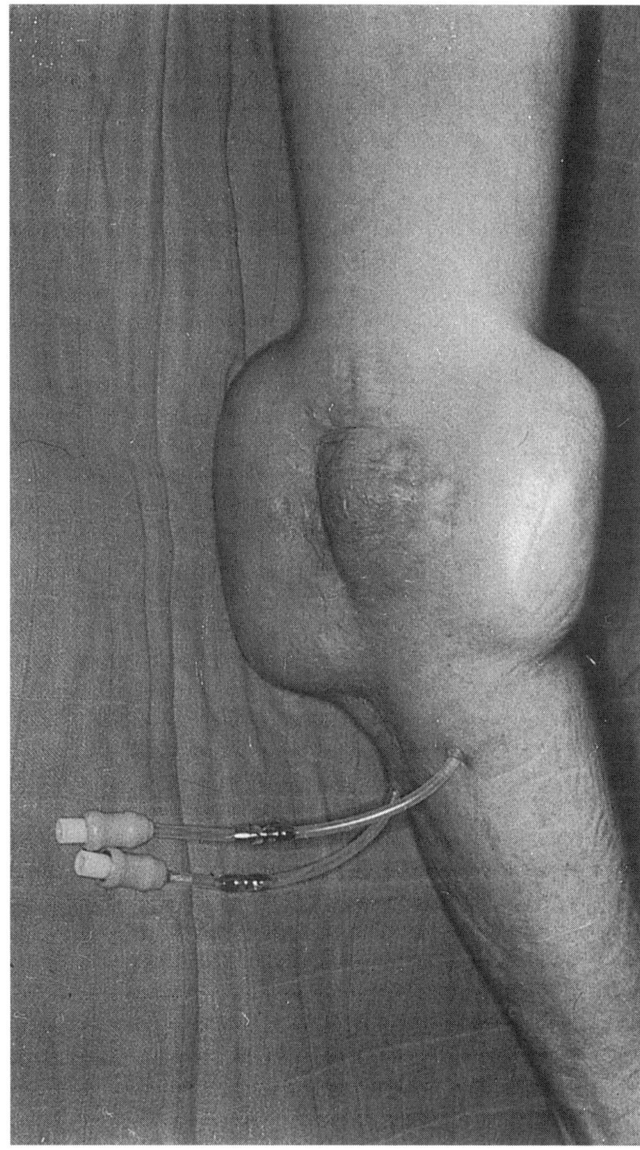

FIGURE 6-7
End of expansion with external valves.

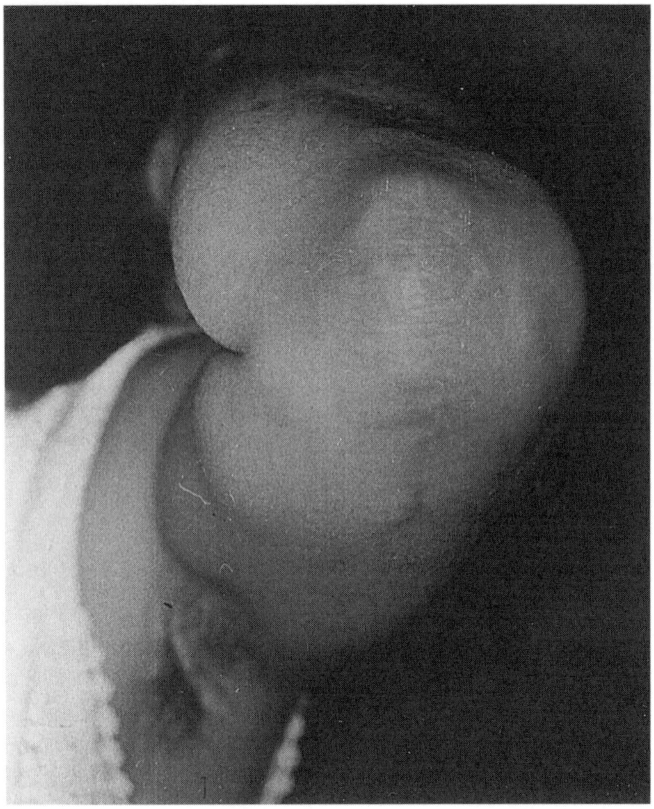

FIGURE 6-8
Final result 6 months later.

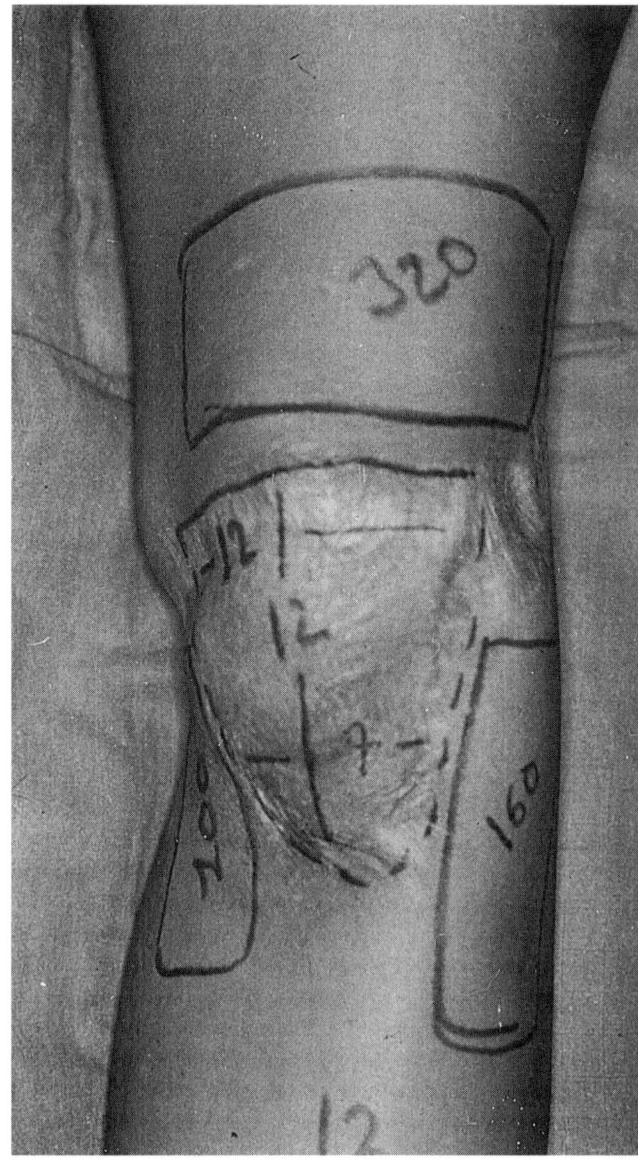

FIGURE 6-9
Uncosmetic skin graft on the knee.

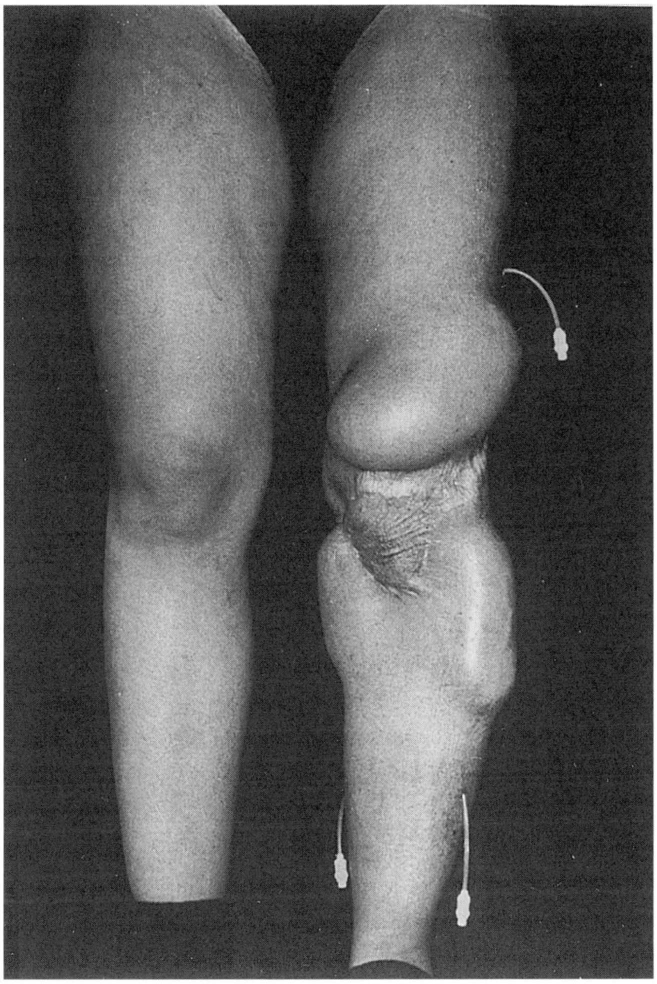

FIGURE 6-10
End of expansion: three tissue expanders, three filling valves.

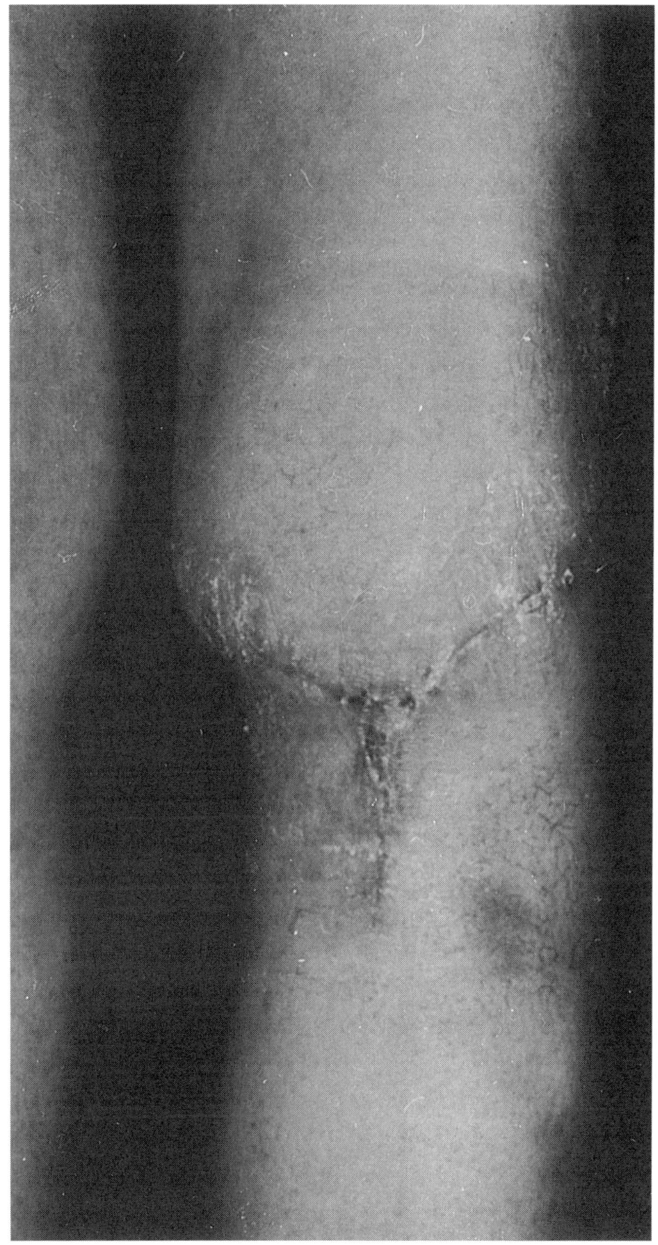

FIGURE 6-11
Final scar.

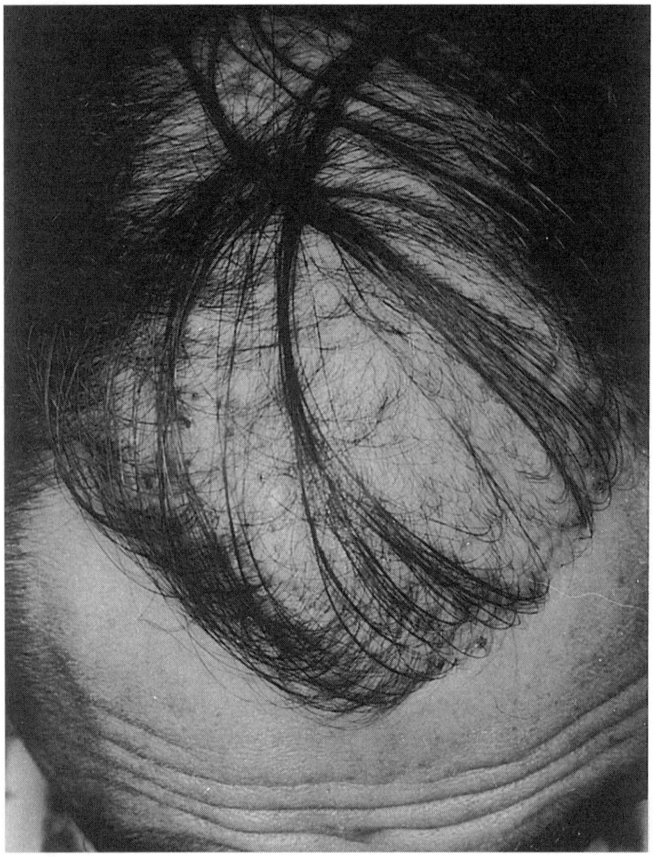

FIGURE 6-12
Male pattern baldness.

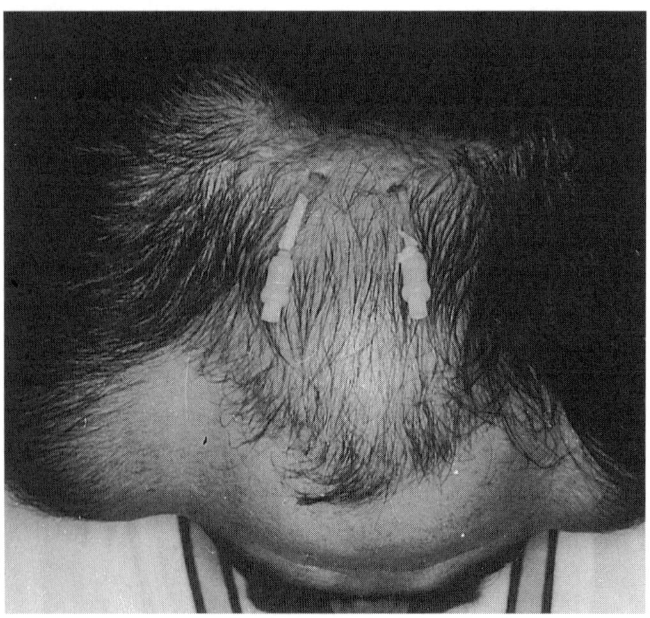

FIGURE 6-13
End of expansion with external valves.

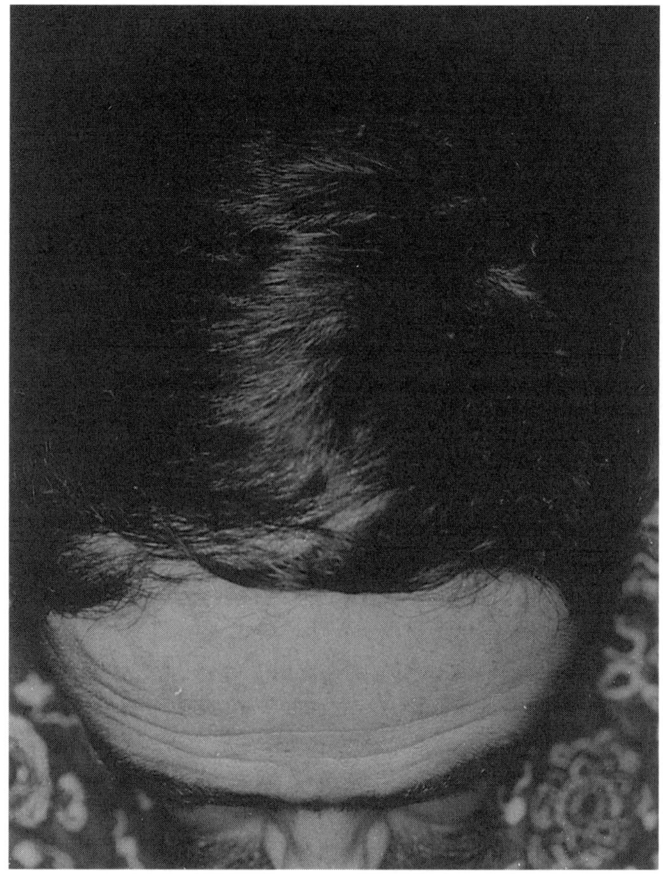

FIGURE 6-14
Final result 6 months later.

REFERENCES

1. Jackson IT, Sharpe DT, Polley J, Costanzo C, Rosenbert L. Use of external reservoirs in tissue expansion. *Plast Reconstr Surg* 1987;80:266–271.
2. Aubert JP, Paulhe P, Bardot J, Magalon G. L'apport des valves externes dans l'expansion cutanée. *Ann Chir Plast Esthet* 1991;38(3):218–222.
3. Aubert JP. *Etude clinique et expérimentale de l'expansion tissulaire.* Marseille: Thèse de Médecine, 1987.

7

Filling Policy

JINCAI FAN, EDOARDO RAPOSIO,
AND ROLF E. A. NORDSTRÖM

Obtaining a suitable flap of expanded skin (fitting the area to be reconstructed) depends mostly on adequate preoperative geometrical planning and on the outcome of sufficient filling of the implant. The most effective and safe way to fill tissue expanders is still unknown.[1] However, even an inexperienced surgeon will be able to successfully complete the filling of an implant if he or she is aware of a number of basic and simple rules.

FACILITIES AND EQUIPMENT

Usually, the filling is performed as an outpatient procedure by a physician or a trained nurse. If the patient lives far from the hospital, the local physician, after receiving appropriate instructions, can perform the filling. In particular cases, if the patient or a member of his/her family is trained and understands the tissue expansion procedure sufficiently, he/she can perform the filling at home. Normal saline is used for the filling of the expander. At present, air or other gases are not used because of disadvantages such as uncontrolled volume, unstable intraluminal pressure, and risk of air embolism. The instruments used for filling the expander are quite simple. A normal syringe with a 23- or 25-gauge needle is used for the routine filling of the expander. We prefer to use a 25-gauge butterfly needle to diminish dome leakage. If several expanders have to be inflated the same day, the filling is made easier with a filling kit. This kit, (McGhan Medical Corporation) consists of a 10-milliliter Luer lock syringe with automatic spring return, a transfer set with a vented piercing device and two-way check valve, and a No. 21 needle infusion set. The syringe is filled automatically by a returning spring, thus sparing time and energy, making the filling procedure quicker and easier. The filling procedure is simplified by the use of expanders with external ports because they allow easy and pain-free filling injections.

PLANNING OF THE FILLING

The first filling of the expander usually is performed in the operating room in conjunction with the insertion of the expander. This obliterates the dead space caused by the undermining of the pocket. The risk of hematoma, seroma, and infection is thus decreased. It also enables one to check that the implant does not leak and that there is no kinking of the connecting tube. The amount of saline injected at this stage is about 10 percent of the expander nominal volume, and one should take great care not to fill the implant too much as this can compromise the circulation of the overlying skin.

Subsequent fillings are usually about 10% to 20% of the nominal expander volume, with intervals between fillings ranging from one-half to two weeks, the length of time depending on various factors, such as the age of the patient and the position of the implant. The start of the expansion very much depends on the location of the incision that is made for the placement of the implant. Thus, if the expander is inserted through an incision that will be pulled open by the expansion, it will be necessary to wait until the tensile strength of the scar is sufficient to tolerate the stretching caused by the expander inflation. To avoid this delay, it is advisable to try to insert the expander through a radial incision, as described in Chapter 3 of this book (Geometric Planning). This method prevents the incision from being forced open during the filling and often allows one to start the expansion immediately.

FILLING PROCEDURE

The filling should be performed under aseptic conditions. First, the filling dome must be located. When dealing with children, it might be helpful to apply a local anesthetic ointment, such as Eutectic Mixture of Local Anaesthetics (EMLA, Astra Pharmaceuticals, Westborough, MA), on the skin overlying the injection dome 30 minutes before each filling session. This decreases the pain associated with needle penetration.

To start the filling, one hand fixes the injection port, while the other hand holds the butterfly needle, which penetrates the skin into the cavity of the dome at a 90° angle. When the tip of the needle reaches the metal base of the dome, we usually withdraw some saline to verify proper placement of the needle tip. Thereafter, the injection of the saline solution can begin. Usually, the injection meets less resistance in the beginning. If no saline can be aspirated or if there is much resistance, the needle is probably in the wall of the dome and not in the cavity. If, after reinserting the needle, the resistance persists, then the whole expander system must be checked to find the source of the obstruction (for example, kinking of the connecting tube). During the filling, it is important to palpate the expander to feel the tension in it. One should also communicate with the patient regarding pain and discomfort. The filling is continued until the patient complains about pain or the skin begins to blanch. Finally, a small volume of saline is withdrawn from the expander to return the capillary refill and to minimize pain and discomfort.

FACTORS INFLUENCING THE FILLING

Capsular contracture, compression of adjacent sensitive structures or other factors that diminish the circulation of the overlying skin require longer intervals between fillings. Major complications—such as infection, expander exposure, expander rupture, and expanded skin necrosis—usually lead to a cessation of the expansion procedure and removal of the implant. Special care must be taken when the expanded area has been previously irradiated or burned, or when there is scar tissue, because, in these cases, the expanded skin is more prone to most of the major complications.

Pain and Discomfort

The discomfort that usually occurs at the end of each filling session is related to the tenseness of the expanded skin. While breast reconstruction with tissue expansion is not usually painful, a considerable amount of discomfort may occur after filling on the forehead and scalp or after filling distally, in the extremities. The filling on the trunk, back, neck, and face may cause some discomfort, but it generally disappears in some hours after each filling. However, the pain is usually the first signal of overzealous inflation of the expander, and it should alert the physician to end the filling session. This is particularly important in patients with dark skin, because it is particularly difficult to check the capillary refill in these patients.

Leakage and Rupture

A considerable discrepancy between the total injected saline volume and the actual volume of the expander may sometimes be noticed. This discrepancy is almost always due to filling port leakage, and it may have clinical significance if it reaches disturbing proportions.[2]

Acute rupture of the implant usually results from the perforation of the expander sack (or of the connecting tube) during the filling procedure. Perforation of the sack occurs more often with integrated-dome expanders, but it also occurs with internal remote port expanders when the dome is too close to the expander sack. Thus, when the dome location is in question, ultrasound or X-ray techniques should be used for assistance.

Infection

Infection caused by a filling injection is rare, because aseptic techniques should be used. If an infection develops, antibiotic therapy, drainage, and irrigation can be tried. If these are unsuccessful, the implant should be removed promptly.

Psychological Reactions

Patients invariably respond to the obvious distortion of body image in a number of ways; negative reactions range from depression to withdrawal and even to exhibitionist behavior.[3] Some patients may simply be unable to tolerate tissue expansion because of unacceptable body image distortion related to the cosmetic deformity. Thus, it is advisable that patients be well informed, motivated, mentally competent, and completely aware of the nature of the procedure.

HOW TO MONITOR THE EXPANDED SKIN DURING THE FILLING PROCESS

The major parameters to be monitored during tissue expansion are the intraluminal pressure and the circulation of the expanded tissue. Subjective evaluations, such as the patient's complaint of pain, skin blanching, implant pressure estimated by palpation, and capillary refill, are usually reliable. But, in particular circumstances, an objective monitoring device may be helpful to better evaluate the perfusion of the expanded skin, and thus decrease the risk of tissue necrosis and implant exposure.

Many techniques for monitoring tissue expansion have been developed and evaluated. An ideal monitoring technique should be noninvasive, simple, able to be continuously recorded, safe, portable, inexpensive, and easily interpretable by medical staff or even by patients. Unfortunately, such a device does not currently exist.

Intraluminal Pressure Monitoring

In the Department of Plastic, Reconstructive, and Hand Surgery, University Hospital, Tromsø, Norway, we have a Stryker Intra-Compartmental Pressure Monitor System (Stryker Surgical, Chicago, IL) to test the intraluminal pressure when the expander is filled. This system is a hand-held, portable pressure monitor system designed for measuring intracompartment pressure. The Stryker system has been developed to produce simple, quick responses to intraluminal pressure changes. It consists of two major components, the hand-held pressure monitor system, which has an auto-zeroing device that eliminates manual adjustments, and a sterile, disposable, tissue expander monitor set. Other components are Luer lock connectors for syringe and needle, a diaphragm, a 30-cm (12-in.) tube to the needle, and a 75-cm (30-in.) tube to the monitor. Although this device is simple to use, one very seldom has the need to use it.

Skin Tensionmetry

Devices such as Dermaflex (Cortex Technology, Hadsund, Denmark) and ballistometer have been investigated[4,5] to measure the biomechanical properties of human skin. Unfortunately, such devices are of no real clinical value in monitoring dynamic changes of the skin during the filling process.

Transcutaneous pO_2 Measurement

Tissue pO_2 seems to be a suitable parameter of skin blood flow during tissue expansion. The method has been used to evaluate the blood supply and viability of expanded skin both experimentally and clinically. To take the measurement, an oxygen sensor is taped to the skin. To induce maximal vasodilatation, the device must be heated to 44°C to 45°C. Therefore, a certain amount of time is required to establish equilibrium. However, if the measurement is carried out many times over a short period, the skin may be burned by the local hyperthermia. These kinds of devices cannot, therefore, be used to measure the skin continuously.

Laser Doppler Flowmetry

When a beam of laser light is reflected from moving blood cells, it undergoes a change in wavelength (Doppler shift) while the wavelength of the light reflected from static objects is unchanged. The magnitude and frequency distribution of these changes in wavelength are directly related to the velocity, number, and direction of movement of the particles. The frequency shift is then received by a return fiber probe and converted into an electronic signal. Two basic parameters may be obtained from the laser Doppler flowmeter. First, a flow value is determined by the movement and concentration of the particles in specific volumes of tissue. Second, a direct current value converts the total amount of light scattered back to a photodetector. Although this kind of device is more valuable than many others for monitoring the filling procedure, care should be taken when using a standard probe because the holder may increase the pressure on the ball-shaped surface skin of the expanded area, causing a lower reading. In our department, we use a multichannel laser Doppler flowmeter system (PeriFlux 4001 Master, Perimed, Sweden) equipped with a diode laser, which allows the optional selection of wavelengths. The measurements obtained using this device are given in an arbitrary Perfusion Unit (PU). The measuring depth depends on several factors, such as the structure and density of the capillary bed, the wavelength of light used, and the distance between the fibers in the probe. In normal skin, the measuring depth is about 0.5 to 1 mm. However, because the degree of light absorption by the skin is influenced by blood content, degree of oxygenation and pigmentation, the recorded values are relative values rather than absolute values of flow. By measuring blood flow, concentration of moving blood cells (CMBC), and blood velocity, one can evaluate microvascular perfusion in real time. The blood flow (perfusion) represents both the movement and concentration of blood cells through the microvasculature (that is, the CMBC in the measured volume multiplied by the mean velocity).

EVALUATION OF DIFFERENT FILLING TECHNIQUES

One of the major disadvantages of conventional tissue expansion is that it takes a considerable amount of time. To shorten the length of the expansion process, many innovations have been developed using different techniques for different purposes.

Rapid Tissue Expansion

This technique was first described by Marks and associates[6] with the purpose of providing early coverage of traumatic defects. The method involves filling the expander to a maximal tolerable volume each day for about 7 to 10 days. In a pig model, during 5 days of filling, the authors found that the intraluminal pressure after each filling decreased from a mean of 71.2 mmHg to 4.1 mmHg in 24 hours. The tissue oxygen tension after the filling was recovered in 72 hours. The flap survival was significantly longer than in the acute flap, which served as a control. Furthermore, the capillary blood flow in the expanded flap was significantly greater than in the nonexpanded flap and in the acute flap after its transposition.[7,8] Clinically, the filling is carried out until the overlying skin feels very tense or the area is blanching. When blanching occurs, the patient must be observed for 30 minutes. If blanching persists, fluid should be withdrawn from the expander until the blanching disappears. However, because patients usually feel pain, the filling should be stopped if the pain becomes significant. Rapid tissue expansion may cause loss of skin sensitivity. Furthermore, the complication rate is usually considerably higher using the rapid expansion technique compared to conventional tissue expansion.

Overfilling Technique

Overexpansion means that the pressure in the expander is increased to a point at which either the dermal capillary flow is zero or the patient experiences so much discomfort that inflation cannot be continued. The pressure in the expander is thereafter gradually decreased by withdrawing saline until both the capillary refill and all other clinical criteria are sufficient for safe dermal circulation. Pietilä, Nordström, et al.[9] have found an average volume gain of about 59 percent with this system. Thus, temporary overexpansion of the tissues at each filling session increases the actual inflation volume by about 59 percent, shortening the process of tissue expansion considerably.

Continuous Tissue Expansion

With this technique, the filling pressure of the expander is maintained at a level slightly less than capillary filling pressure for a period of days. This is obtained using a controlled continuous tissue expansion pump system, first developed by Logan and Hayden in 1989.[10] This system includes a continuous filling pump and a controller. When the expander is filled to reach a certain level of intraluminal pressure, the filling pump automatically stops. Conversely, if the pressure within the implant is less than the stated level, the pump starts the filling. The intraluminal pressure is usually set at 35 mmHg. The prototype continuous tissue expansion device has undergone major modifications and improvements, including the incorporation of a minicomputer that allows programming and recording of data as well as the ability to vary and maintain pressures in a programmable fashion. This technique has been applied in a pig model, demonstrating the advantage of rapid (3 days) continuous tissue expansion compared to intraoperative expansion technique.[11]

Cycled Hyperinflation

Cycling hyperinflation and deflation of the expander at each filling session enhances the effects of momentary hyperexpansion. Presumably, this phenomenon is due to the fact that if the skin is "load cycled," the maximal extension is not usually obtained on the first loading, but it may be after three or four fillings.[12] This is similar to the overfilling technique, but in a repeated form. A higher risk of flap necrosis and possible discomfort to the patient should be considered when using this technique.[13]

INTRAOPERATIVE TISSUE EXPANSION

In 1987, Sasaki introduced the concept of intraoperative sustained limited expansion (ISLE). The author observed that this technique, by means of cyclic loading of temporary expanders, allows skin to stretch beyond its inherent limits within minutes, thus gaining extra tissue for use in closing small skin defects with minimal tension. The procedure should be performed as follows.[14]

Under local or general anesthesia, a pocket slightly larger than the base dimensions of the expander is dissected, the implant is positioned, and serial expansion is then performed, with the tubing exiting from the defect site. A finger (or, better, a temporary subdermal suture) is placed over the entry site of the expander to prevent it from bulging out of its pocket during inflation. The expander is filled to tissue tolerance as determined by inspection (capillary refilling) and palpation (firmness). The fill volume is maintained for 2 to 3 minutes, and then the expander is deflated to allow tissue perfusion. This procedure is repeated for two additional stress load cycles. The fill volume will increase after each cycle, indicating effective tissue recruitment. Successful use of intraoperative rapid tissue expansion has also been reported by Abramo et al.[15] in cleft palate repair for closure of large gaps between cleft borders using less tension. Using a 5-milliliter Foley catheter with the distal tip trimmed as the tissue expander, the authors were able to increase surface area by distension of the palate and recruitment of the surrounding soft tissues. Mackay et al.,[16] on the other hand, have tested the efficacy of intraoperative expansion in reducing the tension of wound closure in a pig model, and have reported that, while undermining of the wound edges resulted in a remarkable decrease in the force required to close the wound, intraoperative tissue expansion did not significantly decrease such tension. Similarly, Hochman et al.[17] have reported that, although a statistically significant reduction in the force required to close a wound after intraoperative tissue expansion may be observed, when the degree of undermining achieved by the expander is matched in an equivalent defect, no difference in the resultant wound-closing tensions can be discerned between the two techniques.

In our experience, the intraoperative tissue expansion procedure has some value in harnessing the viscoelastic properties of the skin, even if the reported enhancements in area gain are mainly to be attributed to the increased extent of undermining required to place the expanders under the wound margins.

REFERENCES

1. Hallock GG, Rice DC. Increased sensitivity in objective monitoring of tissue expansion. *Plast Reconstr Surg* 1993; 91:217–222.
2. Nordström REA, Pietilä JP, Voutilainen PEJ, et al. Tissue expander dome leakage. *Plast Reconstr Surg* 1988;81:26–29.
3. Brown MD, Johnson TM. Complications of tissue expansion. *J Dermatol Surg Oncol* 1993;19:1120–1122.
4. Wickman M, Olenius M, Malm M, et al. Alterations in skin properties during rapid and slow tissue expansion for breast reconstruction. *Plast Reconstr Surg* 1992;90:945–950.
5. Tosti A, Compagno G, Fazzini ML, Villardita S. A ballistometer for the study of the plasto-elastic properties of skin. *J Invest Dermatol* 1977;69:315–318.
6. Marks MW, Mackezie JR, Burney RE, et al. Response of random skin flaps to rapid expansion. *J Trauma* 1985;25: 947–952.
7. Marks MW, Burney RE, Mackezie JR, Knight PR. Enhanced capillary blood flow in rapidly expanded random pattern flaps. *J Trauma* 1986;26:913–917.
8. Marks MW, Argenta LC, Thornton JW. Rapid expansion: Experimental and clinical experience. *Clin Plast Surg* 1987;14:455–459.
9. Pietilä JP, Nordström REA, Virkkunen PJ, et al. Accelerated tissue expansion with the "overfilling" technique. *Plast Reconstr Surg* 1988;81:204–207.
10. Logan SE, Hayden JM. A control unit for maximal-rate continuous tissue expansion. *Biomed Sci Instrum* 1989;25: 27–30.
11. Wee SS, Logan SE, Mustoe TA. Continuous versus intraoperative expansion in the pig model. *Plast Reconstr Surg* 1992;90:808–814.
12. Fan J. Tissue expansion. Ph.D. Thesis, Beijing Union Medical University, Beijing, China, 1991.
13. Fukuta K, Jackson IT, Noreldin AA, Pieper DR. Efficacy of cycled hyperinflation for rapid tissue expansion. *Plast Reconstr Surg* 1993;91:846–852.
14. Sasaki GH. Intraoperative sustained limited expansion (ISLE) as an immediate reconstructive technique. *Clin Plast Surg* 1987;14:563–573.
15. Abramo AC, Viola JC, Angelo AJ. Intraoperative rapid expansion in cleft palate repair. *Plast Reconstr Surg* 1993; 91:441–444.
16. Mackay DR, Saggers GC, Kotwal N, Manders EK. Stretching skin: Undermining is more important than intraoperative expansion. *Plast Reconstr Surg* 1990;86:722–730.
17. Hochman M, Branham G, Thomas JR. Relative effects of intraoperative tissue expansion and undermining on wound closing tensions. *Arch Otolaryngol Head Neck Surg* 1992;118:1185–1187.

8

Complications: How to Avoid and Treat Them

JULIEN H. A. VAN RAPPARD

Tissue expansion will not realise its full potential until the complication rate is consistently below 10%.[1]

Skin adaptations to growth, obesity, and pregnancy are physiologic phenomena. The first application of nonphysiologic soft-tissue expansion has been a not very well known but regularly used principle in Africa. Thaw,[2] Burnet,[3] Weeks,[4] and Caputo[5] describe in several issues of *National Geographic* the practice of stretching of lips, earlobes, and necks by means of wooden disks or rings. In none of these articles is the word "complication" mentioned. No infection, no pain whatsoever took place or was bad enough to complain about (Figure 8-1). When tissue expansion reached the "civilized" world and the technique turned out to be a medical reconstructive technique, problems showed up immediately.

In 1957, Neumann[6] described the first clinical use of tissue expansion. In this first medical article about tissue expansion, Neumann already mentions complications such as infections. He did not continue to use tissue expansion techniques because the material he used presented problems. The expanders were made out of a rubber material that was not inert enough and resulted in undesirable reactions and problems.

The second generation of tissue expansion techniques, presented by Radovan,[7] Austad and Rose,[8] and Argenta et al.[9] did not at first mention high percentages of complications. In time, indications for this technique extended as did the population of patients on whom it was performed. At that time, an enormous increase in complication rate took place, which resulted in publication of complications ranging from 0% to 51% (Table 8-1). When the criteria for indications became stricter, the complication rate decreased again to an acceptable level. The basis for complications is twofold. Apart from the usual problems one meets when implanting a foreign body, in tissue expansion one is confronted with conditions that are more or less continuously threatening to tissues.

WHAT SHOULD BE CONSIDERED A COMPLICATION?

When discussing the issue of complications of tissue expansion, it is important to differentiate between major and minor complications. Masser[10] defines major complications as those that cause the expansion to abort and prevent the intended outcome; minor complications are those that require treatment and may slow down the program and incur expenses, but they do not affect the final results. In addition, there are the complications of general or regional anesthesia, common to other surgical procedures. These can be considered separately. The complications are best shown schematically (Table 8-2).

HOW TO AVOID COMPLICATIONS

"To avoid is better than to heal."

To decrease complication rates it is important to use strict criteria for indications and contraindications. In addition, the surgeon's accuracy and experience in the technique has been shown to play a very important role

FIGURE 8-1
"Noncivilized expansion." No complications; moreover the film he was wearing was beautiful.

in diminishing the complication rate.[11] After the rebirth of the tissue expansion technique in the late 1970s, indications that first appeared to be limited to breast reconstruction began to see broader application. A situation similar to that seen shortly after the development of the microsurgical technique took place in tissue expansion. No indication was considered "too wild."

There are several contraindications for tissue expansions. Some of these are more obvious than others, but all are of substantial importance. If a surgeon decides to perform tissue expansion in spite of the existence of one or more of the contraindications, he/she should keep in mind the risks being taken. The following circumstances are clear contraindications:

1. A malignancy that has not been radically excised.

Any form of malignancy should be radically excised. Only then is reconstruction using tissue expansion justified.

2. Infection.

In case of infection, no foreign body, such as a tissue expander, should be implanted.

3. Irradiated tissue.

Irradiation leads to fibrosis of all tissues. This will reduce the potential for cell regeneration and wound healing. These changes do not decrease as time passes. Even after 5, 10 or 20 years, so-called "late irradiation effects" can be seen. At present, doses of radiation are much lower and more localized than in the past. Nevertheless, tissue expansion in irradiated tissue is not a good choice. If tissue expansion is performed, special care should be taken. Inflation should be done over much longer time intervals and with considerably less saline per inflation.

4. Poorly vascularized tissue.

In cases of morbus Bürger or Raynaud syndrome, expansion of distal parts can cause problems. For

TABLE 8-1
Complication Rates Reported in Literature

		Number of Patients	*Overall Complications (Percentage)*	*Compromising Major Complications (Percentage)*
Radovan	1982[12]	18	0	0
Manders et al.	1984[13]	35	40	25
Neale et al.	1987[14]	77	20	11
Dickson	1987[15]	75	40	10
Antonyshyn et al.	1988[16]	66	39	20
Austad	1988[17]	35	51	24
Donelan	1988[18]	37	18	11
van Rappard	1988[11]	65	24	17
Armstrong	1989[19]	49	24	24
Elias et al.	1991[20]	49	25	14

TABLE 8-2
Complications in Tissue Expansion

I. Major Complications (Problems Leading to Failure)

1. During placement:
   —skin damage during undermining
2. During filling:
   —anaphylactic foreign body reaction
   —infection
   —exposure through placement incision
   —exposure by necrosis of overlying tissue
   —invincible capsule
   —unbearable pain
   —psychological problems
   —prosthesis-related problems:
      disconnection between tube and prosthesis
      leaky prosthesis
3. During or after removal:
   —necrosis of the expanded flap
   —inadequate expansion
   —venous air embolism during removal of tissue expanders

II. Minor Complications (Manageable Problems)

1. During placement:
   —apparent injury of the implant
   —abundant bleeding
2. During filling:
   —hematoma
   —seroma
   —pain
   —valve-related problems:
      valve displacement
      capsized valve
      leaky valve
      exposure of valve
   —organ compression:
      bone
   —rib
   —skull
   —other:
      nerves
      vessels
   —alopecia (temporary)
   —striae
3. During or after removal:
   —intubating complication
   —hematoma
   —seroma
   —rupture of suture lines
   —ulceration
   —dog-ears
   —spreading of scars
   —subjective anomalies of expanded skin (numbness)

TABLE 8-3
Body Regions Ranked According to Decrease in Tolerance to Tissue Expansion

* scalp (tolerant)
* breast
* trunk
* face
* neck
* back
* upper leg
* upper arm
* fore-arm
* lower-leg (intolerant)

example, correction of leg or arm stumps after amputation can be risky procedures. Diabetes mellitus is not always a contraindication for tissue expansion. Gradual expansion in such cases will not produce extra problems except in the area below the knee.

5. Chemotherapy.

Patients receiving chemotherapy should not undergo tissue expansion. The fall in the white blood cell count following chemotherapy increases the risk of infection around the expander. Unlike irradiation, chemotherapy does not have long-term effects. When blood counts return to normal, tissue expansion can be undertaken.

6. Psychologically unsuitable patients.

The entire tissue expansion procedure requires the full cooperation of the patient. Therefore, patients who cannot accept temporary disfigurement or who cannot withstand the pain are not suitable candidates for this technique.

It should be emphasized that different anatomical regions have differing sensitivities to tissue expansion. Table 8-3 presents different tissues ranked according to tolerance of tissue expansion.

HOW TO HANDLE COMPLICATIONS

The best way to handle complications with any surgical procedure is to avoid them. But complications can still occur, even if care is taken in proper patient selection, a proper implant selection and placement, a judicious inflation schedule, and a proper use of the additional tissue generated by the expansion. The issue is how to handle these problems if and when they arise.

Pain

Patients may complain of pain during the inflation. Mostly the pain is experienced as an unpleasant tightened sensation. The intensity of this sensation depends on the area of the body affected and the rate and quantity of inflation. That pain can be localized to a specific region is easily demonstrated by the fact that the expansion of breast or scalp is seldom painful. In contrast, however, patients whose backs or arms are expanded frequently complain. The larger the volume of inflation, the higher the pressure on the skin and the more painful the sensation can be. This pain subsides in 1 hour and is over in 12 hours. In a prospective study, Cole et al.[21] demonstrated that the discomfort, which was found to be related to a sharp rise in intraluminal pressure after bolus inflation, could be avoided by using a syringe pump for more gradual and even quicker inflation. De Greef[22] recommended sedation at night to minimize patient discomfort. Cohen and Roberts[23] advised adding 25 milligrams lidocaine (1 percent solution) into the implant through the fill-valve; this provided complete relief of pain. In general, it is sufficient to withdraw some liquid after an inflation that is too painful. The discomfort decreases almost immediately. Another cause for pain is the percutaneous injection itself. This can be very frightening particularly for children, and is a reason to stop the procedure. The use of Eutectic Mixture of Local Anaesthetics® (EMLA) (Astra Pharmaceuticals, Westborough, MA) ointment 1 hour before puncture can avoid this. When possible, the use of external valves as described by Jackson et al.[24] is recommended in children. The advantages of this technique, in addition to the lack of puncture pain, are reduced operating time, ease of injection into the valve, early detection of valve or filling tube junction leakage, and lack of complications associated with the valve.

Hematoma

Hematomas are almost always the result of insufficient hemostasis, although a disturbed clotting mechanism can also be the cause. After implantation of the expander, usually one or two suction drains are inserted. Initial filling of the expander during the first stage operation causes increased pressure on the surrounding tissues and probably diminishes the occurrence of hematomas. Ashall and Quaba[25] presented a case in which a delayed acute hemorrhage took place. The likely mechanism was erosion of an artery by the edge of the hard reinforced backing of the expander.

Seroma

The body reacts to an expander as it would to any foreign body by producing extra tissue fluid. In general, this produces no problems. Goldstein and Schuster[26] recommend the addition of 10 milliliters methylene blue to the expander. This small addition of dye is helpful in distinguishing postoperative seroma from leaks in the expander and aids in locating the fill port when it is placed in heavily scarred areas.

Capsule Formation

Capsule formation or contracture can limit the tissue expansion procedure. The finding that tissue expansion reduces the incidence of capsular contracture as described by Radovan[27] and Argenta[28] is not substantiated by the study of Holmes.[29] In 29 percent of expansions, capsule contracture of Baker III or IV took place. In 2 percent of the procedures, capsulectomy and reinsertion of the expander was necessary. To avoid many problems concerning capsule formation, expansion should be carried out as quickly as possible. Routinely, we start about 4 or 5 days after the first-stage operation and inflate as much as the condition permits.

Infection

Every implantation and the subsequent inflations carry the risk of infection. Therefore, optimal sterile conditions are necessary. Disinfection before inflation can minimize the problem. In general, the use of antibiotics is not indicated and are only prescribed when infection is suspected. In some cases, interruption of the procedure can be prevented in this way. Armstrong et al.[18] showed that the use of a tissue expander at the time of immediate reconstruction, especially in cases where there is to be surgery of the contralateral breast, has a much higher risk of infection. We do not use prophylactic antibiotics in the tissue expander to decrease the rate of infection, as mentioned by Nordström.[30]

Induced Ischemia

If too much saline is injected, the skin is exposed to an unacceptably high pressure. As a result, circulation will be reduced and necrosis may follow. Blanching of the skin over the tissue expander is a sign of possible trouble. However, should this occur, it is advisable to wait for half an hour to see if there is a return of color and capillary refill. If blanching continues, saline should be withdrawn until normal color returns. Maximal caution for pressure-induced ischemia is necessary in irradiated areas. In rapid tissue expansion, isolated necrosis of deep tissue under the expander was described by Lawrence[31] but was never seen in our practice. When flap necrosis takes place in breast reconstruction by means of tissue expansion, early and aggressive excision of the ischemic tissue with immediate primary closure often can achieve rapid primary healing (Kroll[32]). In our opinion, noninvasive, objective monitoring that utilizes oxygen levels measured transcutaneously, as recommended by Hallock and Rice[33], is too cumbersome for routine use.

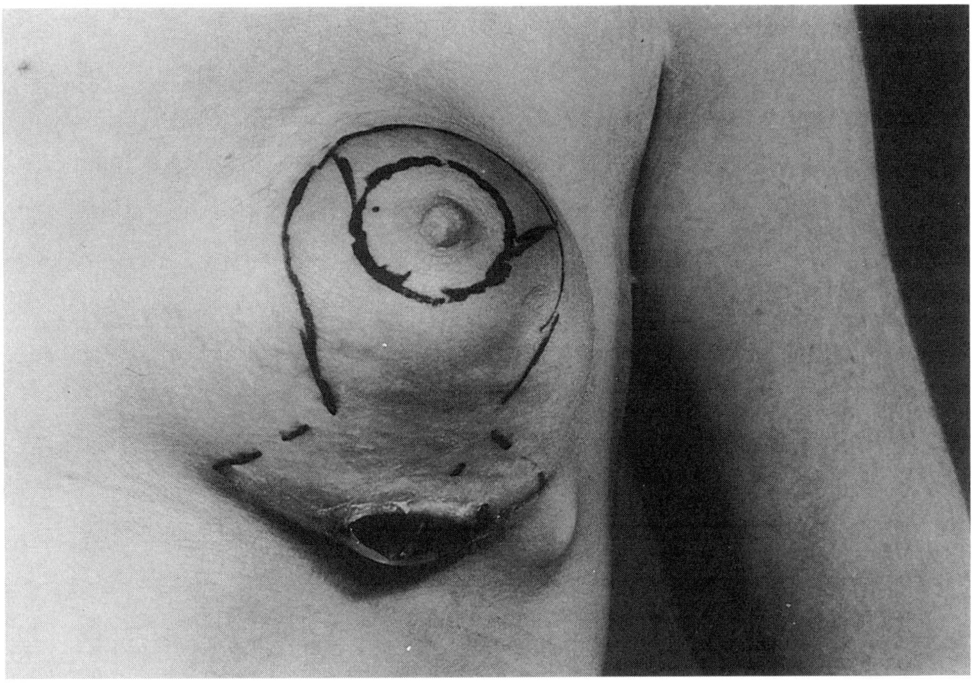

FIGURE 8-2
"Civilized expansion." Major complication; exposure of expander because of bad positioning of the implant; a fold in the expander is visible.

Expander Exposure
This problem can result from wound dehiscence, infection, or necrosis (Figure 8-2). Initially, 10% to 20% of the expander is filled during the first-stage operation to prevent folds in the envelope, which can cause erosions and exposure of the expander. To prevent rupture of suture lines or wound dehiscence, the initial incision should be made, when possible, "radial" to the direction of expansion (see Chapter 3), and the expander must be placed more than 2 to 3 cm away from the incision (Kawashima et al.[34]).

If expander exposure occurs, there are several things one can do. The most conventional solution is to take the implant out. Another option is to proceed with the intended expansion. Most of the time, the area of erosion tends to enlarge as the implant is inflated further. Baron et al.[35] suggest closing the dehiscence with surgical adhesive film, which allows continuation of the expansion without enlarging the defect. Saxby[36] describes the use of inelastic tape, this time to invert the "knuckle" or fold of an unfolded expander and to avoid the exposure. Harving et al.[37] recommended salvaging the exposed implant either by reducing the implant volume or by using a skin flap. If the procedure is not successful, a removal of the implant must be performed and the expansion process will be delayed for some time. In case of valve exposure, one should consider completely detaching the valve from the soft tissue and using it as an external port.

Organ Compression
Our extensive histomorphological studies[38] demonstrate that, with exception of the epidermis, tissue expansion is actually the stretching rather than the formation of new tissue. More tissue than just the cutis, subcutis, and muscle can be stretched or compressed. As described by Dickson and Sharpe,[13] McKinney et al.,[39] and Sinow et al.,[40] chest wall deformity is a common sequel to tissue expansion for breast reconstruction in a variety of clinical settings. Resolution of the deformity proceeded rapidly after placing permanent implants, and only minor abnormalities persisted for any length of time. The same situation can take place in the skull, as decribed by Cullen et al.,[41] Hemmer et al.,[42] Fudem and Orgel,[43] Paletta et al.,[44] and Penoff.[45]

Prosthesis-Related Problems
Sometimes, the expander is damaged accidentally (Figure 8-3). The envelope may be perforated and liquid may slowly leak out into the tissue. Although this will do no harm, a new expander must be placed. However, when perforation occurs at the end of the expansion

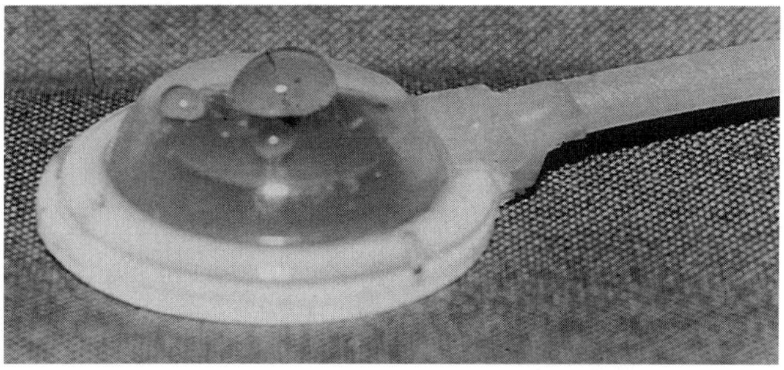

FIGURE 8-3
Valve leakage. Do not use thick needles. A gauge of 21 is the maximum thickness to use.

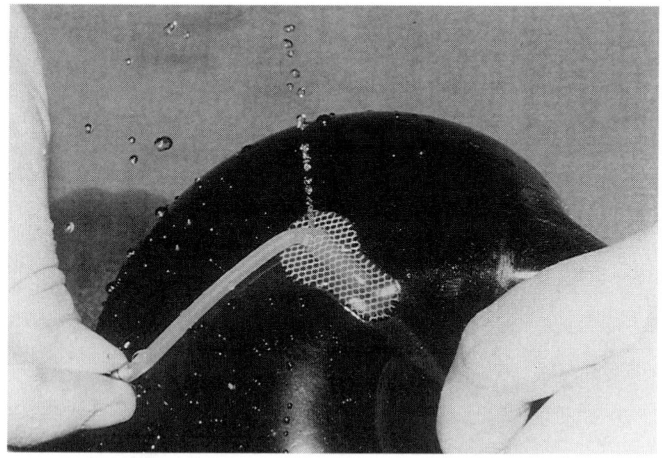

FIGURE 8-4
Expander leakage. Problem in the tubing at the connection of the tubing with the body of the expander.

procedure, there is still the possibility of using the already created skin surplus. To avoid this problem at the outset, the internal valve should be placed at a sufficient distance away from the envelope somewhat closer to the surface. A suture to fix the fill-valve may be helpful. Depending on the area of the body, a low- or high-profiled dome is used. A high-profile dome avoids problems during the filling period, especially in areas of the body where the skin is thick, such as the upper leg. Injection domes can leak, too.[46] To minimize the risk of leakage of the fill-valve, Lawrence[47] recommends using Huber needles (used in Port A Cath-type implantable vascular access devices). The opening of the Huber needle is situated on the lateral aspect of the tip as opposed to standard needles, which have the opening at the bevel on the tip. We use 21-gauge butterfly-needles.

Aside from surgical or other external causes, some problems are due to the device. Manders et al.[11] described a defective injection port that lacked a plastic disk to stop the needle. Chisholm et al.[48] and Levine[49] described problems in the tubing occurring at the connection of the tubing with the body of the expander and at the stainless steel tubing connector (Figure 8-4).

Skin Shortage after Expansion

If insufficient tissue is gained from the expansion procedure, it is impossible to close the entire defect. In such a situation, it is sometimes possible to expand the already expanded tissue during the second-stage operation. This is similar to the technique of intraoperative sustained limited expansion first presented by Sasaki (1986, personal communication). In this way one or two extra centimeters of skin can be gained. The other possibility is to repeat the whole procedure. After excision of only a part of the pathological tissue, the defect is closed, leaving the expander in place. Afterward the same procedure can be started again, now expanding the already expanded skin. However, this requires one extra operation to reach the final goal. These problems can be avoided by adequate handling and good choice of expander. The increase in the surface area by inflating the expander can be calculated mathematically. This is rather complicated and does not indicate exactly which expander should be used. We reported a simple method to convert this rigorous mathematical approach into an easy rule of thumb to help choose the expander needed.[50]

Dog-ears

After expansion, defects are generally closed by advancement or rotation. Dog-ears cannot always be avoided. Immediate correction is contraindicated when too much damage to the circulation is expected. After several months, the dog-ears usually have flattened out spontaneously or can easily be corrected.

Scar Widening

A rather annoying problem is scar widening after the final operation. This can sometimes be avoided by overexpanding the skin to obtain extra tissue to work with. By creating a large skin surplus, one can prevent wound tension and the phenomenon of stretch-back as described by Nordström.[51]

Psychological Problems

The level of psychological stress in patients varies, depending on the area in which the expander is situated. Some patients cannot stand the temporary disfigurement by the procedure, and they stay at home all the time. Goin and Goin[52] studied the psychological discomfort patients experience during tissue expansion procedures. Careful patient selection and extensive preoperative patient education are important factors in preventing psychological problems in patients treated with tissue expansion.

Miscellanea

In addition to the minor and major complications mentioned, some very rare problems should be kept in mind. Anaphylactic reactions to expanders (Aubert, personal communication 1989), complicated intubation associated with tissue expanders (Ferguson[53]), bradycardia associated with insertion and inflation of tissue expanders (Francis and Jayarajah[54]), and venous air embolism during removal of expanders (Roth and Dado[55]) are all extremely rare situations.

Studies by Fenton[1] give a complication rate of 10 percent. The complication rate in our studies and in those of most of our colleagues is below 10 percent. Less than half of this number concerns major complications. On the whole, the outcome of reconstruction using tissue expansion is more dependent on experience than would be expected considering the simplicity of the technique. When using stringent indications, the tissue expansion technique can be considered a very valuable contribution to reconstructive surgery.

ACKNOWLEDGMENTS

Ph.A. van Damme, M.D., D.M.D., member of the scientific library committee of OCITES is gratefully acknowledged for this literature survey.

REFERENCES

1. Fenton O. Complications of soft-tissue expansion (discussion). *Br J Plast Surg* 1988;41:249.
2. Thaw LC, Stout Thaw M. Trans-Africa safari. *National Geographic* 1938;74:327–364.
3. Burnet W. Yank meets native. *National Geographic* 1945;88: 105–128.
4. Weeks GS. Into the heart of Africa. *National Geographic* 1956;110:257–263.
5. Caputo, R. Ethiopia: Revolution in an ancient empire. *National Geographic* 1983;163:614–644.
6. Neumann CG. The expansion of an area of skin by progressive distention of a subcutaneous balloon. *Plast Reconstr Surg* 1957;19:124–130.
7. Radovan C. Development of adjacent flaps using a temporary expander. *Plast Surg Forum* 1979;2:62.
8. Austad ED, Rose GL. A self-inflating tissue expander. *Plast Reconstr Surg* 1982;70:588–593.
9. Argenta LC, Watanabe MJ, Grabb WC. The use of tissue expansion in head and neck surgery. *Ann Plast Surg* 1983; 11:31–37.
10. Masser MR. Tissue-expansion: A reconstructive revolution or a cornucopia of complications? *Br J Plast Surg* 1990;43:344–348.
11. Van Rappard JHA. Controlled tissue expansion in reconstructive surgery. *Thesis* ScD, University of Groningen, The Netherlands, 1988.
12. Radovan C. Breast reconstruction after mastectomy using the temporary expander. *Plast Reconstr Surg* 1982;69: 195–206.
13. Manders EK, Schenden MJ, Furrey JA, Hetzler PT. Soft tissue expansion: Concept and complications. *Plast Reconstr Surg* 1984;74:493–507.
14. Neale HW, High RM, Billmire DA, Carey JP, Smith D, Warden G. Complications of controlled tissue expansion in the pediatric burn patient. *Plast Reconstr Surg* 1988;82: 840–845.
15. Dickson MG, Sharpe DT. The complications of tissue expansion in breast reconstruction: A review of 75 cases. *Br J Plast Surg* 1987;40:629–635.
16. Antonyshyn O, Gruss JS, Mackinnon SE, Zuker R. Complications of soft tissue expansion. *Br J Plast Surg* 1988;41: 239–250.
17. Austad ED Contraindications and complications in tissue-expansion. *Fac Plast Surg* 1988;5:379–382.
18. Donelan MB Jr. Complications of controlled tissue expansion in the pediatric burn patient (discussion). *Plast Reconstr Surg* 1988;82:846.
19. Armstrong RW, Berkowitz RL, Bolding F. Infection following breast reconstruction. *Ann Plast Surg* 1989;23: 284–288.
20. Elias DL, Baird WL, Zubowicz VN. Applications and complications of tissue expansion in pediatric patients. *J Pediatr Surg* 1991;26:15–21.
21. Cole RP, Gault DT, Mayou BJ, Davis PKB. Pain and forehead expansion. *Br J Plast Surg* 1991;44:41–43.
22. De Greef A. Tissue-expansion in reconstructive surgery. *S Afr Med J* 1986;69:705–707.
23. Cohen IK, Roberts C. Lidocaine to relieve pain with tissue expansion of the breast. *Plast Reconstr Surg* 1987; 79:489.
24. Jackson JT, Sharpe DT, Polley J, Constanzo C, Rosenberg L. Use of external reservoirs in tissue expansion. *Plast Reconstr Surg* 1987;80:266–271.
25. Ashall G, Quaba A. A hemorrhagic hazard of tissue expansion. *Plast Reconstr Surg* 1987;79:627–630.
26. Goldstein RD, Schuster SH. Methylene blue, a simple adjunct to aid in soft tissue expansion. *Plast Reconstr Surg* 1986;77:452.
27. Radovan C. A self-inflating tissue expander (discussion). *Plast Reconstr Surg* 1982;70:594.
28. Argenta LC. Controlled tissue expansion in reconstructive surgery. *Br J Plast Surg* 1984;37:520–529.
29. Holmes JD. Capsular contracture after breast reconstruction with tissue expansion. *Br J Plast Surg* 1989;42: 591–594.
30. Nordström REA. Antibiotics in the tissue expander to decrease the rate of infection (letter). *Plast Reconstr Surg* 1988;81:137–138.

31. Lawrence WT. Isolated necrosis of deep tissue as a complication of rapid tissue expansion (letter). *Plast Reconstr Surg* 1989;84:172.
32. Kroll SS. The early management of flap necrosis in breast reconstruction. *Plast Reconstr Surg* 1991;87:893–901.
33. Hallock GG, Rice DC. Objective monitoring for safe tissue expansion. *Plast Reconstr Surg* 1986;77:416–420.
34. Kawashima T, Yamada A, Hashimoto M. Problems in the use of tissue expanders for scalp reconstruction (abstract). *Plast Reconstr Surg* 1990;85:834.
35. Baron JL, Reynaud JP, Bousquet P, Quillot M. Dehiscences cutanees en cours d'expansion: Utilisation d'un film plastique chirurgical adhesif. (Cutaneous dehiscences during tissue expansion: Use of surgical adhesive plastic film). *Ann Chir Plast Esthet* 1991;36: 167–168.
36. Saxby PJ. Inelastic tape to treat problems of tissue expansion (letter). *Br J Plast Surg* 1988;41:666–667.
37. Harving S, Ravnsbaek J, Johansen LV. Salvage of exposed implants. *Scand J Plast Reconstr Surg Hand Surg* 1989;23: 143–144.
38. Van Rappard JHA, Jerusalem C, Sonneveld GJ, Borghouts JMHM. Histologic changes in soft tissue due to tissue expansion. *Fac Plast Surg* 1988;5:280–286.
39. McKinney P, Edelson R, Terrasse A, Zukowski M. Chestwall deformity following soft tissue expansion for breast reconstruction. *Plast Reconstr Surg* 1987;80:442–444.
40. Sinow JD, Halvorsen RA Jr., Matts JP, Schubert W, Letourneau JG, Cunningham BL. Chestwall deformity after tissue expansion for breast reconstruction. *Plast Reconstr Surg* 1991;88:998–1004.
41. Cullen KW, Clarke JA, McLean NR. The complications of tissue expansion in the burned scalp. *Burns Incl Therm Inj* 1986;12:273–276.
42. Hemmer KM, Marsh JL, Picker S. Calvarial erosion after scalp expansion. *Ann Plast Surg* 1987;19:454–459.
43. Fudem GM, Orgel MG. Full-thickness erosion of the skull secondary to tissue expansion for scalp reconstruction (letter). *Plast Reconstr Surg* 1988;82:368–369.
44. Paletta CE, Bass J, Shehadi SI. Outer table skull erosion causing rupture of scalp expander. *Ann Plast Surg* 1989; 23:538–542.
45. Penoff, J. Skin expansion—a sword that "stretches" two ways: Scalp expansion and bone erosion. *J Craniofac Surg* 1990;1:103–105.
46. Nordström REA, Pietilä FP, Voutilainen PE, Lilius GPS, Virkkunen PF, Rintala AE. Tissue expander injection dome leakage. *Plast Reconstr Surg* 1988;81:26–29.
47. Lawrence WT. Huber needles for tissue expander inflation. *Plast Reconstr Surg* 1986;78:541–542.
48. Chisholm EM, Marr S, Macfie J, Broughton AC, Brennan TG. Post-mastectomy breast reconstruction using the inflatable tissue expander. *Br J Surg* 1986;73:817–820.
49. Levine RA. Surgical product failures (letter). *Plast Reconstr Surg* 1991;88:916.
50. Van Rappard JHA, Sonneveld GJ, Borghouts JMHM. Geometric planning and the shape of the expander. *Fac Plast Surg* 1988;5:287–290.
51. Nordström REA "Stretch-back" in scalp reductions for male pattern baldness. *Plast Reconstr Surg* 1984;73: 422–426.
52. Goin MK, Goin JM. Growing pains: The psychological experience of breast reconstruction with tissue expansion. *Ann Plast Surg* 1988;21:217–222.
53. Ferguson DJ. Tissue-expanders complicating intubation (letter). *Anaesthesia* 1985;40:822.
54. Francis GA, Jayarajah M. Bradycardia associated with insertion and inflation of tissue expanders under the scalp. *Acta Anaesthesiol Scand* 1991;35:366–367.
55. Roth AG, Dado DV. Venous air embolism during removal of tissue expander in a supine child. *Anaesthesiology* 1988;68:457–459.

9

Face and Neck

ERIC DAVID AUSTAD

Functional and aesthetic reconstruction of the face and neck is limited by the availability of adjacent donor tissue. Historically, reconstructive plastic surgery has directed much of its energy toward the design and execution of increasingly ingenious methods of tissue transfer. It is not coincidental that our specialty traces its origins to the most prominent structure of the face, the nose. Tagliacozzi (1597),[1] of Bologna, described the first transposition of a delayed flap from the arm for nasal reconstruction. This work was preceded by the "Indian" method of nasal reconstruction, which used a forehead flap, as described by Sushruta circa 600 B.C.[2] While other types of acquired or congenital deformities certainly caused comparable or greater functional disability, nasal reconstruction was the first to require a level of aesthetic expertise so challenging that a surgical specialty could evolve from it. In more modern times, the energy and imagination devoted to the reconstruction of congenital facial clefts bears witness to the difficulty of reconstructing facial features and the importance of restoring both function and aesthetic acceptability to patients afflicted with facial deformity. It is not original, then, to observe that facial reconstruction and, to a lesser extent, reconstruction of the neck, are difficult and technically demanding. It may be a new perspective to some, however, to consider that most of the *functional* requirements of facial and neck reconstruction can be met by transfer of tissue from local or distant donor sites. Free-tissue transfer has immensely enlarged this capability. It is now very difficult to imagine a soft-tissue or skeletal injury or deformity that cannot be closed, stabilized, or functionally restored with these techniques. With this ability, what is the value of tissue expansion? In its simplest and, perhaps, its only major clinical application, tissue expansion can provide additional, innervated, matching skin or mucous membrane. Specifically, it provides additional epidermis and "secondary" epidermal structures such as hair follicles and sweat glands. When expansion occurs adjacent to a defect, it provides an elegant match of texture, color, hair-bearing characteristics, and sebaceous function. No anatomical region is more demanding, or deserving, of this technique than the face; it can be resurfaced with tissue derived from itself.

ANATOMICAL CONSIDERATIONS

As described by Argenta and Austad,[3] facial and neck skin can be separated into three distinct types. The forehead and nose are a distinct anatomical unit, covered by thick skin that has few hair follicles and numerous sebaceous glands. The periorbital region is covered by relatively thin and pliable skin, containing few sebaceous glands. The cheeks, upper lip, and neck comprise the third anatomical region: the skin is relatively thin, hair follicles are numerous, and relatively few sebaceous elements are present. Optimally, defects are reconstructed with expansion using skin derived from corresponding donor sites—a concept that is not new. The size or location of the defect in question may obviate the expansion of well-matched local tissue; in this case, it is appropriate to consider the use of expanded full-thickness grafts, as described in chapter 22 in this text by Bauer. He has reported superb results in the resurfacing of massive congenital and acquired deformities, both in the face and elsewhere.

In addition to the physical/histological qualities of the donor site, important structures underlying or adjacent to the area of anticipated expansion must be considered. In some cases, this will facilitate the reconstruction. For example, inclusion of a major neurovascular bundle into the expanded flap may add an extra margin of reliability to the flap transfer. Conversely, an awareness of important anatomical structures in the region of the planned expansion should lead to a more complete appreciation of potential complications, such as facial nerve injury or ectropion. It should be noted that expansion has been successfully performed over or adjacent to virtually every major nerve, including the facial nerve. Manders et al.[4] have extensively studied the response of nerve and muscle to expansion. Their findings are particularly pertinent to expansion in the face and neck, and are summarized below:

1. Skeletal muscle may be expanded, with function retained. This was observed specifically with frontalis muscle, but generalization to other muscle groups appears appropriate if denervation does not occur during implant placement or subsequent mobilization, and this has been the clinical experience of multiple surgeons.

2. The human median nerve and, presumably, other nerves can be expanded with no adverse response. Intraluminal pressure in this study remained below 40 mmHg. Higher pressures resulted in immediate reduction of axon potentials.

3. A definite increase in nerve length can be elicited by underlying expansion, although this appears to involve a significant component of recruitment from adjacent nerve.

In related work, Stark et al.[5] have confirmed the ability of axial blood vessels to elongate in response to expansion. While their initial report focused upon the vascular sequelae of expanding the femoral neurovascular bundle, elongation of the adjacent nerve is obvious in their illustrations, and a detailed study of nerve-related expansion will surely follow. While the location of major blood vessels in the face and neck is of importance in preoperative planning, expansion per se has not been reported as a cause of arterial occlusion or venous obstruction. In fact, the net effect of expansion upon donor tissue vascularity appears to be highly beneficial. Radovan was the first to note that expanded flaps appear inordinately robust with regard to their perfusion and ability to tolerate transfer without vascular compromise.[6] He noted, quite insightfully, that expanded flaps appeared comparable to classically delayed flaps in these characteristics. This phenomenon has been subsequently studied by Cherry et al.[7] and by Sasaki and Pang.[8] Although these two studies differ somewhat in their proposed mechanisms for increased vascularity, they concur in the conclusion that expansion results in a major enhancement of overlying and adjacent tissue vascularity.

GENERAL PRINCIPLES OF TISSUE EXPANSION IN THE FACE AND NECK

Ear and nasal reconstruction are discussed in chapters 10 and 11 in detail, as is appropriate for these specialized structures. In addition, it should be noted here that many defects of the face and neck are amenable to primary closure or to closure with Sasaki's method of intraoperative rapid expansion: the intraoperative sustained limited expansion (ISLE) technique. As noted previously, expanded full-thickness skin grafts, as described by Bauer, are also very useful in this region. The following is a discussion of the benefits and limitations of tissue expansion adjacent to defects not amenable to those techniques.

Forehead

The scalp and forehead are histologically similar, and acquired or congenital deformities of the frontal region often involve components of both. Fortunately, the convex surfaces of this region are highly receptive to the advancement or rotation of hemispherical flaps, as result with expansion. The reconstruction is approached by first establishing a complete definition of the defect in need of resurfacing. This may appear to be a gratuitous comment, but in reality the margins of the defect may not be clearly defined; as, for example, in the case of nevi, which may be extensive but relatively inconspicuous behind the hairline. Accepting the presence of some residual nevus-bearing tissue may be appropriate in the interest of the reconstructive result. Moreover, multistaged expansions should be considered an acceptable approach in all expansion procedures involving large defects, and this eventuality must be a component of preoperative consultations. Because of the deformity and discomfort associated with expansion in this region, it is often most straightforward to proceed with expansion according to the tolerance of the patient, advancing the expanded tissue and reexpanding as appropriate.

After definition of the defect at issue, a consideration of functional requirements should be performed. In the forehead, these are relatively few: normal frontalis function and sensibility are surgical goals, and it is es-

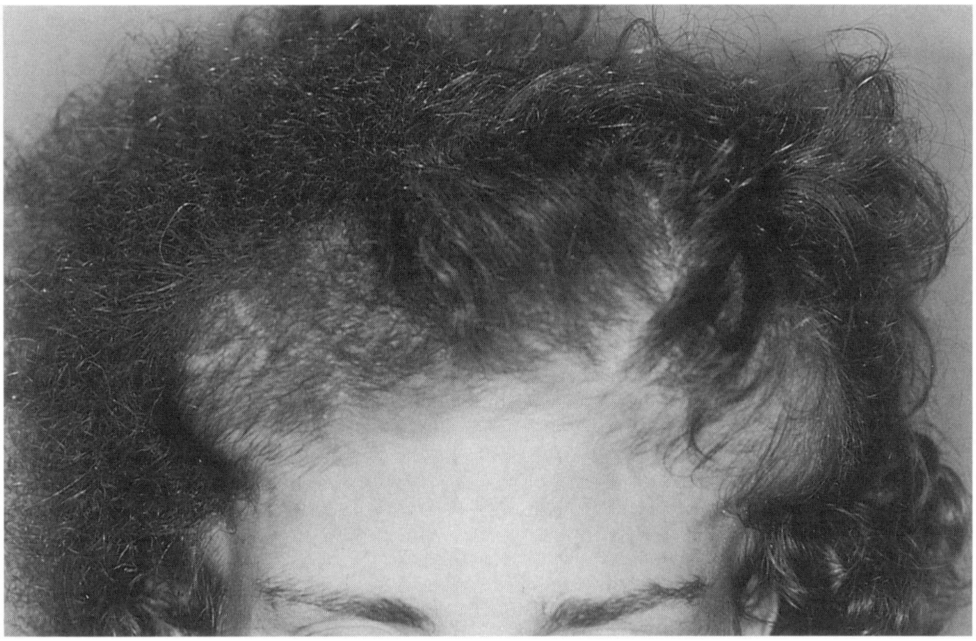

FIGURE 9-1
Scalp/forehead expansion for excision of congenital pigmented nevus.

sential to protect the contralateral (donor) frontalis and supraorbital/supratrochlear neurovascular bundles during expander placement and subsequent expanded flap advancement. This usually leads to the optimal placement of the initial incision either along the anterior hairline or within the defect, and this incision will ultimately be resected or resurfaced. In the forehead, skin grafts are often thin and adhere to bone, and it is my preference to use an anterior hairline incision in normal tissue rather than placing the initial incision at the interface of the graft with normal tissue. Nevus-containing skin or scalp appears to tolerate undermining quite well, and subsequent healing of incisions made in this tissue has been uneventful in my experience. However, in contrast to scar or split-thickness graft, nevus-bearing tissue appears to expand at least as readily as normal skin or scalp. It is disconcerting to find that a small island of nevus in the expanded flap has become significantly enlarged, and this should be considered.

Another possibility of forehead expansion that should be considered preoperatively is that of unusually severe pain, which is more frequently associated with implant inflation in this region. Cole et al.[9] have studied this specifically after observing three patients whose forehead expansions were noted to be "extremely painful." The authors subsequently studied four patients who displayed sharp elevations in intraluminal pressure after injections of volumes less than 10 milliliters. These pressure increases were associated with severe pain. To address this problem, they developed a constant-flow pump that eliminated the sharp pressure increases associated with bolus injections and found that two of the patients studied could be rapidly expanded (mean expansion time of 8.5 days) with significantly less pain. Although there are obvious practical considerations when tethering a patient to a constant-flow pump for a period of days or weeks, this work is being carried out in several medical centers. Physiologically, the procedure is probably comparable to the osmotically driven self-inflating tissue expander we worked with in the late 1970s.[10] In those devices, as in the constant-flow pump, inflation occurred at a constant and continuous rate, and was well tolerated by the two patients who received them.

Figure 9-1 describes the beginning of a straightforward and uneventful forehead/frontal scalp nevus excision, using expansion. The patient is a 27-year-old caucasian woman with a congenital 9 cm × 18 cm pigmented nevus, which begins below the anterior hairline on the right and extends to the vertex. It had become nodular in several areas and had been biopsied twice by her dermatologist. Histologic evaluation revealed no evidence of malignancy, but in the interim between biopsies her mother developed melanoma in a similar lesion on her trunk, and this increased the

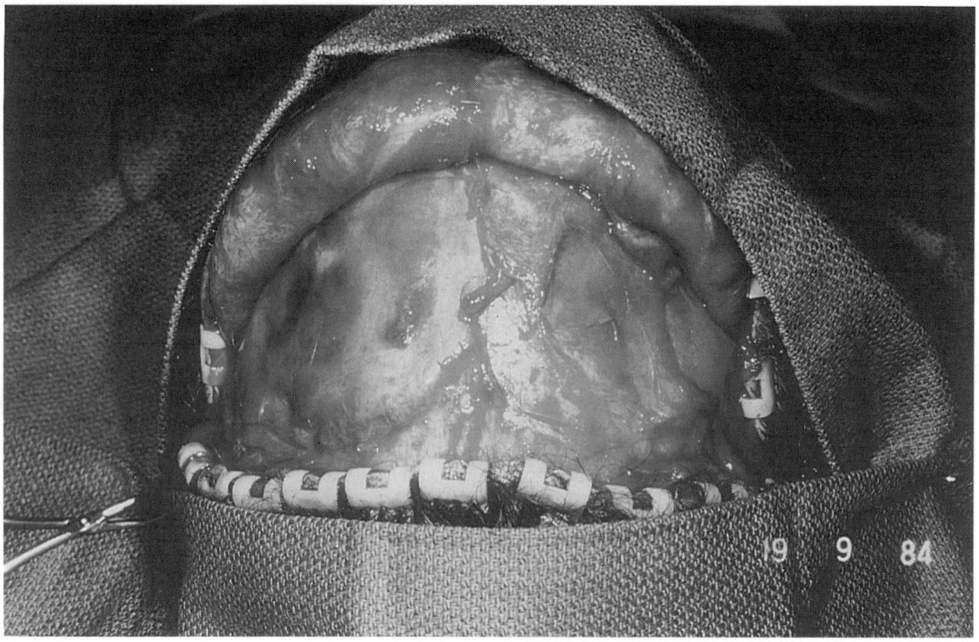

FIGURE 9-2
Skull erosion and periosteal saucerization below tissue expanders.

patient's level of anxiety significantly. Two crescent-shaped 1200-milliliter expanders were placed via an intralesional parasagittal incision, dissecting beneath the nevus to create pockets for the implants at both the right and left margin of normal forehead skin and scalp. Implant inflation began at 2 weeks and proceeded uneventfully over the next 6 weeks.

Upon implant removal, it was apparent that some saucerization or recontouring of the frontal periosteum and skull had occurred (Figure 9-2). This is a frequent finding when the expander is placed immediately adjacent to bone, and it does not appear to be of clinical significance in the scalp or face. It should be noted, however, that dramatic rib deformities were reported during expansion for breast reconstruction by Sinow et al. in 1991.[11] The nevus was subsequently excised uneventfully, and primary closure of the resulting defect was possible (Figure 9-3). The patient was followed postoperatively for 1 year, displaying good healing and no late alopecia or other difficulties.

The patient described in Figures 9-4 through 9-8 is a 14-year-old caucasian girl whose left forehead reconstruction remains a work-in-progress. She illustrates many of the benefits and limitations of tissue expansion in the face. Her congenital nevus involved virtually the entire left frontal region and extended posteriorly to the vertex. At age 8 she underwent gross total excision of her nevus with placement of split-thickness graft. Frontalis was apparently resected with her initial procedure, and her graft was firmly adherent to frontal skull (Figure 9-4). The patient's displeasure with the appearance of her graft became increasingly apparent with the onset of adolescence, and her reconstruction began with placement of a rectangular incorporated-valve expander at age 13. The 500-milliliter expander was placed via an anterior hairline incision, and implant inflation began 2 weeks after placement. These were tolerated relatively well until a volume of 340 milliliters had been attained, requiring a total of 6 inflations over six weeks (Figure 9-5). In spite of multiple preoperative discussions of the expansion process, with its resulting discomfort and grotesque appearance, it became clear at this point that she would not readily tolerate further expansion. When this difficulty with patient compliance occurs, it is essential that rapport between patient and surgeon remain strong. This is not a failure of the technique, but rather is a tribute to its versatility. The implant was simply removed, and the forehead was advanced to the extent possible (Figure 9-6). Capsulectomy or capsulotomy was not performed in this situation, although it might have added some additional length to the advancement, as I wished to preserve all elements of expanded frontalis muscle in the expanded flap. Prior to this second procedure, which entailed implant removal, the option of leaving a deflated expander in place, anticipating a second expansion, was discussed. The patient did not appear psychologically prepared to embark upon this and was

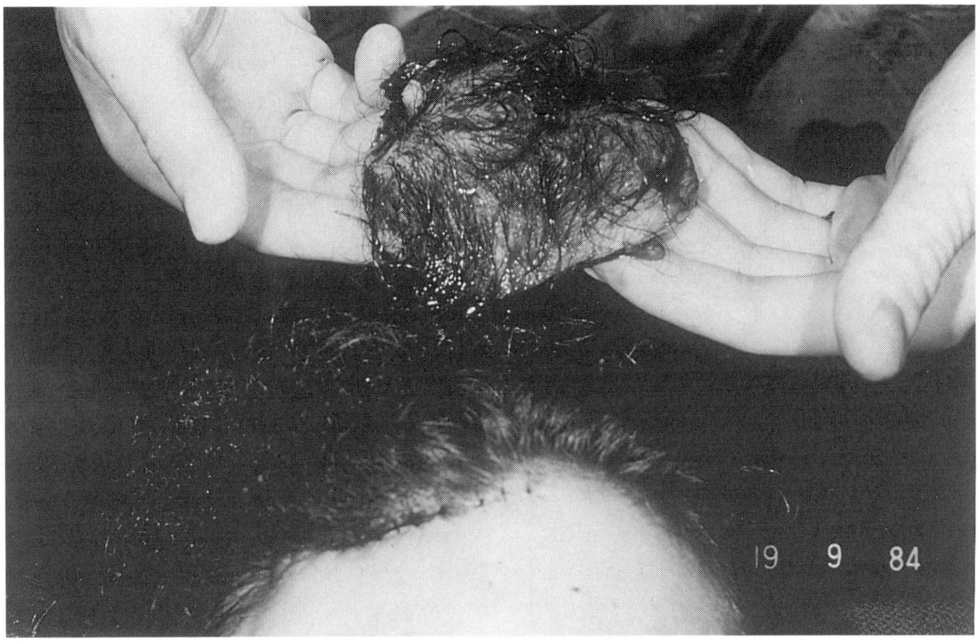

FIGURE 9-3
Excision of congenital nevus after expansion with primary closure.

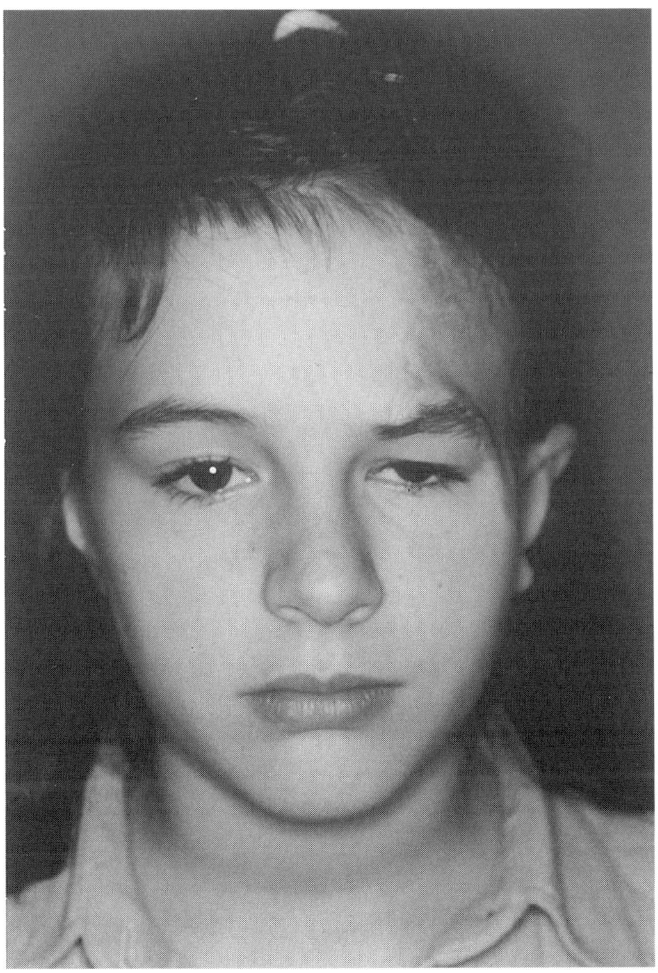

FIGURE 9-4
Eleven-year-old child with hyperpigmented split-thickness graft, left forehead and malar region.

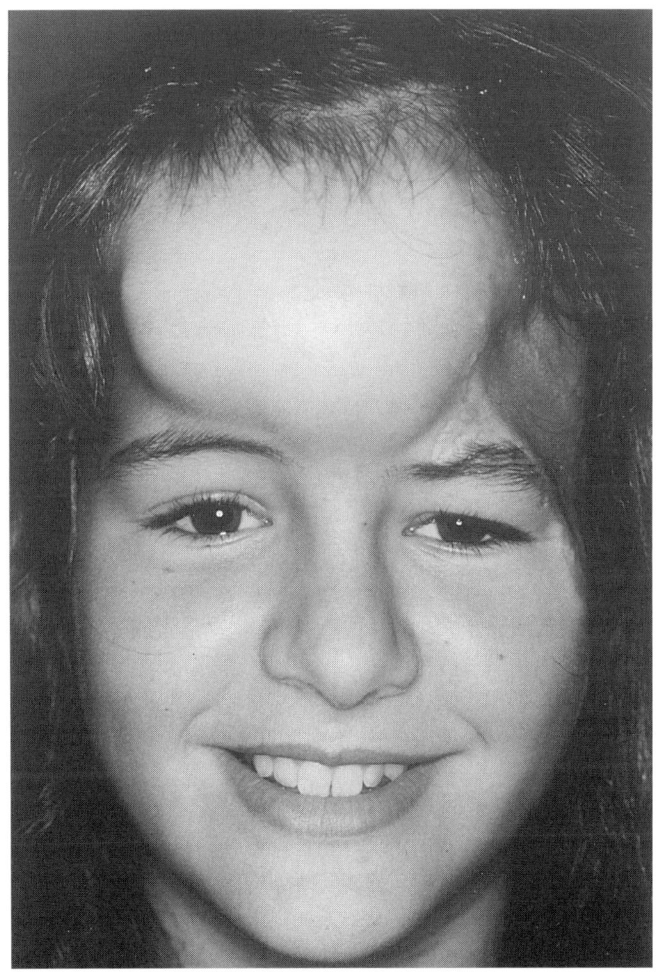

FIGURE 9-5
Midforehead expansion at 340 milliliters volume.

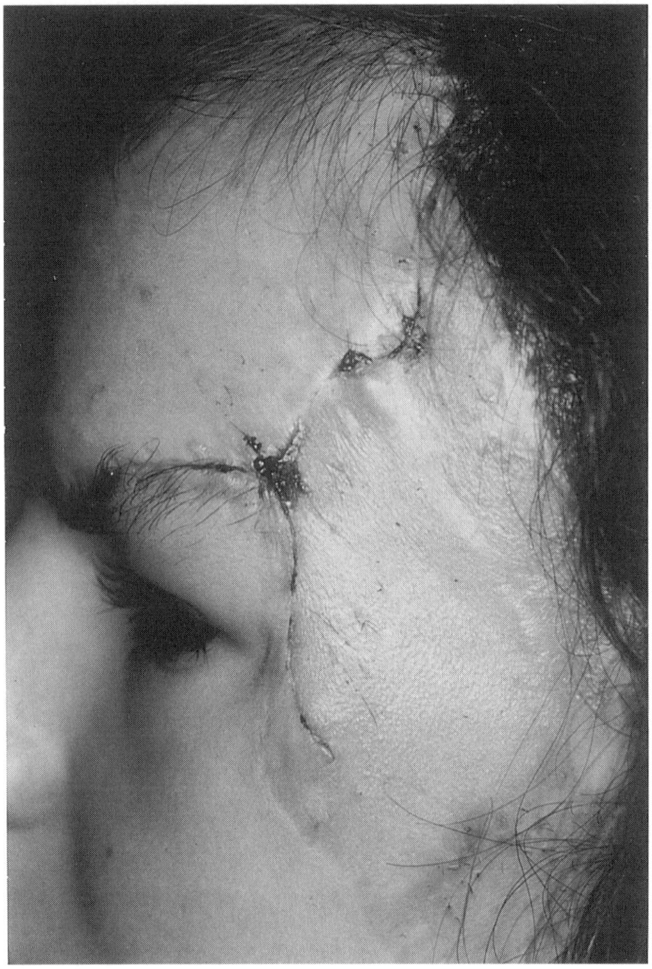

FIGURE 9-6
Early result after subtotal graft excision.

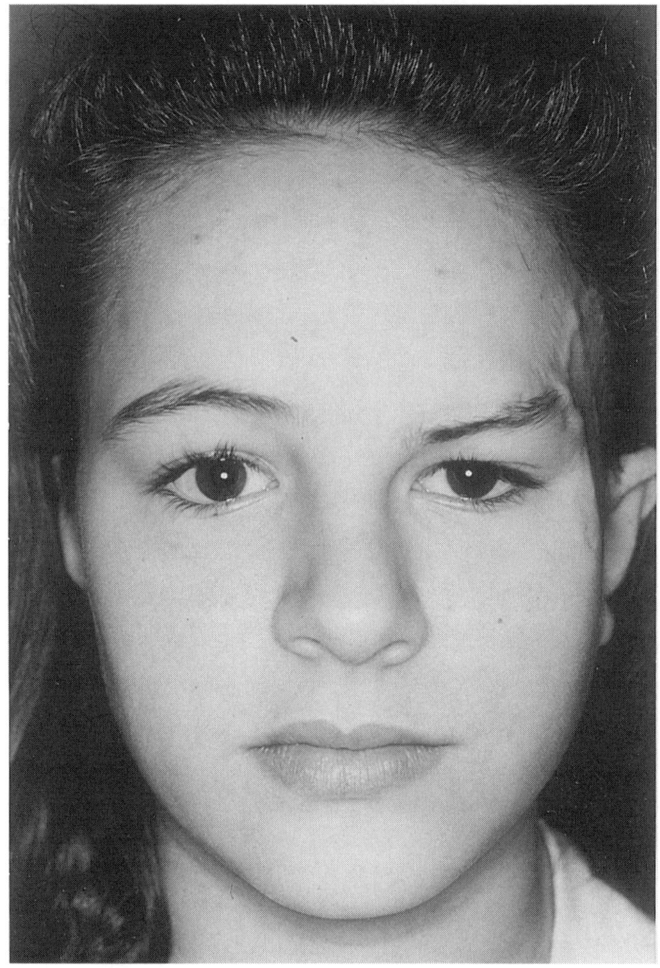

FIGURE 9-7
One-year result, with normal sensate skin of left forehead.

not urged to do so. Her result at one year shows some significant improvement. She has gained approximately 20 square centimeters of normal sensate skin over her left forehead, and she shows some return of frontalis function in this region (Figure 9-7). One year postexpansion, again after extensive discussion with the patient and her parents, it was decided to proceed with intraoperative expansion of the advanced forehead flap, anticipating a subtotal resection of residual graft. The patient tolerated this single-stage procedure much more readily, although her improvement was not dramatic. For the present, we have agreed to continue with this approach of intraoperative expansion/serial resection of graft. As she matures, long-term (6 to 8 weeks) expansion may be more appropriate.

As a final point of interest regarding this patient, it should be noted that the graft overlying her left zygoma and anterior temporal region will ultimately require excision and resurfacing (Figure 9-8). This is an ambitious goal if expansion is to remain confined to the forehead skin. Ultimately, it will require that tissue originally located to the right of midline must reach that region. The temptation, of course, has been to expand the left infraorbital region horizontally and rotate the resultant flap 90° into the lateral brow/zygomatic region. I have refrained from doing so because the cheek and infraorbital region are pristine, and the patient has no ectropion at present. An infraorbital scar is not a disaster, and this may ultimately prove to be the only means of adequately addressing the most lateral extent of her graft and scar. For the present, I will continue to work within the anatomical unit comprising the majority of her deformity: her forehead.

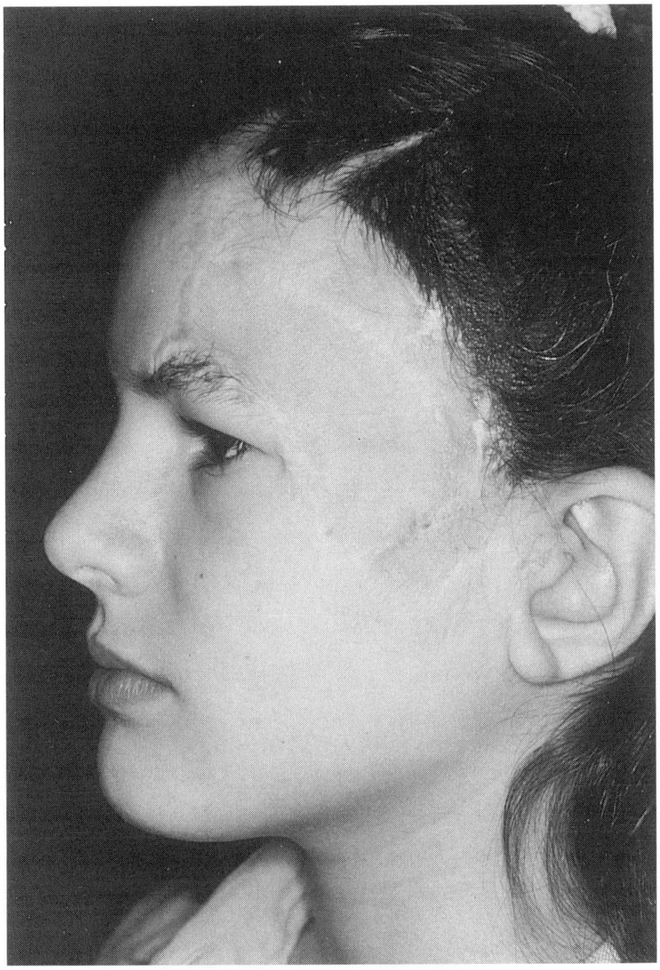

FIGURE 9-8
Residual left forehead and zygomatic defect.

Midface/Cheek

In the spring of 1977 an 8-year-old caucasian girl was referred to William C. Grabb, M.D., Chairman of the Section of Plastic Surgery at the University of Michigan Medical Center. She displayed a hyperpigmented 4 cm × 12 cm split-thickness graft of the left infraorbital region and cheek, the result of a full-thickness burn sustained in an automobile accident 2 years prior to this evaluation (Figure 9-9). The patient had subsequently undergone the first stage of an anticipated serial excision of the graft, which had resulted in a mild ectropion. Approximately 8 months prior to her initial evaluation, my research group had begun studying the concept of tissue expansion in animals, and Dr. Grabb suggested that this child might be a candidate for that technique. Moreover, Radovan had presented his initial clinical experience with breast reconstruction at the annual meeting of the American Society of Plastic and Reconstructive Surgeons in Boston[6] only a few months prior to this child's first visit to us, and there was little doubt that the technique had at least some clinical validity. This child exemplified a situation we often encountered: an otherwise intact patient with no functional loss, who displayed a prominent and permanent deformity that would obviously impact her entire life, and for whom no conventional means of major improvement existed. Further serial excisions appeared futile in view of her ectropion, overgrafting with split-thickness skin did not appear likely to benefit her significantly, and neither free-tissue transfer nor the various types of flaps developed and/or catalogued by Dr. Grabb appeared to have pertinence. We deferred surgical treatment of this patient until 1981, wishing to gain additional experience with the technique in the laboratory and in more straightforward clinical situations. This is perhaps an opportunity to emphasize one of the major observations of virtually every surgeon working with tissue expansion techniques: although the technique is generally straightforward, there are many opportunities for error or misfortune, and early experience is best gained with tattoo removals, "uncomplicated" breast reconstruction, and other types of less-demanding reconstructions. The separation of conjoined twins, facial reconstruction in children, and other similar endeavors deserve a surgeon who is experienced with the technique.

In 1981, after multiple discussions with the patient and her parents, a rectangular 300-milliliter Radovan-type tissue expander was placed below the graft via a preauricular face lift-type incision, undermining virtually the entire left cheek for use as a presumptive donor site (Figure 9-10). In the early preoperative weeks, edema was prominent, and implant inflation did not commence until week 4. Although the patient developed prominent erythema over the implant and subsequently developed a wound dehiscence where the connector tube became displaced and eroded the preauricular incision, the implant was ultimately inflated to a volume of 220 milliliters before an impending implant erosion became obvious. Abandoning the procedure at that point was one reasonable alternative; the implant could have been removed with replacement occurring at some later date. This would have left the child with an investment of approximately 2 months in a procedure that left her with redundant facial skin and no improvement of her original problem. After discussing the situation at length with the patient and her family, we elected to proceed with advancement of the expanded flap, realizing that we had not reached the optimal degree of expansion. Intraoperatively, wide

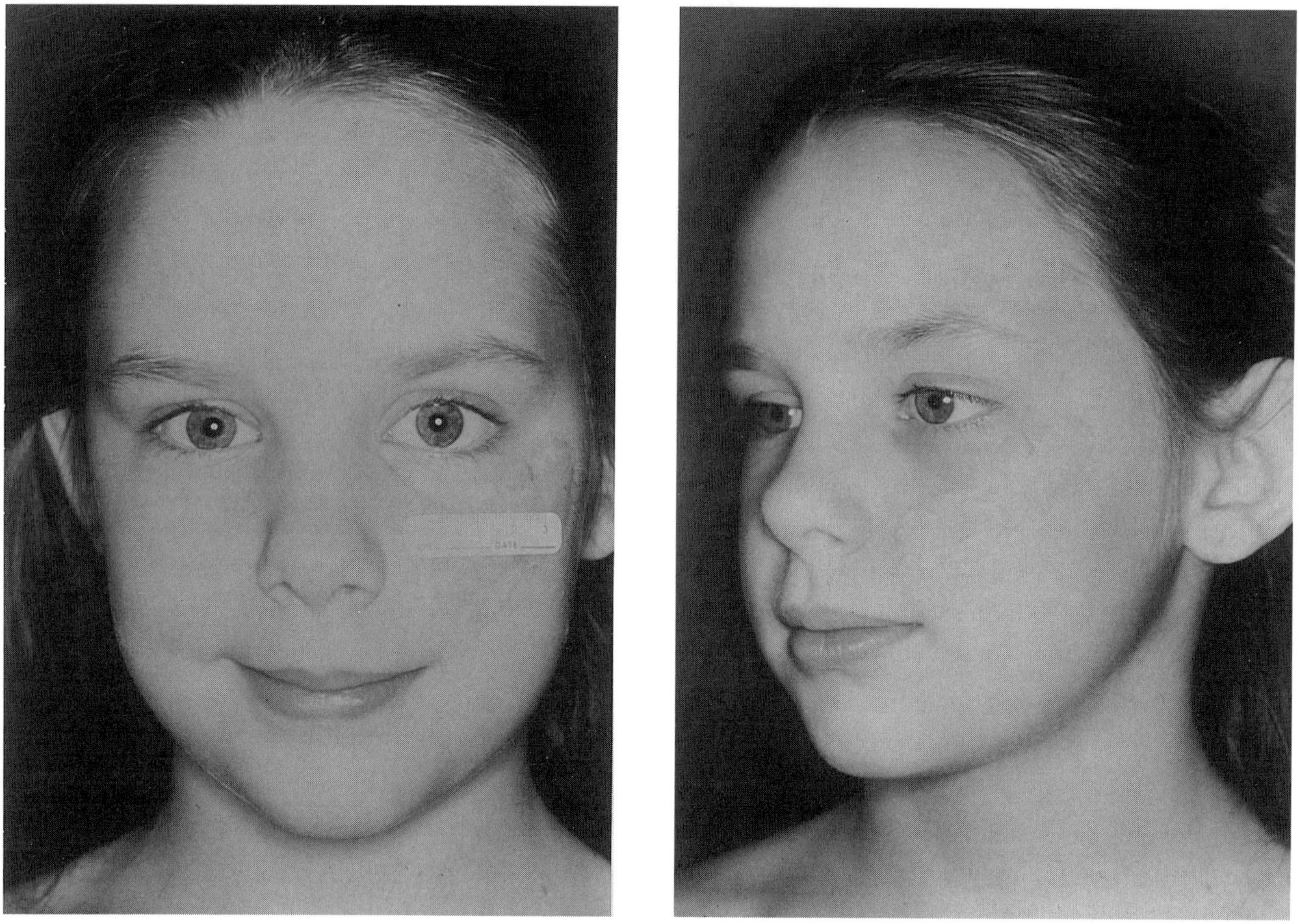

A B

FIGURE 9-9
Eight-year-old child with 4 cm × 12 cm split-thickness graft, left cheek.

undermining of cheek skin adjacent to the expansion allowed complete excision of the graft with what appeared to be an acceptable level of tension on the wound (Figure 9-11). As is usually the case when a graft is excised, the defect became significantly larger than it appeared preoperatively. Closure was performed in layers, and wound healing was uneventful, but the patient proceeded to develop a significant ectropion postoperatively. This might have been prevented by placement of permanent sutures fixing the deep surface of the expanded flap to the infraorbital region, eliminating tension on the lid itself. I would strongly recommend this in similar situations. A canthoplasty at 1 year has restored good function and contour to this patient's lower lid, and it is difficult to discern her scars with minimal makeup. She is now a successful college student, pleased with her result and obviously devoting her full attention to other, more important, aspects of her life (Figure 9-12).

Neck

Tissue expansion in the lower face and neck is comparable in most regards to expansion in other regions of the face. If nerve and vascular structures are not injured by implant placement, they appear to tolerate expansion exceedingly well. Because this skin is relatively thin, placement of the implant below the platysma muscle should be routinely considered. The geometry of the neck and lower face may add an additional element of difficulty to the advancement of an expanded

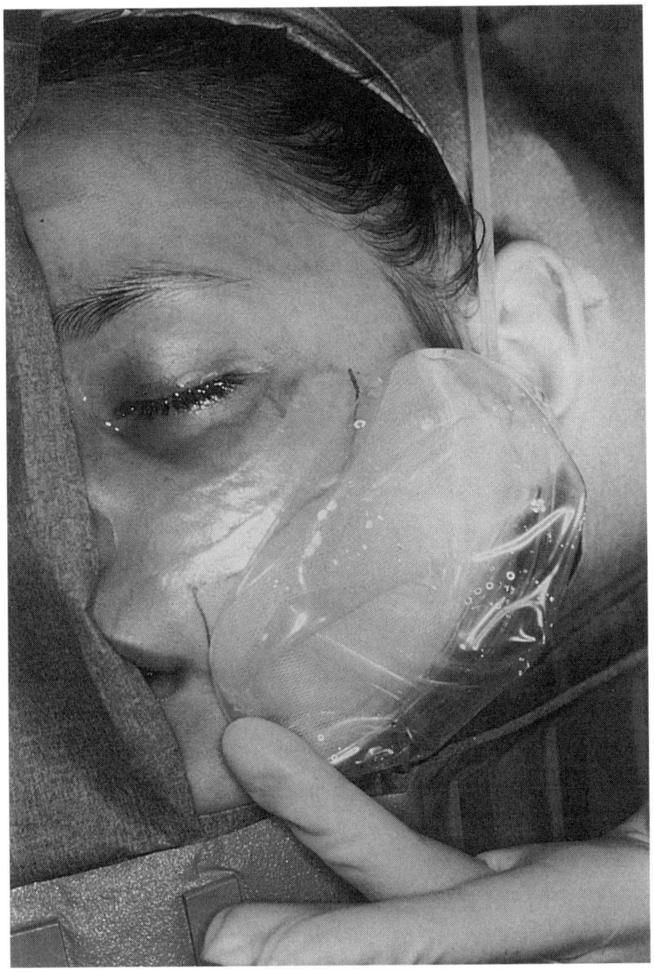

FIGURE 9-10
Placement of 300 milliliters Radovan-type tissue expander, left face.

flap, however, and this should be considered preoperatively. Advancement across the midline is relatively straightforward, as the hemispheric flap adapts quite readily to the convex surface of the contralateral neck. However, expansion of the submandibular region in anticipation of advancement to the lower face is much less straightforward. In this situation, the skin of the neck expands readily, but the craniocaudad dimension is expended to a large extent when it is replaced into the concave high cervical and submandibular region. Actual advancement over the mandibular margin is frequently less extensive than would be expected, and the patient must be prepared preoperatively for a series of expansions.

The hair-bearing characteristics of the lower face and neck are preserved, or essentially so, with expansion. With major soft tissue trauma or ablative surgery to the mandibular region, particularly in males, the ability to mask residual scarring with luxuriant beard growth has been extremely beneficial. Argenta[12] has reported a superb example of this, using expansion of the lateral mandibular regions to obtain oral closure and continence in treating the victim of a self-inflicted shotgun injury. The expanded tissue supported bone grafts to the central mandible without difficulty, and the patient's subsequent beard growth made his massive injury virtually impossible to discern at conversational distances.

Specialized Applications of Tissue Expansion in the Face and Neck

Face-lift. In 1989, Man[13] described the use of a custom-fabricated tissue expander for intraoperative use during face-lift. In sixty-seven patients undergoing some variation of this procedure, he reported excellent clinical results, and observed that he was able to excise "significantly more skin." Publication of this article prompted extensive debate regarding the merits of this technique. Surgeons considering this approach are urged to review this literature; I have not used this approach, but would acknowledge that Dr. Man's results are excellent.

Microorbitalism/Ectropion. Elisevich et al.[14] have described the use of tissue expanders for orbital expansion in a series of patients with unilateral or bilateral congenital anophthalmia, and their results are superb. The study is well documented with preoperative and postoperative computerized tomography (CT) scans and provides a new and ingenious approach to this difficult problem. Victor and Hurwitz[15] described the use of tissue expansion in the treatment of cicatricial ectropion after blepharoplasty but this application has not become widespread.

Orofacial Clefts. In 1989, Lew et al.,[16] described the use of tissue expanders in the treatment of a patient with congenital rubella, in whom a severe perioral and submental defect was beautifully reconstructed. In 1990, Toth et al.[17] reported an excellent result with the use of a forehead tissue expander in treating a child with an atypical Tessier No. 3 craniofacial cleft.

Webbed Neck/Turner's Syndrome. In December 1989, Niranjan published the first application of tissue expansion in the webbed-neck deformity of Turner's syndrome.[18] In his approach, the lateral neck is expanded bilaterally with subsequent posterior advancement. The webbed deformity is eliminated, and the low hairline is elevated. Miller et al.[19] and Thomson et al.[20] have subsequently confirmed the value of this approach.

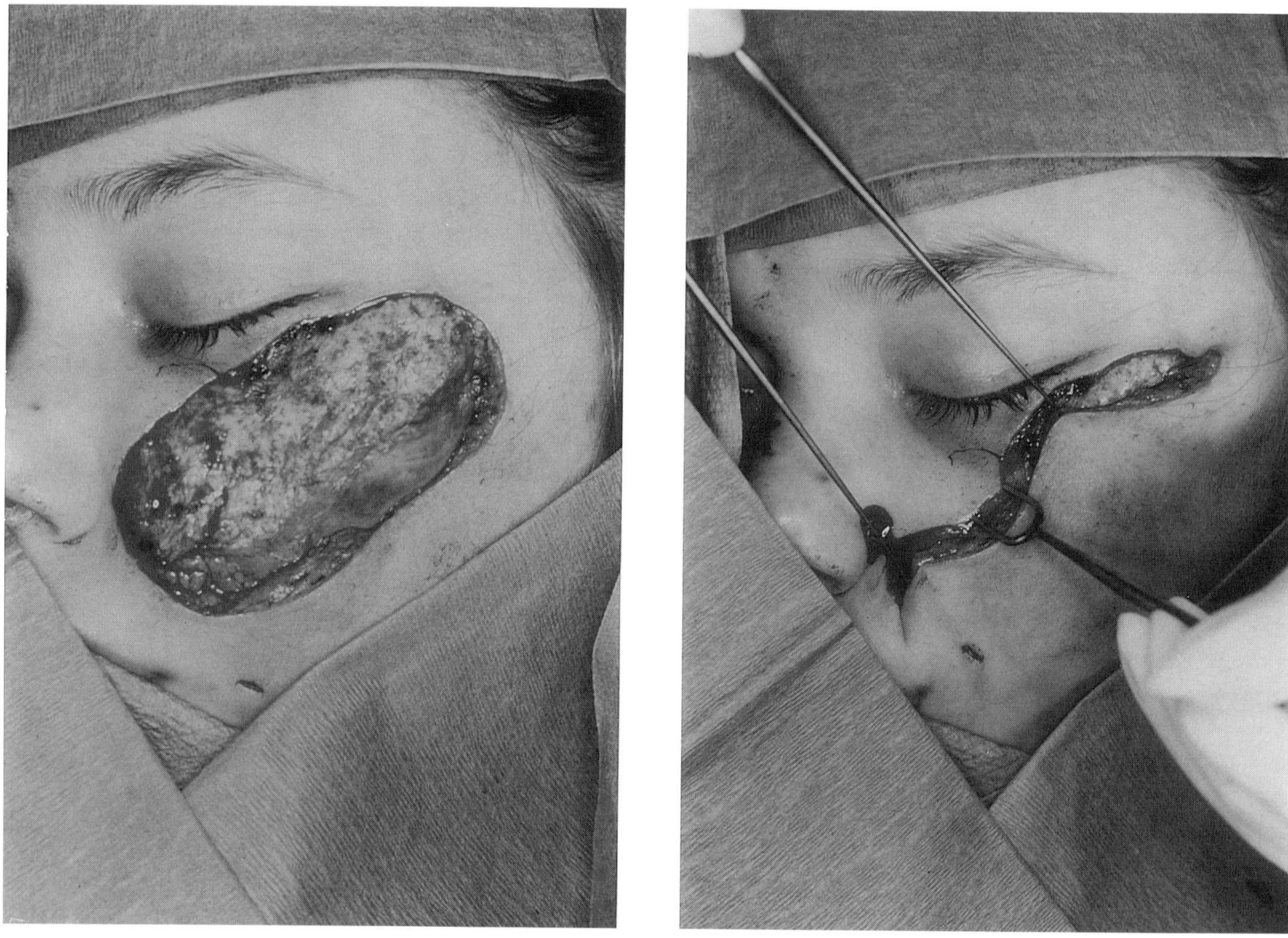

A B

FIGURE 9-11
Resulting skin defect after graft excision, (A) with subsequent closure, (B) under moderate tension.

Psychological Aspects of Expansion in the Face and Neck
Goin and Goin[21] have studied the psychological reaction of patients to tissue expansion in various anatomical locations, and they have found that virtually all of the patients experienced a significant alteration of "body image." This may be transient or quite profound. During preoperative consultations, it is imperative to be aware that each patient is psychologically unique and complex and that the surgical technique being discussed is protracted and technically complicated. Many surgeons who use the technique on a regular basis note that an unusually strong bond is formed with these patients. Tissue expansion is obviously a cooperative effort, in which the patient is entrusted not only to protect his/her implant during expansion, but also is charged to biologically create the "raw material" for the reconstruction itself. The intensity of both the patient's response and the surgeon's involvement is obviously dependent, to some degree, upon the complexity and location of the reconstruction. A small expander placed upon the forearm for ablation of a tattoo is not likely to precipitate a great deal of psychological response or interaction. However, reconstructions involving the face or neck are virtually never subtle—most often they involve a period of some weeks during which an already disfigured patient becomes truly grotesque. Memories of original disfigurement and subsequent painful treatment may be rekindled, and regular routines are usually interrupted to some degree. Some patients are simply not candidates for this procedure, and some deformities do not warrant its undertaking. This is vir-

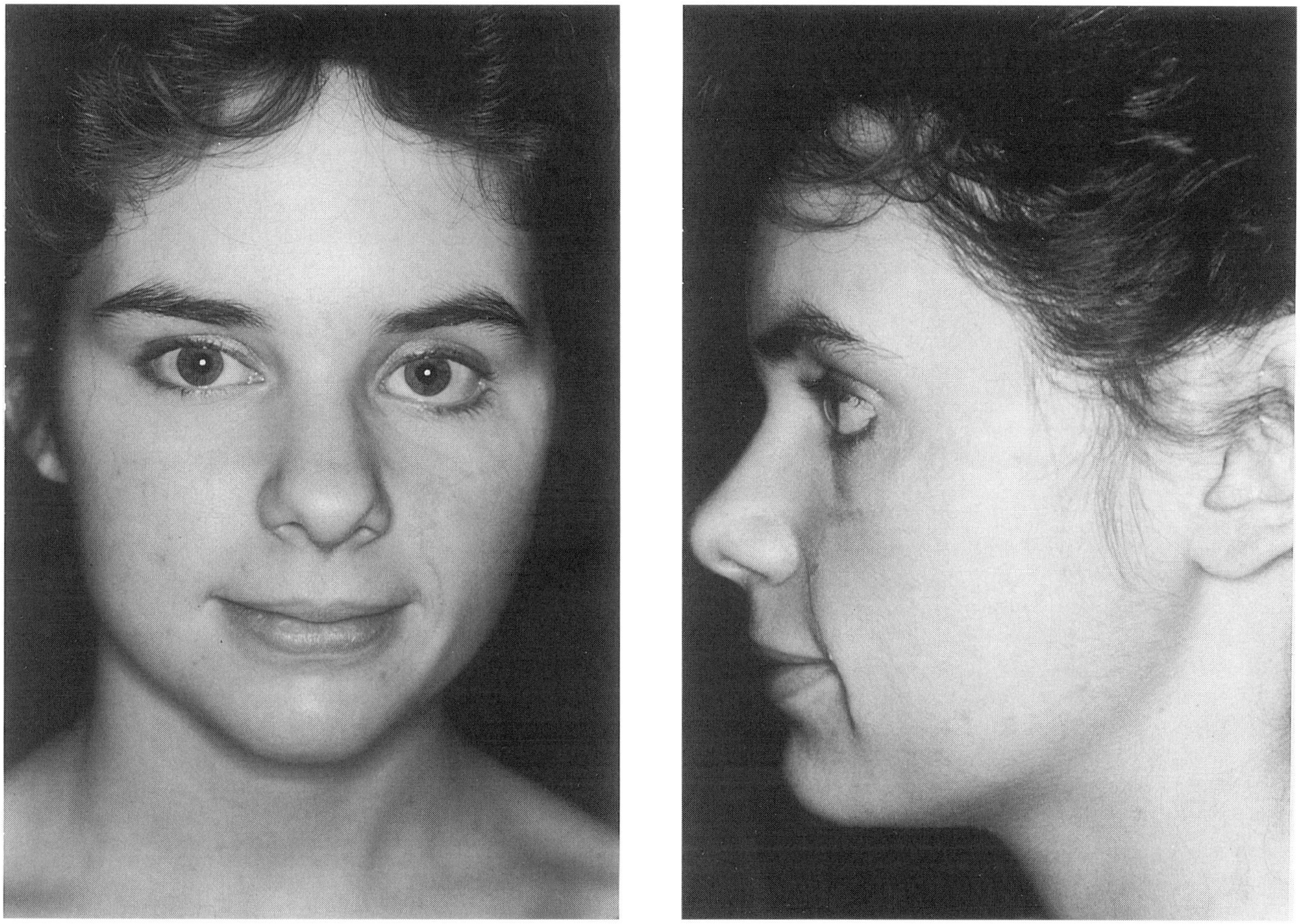

A B

FIGURE 9-12
Seven years postexpansion, without makeup.

tually always an elective procedure, and there is adequate time for preoperative consultation and patient comprehension. I do not routinely have these patients undergo preoperative psychiatric consultation, but I have found this useful on occasion, and it can be the basis for additional psychological support during the postoperative phase. Fortunately, most patients tolerate expansion in the face and neck surprisingly well, viewing it as a transient disruption, or another step toward eliminating their disabilities, or as a remarkable means of restoration.

REFERENCES

1. Tagliacozzi G. *De curtorum chirurgia per insitionem.* Venice, Italy: B. Bindonus, 1597.
2. Sushruta S, as reported by Nichter LS, Morgan RF, Nichter MA. The impact of Indian methods for total nasal reconstruction. *Clin Plast Surg* 1983;10:635.
3. Argenta L, Austad E. Priniciples and techniques of tissue expansion. In *Plastic Surgery.* Philadelphia: W. B. Saunders, 1990, pp. 475–507.
4. Manders E, Saggers G, Diaz-Alonso P, et al. Elongation of peripheral nerve and viscera containing smooth muscle. *Clin Plast Surg* 1987;14:551–562.
5. Stark GB, Hong C, Futrell JW. Rapid elongation of arteries and veins in rats with a tissue expander. *Plast Reconstr Surg* 1987;80:570–581.
6. Radovan, C. Adjacent flap development using expandable silastic implants. Presented at the Annual Meeting of the American Society of Plastic and Reconstructive Surgeons, Boston, MA, September 13, 1976.
7. Cherry G, Austad E, Pasyk K, McClatchey K, Rohrich RJ. Increased survival and vascularity of random pattern skin

flaps elevated in controlled, expanded skin, *Plast Reconstr Surg* 1983;72:680–685.
8. Sasaki G, Pang C. Pathophysiology of skin flaps raised on expanded pig skin. *Plast Reconstr Surg* 1984;74: 59–65.
9. Cole RP, Gault DT, Mayou GJ, Davis PK. Pain and forehead expansion. *Br J Plast Surg* 1991;44:41–43–46.
10. Austad ED, Rose GL. A self-inflating tissue expander. *Plast Reconstr Surg* 1982;70:588–593.
11. Sinow JD, Halvorsen RA, Jr., Matts JP, Schubert W, Letourneau JG, Cunningham BL. Chest-wall deformity after tissue expansion for breast reconstruction. *Plast Reconstr Surg* 1991;88:998–1004.
12. Argenta L. Controlled tissue expansion. *Surgical Rounds,* 1986;February:70.
13. Man D. Stretching and tissue expansion for rhytidectomy: An improved approach [comment]. *Plast Reconstr Surg* 1989;84:561–571.
14. Elisevich K, Bite U, Colcleugh R. Microorbitalism: A technique for orbital rim expansion. *Plast Reconstr Surg* 1991;88:609–612.
15. Victor WH, Hurwitz JJ. Cicatricial ectropion following blepharoplasty: Treatment by tissue expansion. *Can J Ophthalmol* 1984;19:317–319.
16. Lew D, Shroyer JV Jr., Unhold GP, Stutes RD. The use of tissue expanders in the reconstruction of orofacial defects secondary to congenital rubella: Case report. *J Oral Maxillofac Surg* 1989;47:1202–1207.
17. Toth BA, Glafkides MC, Wandel A. The role of tissue expansion in the treatment of atypical facial clefting. *Plast Reconstr Surg* 1990;86:119–122.
18. Niranjan NS. Webbing of the neck: Correction by tissue expansion. *Plast Reconstr Surg* 1989;84:985–988.
19. Miller LB, Kanter M, Wolfort F. Treatment of webbed neck in Turner's syndrome with tissue expansion. *Ann Plast Surg* 1990;24:447–50.
20. Thomson SJ, Tanner NS, Mercer DM. Web neck deformity, anatomical consideration and option in surgical management. *Br J Plast Surg* 1990;43:94–100.
21. Goin M, Goin J. Growing pains: The psychological experience of breast reconstruction with tissue expansion. *Ann Plast Surg* 1988;21:217–222.

10

Nose

ROLF MÜNKER

Since the introduction of soft tissue expansion in head and neck surgery by Argenta et al. in 1981,[1] subtotal and total nasal reconstruction has reached a new level. A wide range of indications and different techniques have become available regarding the expander type, the various shapes and volumes of the devices, and finally the option for chronic, intermediate, short, and intraoperative rapid expansion. The most significant advantage of tissue expansion, particularly in facial reconstruction, is the replacement of cutaneous defects with adequate tissue.

Nasal reconstruction requires the use of thin hairless skin that will not conceal the contour of the underlying structures. Major full-thickness nasal defects need greater amounts of skin to provide both cover and lining. Resurfacing cartilage grafts that have been put in place simultaneously requires flaps to be adapted and folded in without tension.

Classical procedures using forehead and scalp flaps may provide sufficient tissue but leave donor site defects that have to be covered by skin grafting in the visible area of the forehead or hair-bearing scalp. John Conley,[2] one of the masters of the use of regional flaps in head and neck reconstruction, was aware of this unfavorable condition when he stated: "The forehead flap has the appeal of providing approximately 90 sq. cm of tissue for resurfacing and reconstruction purposes in a nondelayed technique. Its principal disadvantage is the defect of the donor area of the forehead."

The application of chronic tissue expansion in preparing a forehead flap provides additional length and width for the flap and enables primary donor site closure at the same time. This substantial advantage was outlined by Manders who remarked:

> Following soft tissue expansion of forehead tissue, flaps of remarkable proportion may be raised with primary closure of the donor site. The flaps prepared with expansion have an increased vascularity and are quite hardy. Thus more reliable flaps of dimensions greater than those readily obtained in the past are available for large nasal reconstructions. A special advantage is the minimizing of the donor defects.[3]

The startling contrast between these two statements indicates the dramatic evolution in nasal reconstruction within a single decade. Due to the use of soft tissue expansion, the disadvantage of secondary skin grafts to cover the donor area of a forehead flap is no longer a concern. Ideal skin texture, thickness, color match, and vascularity of the forehead flap provide the best conditions for nasal reconstruction compared with other methods, such as scalp flaps, brachial flaps (the Tagliacotian or Italian technique), cervical flaps, acromiopectoral, and deltopectoral flaps. Generally, it can be stated that soft tissue expansion generated a revival of local tissue flaps. In nasal reconstruction, local tissue refers to direct nasal skin or to structures immediately adjacent. The preexpanded forehead flap is currently the method of choice for providing large soft tissue coverage of the nose, while reconstruction of smaller nasal defects is an ideal field for rapid intraoperative expansion using miniexpanders.

Harbon et al.[4] reviewed 16 cases in a trial comparing Converse[5] scalping flaps to expanded medium forehead flaps. The authors discussed the respective advantages of the two methods, which both gave good results in subtotal nasal reconstruction. The Converse scalping flap was advocated as a simple and quick solution, especially after cancer resection in elderly patients. At present, the authors favor forehead expansion, despite its duration and complexity, for the very significant aesthetic benefits in terms of forehead scarring.

THE FOREHEAD FLAP

Anatomy

Blood supply to the forehead is excellent and allows development of a variety of axial pattern flaps. The medial forehead is supplied by supratrochlear and supraorbital branches deriving from the internal carotid artery. The lateral frontal area is vascularized by the zygomaticoorbital and superficial temporal branches of the external carotid artery. Both vascular systems are connected by an extensive network of anastomoses. The various types of horizontal, vertical, and oblique axial forehead flaps described for nasal reconstruction are based on this rich vascularization.

For our purposes of tissue expansion of the forehead, the oblique paramedian flap provides ideal anatomic conditions. Its blood supply is derived from the supratrochlear vascular bundle accompanied by the affiliated sensory nerve branch (Figure 10-1). The anatomic advantage of this supratrochlear flap, compared with other forehead flap variations, is the very central and deep inserting axial vascularization. The skull perforating point of the supratrochlear vessels is situated in the middle of a line between the inner canthus and inner eyebrow border. This location allows the surgeon to position the flap at the pivot point nearest to the nasal defect to be covered. The supratrochlear vascular bundle is running as a vertical dermal plexus, anastomosing at the hair-bearing area with the frontal branches of the superficial temporal artery and vein. This system nourishes the adjacent skin area and, through small penetrating vessels, the underlying frontal muscle. Because there is a definite cutaneous axial blood supply, we feel there is no need to incorporate frontalis muscle in a forehead flap designed for nasal reconstruction. The reliability of a myocutaneous flap, as used by other investigators[6,7] has to result in unnatural thickness and hardness. On the other hand, meticulous surgical dissection of a subcutaneous plane directly over the frontalis muscle is mandatory to protect the well-defined axial blood supply. A reliable cutaneous flap can be raised if knowledge and use of bedside and intraoperative Doppler examination are considered as absolute requirements to record possible expander-specific alterations of the vascular anatomy. If a round expander has been used or if the lateral edge of a smaller rectangular expander has been placed near the vascular bundle, then we would verify a movement of the supratrochlear bundle from a paramedian line laterally. As expansion proceeds, the stretched skin—including the vessels—moves from the apex to the base of the expander hemisphere. This is necessary to create a reliable forehead flap. There was no appreciable vascular lateroposition found if large rectangular expanders (160 milliliters and more) for the entire forehead have been used.

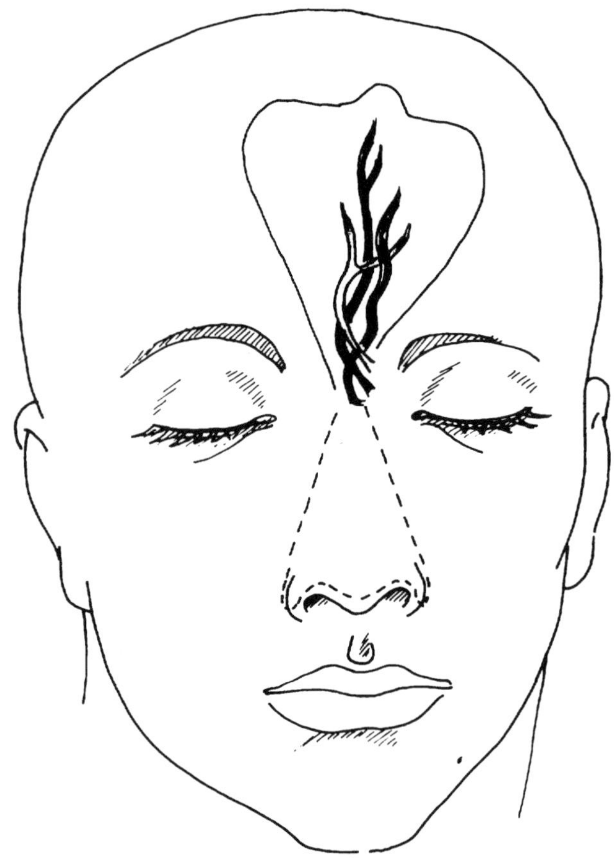

FIGURE 10-1
Diagram of the paramedian forehead flap. The rotation flap is based on a supratrochlear neurovascular axial pattern for nasal reconstruction. There is flap rotation for defect coverage of 170 degrees. Münker R. Nasal reconstruction with tissue expansion. Fac Plast Surg *1988;5:328–337. (Used with permission from Thieme Med. Publ., New York.)*

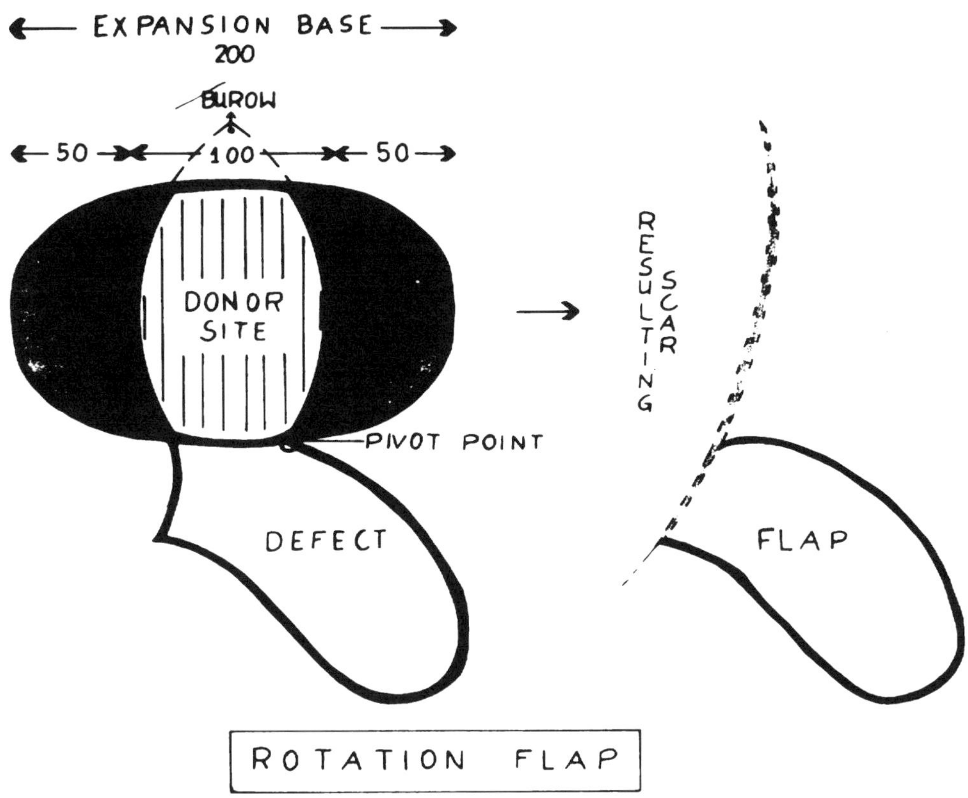

FIGURE 10-2
Geometric conditions of donor site and flap design in a preexpanded rotation flap. Münker R. Nasal reconstruction with tissue expansion. Fac Plast Surg *1988;5:328–337. (Used with permission from Thieme Med. Publ., New York.)*

Geometric Conditions
The forehead provides an area of homogenously structured skin extending approximately 18 cm × 7 cm in an adult and 14.5 cm × 5 cm in a 10-year-old child. A total nasal reconstruction, including tissue for infolding and inner lining, requires a triangular flap measuring 7 centimeters in the horizontal axis and 8 centimeters in the vertical axis. General principles of tissue expansion have shown almost the same geometric conditions for the development of either advancement or rotation flaps. For tensionless per primam closure of a rotation flap donor site, the meridian over the expanded area must be at least double the size of the greatest horizontal diameter of the rotation flap design (Figure 10-2). In the forehead, which is assumed to provide a horizontal extension of 18 centimeters, it is not a problem to create vertical or oblique flaps with a maximum horizontal axis of 8 centimeters, or to perform (at the same time) direct wound closure to the donor site if adequate tissue expansion has been performed.[8]

Expander Choice
Expander size and shape is primarily dependent on the amount of nasal tissue to be replaced by forehead skin. Usually a 160 milliliter rectangular expander provides enough tissue increase to cover subtotal nasal defects (Figure 10-3). There are recommendations in the literature for using 300 to 400 mL expanders;[3,9] until recently, however, we had no situation that exceeded the 250 milliliter size. A rectangular device of 250 milliliters has a base of 13 cm × 6 cm and covers the forehead totally. The deformity resulting from expansion is less conspicuous with this device (Figures 10-4A and 4F) than after full expansion of a smaller round expander that is placed very protuberantly in the middle of the forehead (Figure 10-5C).

The expansion period usually is 6 to 8 weeks and the patient has to tolerate an increasing deformity, which is an important psychological factor. Furthermore, a tissue increase induced by a rectangular expander is higher than that induced by a round expander.[10] We

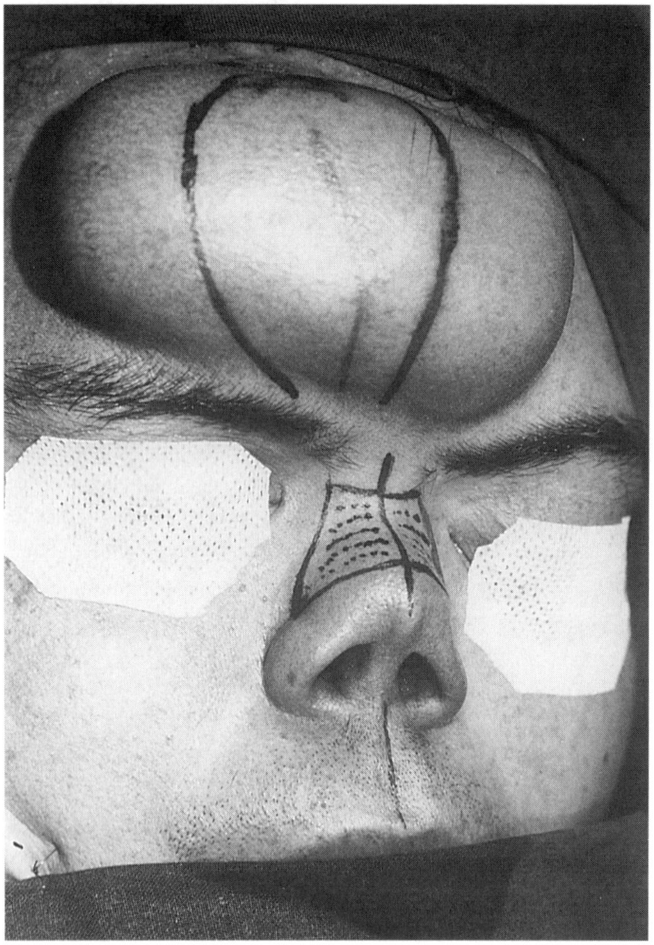

A

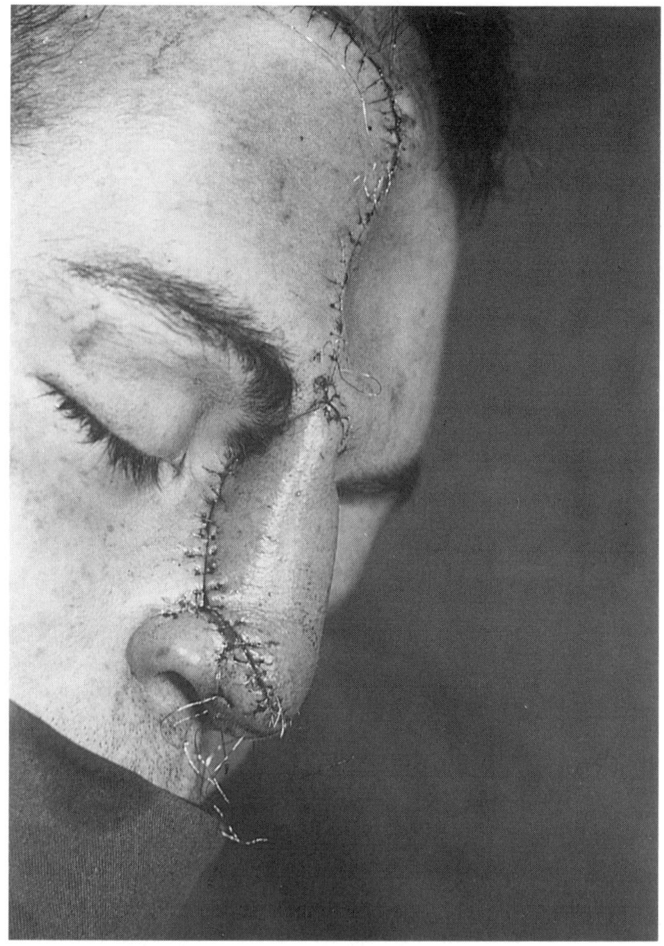

B

FIGURE 10-3

Patient with Binder's Syndrome. There is extreme shortness of the nose and nasomaxillary retrusion, together with exophthalmus, divergent strabismus, optic nerve involvement, and macrostomia. (A) Subcutaneous expander, 100 milliliters, rectangular, incorporated valve on the forehead to develop a paramedian forehead rotation flap based on the supratrochlear vessels mapped by Doppler sonography. The hatched area on nasal dorsum is de-epithelialized, and there are two hinge flaps based on the sagittal midline for inner nasal lining of detelescoped cartilaginous structures. (B) Inner and external elongation of nasal soft tissue by use of the above technique. The paramedian forehead flap is sutured in place. There is a primary donor site closure according to the diagram in Figure 10-2. Skeletal nasal lining is via a costal cartilage graft. (C) Before, front view. (D) Four months after nasal reconstruction and bone grafting of the midface, front view. Later flap trimming, correction of the nasal lining, and chin augmentation are planned. (E) Before, lateral view. (F) After, lateral view. (G) Before, oblique view. (H) After, oblique view. Münker R. Nasal reconstruction with tissue expansion. Fac Plast Surg *1988;5:328–337. (Used with permission from Thieme Med. Publ., New York.)*

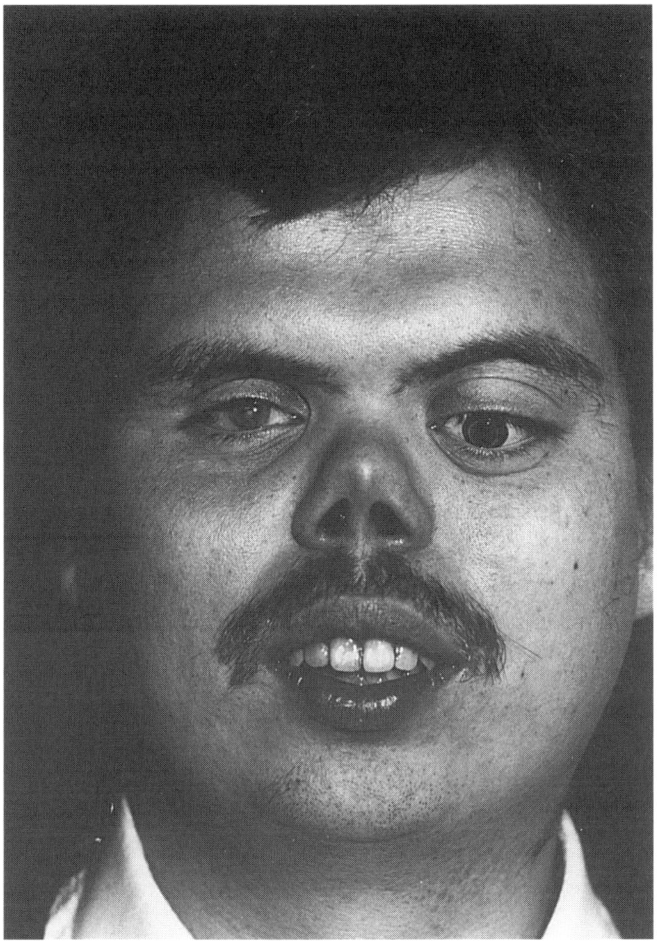

C

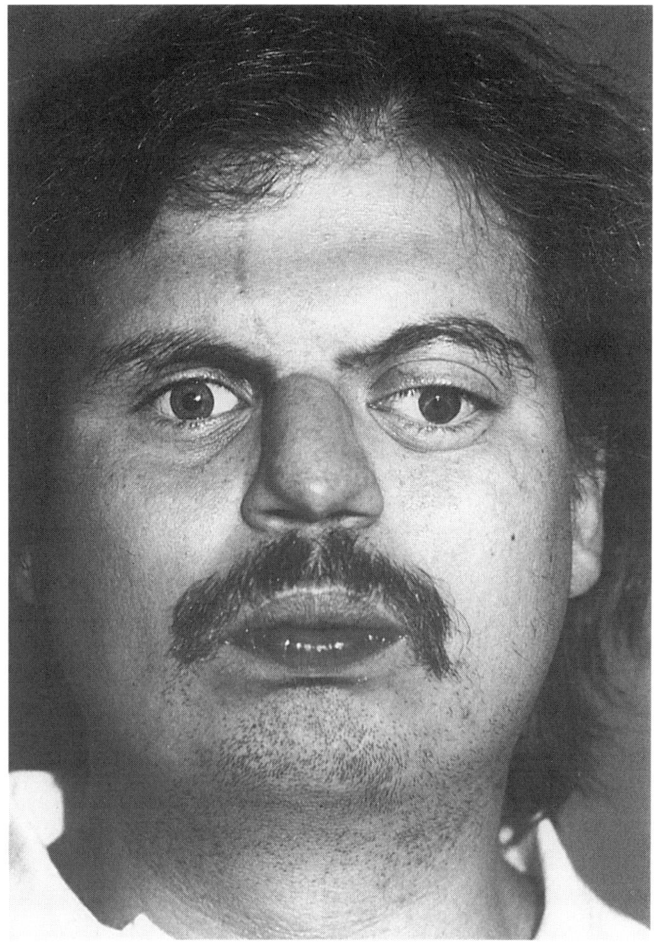

D

FIGURE 10-3 *(Continued)*

first started nasal reconstruction with the original round Radovan expander, supplied with a distant filling valve that was placed in a separate pocket in the opposite frontotemporal area.[11] The volume and stiffness of most of the currently used incorporated filling valves implanted over the frontal bone cause patients discomfort in the early expansion period. To some extent, this discomfort can be diminished by creating a wide expander pocket. The location of the incorporated valve presents more of a problem than does a distant filling valve.[12] There is a magnetic locating device on the market (Magna-finder, McGhan Medical Corporation) that is useful in identifying the integral injection dome.

Indications for Chronic Expansion

Forehead flaps designed to cover smaller nasal defects do not need chronic expansion as long as the width of the flaps does not exceed approximately 3 centimeters. Wide undermining of the wound edges is a sufficient maneuver to close this donor defect. Chronic tissue expansion with its several disadvantages (namely, multiple procedures, visible deformity, psychological impact) should be limited to problematic cases of subtotal and total nasal reconstruction that require the creation of larger amounts of skin for coverage and lining. For this purpose, expanded forehead flaps that are 7 to 8 cm wide present no problem for immediate closure of the donor site[7,8]

Different degrees of subtotal and total nasal defects due to either congenital deformities, traumatic or iatrogenic lesions, and cancer have been treated by chronic forehead expansion.

Complete nasal aplasia, as the most serious malformation, has need for simultaneous construction of the internal airways (choanal atresia), the inner lining, the bony nasal structures, and the external skin. Treatment

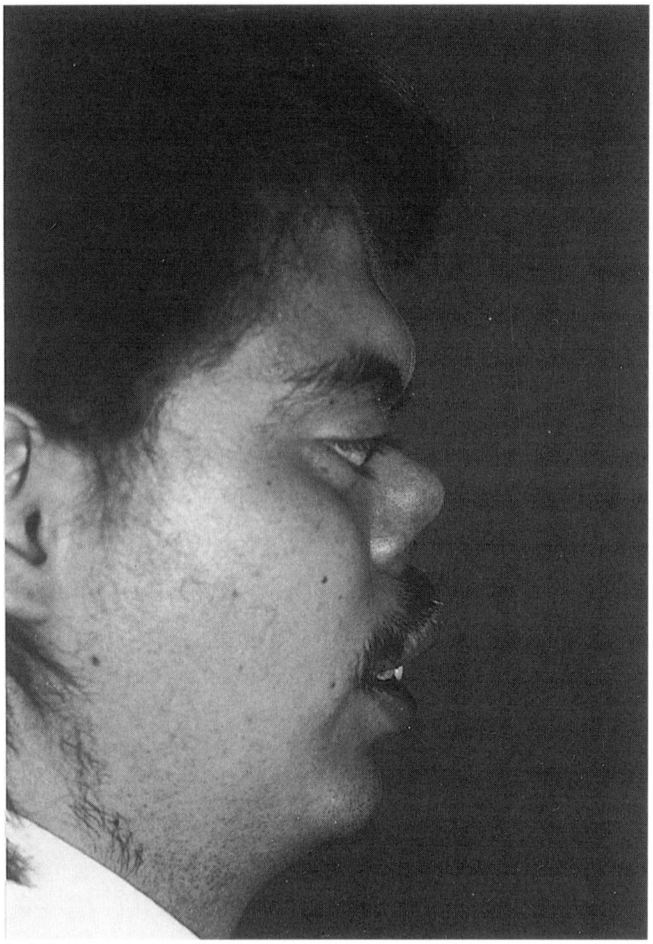

E

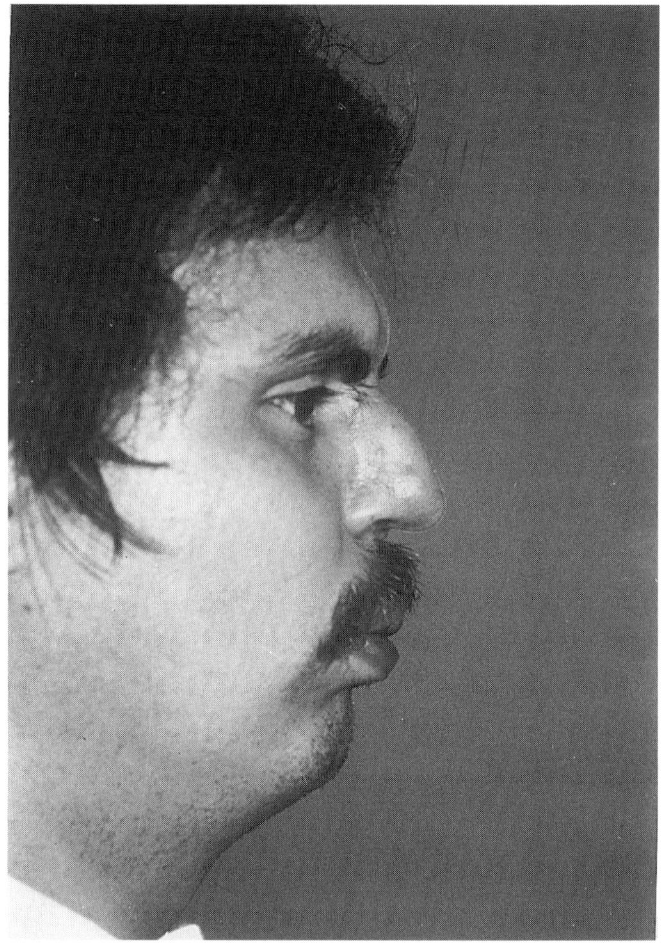

F

FIGURE 10-3 *(Continued)*

in early childhood using a preexpanded forehead flap has been reported by Lampe et al.[13]

A different method for treating complete nasal aplasia at a very early age has been advocated by Möhlbauer et al.[14] These authors preexpanded the midfacial skin of a 4-week-old female infant using two mini-expanders and created skeletal support with an osteoperiosteal flap of the front.

In 1985, we reconstructed an extremely short nose in an adult patient with Binder's Syndrome by using a preexpanded forehead flap at a time when miniexpanders were not available (Figure 10-3). At present, based on the experiences of Mühlbauer, we would prefer local skin expansion of the nose and midface using miniexpanders in this case.

Chronic expansion of the forehead for coverage of nasal cancer defects has some limitations. It is a delayed technique and an expander should never be used before extirpation of a malignancy so that surgical eradication is not compromised. Due to its stretchability, the skin of the predominantly older cancer patient is suitable in most of the cases for an abbreviated expansion procedure. Although prolonged forehead flap expansion usually consists of injections performed once or twice weekly over a period of 6 to 8 weeks, Marks et al.[15] reported on a more rapid expansion technique that uses daily injections over a period of one week. The advantage of shorter morbidity is one important aspect of this intermediate expansion. Furthermore, there is a reduced flap thickness and viability compared with flaps expanded over a couple of weeks.

Technique of Chronic Expansion

Expander implantation for developing an oblique paramedian forehead flap may be performed under local or general anesthesia. Before initiating surgery, the

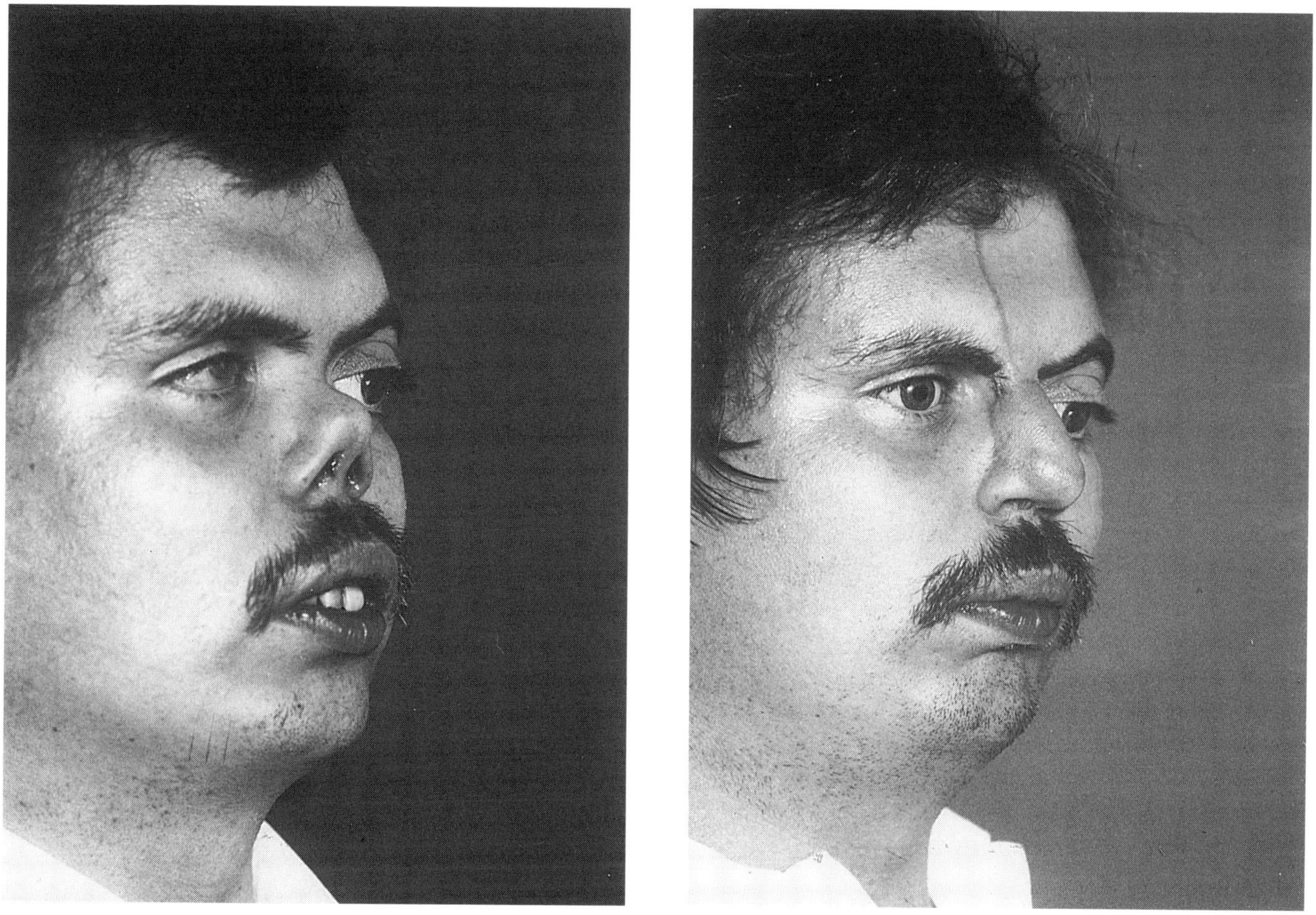

G H

FIGURE 10-3 *(Continued)*

amount of tissue needed for nasal reconstruction is estimated by the use of a rubber model taken from the defect. In the donor area, some additional tissue should be planned for the flap, considering that a cartilage or bone graft incorporated for nasal lining needs extra soft tissue to provide tensionless defect coverage. Furthermore, a distinct amount of stretch-back of the transposed flap has to be precalculated.[16] Before the tissue expansion technique came into use, flap contracture was an evident problem in nasal reconstruction, especially in the vertical axis.

Recording of the supratrochlear vessels via Doppler sonography before surgery (Figure 10-4D) is the next step. To implant the expander, the incision is made 1 centimeter behind the hair-bearing border. Dissection to create an adequate expander pocket is performed in an epigaleal plane. A pure dermal flap is raised, which includes the subcutaneous tissue but preserves the frontalis muscle. Special care has to be taken not to dissect the supratrochlear vascular bundle and its entrance into the dermal layer. To secure this, the expander pocket preparation is performed by "blunt" dissection that is caudally limited by a line connecting the upper eyebrow margins (Figures 10-3A and 10-4A). The expander pocket should exceed the expander base approximately 1 centimeter in the horizontal and vertical directions to ensure tensionless implantation. Some amount of intraoperative expansion of the elevated frontal skin can be gained by inner manual massage. This facilitates accommodation of the rigid incorporated filling valve and reduces postoperative tension and patient discomfort. Suction drainage is routinely used and continued for 6 to 8 days until drainage is minimal.

Initial saline instillation is started at the time of surgery with 10% to 15% of the final expander volume.

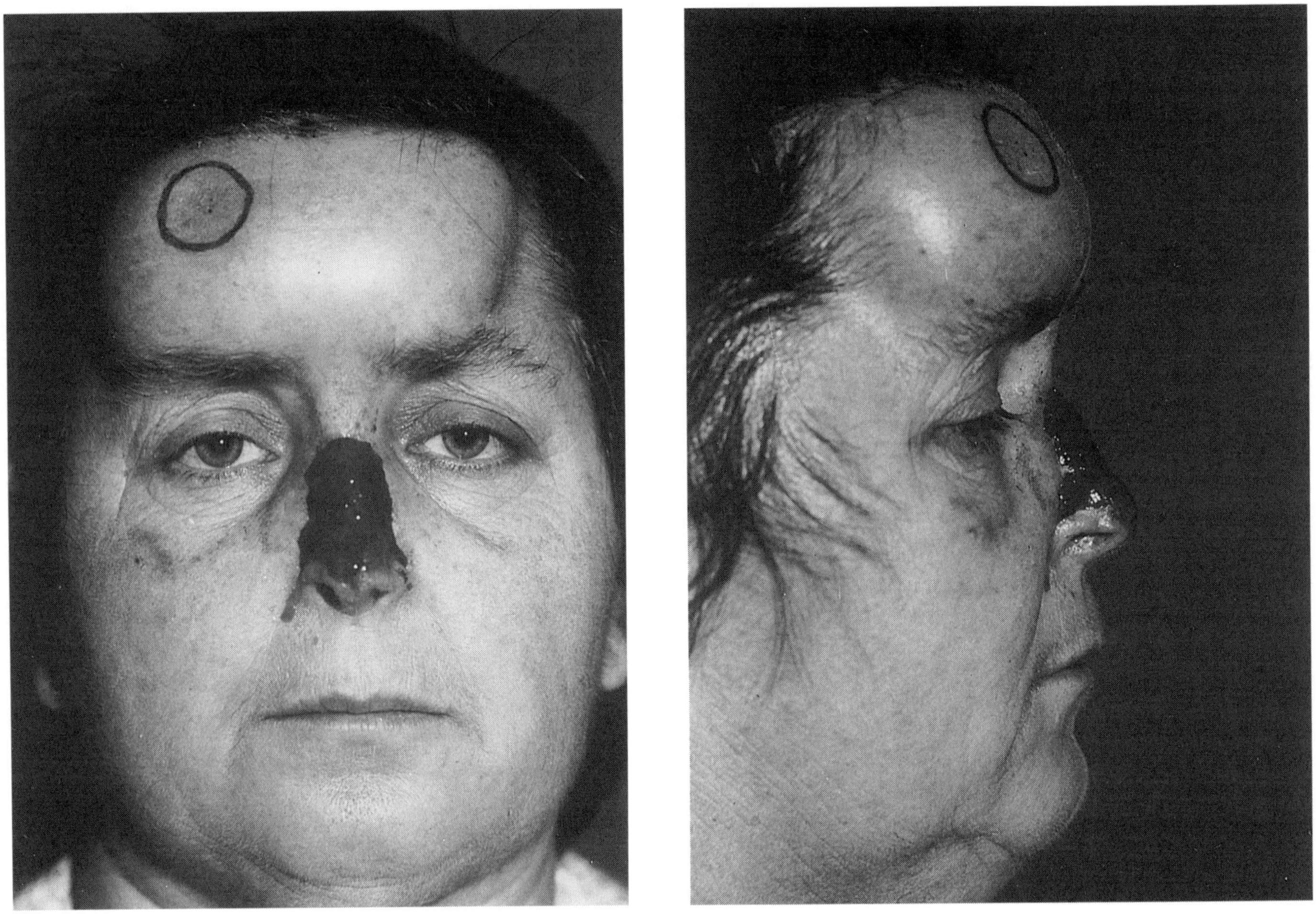

A B

FIGURE 10-4

Recurrent basal cell carcinoma of the nasal dorsum after previous surgeries. (A) Secondary granulation of the raw surface following tumorectomy. Development of cutaneous forehead expansion by use of a rectangular device, 160 milliliters, with incorporated valve, front view. (B) Lateral view. (C) Different expanders and miniexpanders (Dow Corning Wright). Volumes from left to right: 480 mL, 160 mL (as used in this case), 25 mL, 10 mL, and 5 mL. (D) Doppler sonography for marking the right supratrochlear vascular bundle at the end of expansion period. (E) Incision and elevation of the cutaneous paramedian forehead flap. (F) Flap rotation of approximately 170° into the nasal defect. Expander exposed before removal. (G) Flap sutured in place. Donor site defect primarily closed. (H) Preliminary result 2 months postoperatively. Final flap trimming procedure being planned. Münker R. Nasal reconstruction with tissue expansion. Fac Plast Surg *1988;5:328–337. (Used with permission from Thieme Med. Publ., New York.)*

Two weeks postoperatively, saline inflations are continued in weekly intervals with 5 to 10 mL each session, adjusted to the patient's tolerance. Usually, an obvious weakness appears within 2 to 3 days after each saline instillation, and the intervals may be accelerated if needed. When expansion is complete according to the geometric conditions, definite reconstruction by means of flap transfer can be performed. The paramedian forehead flap is planned with its vertical axis adjusted to the course of the supratrochlear vascular bundle. Expansion-related movements of the vessels, as described previously, must be considered. Therefore, consecutive bedside and intraoperative Doppler sonography is mandatory. Elevation of the dermal flap can be performed over the expander without any problems. The small cutaneous flap base measuring 2 centimeters in

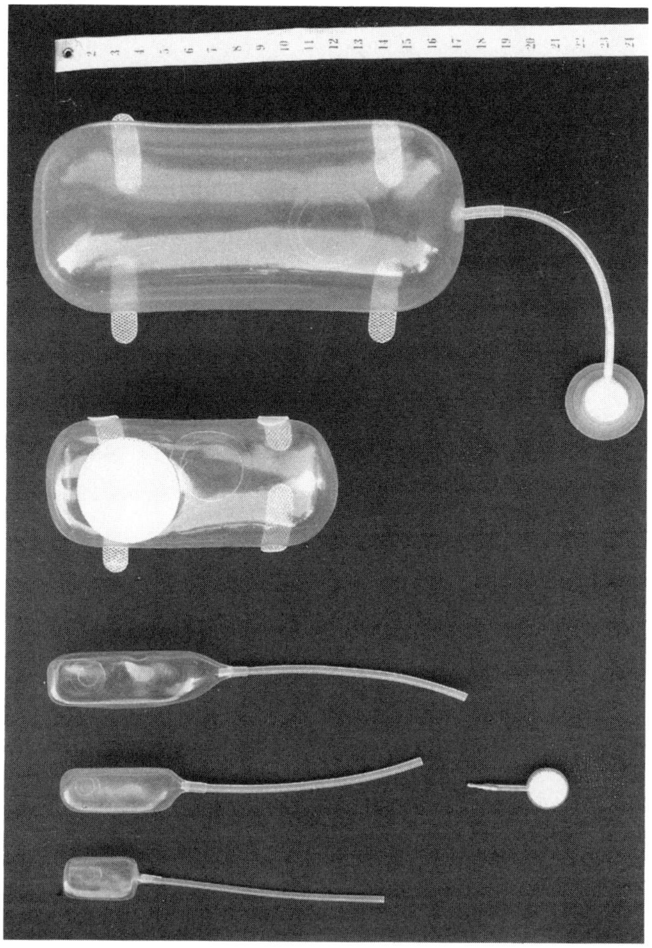

C

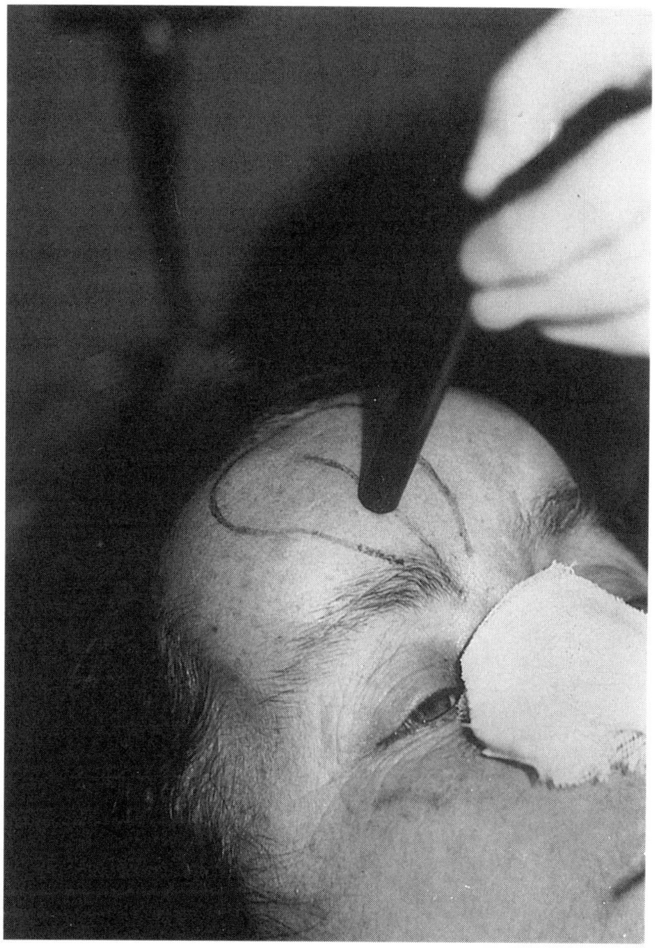

D

FIGURE 10-4 *(Continued)*

width at the pivot point must be developed very carefully. The anatomic landmark is the cross-point of the supratrochlear vessels penetrating the skull and dermal layer (Figures 10-4E and 10-5C). Flap rotation of approximately 170 degrees into the nasal defect is performed (Figures 10-4F and 10-5D). If any horizontal or vertical flap shortness is evident, we do not hesitate to perform multiple capsulotomies for flap lengthening. The direct area of the axial blood supply is strictly preserved. Simultaneous incorporation of autogenous cartilage from the rib or concha, if necessary, is performed (Figures 10-3 and 10-5). A small suction drainage, Charriere No. 8, is placed under the flap for a few days. Primary closure of the donor site should be possible in every case and is considered the major advantage of expansion of the forehead.

When postoperative swelling after 4 to 6 months has disappeared, a remaining dog-ear and lateroposition of the flap base at the root of the nose can be removed and combined with a final scar revision and cartilage graft to support the tip of the nose, depending on the individual situation (Figures 10-5D and H).

Indications for Intraoperative Rapid Expansion. Since the introduction of miniexpanders, a technique for intraoperative rapid expansion has been developed by Sasaki[17] to gain additional tissue. For reconstructive purposes in nasal defects, intraoperative rapid expansion can be used for midline or paramedian forehead flaps and for local flaps of the nasal area itself. Cyclic constant loading followed by relaxation periods induces skin stretching that is more effective than simply applying traction on the skin. Using guinea pig skin, Machida et al.[18] found that chronic expansion increased the skin surface area 137 percent, rapid expansion resulted in a 31 percent increase, and

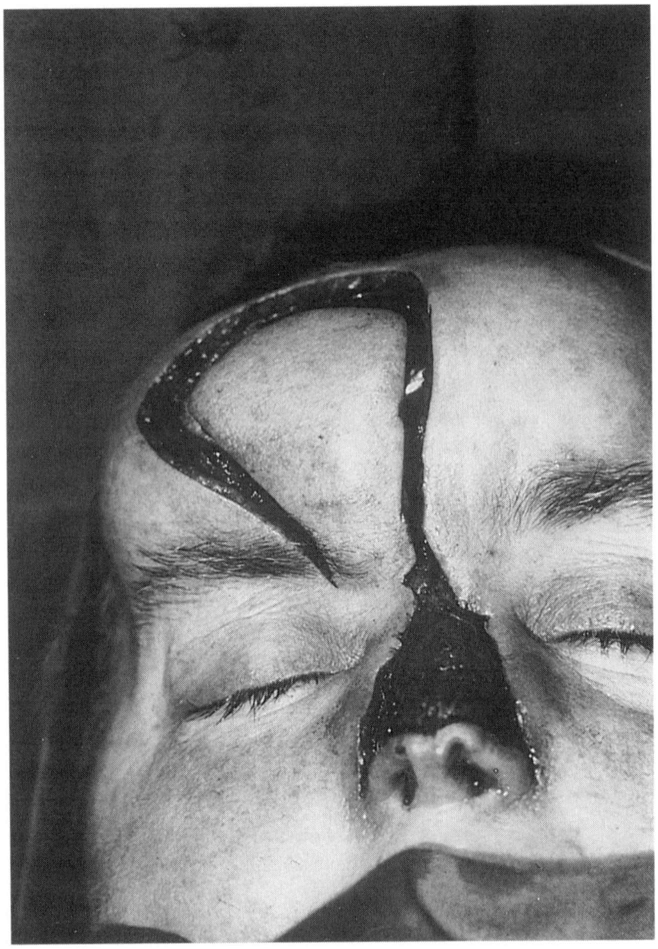

E

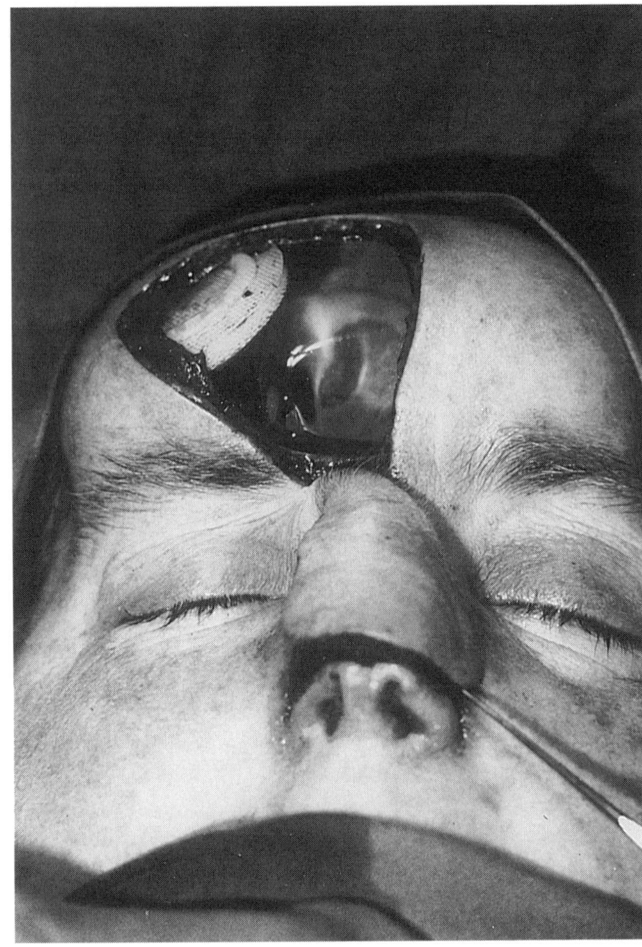

F

FIGURE 10-4 *(Continued)*

simple traction showed an almost negligible amount of skin increase.

Rapid intraoperative expansion is effective primarily for limited expansion to repair small defects. Baker and Swanson[19] reported that intraoperative expansion of midline forehead flaps for nasal reconstruction has been successful in gaining 1.5 centimeters in width of additional forehead skin for utilization in the flap. Two or more expanders can be used, if applicable, to increase this limited amount of tissue gained by intraoperative rapid expansion. For reconstruction in smaller full-thickness nasal tissue loss, a great number of local flaps have been advocated. Among the most sophisticated are the bilobed flap, the subcutaneously sliding transposition flap, and the island flap. Secondary scarring is unavoidable in all of these methods. On the other hand, direct wound closure or simple advancement flaps in the lower third and tip of the nose have been limited to defects no wider than 0.5 centimeter.

The nasal tip and alae are the parts of the nose most frequently mutilated by trauma and tumors. Even this area presents the greatest reconstructive difficulties due to its specific skin rigidity and inelasticity. If coverage of a smaller nasal defect is planned by using a local advancement or rotation flap, intraoperative sustained limited expansion (ISLE) according to Sasaki should be considered as a technique to gain 1 to 2 cm of additional skin for tensionless defect closure. Figure 10-6 demonstrates a recurrent basal cell carcinoma of the nasal tip and dorsum and its treatment by means of rapid intraoperative expansion.

CONCLUSIONS

The application of tissue expansion to established nasal reconstruction procedures has enlarged the possibili-

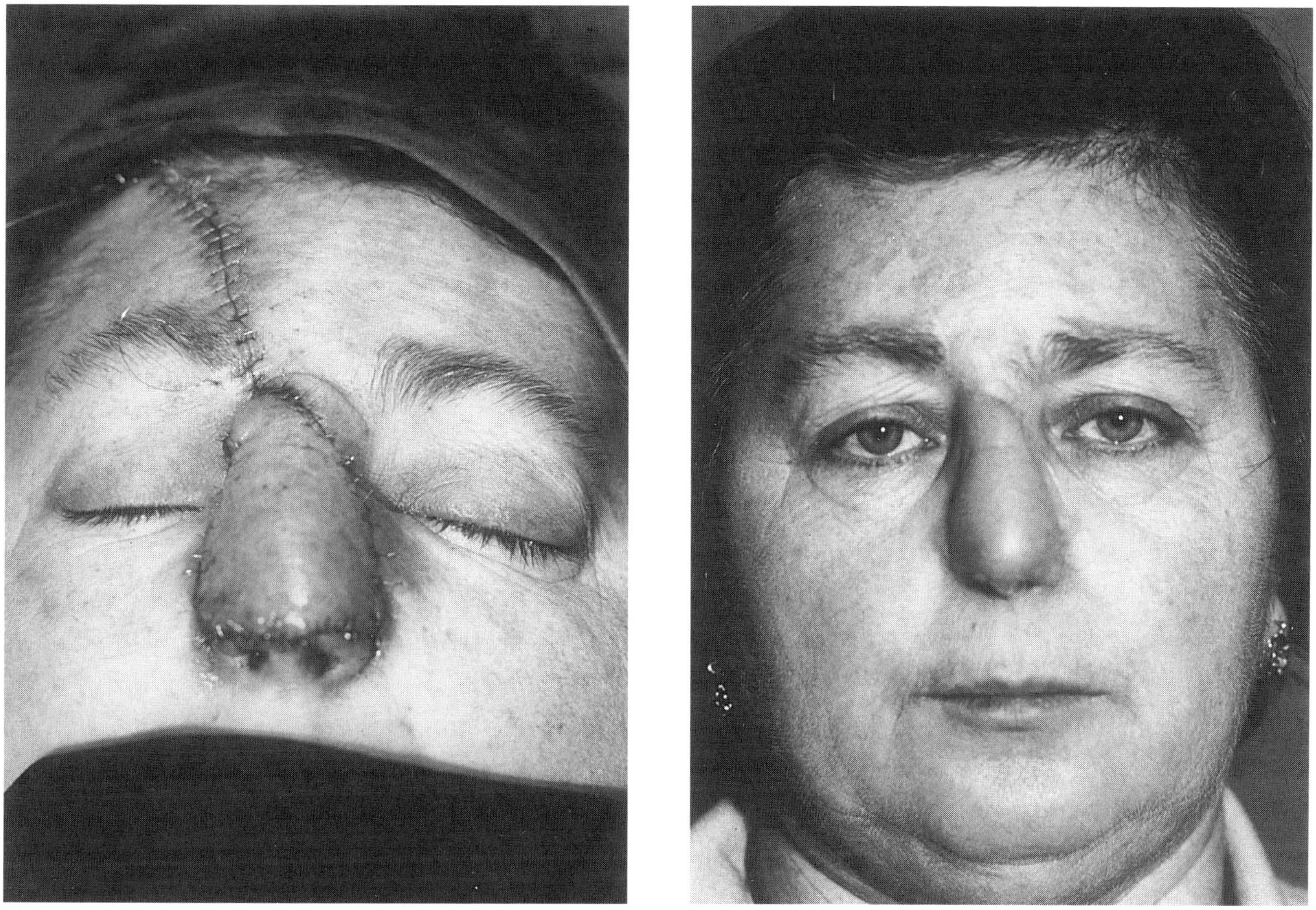

G H

FIGURE 10-4 *(Continued)*

ties for nasal defect coverage with ideal skin match without the disadvantages of major donor site defects. We have seen that the paramedian forehead flap—pre-expanded by a 250 milliliter rectangular expander with an incorporated filling valve—is the ideal solution for resurfacing large nasal soft tissue defects. Reliability of the flap is derived from the supratrochlear axial blood supply. Doppler sonography is mandatory to identify normal vascular anatomy and lateroposition of the vascular bundle due to the expansion process. It should be noted that we have stretched the dermal flap over the galea and not in a subgaleal plane, as reported by others.[3,9]

Some degree of late vertical tissue contracture should be counteracted by means of adequate cartilage support of the dorsum and tip structures. Dog-ears at the forehead flap pivot point have usually been removed at 4 to 6 months postoperatively. The expansion-sustained paramedian forehead flap demands surgical skill as well as comprehension of all circumstances of the procedure by the patient. This is one of the points that makes chronic tissue expansion on the face a delicate procedure. The intraoperative rapid expansion technique (ISLE) avoids the psychological problem, but it is limited to the development of local flaps that are designed for closure of smaller nasal defects.

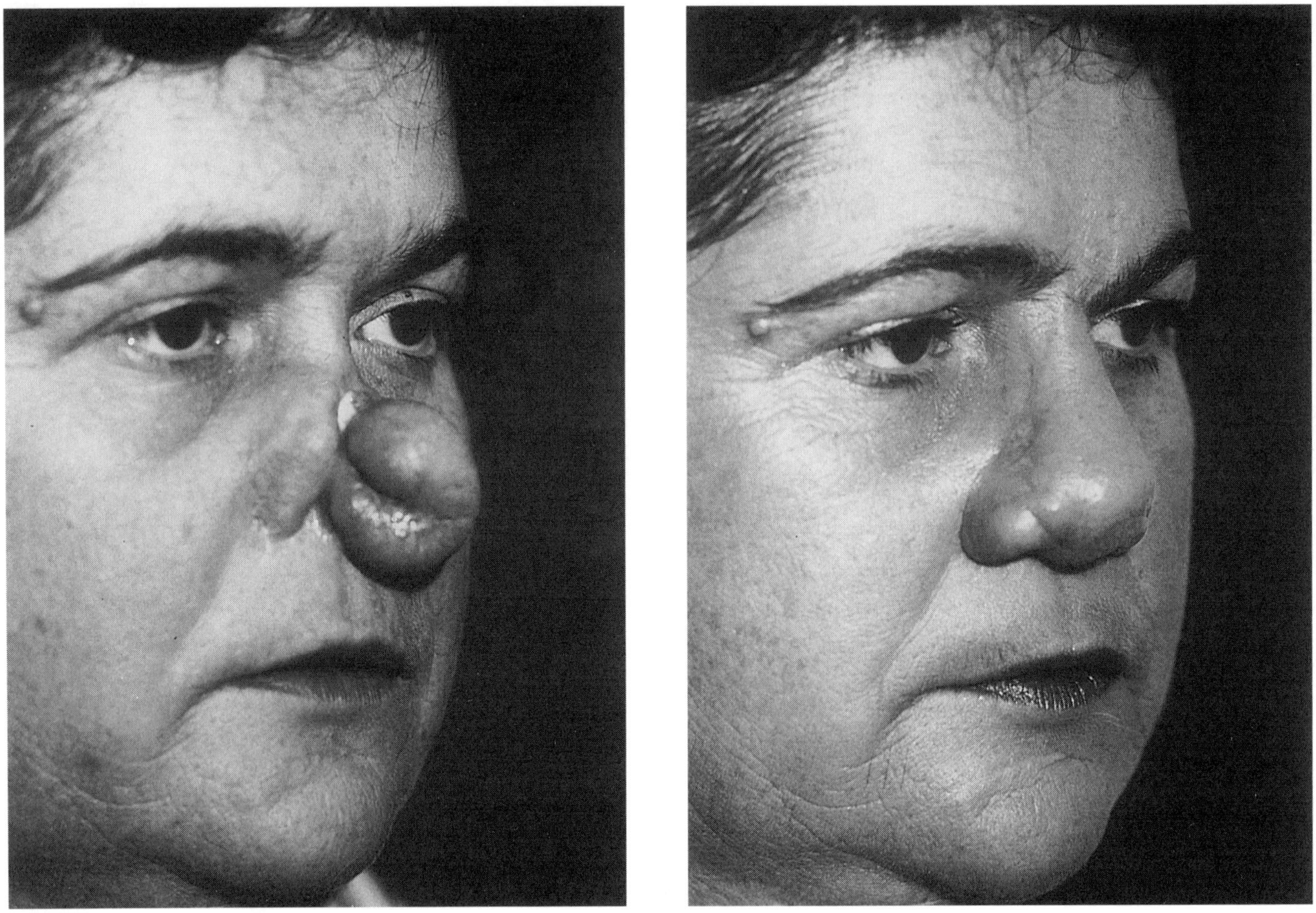

A B

FIGURE 10-5

Continuation of an unsuccessful nasal reconstruction by use of an expanded paramedian forehead flap. (A, E, G) Situation after 30 previous procedures for repeated reconstruction. A delayed brachial flap by means of the "Italian" technique had failed. After that, a tubed pedicle flap from the submental area had been developed but could not produce sufficient coverage. In this situation we took over the further reconstruction. (B, F, H) Final result within three procedures by means of an expanded paramedian forehead flap. (C) Miniflaps have been developed by use of a tubed pedicle flap to provide columellar and alar reconstruction. The paramedian left side forehead flap is elevated over a round expander, 100 milliliters, with a distant filling valve. The dotted line indicates a supratrochlear vascular bundle. (D) Removal of scar tissue on the nasal bridge. There is flap rotation for complete external nasal soft tissue replacement. Also shown are L-shaped costal cartilage and small bent cartilage grafts for dorsum, columella, and alar support, followed by primary wound closure of the donor site. Münker R. Nasal reconstruction with tissue expansion. Fac Plast Surg *1988;5:328–337. (Used with permission from Thieme Med. Publ., New York.)*

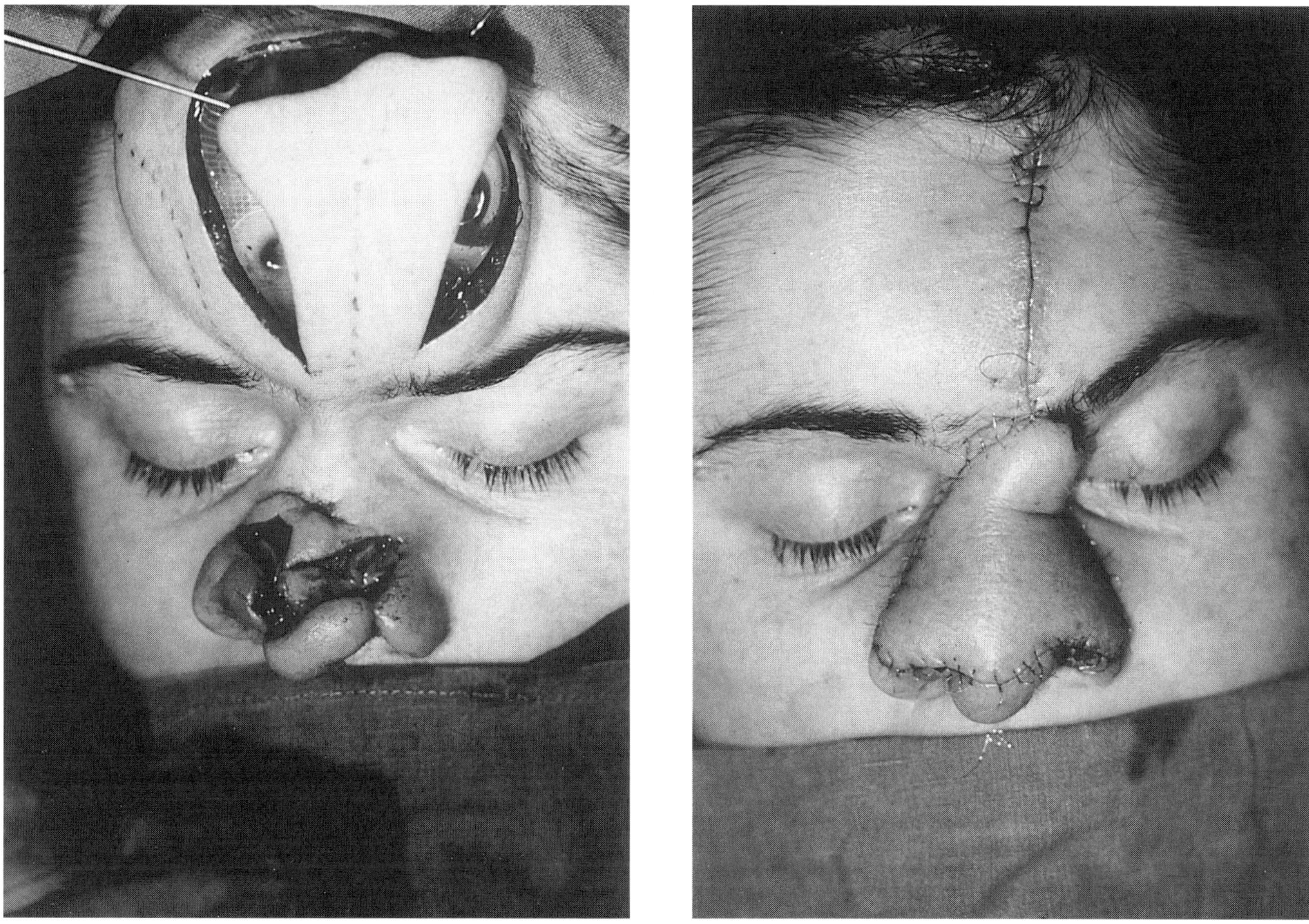

C D

FIGURE 10-5 *(Continued)*

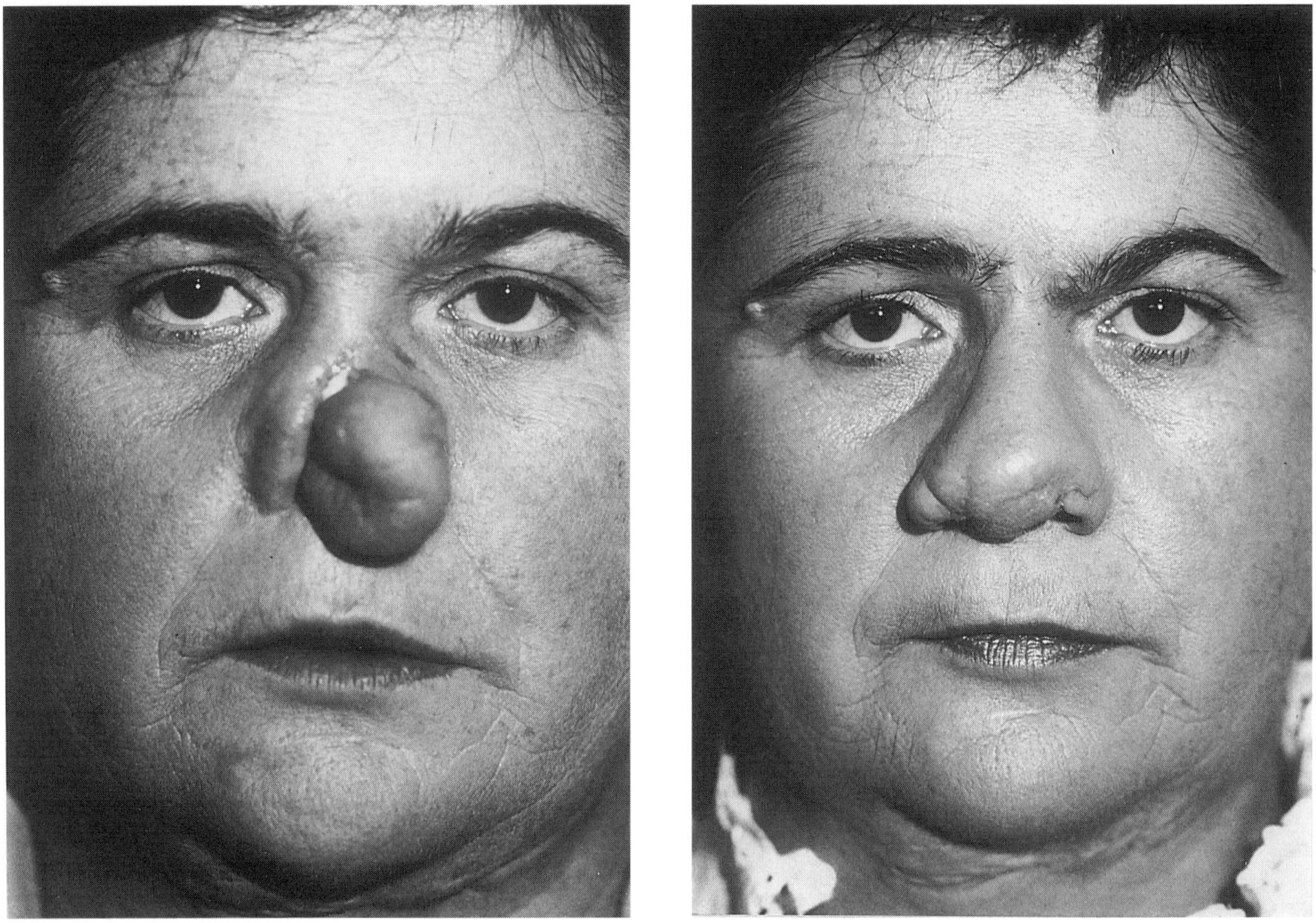

E F

FIGURE 10-5 *(Continued)*

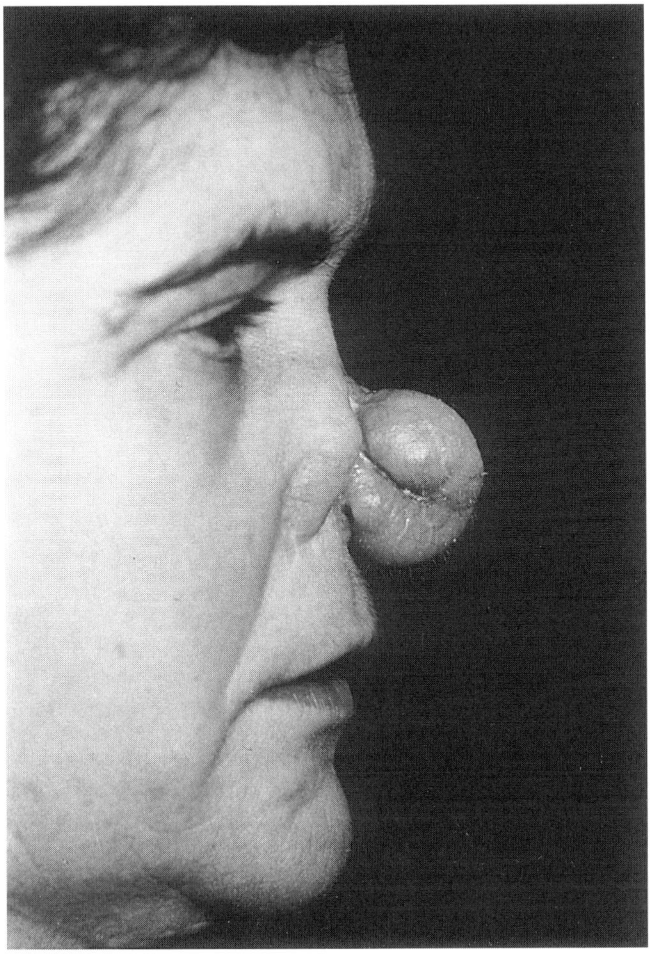

G

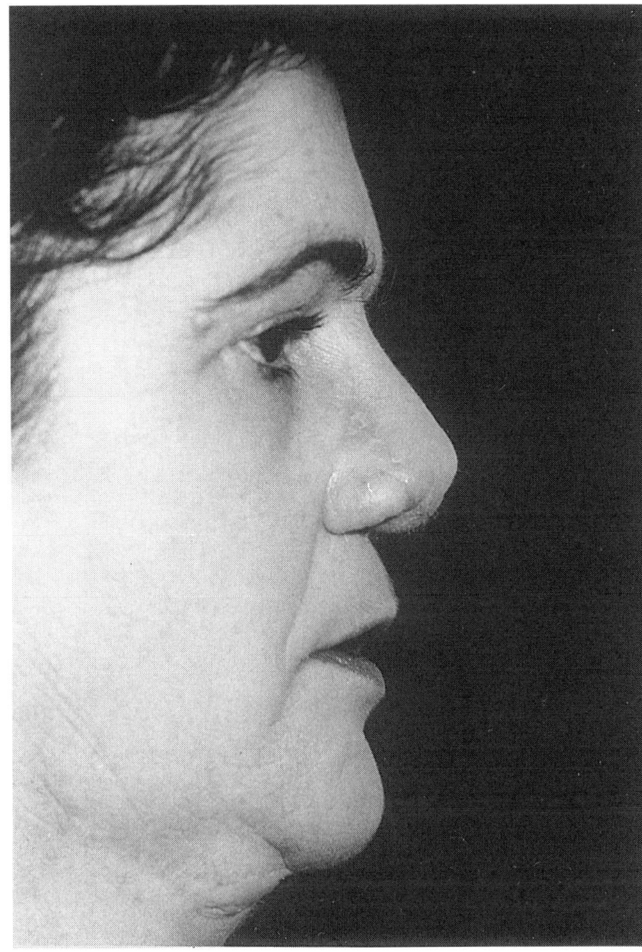

H

FIGURE 10-5 *(Continued)*

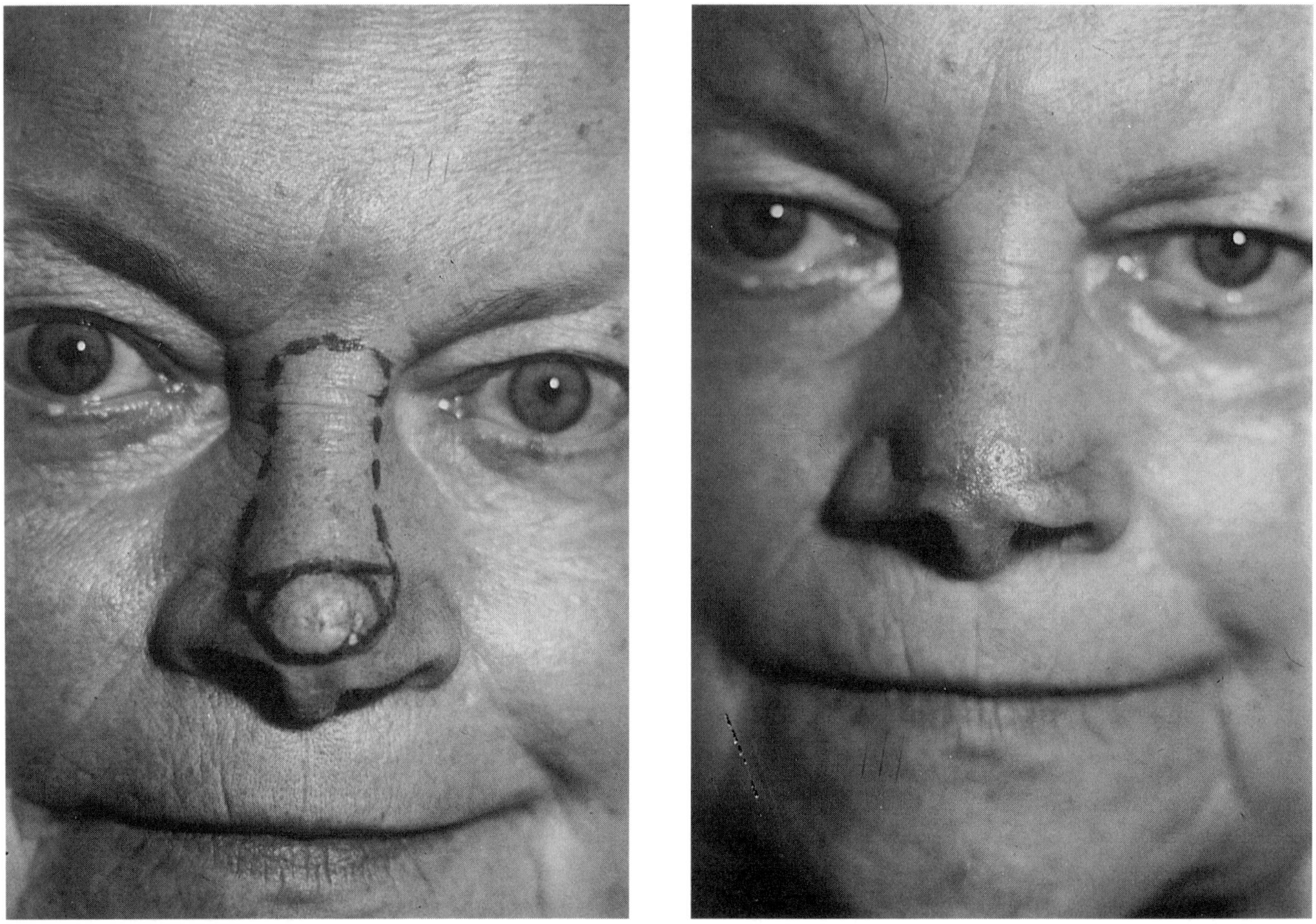

A B

FIGURE 10-6

Recurrent basal cell carcinoma of the tip of the nose. Shown is reconstruction by means of rapid intraoperative expansion. (A) Preoperative markings of tumorectomy and area of intraoperative expansion (dotted line) are shown. (B) Early result, 4 weeks postoperatively. (C) Intraoperative demonstration of a 5-milliliter expander connected with a filling tube and a three-way stopper. (D) Following tumor resection down to the cartilage and frozen section examination, the nasal tip defect measured 2.5 cm × 0.5 cm. Intraoperative repeated rapid expansion of the 5-milliliter expander inserted subcutaneously on the nasal dorsum. Blanching—a phenomenon due to controlled tissue expansion—is visible. (E) Intraoperative result of tip reconstruction by means of the expanded advancement flap. Münker R. Nasal reconstruction with tissue expansion. Fac Plast Surg *1988;5:328–337. (Used with permission from Thieme Med. Publ., New York.)*

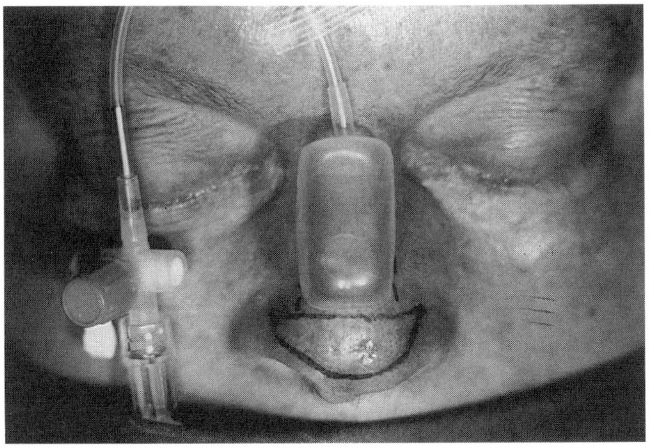

C

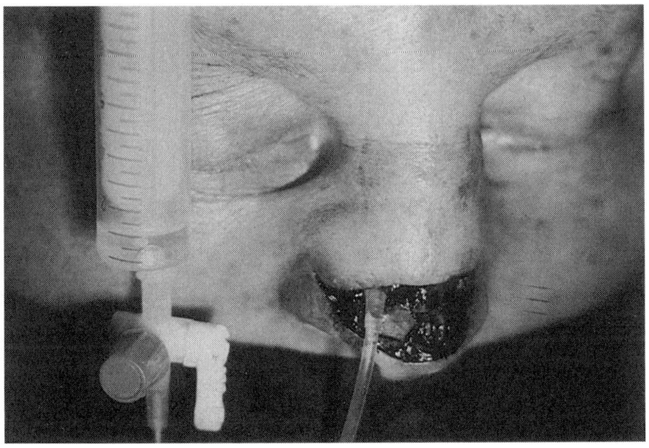

D

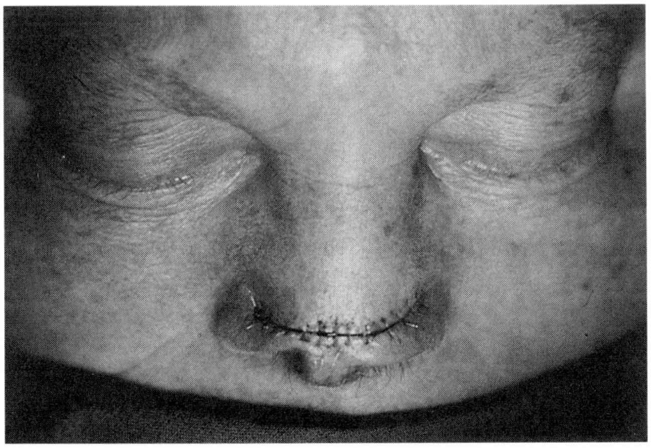

E

FIGURE 10-6 *(Continued)*

REFERENCES

1. Argenta LC, Watanabe MJ, Grabb WC, et al. Soft tissue expanders in head and neck surgery: A new method of reconstruction. *Plast Surg Forum* 1981;68:723–730.
2. Conley J. *Regional flaps of the head and neck.* Stuttgart: Georg Thieme Publishers, 1976, p. 20.
3. Manders EK. Nasal reconstruction aided by soft tissue expansion. Presented at the International Tissue Expansion Symposium of the Plastic Surgery Educational Foundation, San Francisco, 1987.
4. Harbon S, Labbe D, Compere JF, Menard P, Ricbourg B. Rhinopoiese: Essaie comparatif entre lambeau scalpant de Converse et lambeau medio-frontal expanse. A propos de seize cas. *Ann Chir Plast Esthet* 1990;35:351–358.
5. Converse JM. Reconstruction of the nose by the scalping flap technique. *Surg Clin North Am* 959;39,235–239.
6. Marks MW, Argenta LC. Skin expansion in reconstructive surgery. *Fac Plast Surg* 1988;5:301–311.
7. Manders EK. Reconstruction by expansion in the patient with cancer of the head and neck. *Surg Clin of North America* 1986;66:201–212.
8. Münker R. Nasal reconstruction with tissue expansion. *Fac Plast Surg* 1988;5:328–337.
9. Sasaki GH. Total nasal reconstruction using expanded forehead flaps. Presented at the International Tissue Expansion Symposium of the Plastic Surgery Educational Foundation, San Francisco, 1987.
10. Brobmann GF, Huber J. Effects of different-shaped tissue expanders on transluminal pressure, oxygen tension, histopathologic changes, and skin expansion in pigs. *Plast Reconstr Surg* 1985;76:731–736.
11. Münker R. Weichteilrekonstruktion mit dem Gewebeexpander. Presented at the Annual Meeting of the Deutsche Gesellschaft für Plastische und Wiederherstellungschirurgie, Berlin, 1986.
12. Münker R. Various devices available for tissue expansion and clinical experience. *Fac Plast Surg* 1988;5:291–300.
13. Lampe HJ, Lemperle G, Exner K. Der hautexpander: Technik und klinik. *Chirurg* 1985;56:773–778.
14. Mühlbauer W, Schmidt A, Fairley J. Simultaneous construction of an internal and external nose in an infant with arhinia. *Plast Reconstr Surg* 1993;91:720–725.
15. Marks MW, Argenta LC, Thornton JW. Rapid expansion: Experimental and clinical experience. *Clin Plast Surg* 1987;14:455–463.
16. Nordström REA. "Stretch-back" in scalp reductions for male pattern baldness. *Plast Reconstr Surg* 1984;73:422–426.
17. Sasaki GH. Intraoperative sustained limited expansion (ISLE) as an immediate reconstructive technique. *Clin Plast Surg* 1987;14:563–573.
18. Machida BK, Liu-Shindo M, Sasaki GH, Rich DH, Chandrasoma P. Immediate vs. chronic tissue expansion. Presented at the Fifth International Symposium on Facial Plastic and Reconstructive Surgery, Toronto, Canada, June 1989.
19. Baker SR, Swanson NA. Rapid intraoperative tissue expansion in reconstruction of the head and neck. *Arch Otolaryngol Head Neck Surg* 1990;116:1431–1434.

11

Auricle

Rolf E. A. Nordström

The morphology of the auricle is very complex; therefore, its construction is a challenge for the reconstructive plastic surgeon. The most common reason for auricle construction and reconstruction in our patient population is microtia, frequently seen in conjunction with hemifacial microsomia. Bennun considers isolated microtia to be a microform of hemifacial microsomia.[1] The incidence of hemifacial microsomia is estimated by Grabb to be at least 1 in 5600 births;[2] by Poswillo, 1 in 3500 births.[3]

Until 1984, we have used methods for auricle reconstruction advocated by Tanzer[4] and Brent.[5] The lack of adequate amounts of pliable, well-vascularized, and hairless skin to cover the reconstructed framework has been a problem, however, and the cosmetic results have often been disappointing. To overcome these difficulties, we decided to try expansion as an adjunctive treatment modality for auricle reconstruction.[6]

MATERIAL AND METHODS

Since 1985, we have used tissue expansion in 26 patients for auricle construction and reconstruction. Most were cases of microtia, one of total avulsion, two of partial avulsion of the auricle, and two for other auricle deformities.

Mentor (Goleta, CA) 65-mL, 80-mL, and 145-mL expanders, as well as the Cox-Uphoff International 80-mL expanders, all of which are suitably shaped for auricle reconstruction, were applied (Figure 11-1). These expanders all have remote injection ports.

The frame is formed from autologous costal cartilage. However, instead of starting by placing the costal cartilage frame subcutaneously, we insert the tissue expander in the location for the planned auricle. The locations for the pockets for the expander and the injection dome are drawn on the skin (Figure 11-2). A 5-centimeter-long incision through the scalp—about 5 centimeters away obliquely above and behind the planned auricle—is made for insertion of the expander

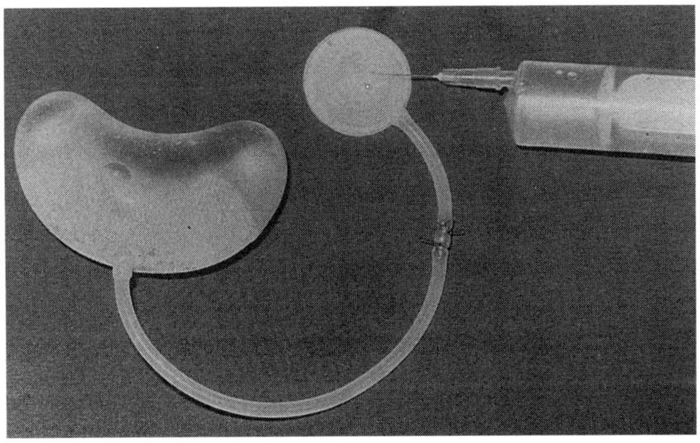

FIGURE 11-1
The expander used in most of these patients was an expander made by the Mentor Corporation, with a nominal filling volume of 65 milliliters. (Reprinted with permission from Nordström REA, Salo HP, Rintala AE. Auricle reconstruction with the help of tissue expansion. Fac Plast Surg *1988;5:338–346.)*

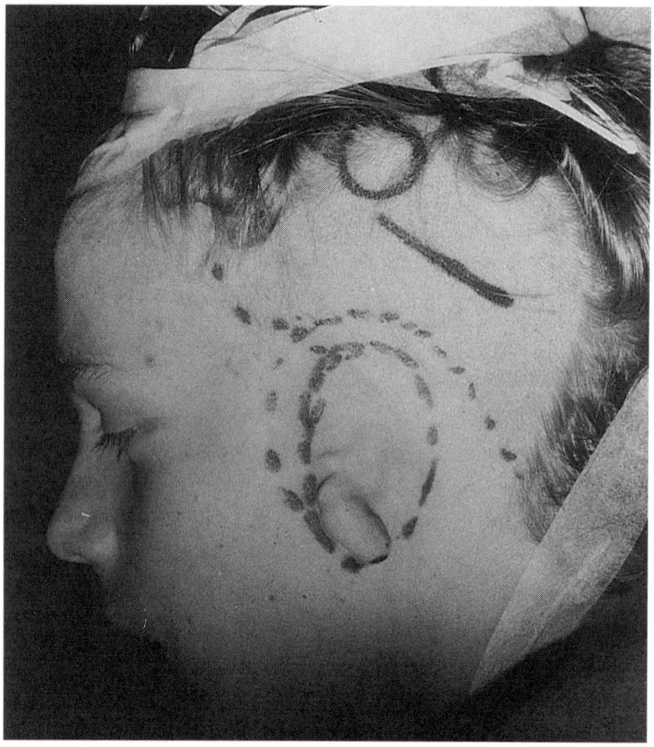

FIGURE 11-2
Typical patient with hemifacial microsomia and concomitant microtia. The circle on the top represents the planned site for the injection dome. The oblique line is the planned incision for the insertion of the expander, and the oval encircling the rudimentary auricle is the planned location for the inflatable sack. (Reprinted with permission from Nordström REA, Salo HP, Rintala AE. Auricle reconstruction with the help of tissue expansion. Fac Plast Surg *1988;5:338–346.*

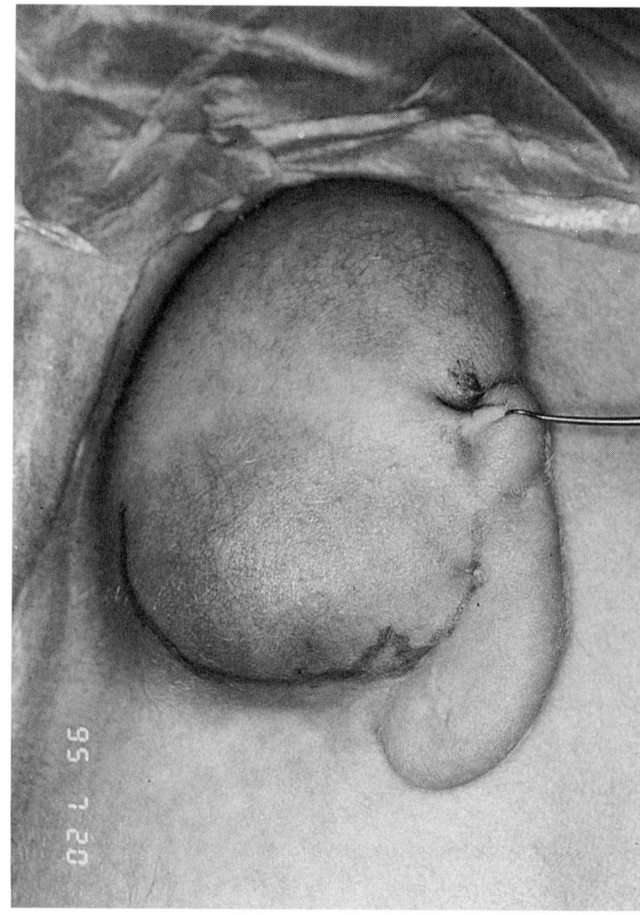

FIGURE 11-3
The skin pocket expanded.

so that it is far enough away from the skin to be expanded. This eliminates the risk for extrusion of the expander through this incision; furthermore, the scar is covered by the hair lying above it. When the incision is this far from the expander, postoperative expansion can be started immediately. Thus, some weeks are saved in the whole program compared to a procedure using an incision close to the expander, which would demand some weeks of healing before expansion. Furthermore, scars close to the auricle later easily interfere with a successful reconstruction of all the details of the auricle.

We feel that the cartilaginous remnants present in most of these auricles are only in the way and are not useful. Thus, most of the remnants are removed from the pocket created for the expander before the expander is inserted. Some additional skin is gained in this manner.

The pocket for the expander is created in a subcutaneous level and must be sufficiently large not to cause any wrinkling of the expander. The injection port is inserted into a separate superior pocket under the scalp (Figure 11-2). A suction drain is applied and is removed 1 to 2 days postoperatively. We have usually filled up the expander to obliterate any dead space, thus diminishing the risk for hematomas or seromas around the expander.

The postoperative expansion is usually initiated at about 2 weeks, and injections are performed once or twice weekly (Figure 11-3). The limiting factor for each filling has been either pain and discomfort or loss of capillary refill, which is a sign of insufficient vitality of the overlying skin.

The second procedure is the removal of the expander and the introduction of the carefully carved autologous rib cartilage (Figure 11-4). This is done af-

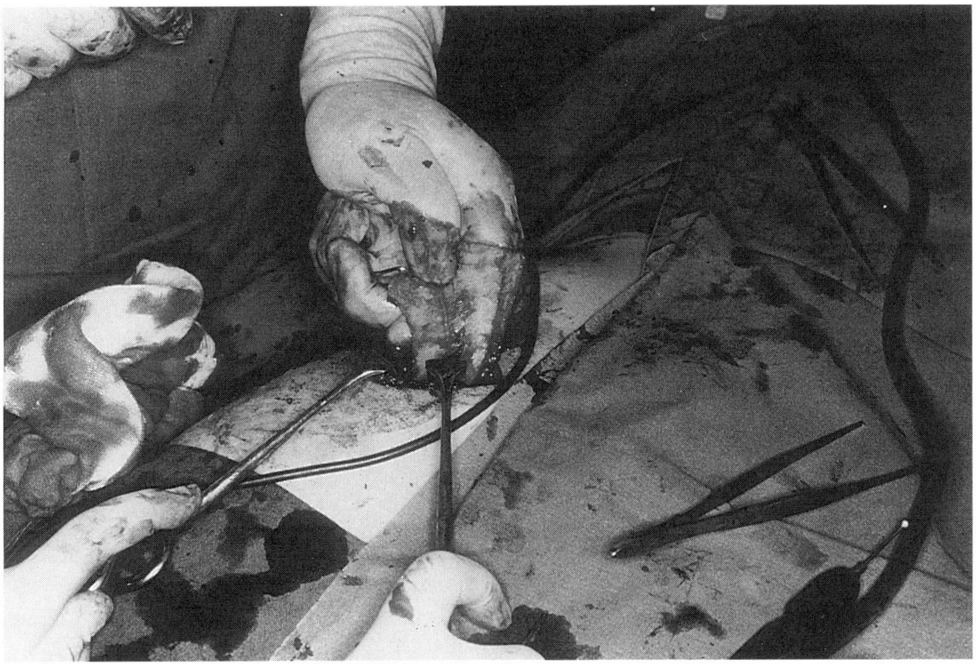

FIGURE 11-4
The costal cartilage is taken from the sixth to eighth rib from the contralateral side. (Reprinted with permission from Nordström REA, Salo HP, Rintala AE. Auricle reconstruction with the help of tissue expansion. Fac Plast Surg *1988;5:338–346.)*

ter the completion of the expansion. The carving of the cartilage frame is carried out according to the guidelines described earlier by Tanzer and Brent[4,5] and later by Nagata.[7]

A high profile of the cartilage frame is important, because the helical rim seems to soften in profile and should be accentuated in this frame (Figures 11-5 and 11-6). An extra piece of cartilage is adjusted under this frame against the skull and serves to create the normal angle of deviation of the auricle. This also helps in creation of the auriculocephalic sulcus. The cartilage frame is introduced into this pocket in exactly the desired position (Figure 11-7). A suction drain will pull the skin to follow the complex shape of the cartilage frame (Figure 11-8). At this point, the value of a sufficiently expanded skin "bag" will be obvious. Retention sutures are inserted to keep the skin folds in place after removal of the suction drain 1 or 2 days postoperatively (Figure 11-9). These sutures are removed about 2 weeks postoperatively.

One of the problems in patients with hemifacial microsomia is that the area of nonhair-bearing skin tends to be insufficient. In some patients, we have tried to epilate the hair-bearing adjacent area by cutting roots of the hairs from below during the procedure in which

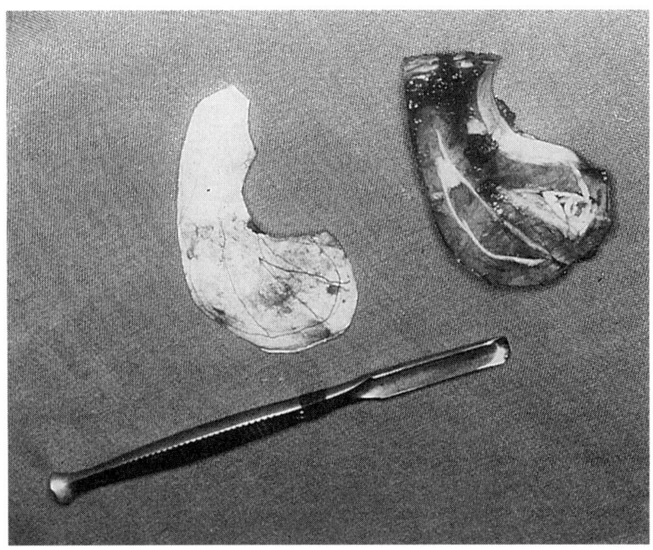

FIGURE 11-5
Carving of the frame is started.

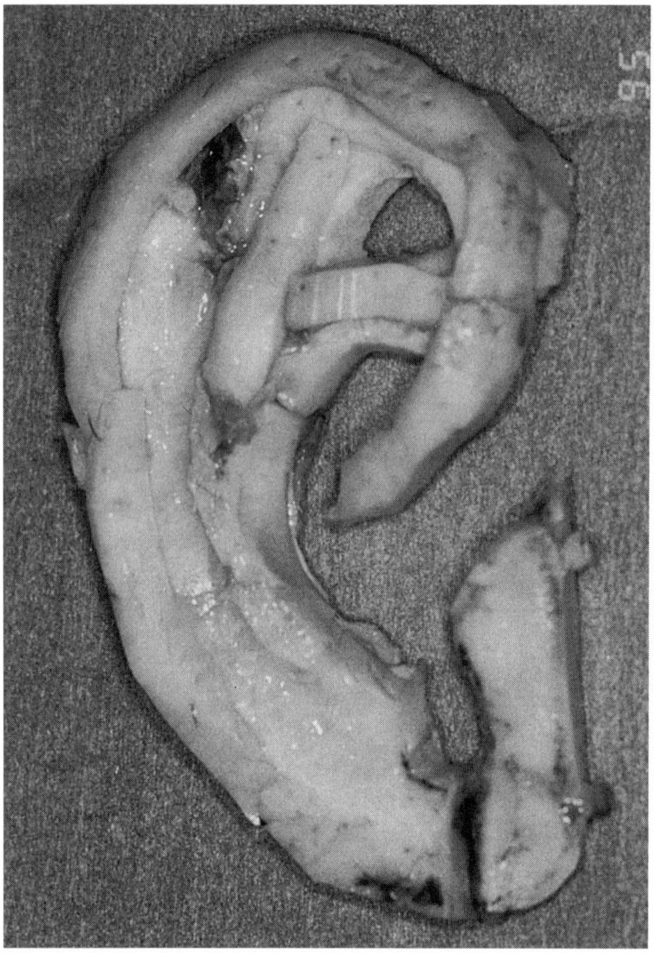

FIGURE 11-6
The frame is carved out to the final shape.

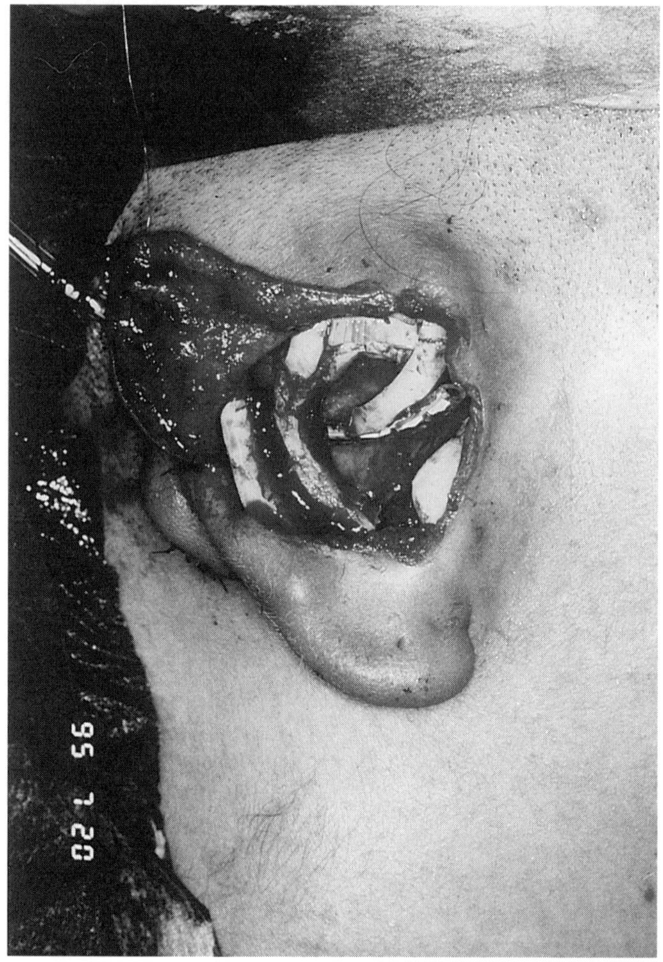

FIGURE 11-7
Insertion of the cartilage frame into the expanded skin pocket.

the expander is introduced. This seems to have some effect, but it is not completely satisfactory. During the procedure in which the cartilage framework is inserted, many of the remaining hair roots can be seen and cut. After surgery, we have used electric epilation for persisting hairs (Figure 11-10).

The third procedure, performed several months after the second one, creates a conchal cavity and tragus and transposes the usually misplaced lobulus (Figures 11-11 and 11-12). Further minor revisions have also been performed on several patients at various stages.

The interval from insertion of the tissue expander until completion of inflation varies from 3 to 16 weeks, with an average of about 8 weeks. The total filling volume should not be less than 130 mL if the intention is to create the auriculocephalic sulcus with the expanded skin.

RESULTS

The cosmetic results have been gratifying compared to those using the previously mentioned techniques and also compared to the results achieved elsewhere using tubed pedicles and other methods (Figures 11-13 through 11-16). It appears that the results are quite permanent when scars have matured and the postoperative edema is gone. (Figure 11-16).

COMPLICATIONS

We have experienced two extrusions of expanders during inflation. In both of these patients, a cartilage frame was introduced during the procedure in which the exposed expander was removed and the reconstruction was performed. In two patients, the implanted cartilage

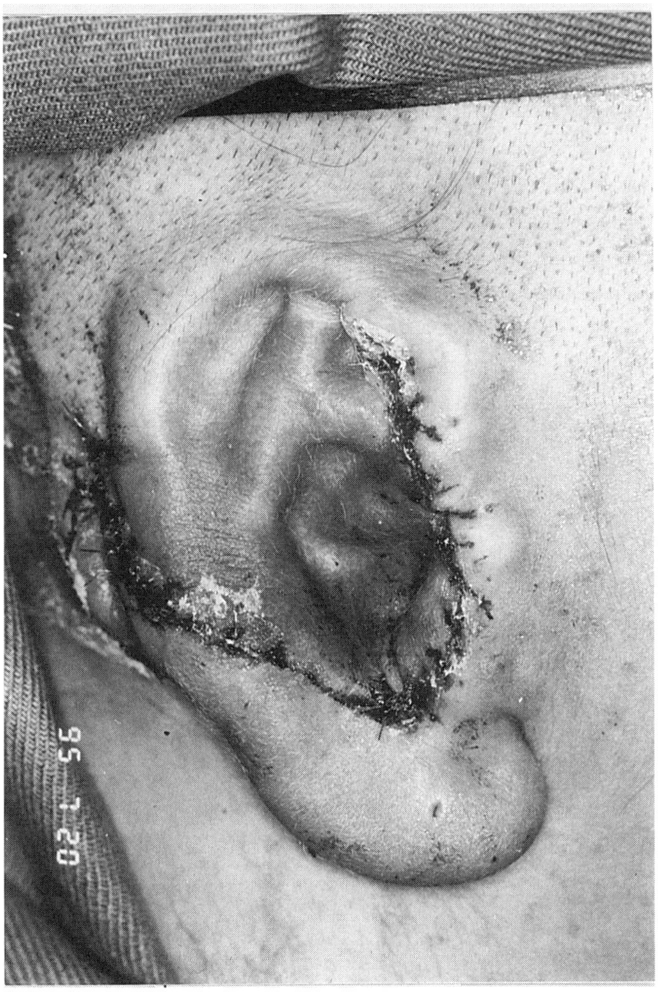

FIGURE 11-8
The suction drain is necessary to pull the skin against the cartilage frame to create the profile of the auricle.

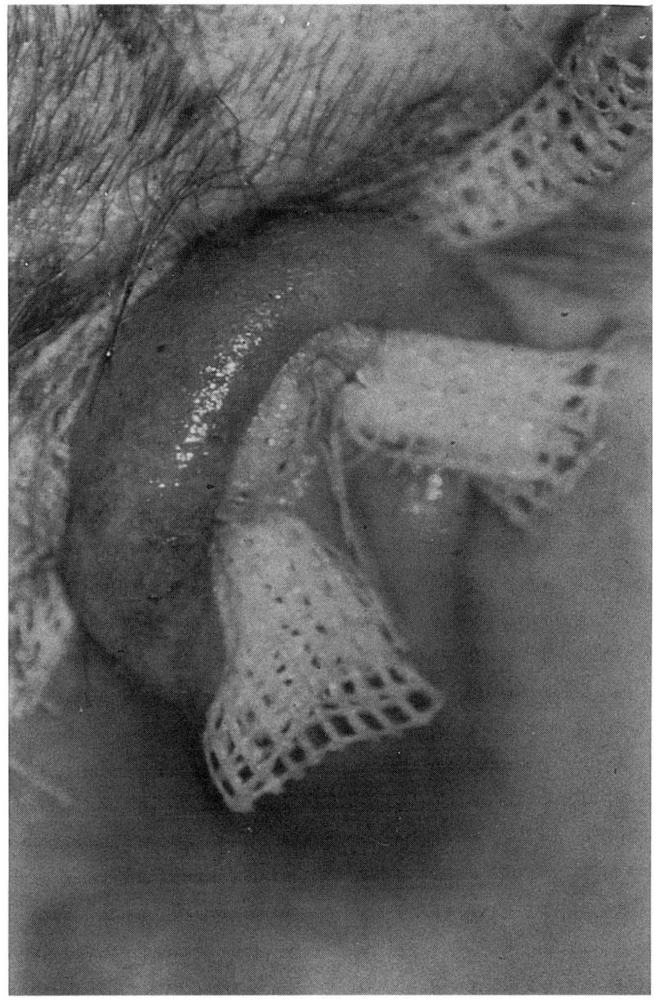

FIGURE 11-9
Retention sutures applied. (Reprinted with permission from Nordström REA, Salo HP, Rintala AE. Auricle reconstruction with the help of tissue expansion. Fac Plast Surg *1988;5: 338–346.)*

became partially exposed; local excision, revision, and suturing of the skin edges salvaged the cartilage. A more extensive removal of the cartilage frame was unnecessary because no infection resulted.

No patient has mentioned any problem with the chest wall donor area of the cartilage. The most common problem has been that the expanded skin has not been large enough to allow for the creation of a good auriculocephalic sulcus. In these cases, a skin graft, skin flap, or a temporal fascia flap covered with a skin graft has been used behind the auricle several weeks later. As mentioned, the expansion should usually not be less than 130 milliliters to enable the surgeon to drape the cartilage frame totally and to avoid using additional skin grafts or flaps.

DISCUSSION

Tissue expansion was for reconstructive plastic surgery purposes first published in 1957 by Neumann[8]. By coincidence, this very first attempt was for auricle reconstruction. Neumann used a balloon that was progressively inflated with air to distend the skin behind the ear to achieve new tissue for the planned partial auricle reconstruction in a patient with a traumatic defect. However, this concept of tissue expansion soon fell into oblivion.

Tissue expansion was later tried for auricle reconstruction in one case published by Brent in 1980,[9] who was disappointed with it. In 1984 O'Neal et al. also

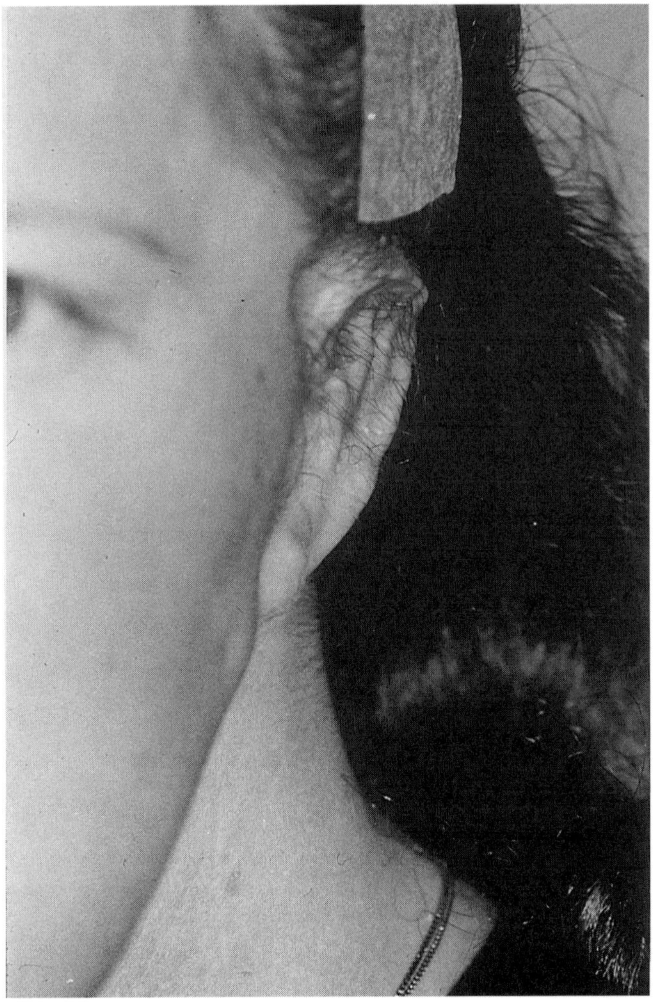

FIGURE 11-10
Another patient with severe hemifacial microsomia with concomitant microtia after reconstruction of the auricle. Due to lack of sufficient amounts of nonhair-bearing scalp in this patient, hair-bearing scalp has been used. Hair roots have been cut from below, but some hairs can still be seen. These will be epilated. (Reprinted with permission from Nordström REA, Salo HP, Rintala AE. Auricle reconstruction with the help of tissue expansion. Fac Plast Surg *1988;5:338–346.)*

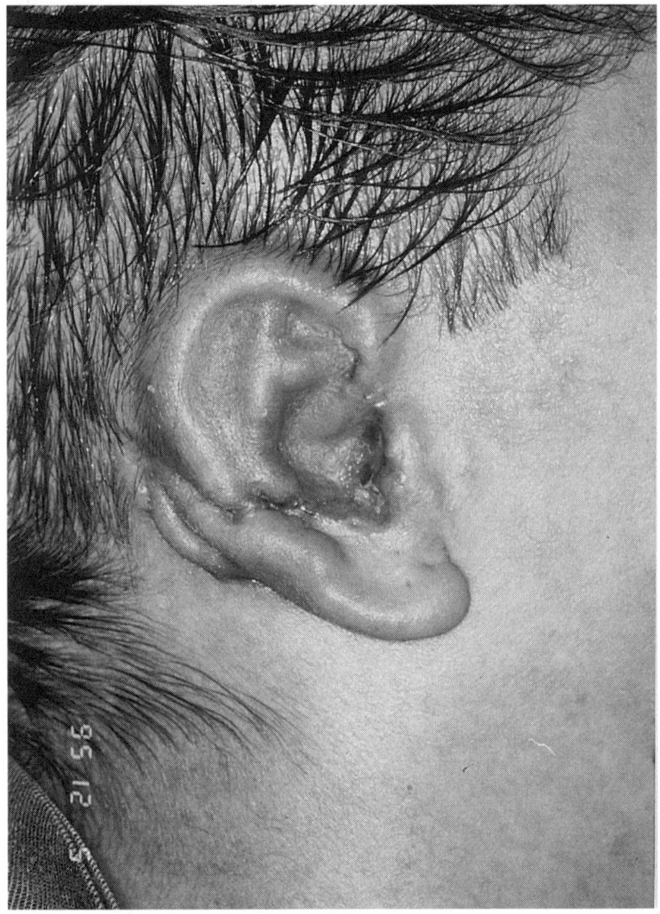

FIGURE 11-11
Five months after the insertion of the cartilage.

reported on the use of a tissue expander in a patient undergoing reconstruction of the upper two-thirds of the auricle after trauma.[10]

One of the goals of tissue expansion has been to gain sufficient amounts of suitable skin to create the auriculocephalic sulcus in the same procedure in which the cartilage frame is inserted. In some patients, this has been achieved; but after some weeks or months, scar contracture appears to significantly diminish the sulcus. This can be created later, for example, by a postauricular flap, as described by Nagata[7] and Isshiki et al,[11] or a skin graft. More expansion than first realized is needed to obtain sufficient skin for the creation of a normal and permanent sulcus. The final volume in the inflated expander should probably be at least 130 milliliters. Furthermore, the bottom area of the expander should not be too large or it may come in contact with hair-bearing skin, and this creates other difficulties previously discussed.

The skin overlying the expander is thinned. In most cases, this is beneficial, as long as its vitality is not jeopardized; thin skin drapes the cartilage frame better.

One further difficulty is the loss of the profile achieved immediately after insertion of the cartilage frame. During the early postoperative months, some of the desired profile is lost, possibly due in part to scar contracture and in part to resorption of cartilage. Six

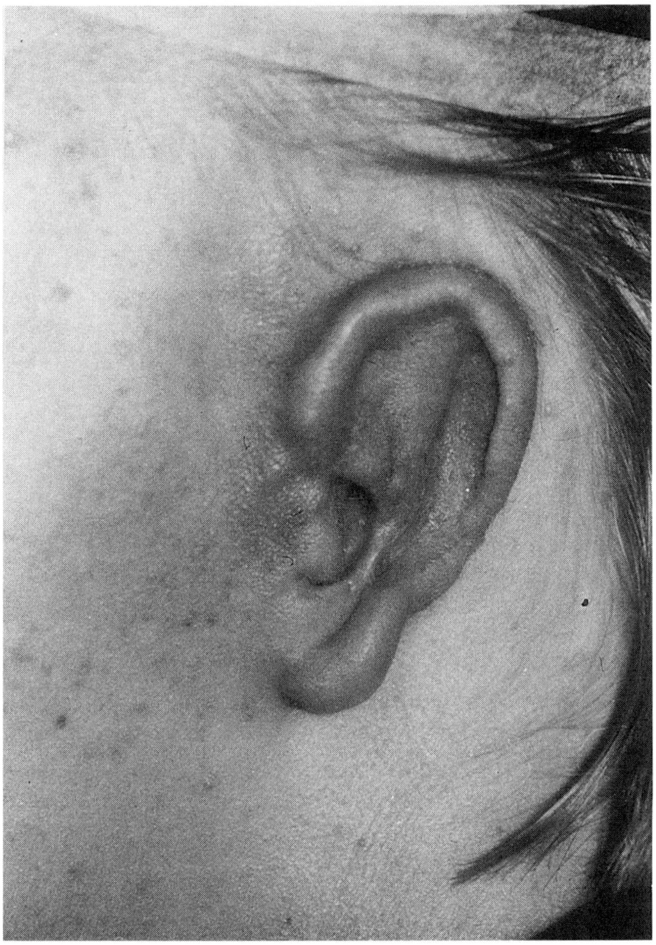

FIGURE 11-12
Another reconstructed auricle. (Reprinted with permission from Nordström REA, Salo HP, Rintala AE. Auricle reconstruction with the help of tissue expansion. Fac Plast Surg *1988;5:338–346.)*

months postoperatively, the scar has softened, and most of the desired profile returns.

Another problem is that the skin is thinnest in the lowest portion of the area to be expanded. Thus, the skin in the lowest portion gives way and expands the most, whereas the biggest demand for new and expanded skin is in the opposite end—the upper pole for reconstruction of the helix.

Many cases of microtia are in fact hemifacial microsomia. Due to this, there are additional deformities and problems, such as a low lateral inferior hairline and a short distance between the lateral corner of the eye and the auricle remnants; if there is any meatus at all, it usually lies below the normal level.

There is a considerable risk for complications with this procedure. As mentioned, we have seen extrusions of the expander. This might be caused by patients lying on the expander and creating a high pressure from the weight of the head. The pressure in such situations has been measured up to about 60 mmHg. Such pressures may be high enough to prevent capillary refill and also create a possibility of injection dome leakage. This has to be taken into consideration when evaluating the real amount of expansion.[12]

The patients with avulsions of the auricle had much thicker skin in this region compared to patients with hemifacial microsomia. However, this skin could be expanded without problems. In the patient with total avulsion, the expansion with a fill of 90 milliliters was not sufficient to enable us to immediately create an auriculocephalic sulcus (Figures 11-17 through 11-19). Because the patient was satisfied with the appearance of the auricle without a sulcus, it was not reconstructed. The expanded skin was considerably thicker, and the delicate structures of the cartilage frame did not show as well as in patients with hemifacial microsomia. In this patient, the avulsed cartilage of the auricle had been successfully buried under the abdominal skin. Still, this was found to be far too soft to give sufficient support to a reconstructed auricle of this type. Thus, we had to use autologous costal cartilage in this case as well.

Our experience has been that a frame made of a nonviable material such as silicone does not give a permanent result. We have seen some of our patients and others in whom the plastic frame has extruded, and the results have been mainly dissatisfactory. Our rough Scandinavian climate, with its cold winters may be different from climates in areas where successful use of these plastic frames has been reported. In two patients as previously mentioned, the cartilage frame was exposed due to skin problems. In both cases, the cartilage remained noninfected after a revision of a part of the cartilage and the skin and after covering the cartilage frame with advanced skin. This is evidence of considerable resistance to infections and trauma of these autologous cartilage grafts. Also, biopsies taken from these cartilage frames have shown them to have living chondrocytes.

In one case, a tubed pedicle had been used in a prior reconstruction several years ago to bring in more tissue; more recently, we used tissue expansion for improvement. The result achieved was satisfactory, although the ear looked somewhat mosaic because of the various colors and textures of skin. Patients with considerable scarring due to previous reconstruction attempts or other types of surgery (for example, to restore hearing) are, in our opinion, not suitable candidates for tissue expansion. In these cases, we recommend and have done

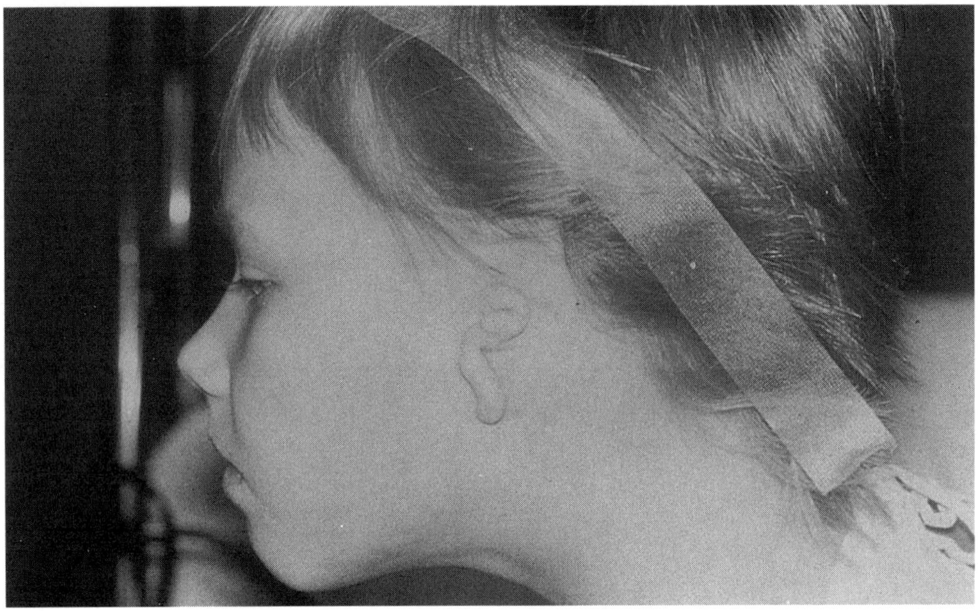

FIGURE 11-13
Hemifacial microsomia and a concomitant microtia. Another patient. (Reprinted with permission from Nordström REA, Salo HP, Rintala AE. Auricle reconstruction with the help of tissue expansion. Fac Plast Surg *1988;5:338–346.)*

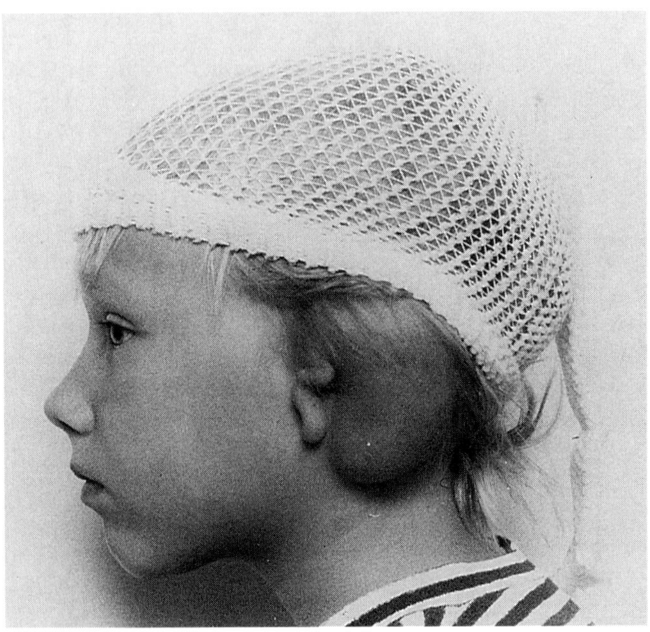

FIGURE 11-14
Patient in Figure 11-13 showing the expanded skin. (Reprinted with permission from Nordström REA, Salo HP, Rintala AE. Auricle reconstruction with the help of tissue expansion. Fac Plast Surg *1988;5:338–346.)*

some titanium implants with a plastic epithesis.[13] Although other methods, such as temporal fascia with a skin graft, may be useful, we have not yet used them.

A frontal view might expose the height of both ears simultaneously. In many cases of hemifacial microsomia, however, it is impossible to position the ears at exactly the same level due to gross asymmetry of the facial skeleton. Still, only one ear is seen at a time in most cases, especially from the lateral aspect. Thus, the viewer does not know the exact shape, size, or inclination of the ear of the opposite side. As a result, the reconstructed ear does not necessarily have to be exactly the same as the auricle of the other side.

SUMMARY AND CONCLUSION

Tissue expansion adds a new dimension to the difficult task of ear construction. Combining tissue expansion with other methods gives us the possibility of tackling one of the most difficult problems: the lack of sufficient amounts of suitable skin. This obstacle is now partially or totally eliminated. The whole process of tissue expansion and auricle reconstruction theoretically seems simple. However, many problems still remain, and experience is necessary to avoid most potential hazards.

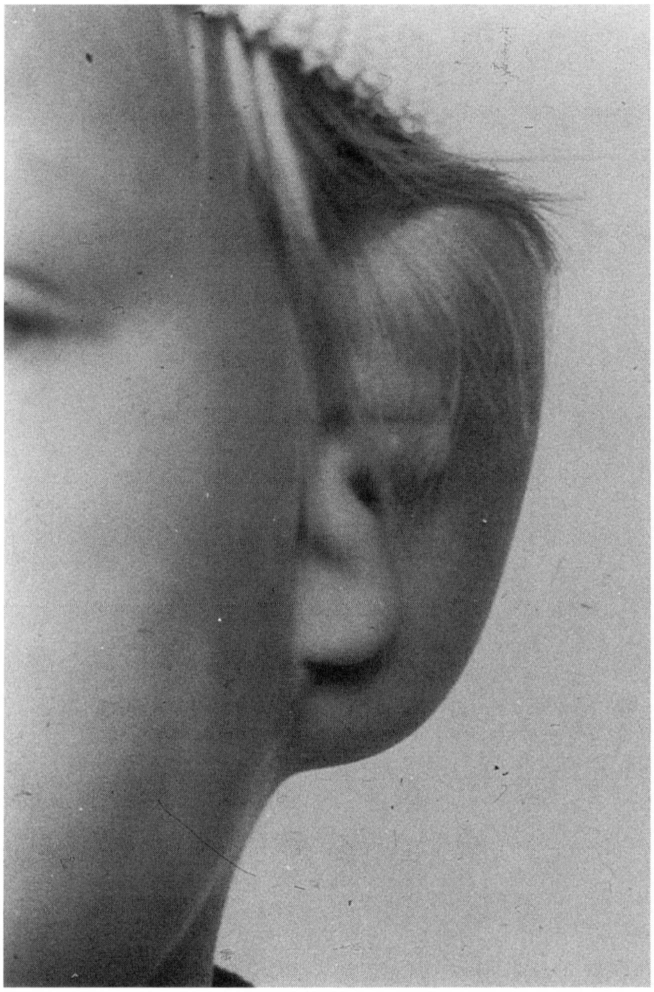

FIGURE 11-14 *(Continued)*

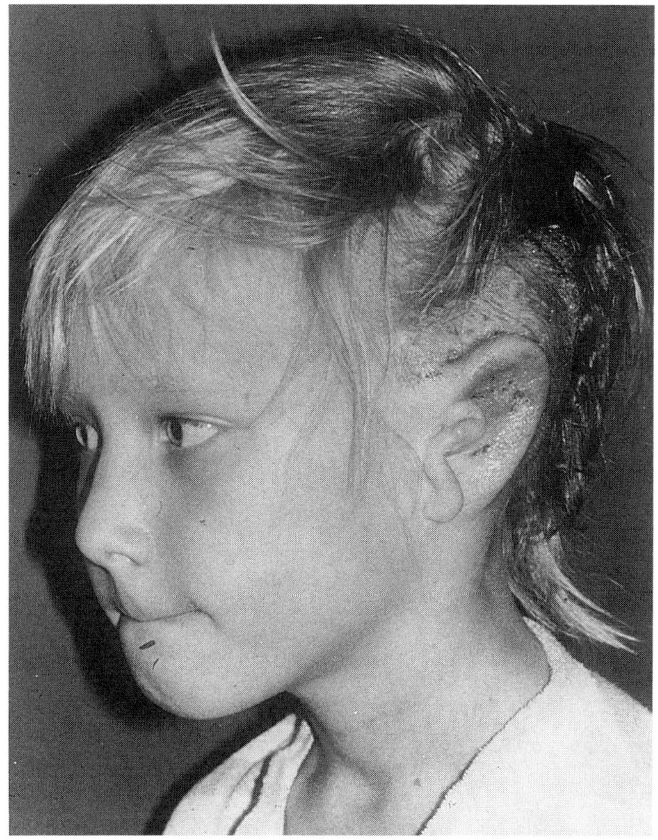

FIGURE 11-15
Immediate postoperative result before the lobulus is transposed to its right position and the conchal cavity and tragus are created. (Reprinted with permission from Nordström REA, Salo HP, Rintala AE. Auricle reconstruction with the help of tissue expansion. Fac Plast Surg *1988;5:338–346.)*

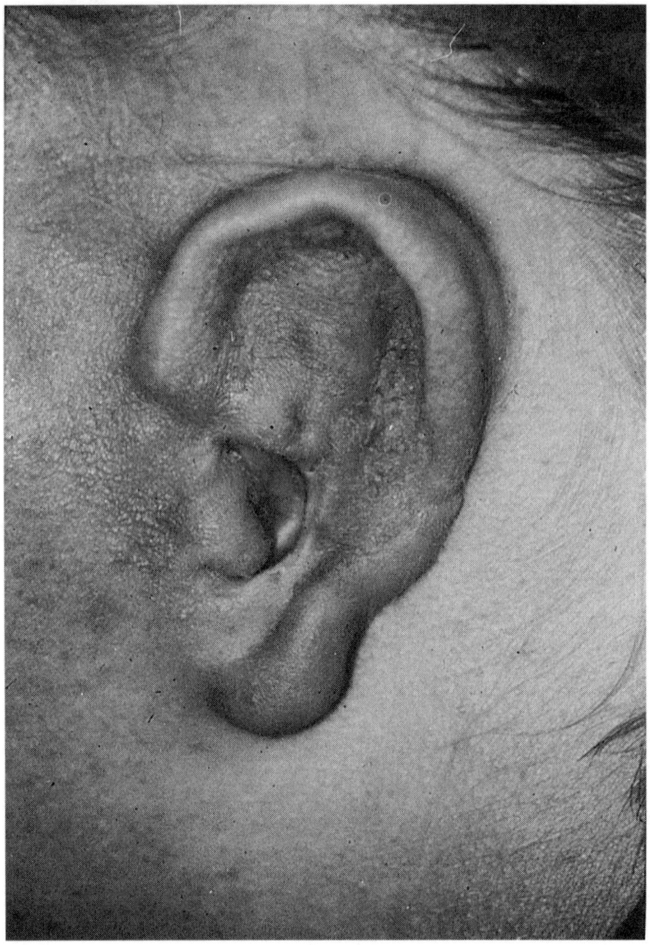

FIGURE 11-16
The auricle in Figure 11-12, 2 years later, showing that the result seems quite stable. (Reprinted with permission from Nordström REA, Salo HP, Rintala AE. Auricle reconstruction with the help of tissue expansion. Fac Plast Surg *1988;5:338–346.)*

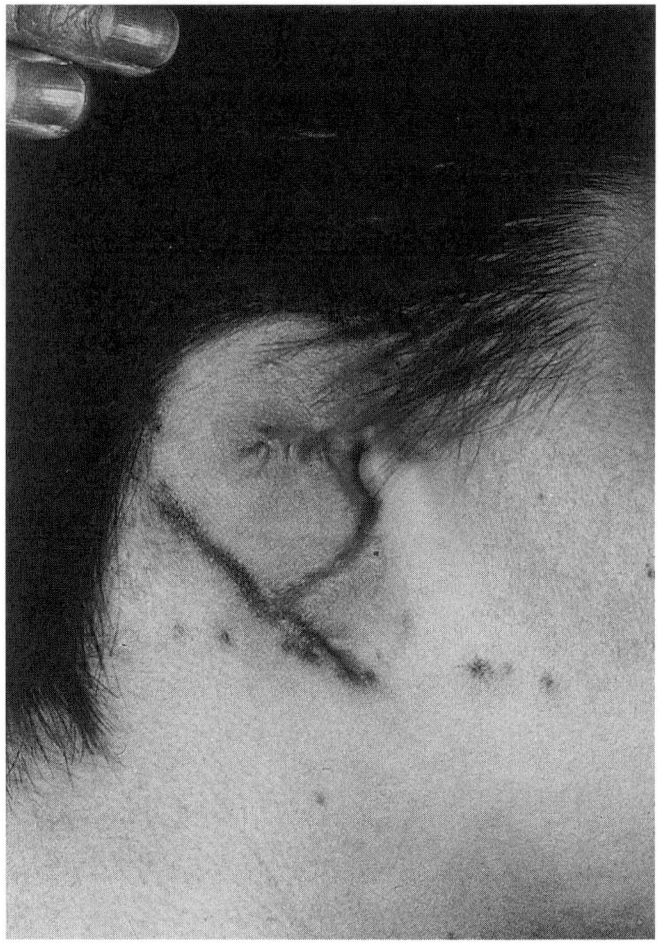

FIGURE 11-17
A patient with an avulsion of the auricle. (Reprinted with permission from Nordström REA, Salo HP, Rintala AE. Auricle reconstruction with the help of tissue expansion. Fac Plast Surg *1988;5:338–346.)*

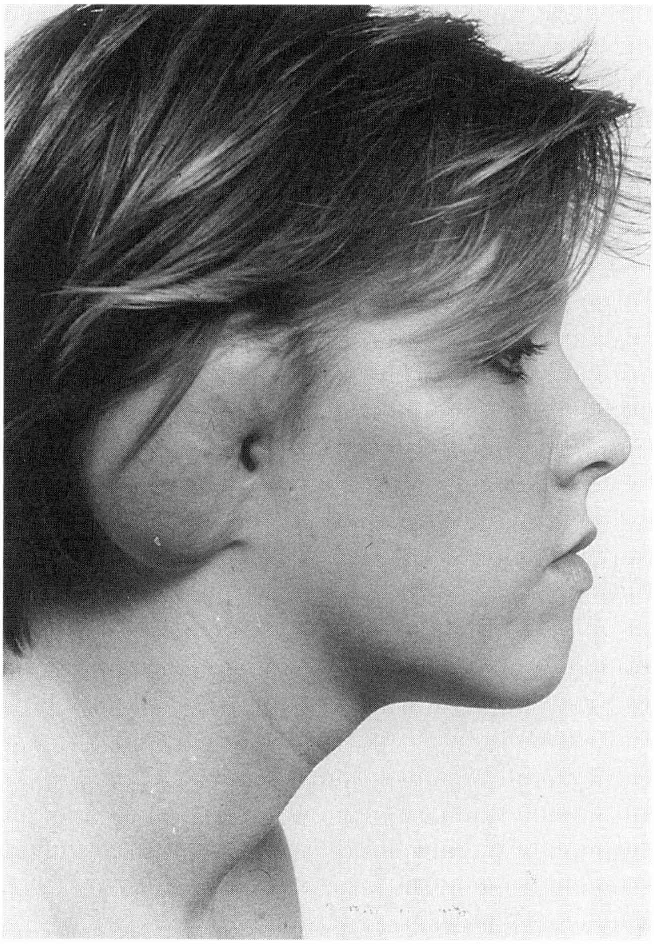

FIGURE 11-18
The patient in Figure 11-19 with the expanded skin pocket. (Reprinted with permission from Nordström REA, Salo HP, Rintala AE. Auricle reconstruction with the help of tissue expansion. Fac Plast Surg *1988;5:338–346.)*

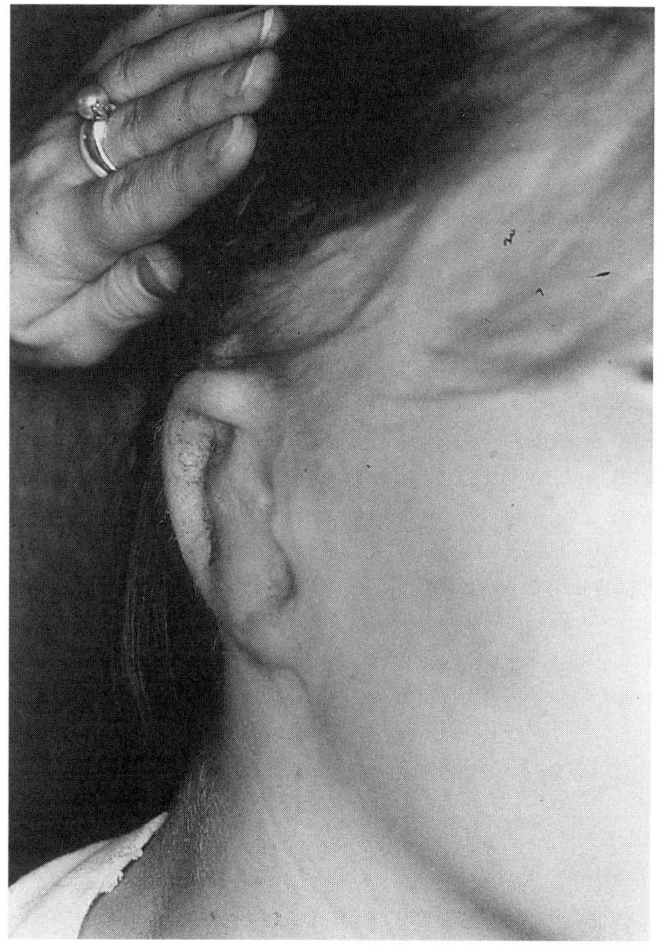

FIGURE 11-19
The result achieved with the cartilage frame inserted. The patient was satisfied with this and did not want a reconstruction of the lobulus. She also did not want to have the auriculocephalic sulcus created, which easily could have been achieved with a flap or graft. (Reprinted with permission from Nordström REA, Salo HP, Rintala AE. Auricle reconstruction with the help of tissue expansion. Fac Plast Surg *1988;5:338–346.)*

REFERENCES

1. Bennun RD. Microtia: A microform of hemifacial microsomia. *Plast Reconstr Surg* 1985;76:859–863.
2. Grabb WC. The first and second brachial arch syndrome. *Plast Reconstr Surg* 1965;36:458–508.
3. Poswillo D. Otomandibular deformity: Pathogenesis as a guide to reconstruction. *J Maxillo Fac Surg* 1974;2:64–72.
4. Tanzer RC. Total reconstruction of the auricle. *Plast Reconstr Surg* 1971;47:523–533.
5. Brent B. The correction of microtia with autogenous cartilage grafts: I, The classic deformity. *Plast Reconstr Surg* 1980;66:1–12.
6. Nordström REA, Salo HP, Rintala AE. Auricle reconstruction with the help of tissue expansion. *Fac Plast Surg* 1988;5:338–346.
7. Nagata S. Secondary reconstruction for unfavorable microtia results utilizing temporoparetial and innominate fascia flaps. *Plast Reconstr Surg* 1994;94:254–265.
8. Neumann CG. The expansion of an area of skin by progressive distension of a subcutaneous balloon. *Plast Reconstr Surg* 1957;19:124–130.
9. Brent B. The correction of microtia with autogenous cartilage grafts: II, Atypical and complex deformities. *Plast Reconstr Surg* 1980;66:13–22.

10. O'Neal RM, Rohrich RJ, Izenberg PH. Skin expansion as an adjunct to reconstruction of the external ear. *Br J Plast Surg* 1984;37:517–519.
11. Isshiki N, Koyama H. Suzuki S, Taira T. Surgical techniques for a deep concha, a pseudomeatus, and high projection in congenital microtia. *Plast Reconstr Surg* 1986;77: 546–555.
12. Nordström REA, Pietilä JP, Voutilainen PEJ, et al. Tissue expander injection dome leakage. *Plast Reconstr Surg* 1988;81:26–29.
13. Tjellström A, Yontchev E, Lindström J, Brånemark PI. Five years experience with bone anchored auricular prosthesis. *Otolaryngol Head Neck Surg* 1985;93: 366–372.

12

Alveolar Ridge

ALBERT R. M. WITTKAMPF

The aim of the different augmentation procedures is to restore the resorbed edentulous alveolar ridge and to create an alveolar process, sufficient in height, width, and firmness so that the upper and lower dentures have support, retention, and stability. The formerly widely accepted augmentation procedures with autologous iliac or rib bone grafts, the Visor-Sandwich osteotomy of the mandible, and the many modifications show a high morbidity. Many augmentation procedures include two surgical interventions with long hospitalizations, two operation sites, and a risk for disturbances of the mental nerve resulting in hypoesthesia of the lower lip in over 80 percent of the patients.[1,2] The accelerated resorption of the bone graft with loss of the initial operation result was another major problem.

The introduction of hydroxylapatite (HA) particles as an onlay augmentation material offered the oral surgeon the opportunity to minimize the above mentioned clinical procedures and a part of their complications. An even greater advantage was that the management of the atrophic mandible could take place on an outpatient basis. As expected, new complications were encountered related to the properties of the material and the new surgical procedure.

While preparing the subperiosteal tunnel in moderate to extreme atrophic ridges (Kent class III and IV;[3] Cawood class V and VI[4]), the unstretchable periosteum can tear, and the result is a loss of control of the HA granules, and their migration into the submucosal tissues. This migration may give rise to complaints of pain due to pressure of the denture rim on the granules, expulsion of granules and loss of height of the augmentated alveolar ridge, subsequent loss of depth of the vestibule due to mucosal perforation, and hypoesthesia of the mental or inferior alveolar nerve (Figure 12-1).

In the literature, many methods are described to prevent the migration of the granules: for example, the use of stents and rubber tubes,[3,5] retention plates,[6,7] modifications in surgical procedures,[7,8] special suture techniques, plaster of paris, collagen, and fibrin glue.[9–14] The fact that so many different solutions have been presented for this problem illustrates the encountered difficulties. Mainly because of the tendency of the HA granules to migrate (no absolute control could be achieved), the inital acceptance of HA decreased. The worldwide introduction and improvement of different implant systems further decreased the use of HA.

In 1985, a "capsule inducing" two-step augmentation procedure was introduced by Krüger.[15] Bonomo[16] then had the brilliant idea of taking advantage of tissue expansion as well as the formation of a fibrous tissue capsule around the expander for use in oral surgery. In this way, he found how to keep the HA granules together. Without any doubt, the subperiosteal tissue expander (STE) method is the only way, at present, to achieve approximately 100 percent control of ridge form and migration tendency.

INDICATIONS FOR A SUBPERIOSTEAL TISSUE EXPANDER IN ORAL AND MAXILLOFACIAL SURGERY

Maxilla

The use of the STE for the maxilla has been advocated in the literature.[17] However, the bony and soft tissue

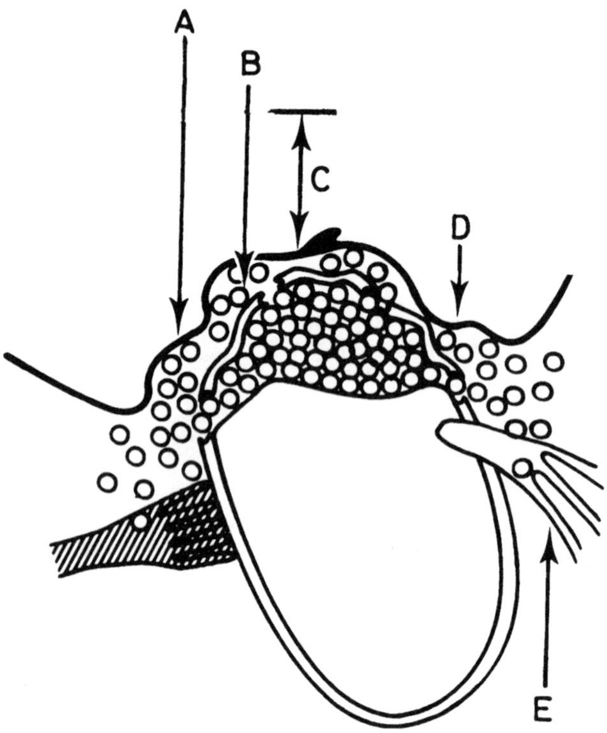

FIGURE 12-1
Migration of the granules as result of tearing of the periosteum. (A) Granules under the dental rim. (B) Migration into the submucosal tissues. (C) Loss of height of the augmentated alveolar ridge. (D) Loss of depth of the vestibule. (E) Sensibility disturbances.

anatomy of the maxilla contraindicates the use of the STE for ridge augmentation procedures. In the first phase of atrophy, the residual alveolar ridge in the maxilla already shows a knife-edge form.[4] Positioning the STE on top of the ridge in these cases is hazardous. Animal experiments show that an STE underlying palatal mucosa will lead partly to expansion of the mucosa but also to extensive resorption of the underlying bone, probably because of the firm palatal mucosal resistance. Also, frequent palatal mucosal perforations occur in these experiments.[18] At present, in the author's opinion, better surgical options are available: lifting the floor of the maxillary sinus combined with bone grafting or autologous onlay grafts combined with permucosal implants; HA fibrin glue reconstruction of the ridge[19] or the use of a vicryl tube filled with HA;[20] and both procedures combined with a submucosal vestibuloplasty.

Mandible

Patients who are referred to a department of oral and maxillofacial surgery because of insufficient retention and stability of the lower denture should be evaluated by a team consisting of an oral surgeon/implantologist and a prosthodontist.

Patients with insufficient retention and stability caused by premature occlusal contacts and improper articulation can be treated by manufacturing a new denture. A class I intermaxillary relationship should be the goal in each patient. Loading capacity of both the upper and lower alveolar processes must be matched properly to avoid pain or denture loosening and iatrogenic resorption of the antagonistic alveolar process.

Knife-edge ridges (Cawood class IV) or local defects can be recontoured using HA granules mixed with fibrin glue. In these cases, the periosteum of the concave defect can be lifted to a convexity, with the HA granules being kept together by the periosteum.[14] In general, permucosal implants are preferable as long as there is good oral hygiene and sufficient bone height.

If bone height is insufficient in the younger patient, I personally would choose an augmentation procedure with autologous bone, combined with permucosal implants. Functional loading of the grafted bone is the first choice in case of alveolar ridge atrophy caused by afunctionality (loss of teeth), denture wearing, and hormonal changes. The STE is indicated in the elderly and patients with a poor oral hygiene. Although the preoperative relationship between soft tissue and mandible is hardly discussed in the literature, it is of significant importance for the clinical results. The need for secondary soft tissue surgery is dictated by this relationship.[21] A preoperative wide denture-bearing area determines the success rate of the STE procedure (Figure 12-2A). If a relatively narrow denture-bearing area is present, the STE—but also any other augmentation procedure—will not lead to the creation of an adequate denture-bearing area (Figure 12-2C). If the floor of the mouth and the vestibule are elevated, additional preprosthetic surgery is mandatory. Patients who previously received extensive preprosthetic surgery, such as Visor-Sandwich osteotomies and free mucosal or split-skin grafts, are prone to complications such as perforations during subperiosteal tunneling, perforations due to expansion, and perforations due to the pressure of the granules.

SUBPERIOSTEAL TISSUE EXPANDER DEVICES

Apart from custom-made tissue expanders with different volumes, three different STE types are available: an STE with a self-sealing filling port in the midline, an STE with a distal filling port (both of these types have a strengthened silicone U-shaped bottom plate with perforations at both ends and an inflatable top with a pre-

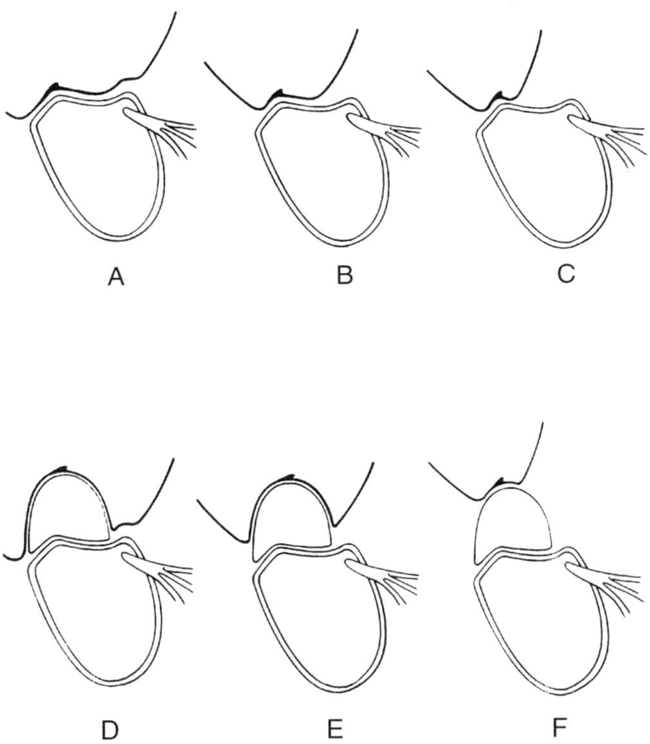

FIGURE 12-2
Subtypes of soft tissue mandibular relationships: (A) broad denture-bearing area potential, (B) intermediate type, and (C) almost no potential denture-bearing area. (D) Extremely good results by expansion of denture-bearing area. (E) Good results by expansion of denture-bearing area using the STE. (F) No gain in denture-bearing area using the STE.

shaped form (Figure 12-3)), and an STE with an integral filling port in the midline. The third type of STE has lateral walls that are reinforced to reduce the tendency of cross-sectional circularity. Additionally, two perforations are present at both ends. The available 3.6 milliliter expanders are suitable for most patients. If the expander is too long, a custom-made STE is preferred because a vestibule in the frontal area is not created as a result of the wrong position of the expander in the midline area.

SURGICAL TECHNIQUE

Insertion of the STE

Prior to the operation, a mouthwash with a 1.2 percent chlorhexidine solution for 1 minute is advised. Local anesthesia is administered at the mandibular foramen at both sides. Infiltrating anesthesia in the buccal sulcus is administered after anesthesia of the lower lip at both sides is evident.

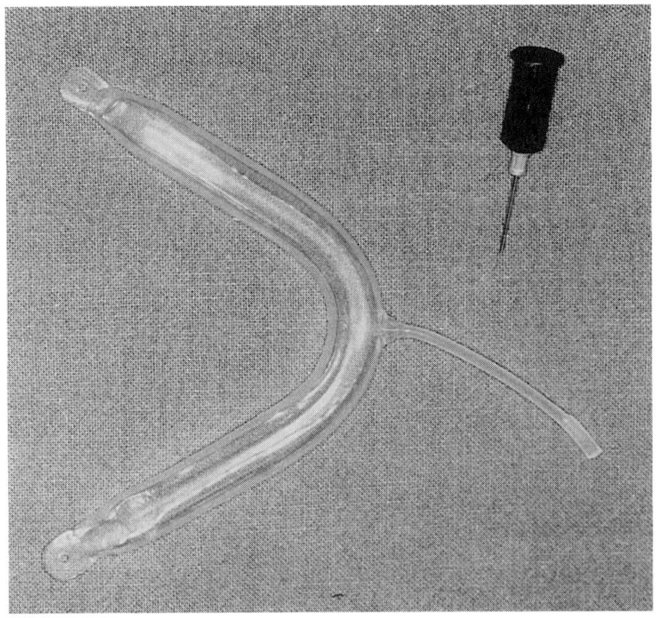

FIGURE 12-3.
Silicone subperiosteal tissue expander with self-sealing filling port and perforations at both ends.

Prior to surgery the available space for insertion of the STE is measured. Many different surgical techniques are recommended in the literature. In plastic surgery, an incision situated above the expander usually is avoided. Undisturbed healing of the incision above a silicone expander in alveolar ridge reconstruction is troublesome because it depends only on the amount of contact between the wound edges of the mucoperiosteum. When insertion of the STE through a distant incision is performed, undisturbed wound healing can take place. The only types of incisions that conform to these criteria are the horizontal incision in the buccal sulcus extending across the midline,[14] and the extraoral sublabial and submental approaches.[22,23] Furthermore, the horizontal incision in the buccal sulcus and the extraoral sublabial and submental approaches have the advantage of a multilayer closure. If the increased risk of wound dehiscence is acceptable, then the vertical midline incision[24] and the two vertical mucosal incisions in the canine/premolar region[25] can be used. A slight curvation of these incisions may be useful to decrease tension. The argument in favor of the vertical incisions, as described by Quayle et al.[25] in the canine/premolar region and the extraoral approaches, is the good visualization of the mental nerve. However, the use of the horizontal incision in the buccal sulcus and the insertion of the Alveolar Retractor (Leibinger GmbH, no: 01-18352/01-18352, Mühlheim-Stetten,

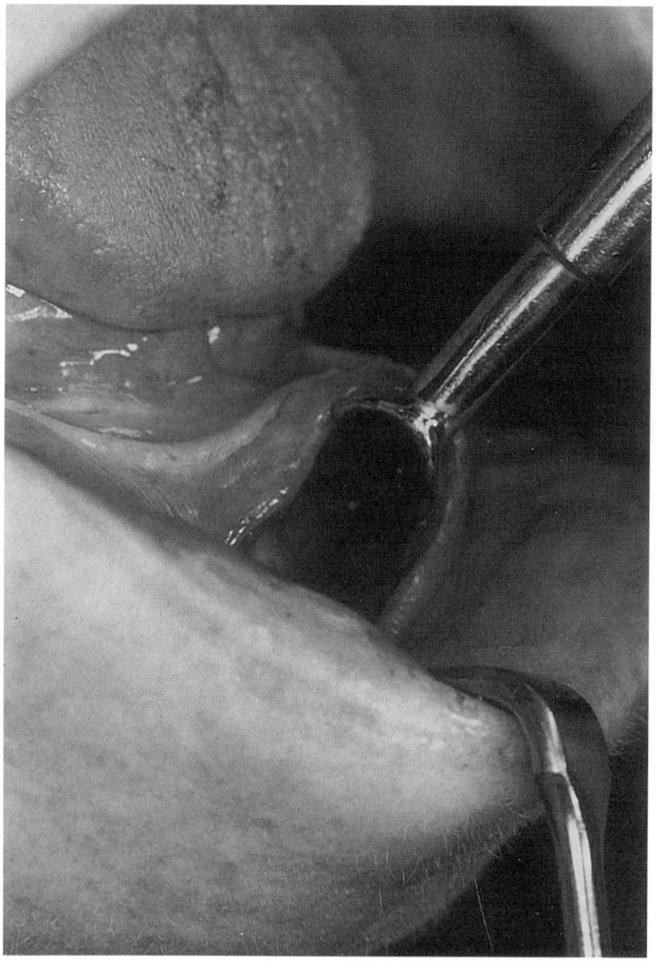

FIGURE 12-4
The alveolar retractor in position to visualize the nerve.

Germany) (Figure 12-4) allow for good visualization of the nerve. The greater risk of scalpel damage to the superficial mental nerve when a vertical incision in the canine/premolar region is performed is not mentioned in the literature. The extraoral approaches specially designed for preserving the mental nerve[25] showed even more hypoesthesia than the horizontal intraoral approach.[21] Another disadvantage of vertical incisions is vertical scar contraction, resulting in loss of depth of the vestibule.

The expander insertion procedure can be performed easily under local anesthesia, but while gaining experience, the surgeon should first perform the STE procedure several times under general anesthesia. A 1.5 to 2 cm horizontal incision is made in the buccal sulcus extending across the midline, 1 centimeter below the residual ridge. The preparation is continued subperiosteally from the incision up to the alveolar crest in the frontal area. Special attention should be given to this area because a sufficient preparation lingual to the alveolar crest must be made to create space for the insertion of the expander just on top of the residual ridge. The tunnel is then continued toward the retromolar pad just on the lingual side of the residual ridge to prevent damage to the alveolar nerve. The elevation of the attached gingiva should be done with great care to prevent perforations.

The mucoperiosteum can then be elevated buccally avoiding injury to the nerve with a subperiosteal elevator. At this stage of the preparation, the alveolar retractor is inserted (Figure 12-4). Residual adhesions from the overlying soft tissue are loosened from the alveolar ridge and the mental foramen and nerve are visualized. The tunnel preparation must be slightly longer than the expander to create enough length on the lateral aspects of the mandible to insert the STE. The tunnel preparation should be equal to the width of the expander or slightly wider to avoid dislocation and the need for perimandibular sutures. It is mandatory to close small mucosal perforations with 4–0 atraumatic suture material before the expander is inserted. In this way, damage to the expander during suturing is avoided. If the desired preparation is achieved, two sutures, which are attached to the Teflon tabs at the end of the expander, are passed through the tunnel using an awl that maintains contact with the mandibular bone. After the awl perforates the mucosal tissue distobuccally at the end of the tunnel, the sutures are grasped at the retromolar pad and the awl is removed. To control the position of the expander during insertion, it is filled with approximately 1.5 mL of sterile saline. By pulling the sutures in a dorsal direction, the expander is easily inserted. As the original alveolar ridge is positioned at the cranial-lingual aspect of the mandibular bone, especially in the molar region, the STE should be positioned slightly more toward the lingual side of the residual ridge than the buccal side. This stimulates the creation of a nice buccal sulcus. The position on the mandible can easily be checked by emptying and refilling the STE. When the expander has been properly positioned, the filling tube is located in the lateral vestibule. This is accomplished by bringing a suture attached to the filling tube (with the help of an awl) from the insertion side to the lateral vestibule in the canine region. The incision is closed in two layers.

Positioning of the filling tube through the vertical midline incision outwards, as originally described by Lew et al.[24] and Bonomo[16] and repeated by several other authors, gives rise to much discomfort for the patient and may increase the percentage of wound dehiscences. For that reason, a specially designed expander was developed.[26] The filling tube placed in the lateral sulcus is very convenient to the patient, but the choice for a distal filling tube may be a matter of personal preference. The latest development, the inflatable STE with an integral filling port as described by Quayle et al.[25] may be superior to the first generations of STEs.

The wound closure should be performed with 4–0 atraumatic suture material in two layers when a horizontal incision is made. However, if vertical incisions are performed, care must be taken that the sutures also pass through the subjacent periosteum to minimize the risk of dehiscences. Care must be taken not to perforate the expander during suturing. An antibiotic drug, taken for 48 hours, and routine analgetics are prescribed after the insertion. The patient is maintained on a soft diet.

INFLATION

Most surgeons fill the expander on the day of surgery to reduce the empty space. Care must be taken to avoid tension on the overlying mucosa. One should be aware of the fact that the patient not only needs an augmented ridge but also a deep vestibule and floor of the mouth. During preparation, a part of the vestibular depth and depth of the floor of the mouth is lost. If one starts the inflations immediately postoperatively, one does not expand the alveolar mucosa but only elevates the buccal vestibule and floor of the mouth. As a consequence, the excellent radiological result contrasts with the poor denture retention. A high percentage of the surgeons who inflate immediately need to perform secondary preprosthetic surgery.[23] Most do not inflate until at least 7 days after surgery; 14 days may be the best choice. The expansion period takes about 2 weeks, during which four to six inflations take place. The maximal inflation in one session should not exceed 0.7 milliliters. The amount of inflation is dictated by the blanching of the overlying mucosa and/or discomfort felt by the patient. If blanching occurs, deflation to a certain extent is advised. The patient should not leave the office with a blanched mucosa, discomfort, or pain. Only minimal tension should be evident. The volume of the expander should be carefully controlled at the end of the inflation period. Overexpansion should be avoided. The expander can be removed 3 to 7 days after the last expansion.

SECOND STAGE

The incision for expander removal is made in the same location as the incision used for the insertion. Although a dehiscence above HA-granules is less serious because of the relatively good granulation tendency, the same arguments against a vertical incision situated above the implant material are valid. A considerable loss of granules is prevented by the horizontal midline incision. Local anesthesia is administered only in the incision area. After deflation, the expander is removed by gentle traction in the midline. The silicone implant, as well as the expansion, induce the formation of a fibrous tissue capsule around the expander.[27] The capsule around the expander should be preserved as much as possible.

Bonomo,[16] Lew et al.,[17,22,24,26] and Bock and Rose[28] advise removal of the capsule between the expander and the mandible by curetting with a sharp instrument before the HA granules are inserted. However, to prevent damage to the nerve caused by the extreme resorption of the alveolar ridge and basal bone in most patients, the capsule on top of the mandible must be preserved. The presence of this tissue does not interfere with stability of the augmentation. According to the literature, the radiological space between the HA granules and the mandible will disappear completely over time in most patients (Figure 12-5). The HA granules are easily inserted into the tunnel with a normal applicator syringe and are compressed to a certain extent by the piston. The firm fibrous tissue capsule prevents migration of the granules. Only 9.5 to 10 g of HA is needed. After filling, the incision is closed in two layers. The same

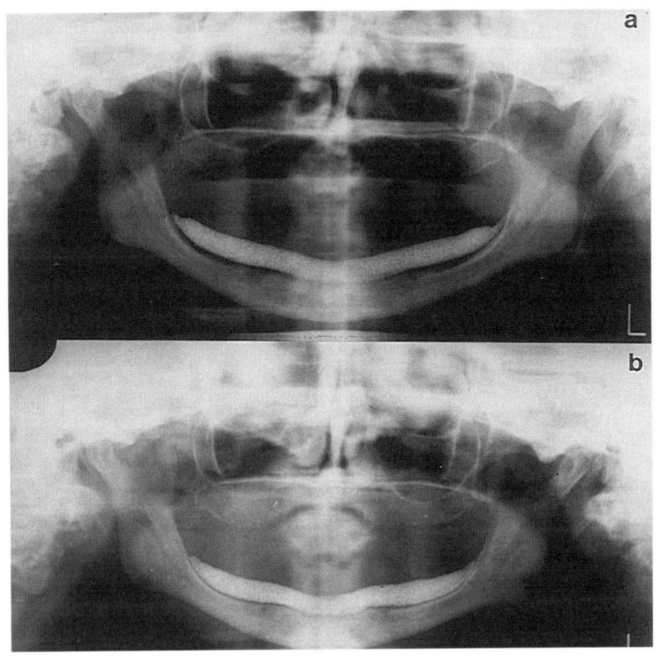

FIGURE 12-5
(A) Radiograph immediately after insertion of HA granules. Note the space between the HA and the bone. (B) Three months later the space is diminished.

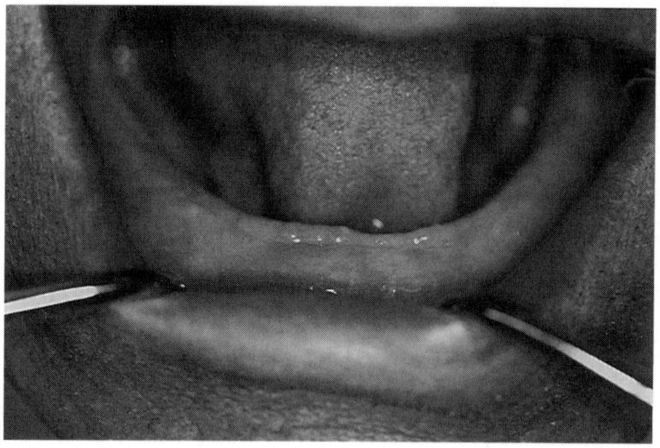

FIGURE 12-6
Final result.

postoperative measures as described following the insertion of the STE are prescribed. Figure 12-6 shows the final result of the STE procedure.

DENTURE LOADING

After several weeks, the granules will be embedded in fibrous tissue. A perforation due to premature pressure of the denture can easily occur and will lead to partial loss of the HA content of the fibrous tissue bag. Denture loading before the 6th week postoperatively is therefore not advised.

Fibrous tissue contraction, settlement of the granules, adaptation of the artificial ridge on the mandible, and resorption of superficial mandibular bone require a careful control of denture fit and stability by a prosthodontist. The reconstructed ridge is very sensitive to occlusal and articulation disturbances. In my experience, mucosal tenderness and redness, often seen in one-stage HA-reconstructed alveolar processes, are less frequent in patients treated with an STE. The fibrous tissue capsule probably acts as a protective shield between the granules and the mucosa loaded by the denture.

If the intraoral retention is still insufficient, a vestibular and floor-of-the-mouth plasty can be performed 4 months after insertion of the granules. The fibrous tissue capsule should then be preserved as much as possible.

COMPLICATIONS

Hypoesthesia due to mental nerve damage occurs and is caused by the preparation. It is important to keep in mind that in extreme atrophy the mental foramen and the inferior alveolar nerve are located on top of the atrophic mandible. Perforations, made during the tunneling procedure, should be closed before the expander is inserted. If they reopen, or if there is a dehiscence of the incision, then a decision must be made whether to continue the expansion or remove the STE. In general, these kinds of dehiscences can easily be closed after removal of the expander before insertion of the granules. A major complication is thinning of the overlying mucosa leading to a perforation. Thinned mucosa can seldom be closed. Although a rapid expansion is the most obvious reason for thinning, it is sometimes hard to avoid. Unknown factors, such as insufficient induction of a fibrous tissue capsule around the STE or, perhaps, an atrophy of the mucosa may influence the resistance to expansion. Infection around the STE has been mentioned only in two cases in the literature.[25,28] Infection of the granules seldom occurs; it should be treated in the usual way—by incision, drainage, and antibiotics.

CONCLUSIONS AND SUMMARY

The STE, used for mandibular alveolar ridge augmentation, is a procedure requiring delicate surgery. However, it permits the oral-maxillofacial surgeon to create an ideal form of the denture-bearing area with a 100 percent control of the granules and with a low morbidity.

The disadvantages of the STE technique are obvious. Two treatments under local anesthesia are necessary, and a 1-month delay, with regular appointments scheduled for inflation of the expander, is also necessary. However, this more than compensates for the lack of secondary preprosthetic procedures when patients are selected on basis of atrophy classifications and of the soft tissue-mandibular relation. The price of the expander is also partly compensated because only 9.5 to 10g of HA are needed.

REFERENCES

1. De Koomen HA. De verhoging van de geresorbeerde mandibula. Thesis, University of Nijmegen, The Netherlands, 1982.
2. Fazili M, Noorman van der Dussen MF, Van Waas, MAJ. Long term results of augmentation of the atrophic mandible. *Int J Oral Maxillofac Surg* 1986;15:513–520.
3. Kent JN, Quinn JH, Zide MF, Guerra LR, Boyne PJ. Alveolar ridge augmentation using non-resorbable hydroxylapatite with or without autogenous cancellous bone. *J Oral Maxillofac Surg* 1983;41:629–642.
4. Cawood JJ, Howell RA. A classification of the edentulous jaws. *Int J Oral Maxillofac Surg* 1988;17:232–236.

5. El Deeb M. Comparison of three methods of stabilisation of particulate hydroxylapatite for augmentation of the mandibular ridge. *J Oral Maxillofac Surg* 1988;46:758.
6. Pham H. The use of an open splint in ridge augmentation with hydroxylapatite. *J Oral Maxillofac Surg* 1986;44: 80–81.
7. Kent JN, Finger IM, Quinn JH, Guerra LR. Hydroxylapatite alveolar ridge reconstruction: Clinical experiences, complications and technical modifications. *J Oral Maxillofac Surg* 1986;44:37–49.
8. Barsan RE, Kent JN. Hydroxylapatite reconstruction of alveolar ridge deficiency with an open mucosal flap technique. *Oral Surg Oral Med Oral Pathol* 1985;59:113–118.
9. Bochlogyros PM, Henser R, Becker R, Zimmermann E. A modified HA implant material. *J Maxillofac Surg* 1985;13: 213–217.
10. Gongloff RK, Montgomery CK, Lee R, Heidereich R. Collagen tubes: Role in subperiosteal contour augmentation. *Int J Oral Maxillofac Surg* 1985;15:669–674.
11. Propper RH. A technique for controlled placement of hydroxylapatite over atrophic mandibular ridges. *J Oral Maxillofac Surg* 1985;45:469–470.
12. Frame JW, Rout PGJ, Browne RM. Ridge augmentation using solid and porous hydroxylapatite particles with and without autogenous bone or plaster. *J Oral Maxillofac Surg* 1987;45:771–777.
13. Mehlisch DR, Taylor TD, Leibold DG, Hiatt R, Waite DE, Waite PD, Laskin DM, Smith ST, Koretz MM. Evaluation of collagen/hydroxylapatite for augmenting deficient alveolar ridges: A preliminary report. *J Oral Maxillofac Surg* 1987;45:408–413.
14. Wittkampf ARM. Fibrin glue as a cement for HA-granules. *J Cranio Maxillofac Surg* 1989;17:179–181.
15. Krüger, E. Alveolarkammaufbau im unterkiefer mit hydroxylapatite. *Deutsches Zeitschrift für Mund-, Kiefer- und Ge-sichtschirurgie* 1985;9:194–195.
16. Bonomo DJ. Subperiosteal soft tissue expansion for ridge augmentation. *Miniclinic C.D.S. Review,* 1986 pp. 34–39.
17. Lew D, Amos EA, Unhold GP. An open procedure for placement of a tissue expander over the atrophic alveolar ridge. *J Oral Maxillofac Surg* 1988;46:161–166.
18. Van Damme PA, Heidbüchel KLWM, Kuijpers-Jagtman AM, Maltha JC, Freihofer HPM. Cranio-maxillo-facial tissue expansion, experimentally based or clinically empiric? A review of the literature. *J Cranio-Maxillo-Fac Surg* 1992;20:61–69.
19. Wittkampf ARM. Augmentation of the maxillary alveolar ridge with hydroxylapatite and fibrin glue. *J Oral Maxillofac Surg* 1988;46:1019–1021.
20. Härle, F. Fortschritte und Schwerpunkte auf dem Gebiet der präprothetische Chirurgie. Fortschritte der Kiefer- und Gesichts-Chirurgie. Stuttgart-New York: Georg Thieme Verlag, 1990, pp. 50–55.
21. Wittkampf ARM. Short-term experience with the subperiosteal tissue expander in reconstruction of the mandibular alveolar ridge. *J Oral Maxillofac Surg* 1989;47: 469–474.
22. Lew D, Amos EL, Shroyer JV. The use of a subperiosteal tissue expander in rib reconstruction of an atrophic mandible. *J Oral Maxillofac Surg* 1988;48:229–232.
23. Schwartz HC, Relle RJ. Extraoral placement of a subperiosteal tissue expander for reconstruction with hydroxylapatite of the severely atrophic mandibular alveolar ridge. *J Oral Maxillofac Surg* 1990;48:157–161.
24. Lew D, Clark R, Shahbazian T. Use of soft tissue expander in alveolar ridge augmentation. *J Oral Maxillofac Surg* 1986;44:516–519.
25. Quayle AA, Marouf H, Holland I. Alveolar ridge augmentation using a new design of inflatable tissue expander: Surgical technique and preliminary results. *Br J Oral Maxillofac Surg* 1990;28:375–382.
26. Lew D, Hinkle RM, Collins SF. Use of subperiosteal implants with distal filling ports in the correction of the atrophic alveolar ridge. *Int J Oral Maxillofac Surg* 1991;20: 15–17.
27. Van Rappard JHA. Controlled tissue expansion in reconstructive surgery. Thesis, University of Groningen, The Netherlands, 1988.
28. Bock NJJ, Rose R. Der subperiostale gewebeexpanderals hilfsmittel bei der alveolarkammplastik. *Deutsches Zeitschrift für Mund-, Kiefer- und Gesichtschirurgie* 1987;11: 443–447.

13

Scalp

ROLF E. A. NORDSTRÖM

The scalp is an excellent source of expanded tissue but it has several special features that should be noted for successful expansion. Furthermore, knowledge of its exact anatomy and the anatomy and physiology of the hair follicles is essential.[1]

THE LAYERS OF THE SCALP

The scalp consists of five layers (Figure 13-1):

1. The skin (epidermis and dermis) is up to 7 millimeters thick in the occipital area.

2. The subcutaneous layer consists of fat and fibrous tissue firmly connecting the skin with the underlying galea aponeurotica or epicranium. This is the space where the main arteries, veins, and also the lymphatics and nerves travel.

3. The epicranium or galea aponeurotica (Figure 13-2) is a very strong tendonlike structure between the frontal muscle in the forehead and the occipital muscle in the neck. Laterally, the temporoparietal and auricular muscles are in conjunction with the galea aponeurotica. The occipital muscle inserts into the superior nuchal line and the frontal muscle inserts into the skin in the region of the supraorbital ridge with the fibers extending into the upper lids and into the subcutaneous tissue at the eyebrows and the nose.

4. The subepicranial layer is a connective tissue layer between the pericranium below and the epicranium above. It is in this very loose, thin, and almost avascular layer that the scalping injuries usually occur. Only a few emissary veins cross this layer. It is important to realize that these emissary veins connect the scalp veins with the intracranial venous system.

5. The pericranium is the periosteum of the skull bone and is the deepest layer of the scalp. It is attached to the skull but can accidentally be ripped off, for example, during surgery of the scalp, but it is regenerated if the skull bone is covered with vascular tissue.

In most avulsions of the scalp, the pericranium is left intact and the denuded area can immediately be covered successfully with a split-thickness skin graft. Often, the full thickness of the avulsed scalp cannot be replaced because there is too much damage to the

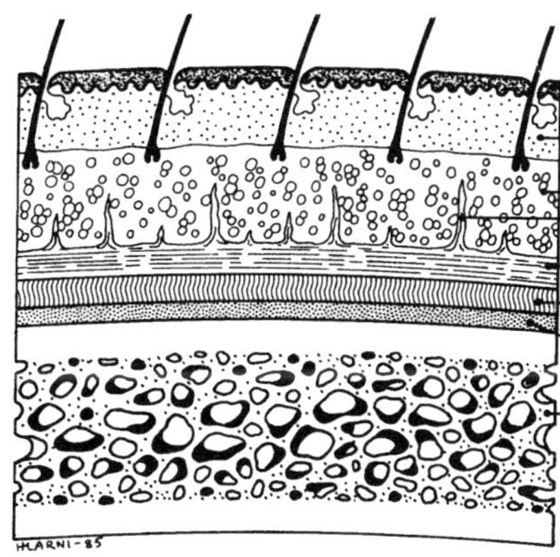

FIGURE 13-1
The layers of the scalp. (Reprinted with permission from Nordström REA. Scalp, hair, baldness and surgery. Fac Plast Surg *1985;2(3):173–177.*

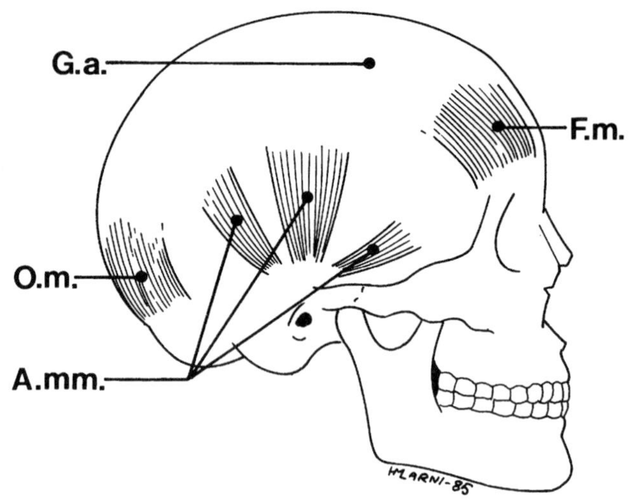

FIGURE 13-2
A schematic drawing of the galea aponeurotica. G.a. = Galea aponeurotica; F.m. = Frontal muscle; O.m. = occipital muscle; A.mm = Auricular muscles. (Reprinted with permission from Nordström REA. Scalp, hair, baldness and surgery. Fac Plast Surg *1985;2(3):173–177.*

avulsed tissue. The avulsed tissue does not survive if it is large or if it is only replaced on the pericranium or bone. It then must be connected by microsurgical techniques to at least one artery and vein. If the pericranium is lost, a split-thickness skin graft does not usually "take," and in these cases the lamina externa of the skull bone must either be removed or holes must be drilled in it so that granulation tissue can form on the bone to provide a good bed for the skin graft.

THE BLOOD CIRCULATION

Especially when designing flaps, but also in other surgery of the scalp, it is important to be familiar with the blood supply of the scalp (Figure 13-3). Three of the five arteries to each side of the scalp are branches of the external carotid artery. These arteries are (1) the superficial temporal artery with its two main branches, the frontal and the parietal; (2) the retroauricular artery, which is small; and (3) the occipital artery, which is almost as large as the superficial temporal artery. In the frontal area, there are two arteries that arise from the internal carotid artery. These are the end-branches of the ophthalmic artery, namely, the supratrochlear and the supraorbital arteries. There are no arteries coming to the scalp through the skull bone; all main arteries travel above the galea in the subcutaneous layer. The location of these main arteries is one of the reasons why scalp flaps survive so well and often make delay procedures unnecessary.

Veins usually follow the arteries. Most of the veins drain into the jugular veins, but emissary veins perforating the skull bone and those following the branches of the ophthalmic artery drain into intracranial veins and thus have the potential of carrying infections inside the skull (although this seems to be extremely rare).

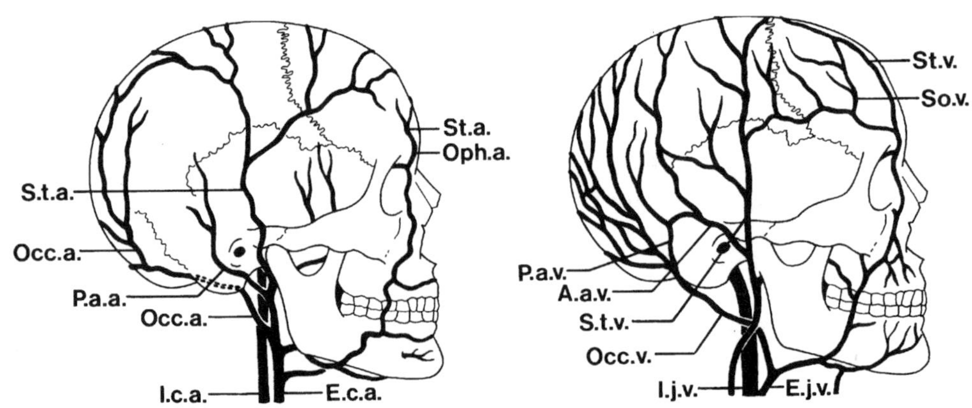

FIGURE 13-3
The main arteries and veins to the scalp. St.a., v. = Supratrochlear a. and v.; So.a., v. = Supraorbital a. and v.; S.t.a., v. = Superficial temporal a. and v.; P.b. = Parietal branch; F.b. = Frontal branch; P.a.a., v. = Postauricular a. and v; Occ.a., v. = Occipital a. and v.; E.c.a. = External carotid a.; A.a.,v. = Anterior auricular v.; I.j.v. = Internal jugular v. (Reprinted with permission from Nordström REA. Scalp, hair, baldness and surgery. Fac Plast Surg *1985;2(3):173–177.*

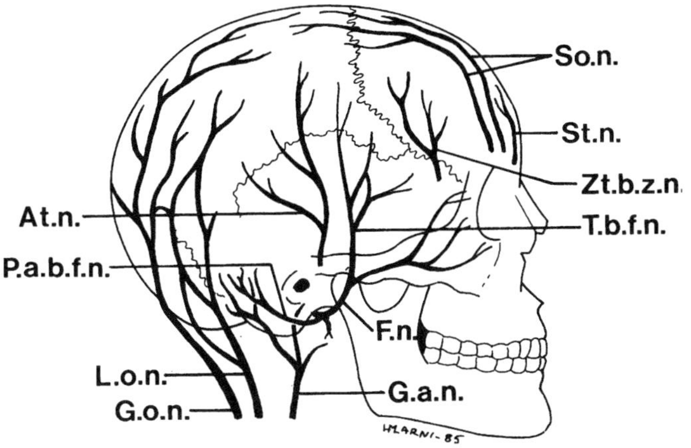

FIGURE 13-4
The nerve supply of the scalp. So.n. = Supraorbital n.; St.n. = Supratrochlear n.; Zt.b.z.n. = Zygomaticotemporal branch of the zygomatic n.; At.n. = Auriculotemporal n.; T.b.f.n. = Temporal branch of the facial n.; G.a.n. = Great auricular n.; L.o.n. = Lesser occipital n.; G.o.n = greater occipital n.; P.a.n. = Posterior auricular n. (Reprinted with permission from Nordström REA. Scalp, hair, baldness and surgery. Fac Plast Surg *1985;2(3): 173–177.*

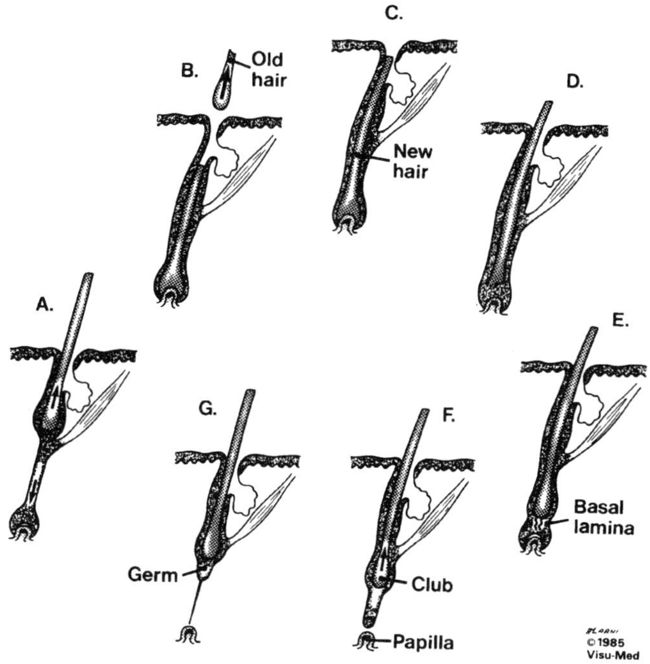

FIGURE 13-5
A hair growth cycle. (A–C) Anagen phase; (D–F) Catagen phase; (G) Telogen phase. (Reprinted with permission from Nordström REA. Scalp, hair, baldness and surgery. Fac Plast Surg *1985;2(3):173–177.*

INNERVATION OF THE SCALP

It is of practical value to know the sensory innervation of the scalp because many surgical procedures can be performed under local anesthesia with nerve blocks. The sensory innervation is directed so that a proximal block creates a field of anesthesia cranially (Figure 13-4).

The lateral and medial branches of the supraorbital nerve and the supratrochlear nerve provide sensory innervation to the forehead and the frontal part of the scalp. These are branches of the frontal nerve, which is the middle branch of the ophthalmic nerve, namely, the superior branch of the trigeminal nerve.

The zygomaticotemporal branch of the zygomatic nerve gives sensory innervation to the temporal region. It is a lateral branch of the maxillary nerve, which is the middle branch of the trigeminal nerve. The lateral part of the scalp just posterior to this receives its sensory innervation from the auriculotemporal nerve. This is a lateral branch of the mandibular nerve, which is the inferior branch of the trigeminal nerve.

The postauricular area is innervated by the great auricular nerve, which is a branch of the cervical plexus. The area just posterior to this is innervated by the lesser occipital nerve, which also branches from the cervical plexus.

The greater occipital nerve provides the sensory innervation to the occipital and posterior crown areas of the scalp. It arises from the posterior ramus of the second cervical nerve. Sometimes the third occipital nerve also provides for a small portion of the sensory innervation close to the midline in a caudal occipital area.

The motor branch to the occipital muscle is the posterior auricular nerve, which is a branch of the facial nerve. The temporal branch of the facial nerve supplies the innervation to the frontalis muscle.

All these nerves travel in the scalp in the subcutaneous layer between the skin and the galea aponeurotica. These nerves are conveniently anesthetized with nerve blocks. The frontal region often requires somewhat more local anesthetic than the other areas.

THE HAIR FOLLICLE

The scalp normally has between 100 000 and 150 000 terminal hairs. In young adults, regardless of sex, about 85 percent of the terminal hairs are in the anagen phase and about 15 percent in the telogen phase. About 1 percent are in the catagen phase, which is a transitional phase between the anagen and telogen phases (Figure 13-5). As a rule of thumb the hairs on the scalp grow for 1000 days and rest thereafter for about 100 days. Also, a maximum of about 100 club hairs, hairs with a club-like root that are in the telogen phase, are normally shed each day. During the anagen phase the scalp hairs grow about 0.3 to 0.4 mm per day.

During human life, the hair cycle is continuously repeated. The same follicle may be transformed into different types of hair. Hairs change from lanugo (fetal hair) to vellus (postnatal) and then to terminal (normal scalp) hair.

The hair consists of a cuticle and a central medulla, with a cortex between the two. The cuticle is a single layer of deeply imbricated scales, with a free margin directed toward the tip of the hair. This anchors the hair in the follicle. The hair bulb surrounding the dermal papilla is made up of rapidly dividing cells, which make up the hair shaft.

DONOR OR RECIPIENT DOMINANCE

Surgical treatment with grafts and flaps can be used in those cases where there is "donor dominance." This means that the properties of the graft or flap prevail in the recipient region and that any disease of the recipient region will not destroy the graft or flap and the hair follicles in it. In recipient dominance, any disease from the recipient region invades the graft or flap and destroys the hair follicles in it.[2]

The following alopecias, for instance, show donor dominance:

1. Androgenetic alopecias (male pattern baldness and female pattern alopecia)
2. All various types of alopecias due to trauma (chemical or physical)
3. All scarring alopecias where the disease is totally quiescent

Chronic discoid lupus erythematosus shows recipient dominance; in 10 of the 12 patients on whom the author has operated, the hair growth in the grafts has diminished over time.[2] Only in two cases has hair growth persisted. Excisions have shown recurrences although, macroscopically, the total lesion has been excised. Alopecia areata is a disease in that one cannot predict whether it will heal spontaneously or progress to the total or universal form. Because of the unstable nature of this disease, I do not recommend surgical treatment except for those unusual cases with a stable small spot of alopecia persisting for several years.

ALOPECIA ANDROGENETICA

Before performing any surgery, it must be ascertained that the patient has an alopecia with donor dominance. By far the most common type of baldness is male pattern baldness (MPB), also known as alopecia androgenetica. It affects, at least to some extent, more than half of all Caucasian men during their lives. Because MPB is so common, its causes will be briefly examined. To become bald, men must have all of these factors: testosterone, genetic makeup, and attainment of a particular age.

The development of MPB requires a testosterone level normal for men. Hippocrates, in the fourth century B.C., recognized that eunuchs did not become bald. On the other hand, for MPB to occur, concentrations of testosterone need not be higher than normal. The previously common misbelief that baldness denoted some kind of supersex men has no verification in hormonal measurements.

MPB is inherited through a dominant gene. A bald man whose father has a full head of hair obtained his hereditary factors for baldness from his mother's side although his mother has a full head of hair. If his father had had one gene for baldness, the father also would have been bald. His mother naturally is not bald, although she had one or two genes for baldness, because she does not meet the first requirement—namely, her level of androgens is not sufficient to cause baldness.

But many females also display a thinning of the hair, though not total baldness, on the top of the head. This is dependent on the same three requirements as MPB and is called female pattern alopecia. It is possible to treat this disease by hair transplantation. On the other hand, a number of these patients do not have a dense growth of hair in the occipital region, making the transplantation less effective.

There seems to be a certain "clock" that times the development of this retrograde metamorphosis, which results in terminal hairs changing in a direction that mimicks the vellus type of minimal, almost invisible hairs. Patients having both high enough levels of testosterone and the hereditary factors for MPB do not develop MPB suddenly; rather, they develop MPB gradually, over a varying number of years. This clock seems to be in the follicle itself or in its very close surroundings. The author did an experiment transplanting hear-bearing scalp from various regions to the forearm in a young man who was becoming bald. The hairs in the grafts on the forearm fell off at the same time that the donor area for the graft became bald.[3] This very clearly demonstrates that MPB is not due to any factor in the scalp such as tension or decreased blood circulation, but that the follicles themselves are programmed to become "bald" at a certain age. This is important to realize as the tension of the scalp increases during tissue expansion.

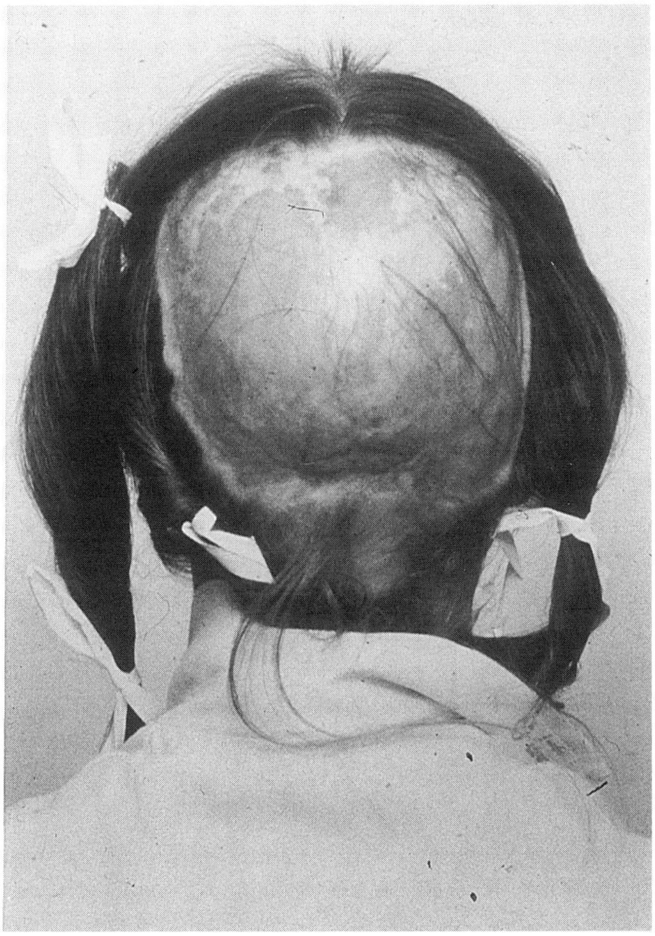

A

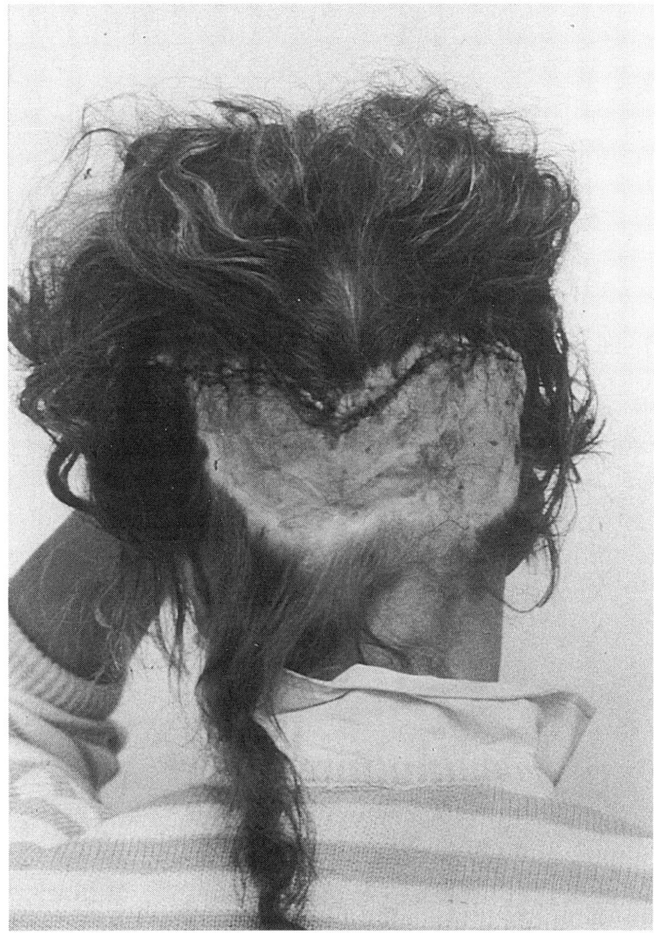

B

FIGURE 13-6
(A) Patient who has lost over 60 percent of her hair-bearing scalp due to an avulsion caused by a dog bite. Primarily covered with a split-thickness skin graft, which is permanently bald. (B) The same patient after correction and expansions completed. (Reprinted with permission from Nordström REA, Pietilä JP, Rintala AE. Clinical experience with tissue expansion. Fac Plast Surg *1988;5:317–327.*

SCALP EXPANSION

The scalp is ideal for tissue expansion (Figure 13-6). The expander is usually placed between the galea aponeurotica and the pericranium. This is an avascular layer, with the rigid galea above protecting the major vessels and nerve of the scalp from the expander. Thus, much higher expansion pressures are tolerated here compared to most other areas.

In most cases, adequate expansion is achieved rapidly despite the rigidity of the galea. In only a few of our scalp expansion patients has such a rigidity of the capsule around the expander developed that continued expansion has been possible only after a capsulotomy. In such patients, the expansion should be started as soon as possible after the operation, before any scar that restricts the tissue expansion has developed. There are also patients who react the opposite way—with a softening of the galea and the development of a thin capsule around the expander.

Accurate planning is essential to scalp expansion; this includes planning the expander shape, size, and location, whereby the flaps to be used should be taken into consideration. Due to the rigid galea, there is a bulging bag of expanded scalp that does not move in the desired direction and position. Due to the shape of the skull, custom-designed expanders for this area are

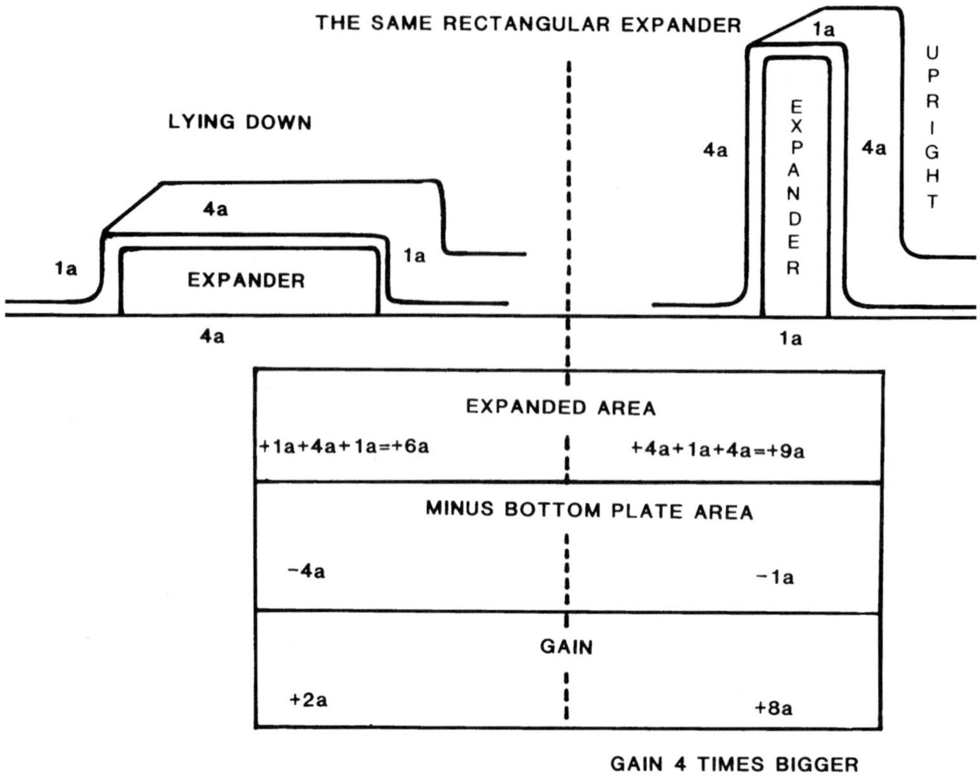

FIGURE 13-7
Schematic drawing showing that the outward projection is the most important direction for yielding new tissue by expansion. Expander of the same size and shape, giving a fourfold result when upright.

usually preferable. We stock a variety of crescent- to horseshoe-shaped expanders, ranging from 12 to 32 cm in length, to allow selection of the most suitable size and shape. A reinforcement on the surface lying against the skull facilitates exact positioning of the expander through a small incision farther away. No wrinkles in the expander are allowed, because they might penetrate the scalp. The reinforcement also helps in achieving a maximal projection of expansion in the optimal direction. The projection is most important in producing as much expanded tissue as possible (see Chapter 3) (Figure 13-7). When the expander bulges out in all directions, the increase in expanded tissue is much less with the same volume fill. A small incision in the right direction permits an immediate postoperative start of expansion and avoids a delay of several weeks for sufficient tensile strength of the incision (see Chapter 3).

Although the scalp is quite resistant to infections, we have seen some low-grade infections. We sometimes inject ampicillin into our expanders,[4] and we believe this reduces the rate of infection.

In one patient with a low-grade *Staphylococcus aureus* infection, most of the capsule and galea disappeared. The patient developed a spotty alopecia and the expander was removed. The hair regrew after 3 months, reaching the anticipated normal density after expansion. Later, the expander was reinserted, and the process of expansion was brought to the planned end point.

Some of the infections we have seen are of a peculiar type. They have started 4 to 8 weeks after the insertion of the expander. In these cases, patients have mentioned that they either had a sore throat, had conducted heavy physical exercise, or had been outside in freezing cold weather. After this, the scalp had become somewhat reddened and sore, and there were palpable tender lymph nodes on the neck and a slight rise in temperature. The patients were treated with antibiotics, but in only one case was there a complete cure of infection. In the remaining patients, the symptoms of infection immediately resolved following removal of the expander. This unusual type of infection indicates that the expanders have a tendency to pick up infections. Bacteria were not introduced during the operation. Bacteria seemed to contaminate the expander via the bloodstream because the expander is a *locus minoris resistentiae*. The expander is a foreign body. Furthermore, especially when higher pressures are used, the circulation is reduced, and there may even be some necrosis of the adjacent tissues that may promote the growth of bacteria.

Another temporary complication seen on occasion is a telogen effluvium, which causes the hair density to diminish. When the expander is removed, hair growth usually returns in 3 months.

In many patients, the process of expansion of the scalp has to be repeated. Partial excision of the lesion is performed and the same or another expander of more suitable size and shape is inserted. We have inserted an expander and repeated the expansion as much as six times in the same patient. The scalp seems to withstand expansion quite well in most patients. It is not only a question of stretching, but also of recruitment of scalp,[5] which has not been clearly pointed out in the literature. This recruitment (migration) of scalp makes it possible to expand the hair-bearing scalp much more than anticipated just from stretching. Furthermore, it is good to realize that in most patients hair density is sufficient to allow for an expansion to double or even quadruple the expanded area. Although this decreases hair density to a half or a quarter of its initial density, it still results in enough hair cover in most patients. Scalp reconstruction where more than half of the scalp has been lost is usually possible.

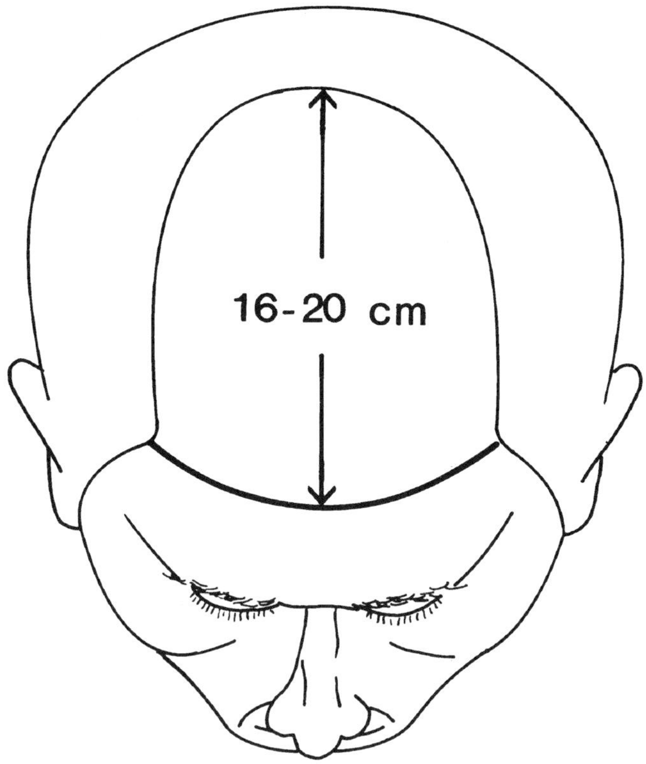

FIGURE 13-8
The characteristics of a suitable case for this system. (Reprinted with permission from Nordström REA. Tissue expanders and flaps for surgical correction of male pattern baldness. Br J Plast Surg *1988;41:154–159.*

A COMBINATION OF TISSUE EXPANSION AND FLAPS FOR SURGICAL CORRECTION OF MALE PATTERN BALDNESS

The combination of tissue expansion and flaps for correction of male pattern baldness has, in certain very select patients, allowed us to achieve a far better result than was previously obtained with punch grafts, alopecia reductions, and flaps. In moderate male pattern baldness limited to the frontal area, it is possible to achieve excellent results with flaps alone; but in patients with more advanced baldness extending over the crown area, it is not usually possible to cover this completely with flaps alone. The sagittal measurement of such baldness is 16 to 20 cm (Figure 13-8). Temporoparieto-occipital flaps are no more than 5 centimeters wide, covering a maximum of 10 centimeters with two flaps and leaving half of the bald area behind them. Leaving bald strips between the flaps often creates a useful result, but these bald strips will show in certain situations. Excision of such remaining bald areas is difficult or unsuccessful because most of the potential laxity of the scalp has already been used for closure of the donor defects.

With punch grafting, a dense and even coverage usually is not achievable, although it is often possible to cover extensive areas and obtain gratifying and useful results. Alopecia reductions, in combination with punch grafting, help to create denser hair growth, especially in patients with a lax scalp; but the result is only rarely as dense as with flaps. It is also not possible to achieve a normal evenness of hair growth, and this unevenness may show, for example, when parting the hair and inspecting the scalp at close range. This is the view that is exposed as a patient emerges from the swimming pool or comes out from the shower.

In patients with a lax scalp, it is possible to do extensive alopecia reductions on top of the scalp. It must be noted, however, that most of the stretching of the scalp occurs in areas close to the line of excision, so that hair density will be least on the top of the scalp and will increase more laterally and occipitally. The goal is the opposite—the densest coverage on the top of the head. A lesser density on the sides and in the neck is usually

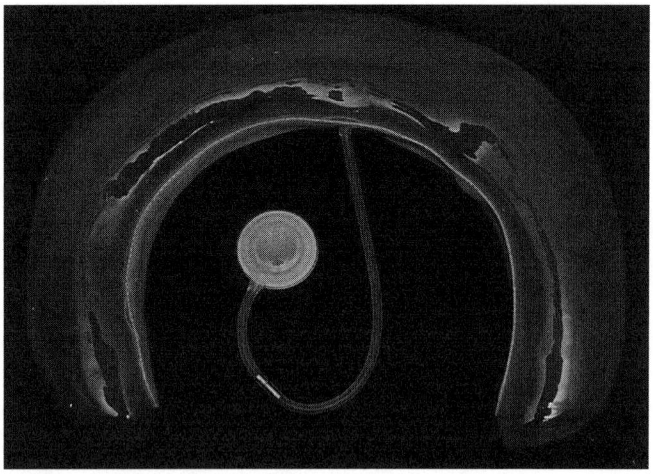

A

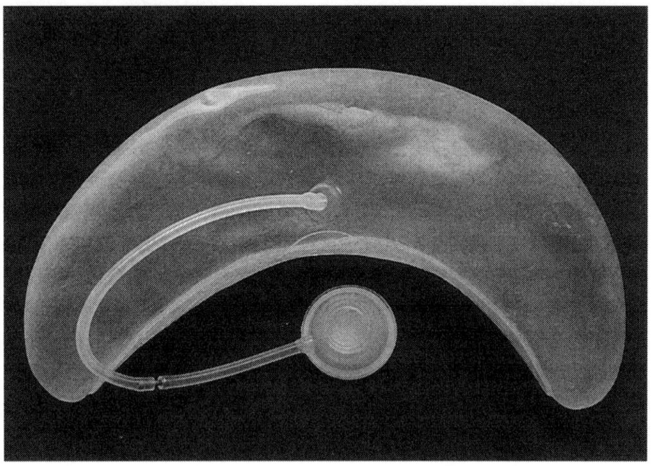

B

FIGURE 13-9
(A,B) The author's two tissue expanders. (Reprinted with permission from Nordström REA. Tissue expanders and flaps for surgical correction of male pattern baldness. Br J Plast Surg *1988;41: 154–159.*

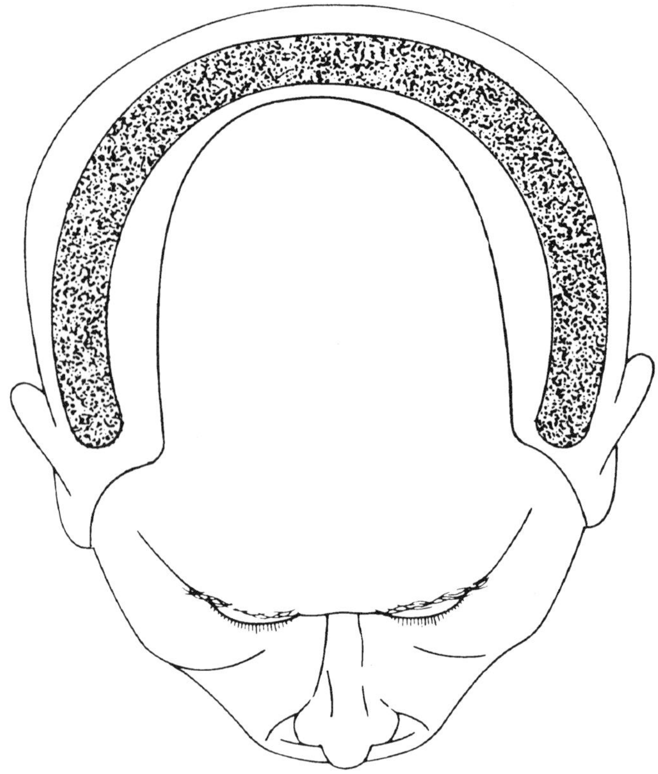

FIGURE 13-10
The first tissue expander in place. (Reprinted with permission from Nordström REA. Tissue expanders and flaps for surgical correction of male pattern baldness. Br J Plast Surg *1988;41: 154–159.*

sufficient. With the system described here, most of these problems are solved.[6]

METHOD

Otherwise healthy patients with baldness (Norwood types IV to VI and IVa to Va), or patients under 30 years of age who supposedly reach these classifications and have a high motivation for the treatment are suitable.[7]

Custom-designed tissue expanders (Figure 13-9) are used. We have in stock U-shaped expanders ranging from 26 to 32 cm in length to suit various patients with different head sizes and shapes. These have reinforced bottom plates that face the skull bone.

The expander is inserted under local anesthesia using a ring block of 1 percent lidocaine with epinephrine around the scalp. It is placed between the galea aponeurotica and the pericranium. The pocket is made as low as possible to produce maximum stretching of the inferior areas of hair-bearing scalp (Figure 13-10).

The sagittal midline incision is closed in two layers, using heavy Maxon® (Davis and Geck, Gosport, UK) running or interrupted horizontal mattress sutures for the galea. The skin is closed with a running thin monofilament suture. Staples can also be used. Before suturing the skin, some saline is injected into the tissue expander to expel the accumulated blood and serous fluid in the undermined areas, and a suction drain is used.

Starting 1 to 2 weeks postoperatively, with one to two weekly filling sessions, the expander is usually sufficiently expanded in 6 to 8 weeks. A sufficient expansion is about 500 milliliters.

The next step is to transpose a temporo-parieto-occipital flap (TPO). The flap to be transposed is one

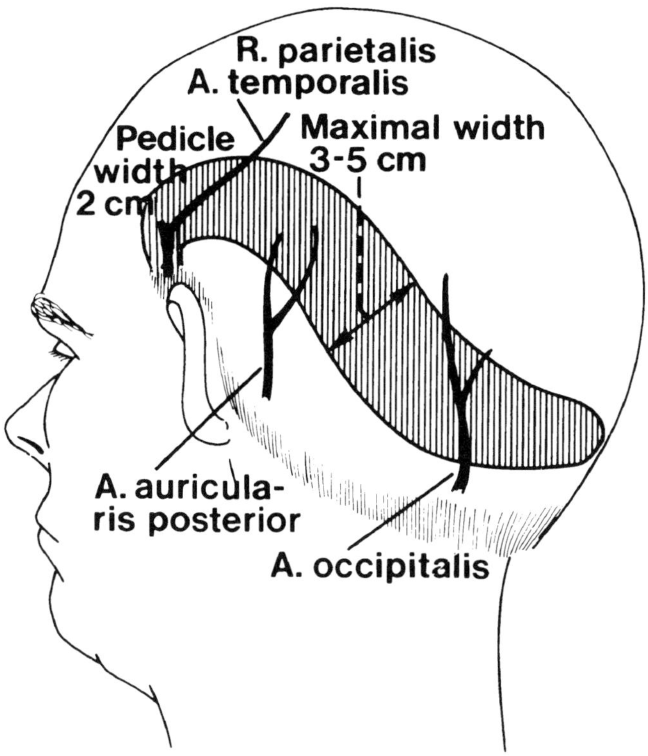

FIGURE 13-11
The author's design of a TPO flap. (Reprinted with permission from Nordström REA. Tissue expanders and flaps for surgical correction of male pattern baldness. Br J Plast Surg *1988;41: 154–159.*

variation of the TPO flaps.[8–10] This is a long flap based on the parietal branch of the superficial temporal artery and is transposed undelayed under local anesthesia (Figure 13-11). The parietal branch of the superficial temporal artery is palpated and/or located with a Doppler scan. The pedicle of the flap is based on this artery, which is kept within the flap usually for a distance of at least 4 centimeters. Very often, the main branch runs too superiorly after this point and has to be left outside the flap. Following this branch too long and too high often includes an area that is already bald or carries an inherent risk for becoming bald. Even when there is little risk for alopecia developing in this area later, the donor scar will be located too high on the scalp to be easily covered and camouflaged by overlying hair. As a result, the donor area for the flap must be curved more inferiorly.

The flap starts with a pedicle width of about 2 centimeters. This narrow pedicle gives less resistance to transposing it about 180 degrees without creating too much of a dog-ear, but it is still sufficient for blood circulation as it includes the superficial temporal artery and vein. The flap then gradually widens to about 6 to 8 centimeters, depending on the laxity of the scalp. The flap is then curved to enclose the occipital artery for a considerable length. I believe that although there will be a retrograde flow with this artery in this flap design, the occipital artery helps considerably in the survival of the very long flap. The trunk of this artery is palpated and/or located with a Doppler scan. The end of the flap turns toward the midline, crosses it, and stops just before the contralateral occipital artery. This is the flap design (although somewhat narrower) that I use also as a routine flap in patients without tissue expansion for coverage of frontal baldness alone. It is performed as an office procedure under local anesthesia and is nondelayed.

Undermining for donor defect closure is carried out all the way to the opposite ear but not extensively to the neck, as this could cut the retroauricular artery. For this reason, there is no risk of the large postauricular necrosis sometimes seen in certain other types of TPO flaps.

Laterally, the flap should not be taken too low either, because all of the area above the flap is going to be discarded as bald or potentially bald when the permanent hair-growing flap is advanced from the contralateral occipital area later during the reconstruction. My impression is also that the collateral circulation of the scalp is better higher up, and this, in turn, improves the safety of such a long flap on a narrow pedicle.

When the inflation is finished, the TPO flap can be transposed (Figure 13-12). The expansion easily creates 6 to 8 centimeters of loose scalp, measured transversely across the scalp, and about 4 centimeters of lax scalp in the occipital area. This occipital laxity is important, because the flap effectively makes use of it as it is taken across the occipital area, leaving only the occipital artery on the contralateral side intact. The occipital donor region is used very effectively for correction of the frontal area as this TPO flap takes advantage of the occipital area, which comprises the largest and most important reserve of permanently growing hair follicles in male pattern baldness. Certain other designs of flaps do not use this important area.[11,12]

Extra laxity of the scalp at this point is used for advancement of the hair-bearing scalp from the contralateral side, as seen in Figure 13-13. Some excision of lax bald scalp can now be performed along the posterior edge of the flap and the donor site.

The pedicle of the TPO flap should be kept rather narrow, about 2 centimeters. This still occasionally creates a considerable dog-ear due to the large transposition angle of the flap, which should be about 180 degrees

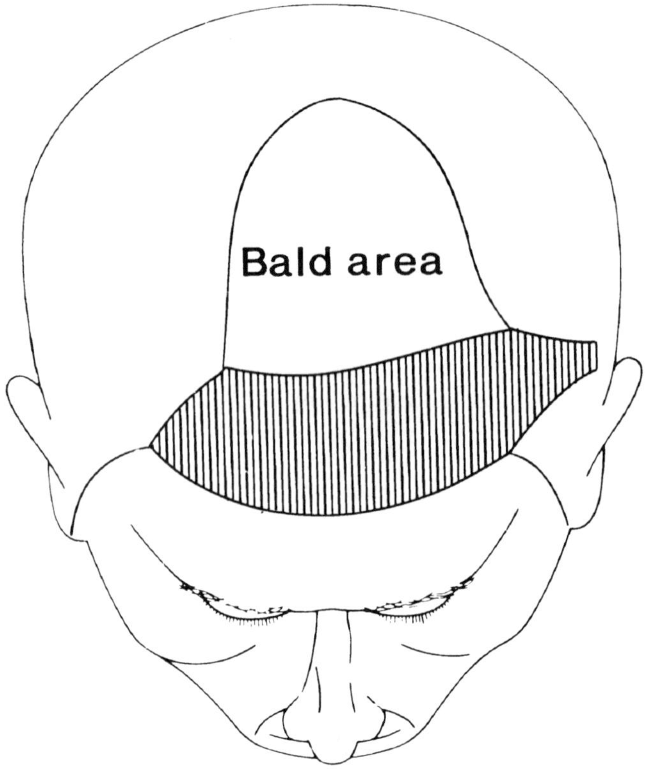

FIGURE 13-12
The situation after transposition of the TPO flap. Note the asymmetry of the bald area, with advancement of the hair-bearing scalp from the contralateral side. (Reprinted with permission from Nordström REA. Tissue expanders and flaps for surgical correction of male pattern baldness. Br J Plast Surg *1988;41:154–159.*

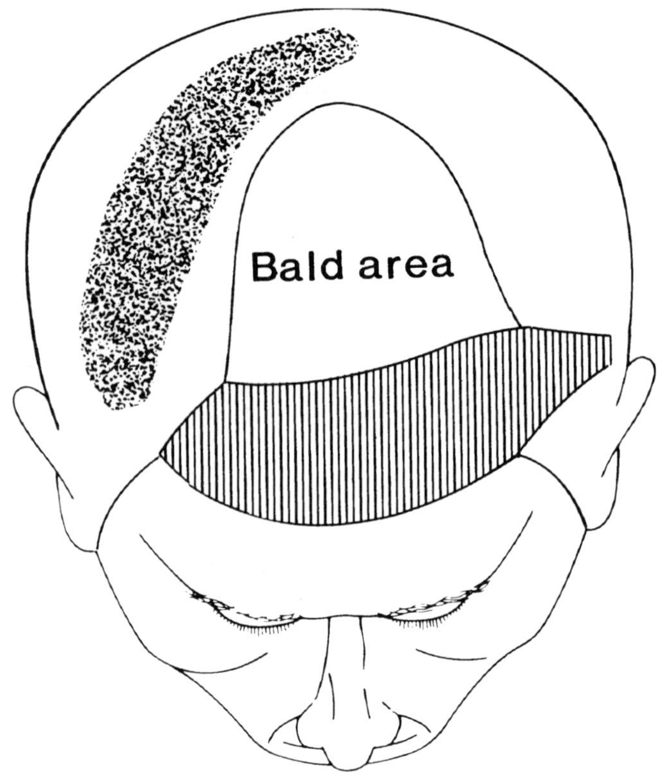

FIGURE 13-13
The second tissue expander in place. (Reprinted with permission from Nordström REA. Tissue expanders and flaps for surgical correction of male pattern baldness. Br J Plast Surg *1988;41: 154–159.*

to achieve a horizontal, natural hairline. If the transposition angle is small, the new hairline will go across the temple, which will look unnatural. This flap is also long enough to reach across the frontal hairline at any height. It allows the possibility of reconstructing the frontal hairline at a level and with a shape that give the best aesthetic result. Some designs of TPO flaps do not extend sufficiently in all patients. The flap should never be placed too low, because correction of such an error is difficult or impossible. The dog-ear usually diminishes rapidly and, if necessary, can be corrected by excision 4 weeks or more postoperatively.

During the procedure in which the TPO flap is transposed, another custom-designed tissue expander, which is 16 to 19 cm long, is inserted under the occipital scalp (Figure 13-14). About 3 weeks should elapse before starting expansion. Inflation is carried out by a once- to twice-weekly expansion to about 500 milliliters.

The expanded retroauricular and occipital hair-bearing scalp is used to cover the top of the scalp and may have to stretch over 10 centimeters (Figure 13-14). This leads to a linear expansion of about 100 percent, decreasing the hair density to about 50 percent. Such a decrease is still quite acceptable, and even a 25 percent hair density satisfactorily covers the scalp in most patients.

Usually about 5 to 10 cm of bald scalp can be excised after one expansion to 500 milliliters, and the procedure may have to be repeated as many as three times. As the expanded tissue is advanced, bald scalp is excised along the posterior border of the previously transposed TPO flap and along the donor area scar of the TPO flap. Thus, no additional scars are created (Figure 13-15). An expansion over 500 milliliters will result in more excision and although it is temporarily more disfiguring for the patient, it is sometimes the choice.

The scalp now has a dense frontal hairline created by the TPO flap; behind this, it is covered by a slightly less dense hair-bearing scalp flap advanced from the contralateral side. There will be no sagittal scar on the top of the scalp; instead, it will lie on the lateral side and be disguised.

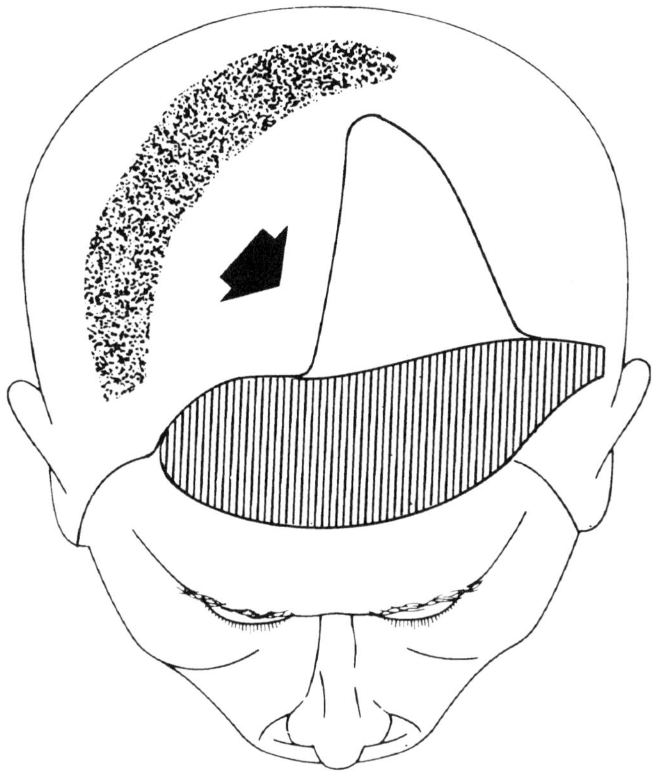

FIGURE 13-14
Advancement of the hair-bearing posterolateral aspect of the scalp. (Reprinted with permission from Nordström REA. Tissue expanders and flaps for surgical correction of male pattern baldness. Br J Plast Surg *1988;41:154–159.*

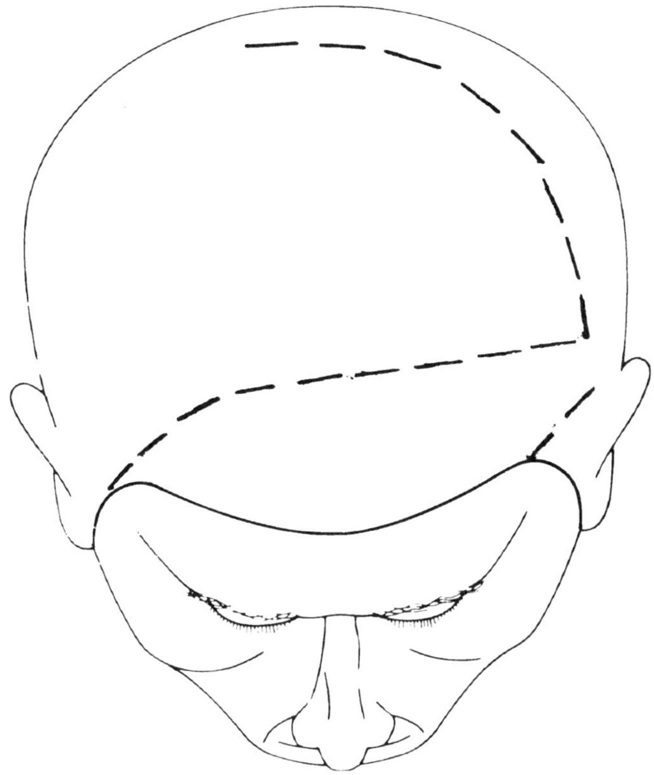

FIGURE 13-15
After total coverage of the bald area by further advancement. There is no sagittal midline scar on the scalp. (Reprinted with permission from Nordström REA. Tissue expanders and flaps for surgical correction of male pattern baldness. Br J Plast Surg *1988;41:154–159.*

RESULTS

Figures 13-16 through 13-19 show typical patients before and after surgery.

DISCUSSION

There have been papers presenting the use of two lateral rectangular expanders. At first, we also used this concept of just lateral expansion in some patients. They were satisfied with the improvement. But the result was a sagittal midline scar that created an artificial part on the top of the head due to the hairs on each side of the scar growing away from the scar, and thus making the scar obvious. This scar is often difficult to disguise. Furthermore, the frontal hairline has an artificial, disturbing part in the middle with hair growing in the opposite direction. It is also usually impossible to sufficiently advance this hairline forward to the aesthetically most pleasing level. Furthermore, as this hairline recedes with age, the scar will be exposed in these patients. The hairline is created by the edge of balding scalp, which almost invariably will recede during the coming years and decades, sometimes even several centimeters. To solve these problems, I have, in most instances, abandoned the use of just lateral expansion with sagittal midline excisions and have proceeded with the procedure described in this chapter.

This system is valuable for baldness classified as Norwood types IVa to Va and moderate type VI alopecia. In Norwood types IIA, III, IIIa, and sometimes IVa, baldness, TPO flaps alone can be used successfully without expanders (one or two flaps).

This is by no means a simple routine system for the average patient. On the contrary, the patient should be highly motivated because of the considerable number of office appointments needed for filling the expanders. Several procedures are needed, resulting in occasional edema of the face, social incapacity, and time off from work after each procedure. The last 2 or 3 weeks of each expansion procedure are sometimes socially difficult for the patient because of the bulging of the expander.

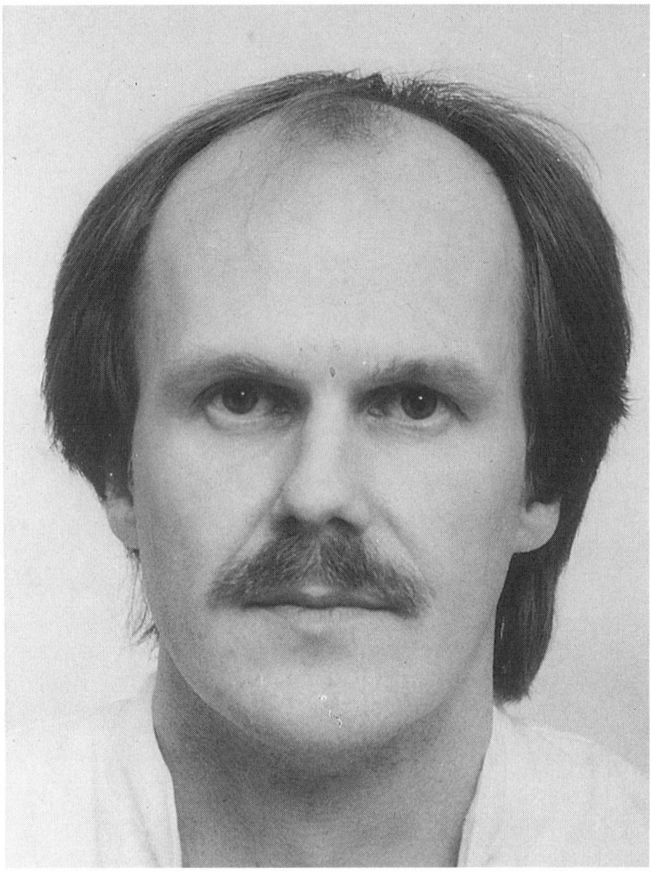

A

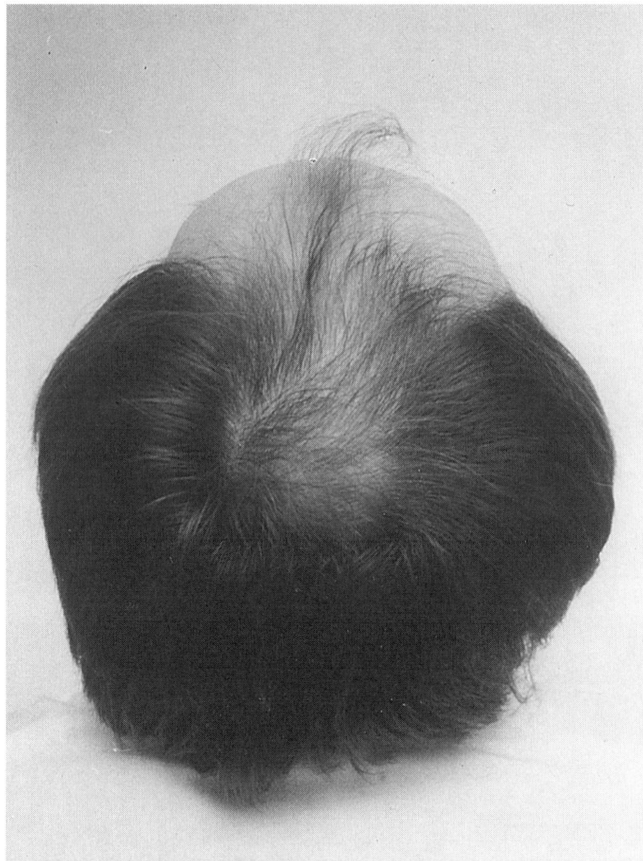

B

FIGURE 13-16

(A,B,C) Patient with male pattern baldness prior to correction. (D,E,F) Same patient after correction . (Reprinted with permission from Nordström REA. Tissue expanders and flaps for surgical correction of male pattern baldness. Br J Plast Surg *1988;41: 154–159.*

Furthermore, this procedure is not recommended for surgeons who are inexperienced with scalp surgery, hair replacement surgery, and tissue expanders. The design of the flap is extremely important, both for aesthetic and flap-survival reasons. The design and exact placement of the expanders is also crucially important. Lack of experience can easily lead to disaster.

This program lasts a minimum of 6 months. However, an advantage is that the hair does not fall out temporarily during the process, as happens for example, in punch grafting, where the hair first falls out from the grafts and then starts to grow about 3 months after transplantation. The patient who undergoes punch grafting has to wait up to 1 year for the hair to grow to the desired length following the last session of punch grafting.

The TPO flap has an upward hair growth direction that is not the most common one in the frontal area, in which the direction of hairgrowth is usually slightly forward. But after performing more than 100 procedures, my impression is that they are very acceptable and usually very good. The frontal hairline scar is generally inconspicuous enough to allow the patients to comb their hair back; if not, it can and should be improved by "micrografts."[13] Minigrafts are excellent to camouflage any disturbing lateral, transverse, or posterior scars. Usually, patients have the option of combing their hair to one side to cover the frontal hairline. A permanent wave helps camouflage the frontal hairline, but it is rarely necessary.

SPECIAL FEATURES OF SCALP EXPANSION

The expander is usually placed between the galea and the pericranium. This layer is very easy to undermine. Furthermore, it is almost avascular. But care must be taken with the bifurcation of the superficial temporal

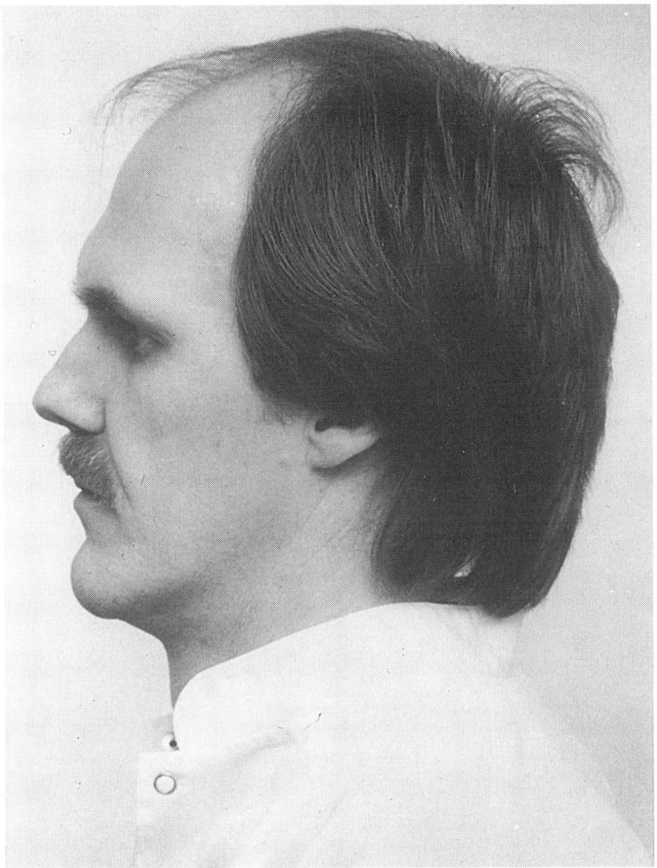

C

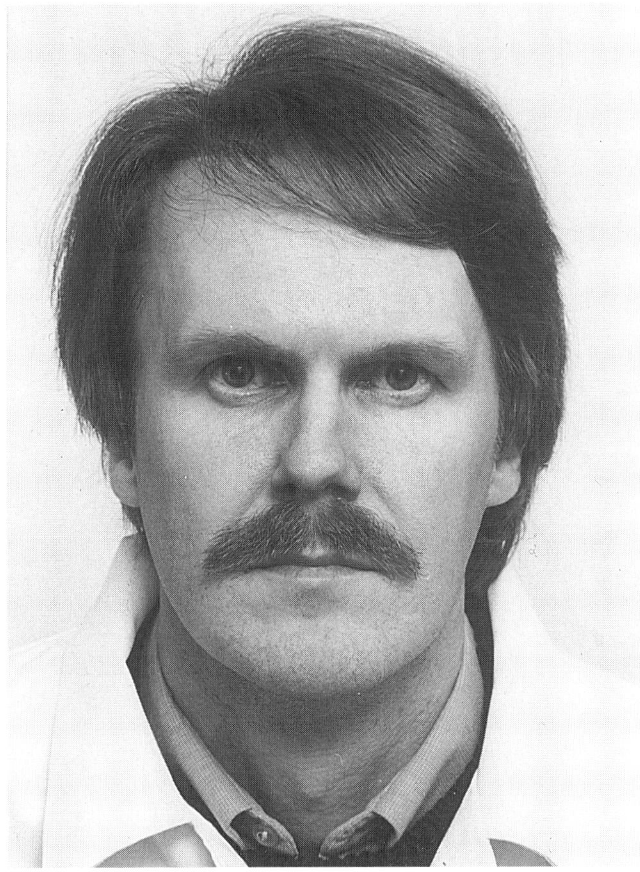

D

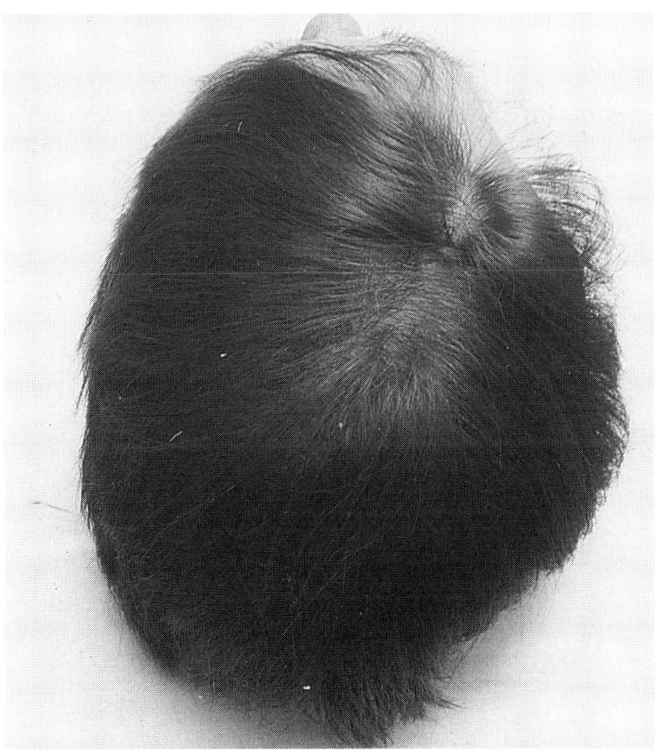

E

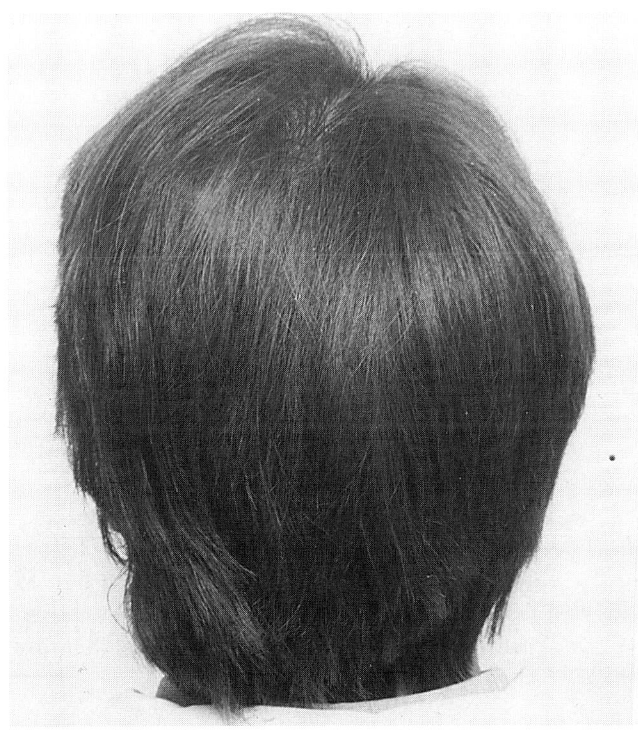

F

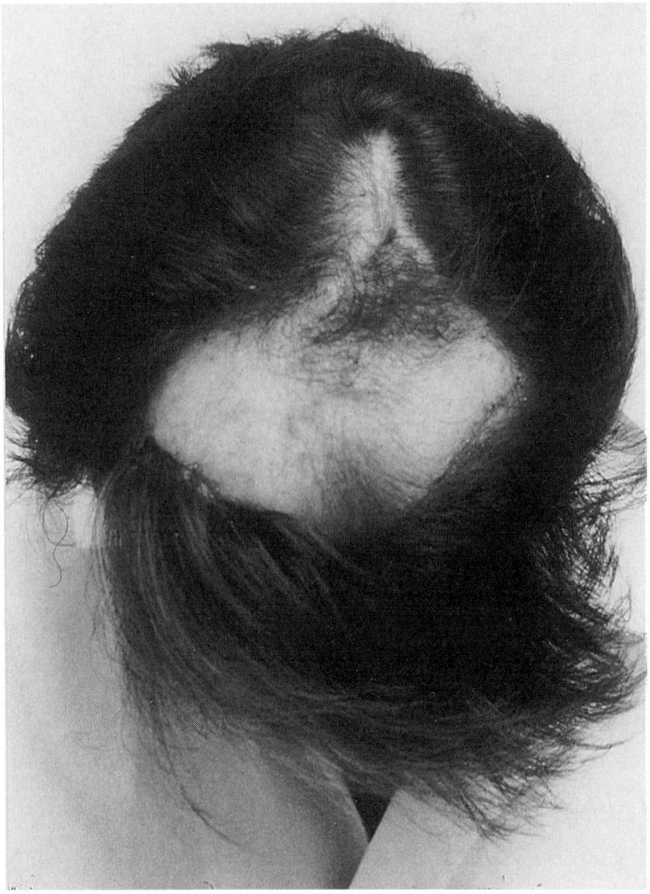

FIGURE 13-17
Intermediate stage showing the TPO flap in place before advancement. (Reprinted with permission from Nordström REA. Tissue expanders and flaps for surgical correction of male pattern baldness. Br J Plast Surg *1988;41:154–159.*

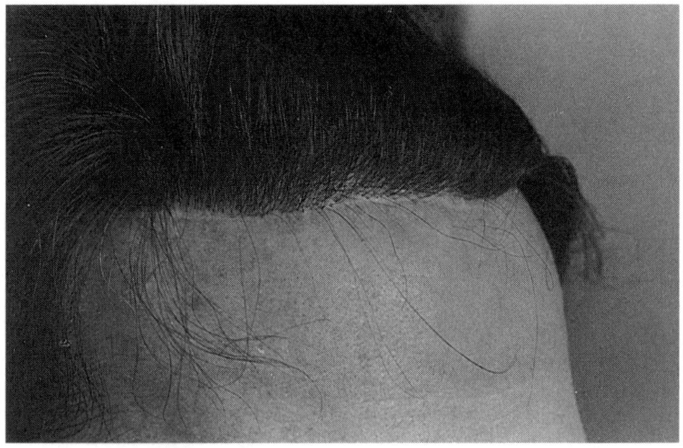

FIGURE 13-18
Close-up of the hairline. (Reprinted with permission from Nordström REA. Tissue expanders and flaps for surgical correction of male pattern baldness. Br J Plast Surg *1988;41:154–159.*

artery into the frontal and the parietal branch because there is a weak area of the galea here and these vessels might be cut. Also notice that the occipital arteries change plane in the region of the nuchal ridge and are easily severed here. Some impressure on the skull bone is possible, but I have never seen any problem with this in my cases. Because a ring of collagenous tissue forms around the base of the expander, quite a depression is often seen as the expander is removed, but this is mostly due to this ring of scar tissue, which can be trimmed somewhat during the operation when the expander is removed. This ring of scar tissue will eventually fade during the postoperative months.

In addition to the major vessels, the hair roots also lie in the subdermal fat layer above the galea and below the dermis. An undermining in this layer would not only be very bloody, but would also destroy hair roots, which are protected when the expander is below the galea.

The skull is convex and thus the expanders have to match this. Furthermore, it is always important to consider the direction the growth of the hair will have in the new location because a wrong direction will create permanent problems for the patient. The hairlines must have a correct, acceptable direction of hairgrowth and they also must start with the normal, soft and sparse growth of hair that is typical for a normal hairline.

Alopecia Due to Trauma

Tissue expansion is excellent for reconstruction of scalp defects. It is important to be familiar with the anatomy described above so as not to go wrong. Defects affecting up to about 70 percent of the hair-bearing scalp have been successfully corrected, as in the case shown in Figure 13-20.

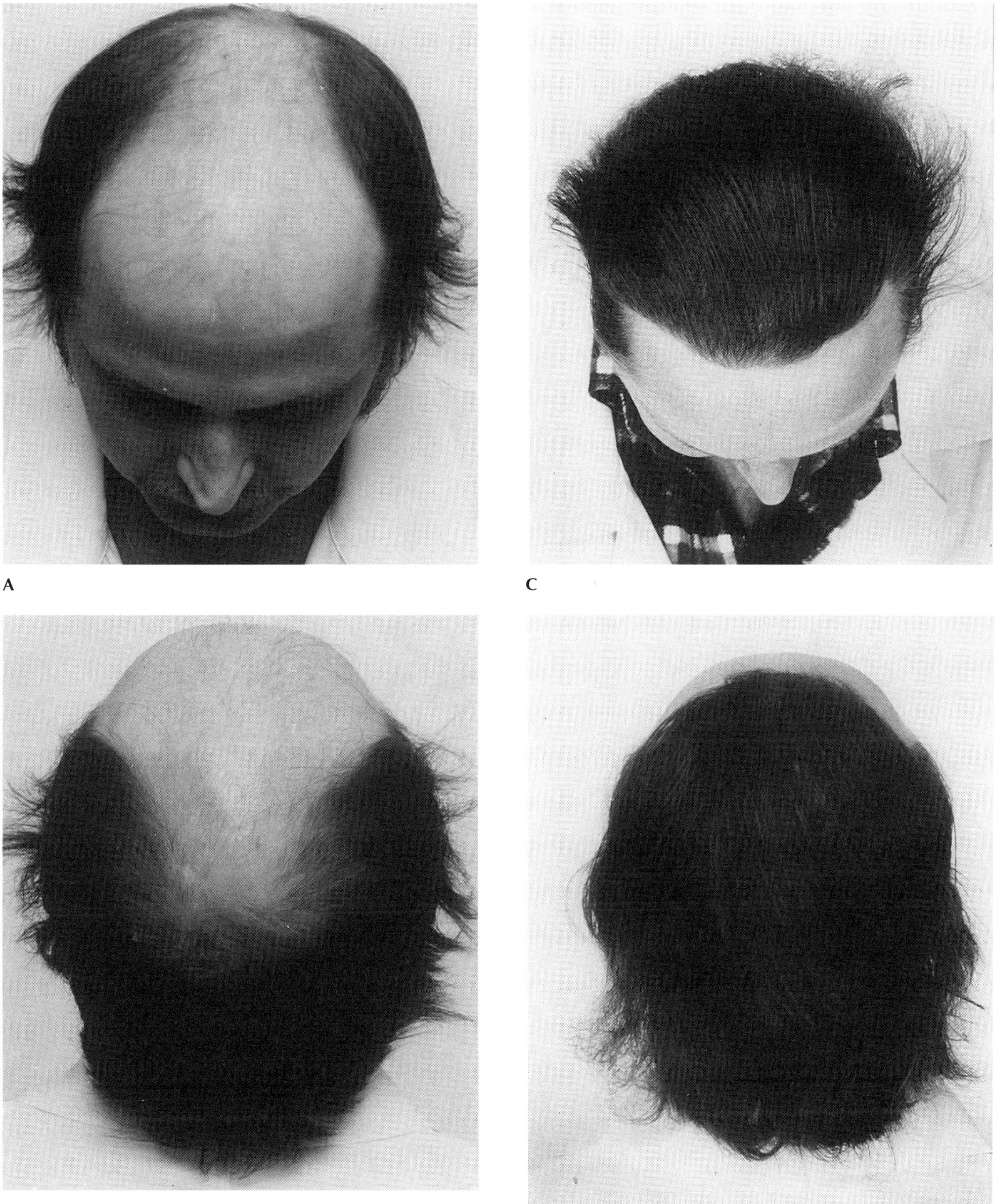

A C B D

FIGURE 13-19

(A,B) Another typical male pattern baldness patient. (C,D) Patient after surgery. (Reprinted with permission from Nordström REA. Tissue expanders and flaps for surgical correction of male pattern baldness. Br J Plast Surg *1988;41:154–159.*

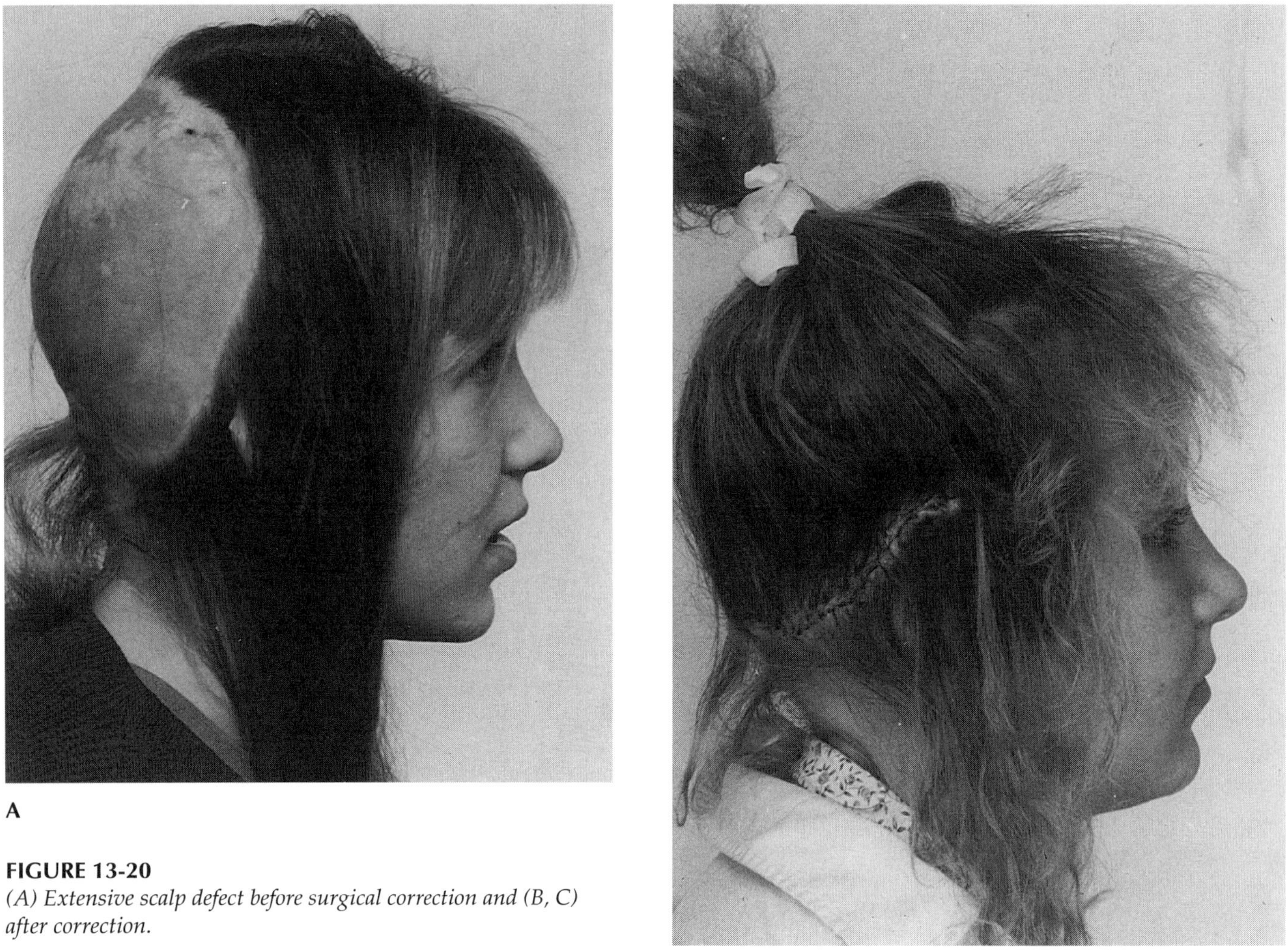

A

B

FIGURE 13-20
(A) Extensive scalp defect before surgical correction and (B, C) after correction.

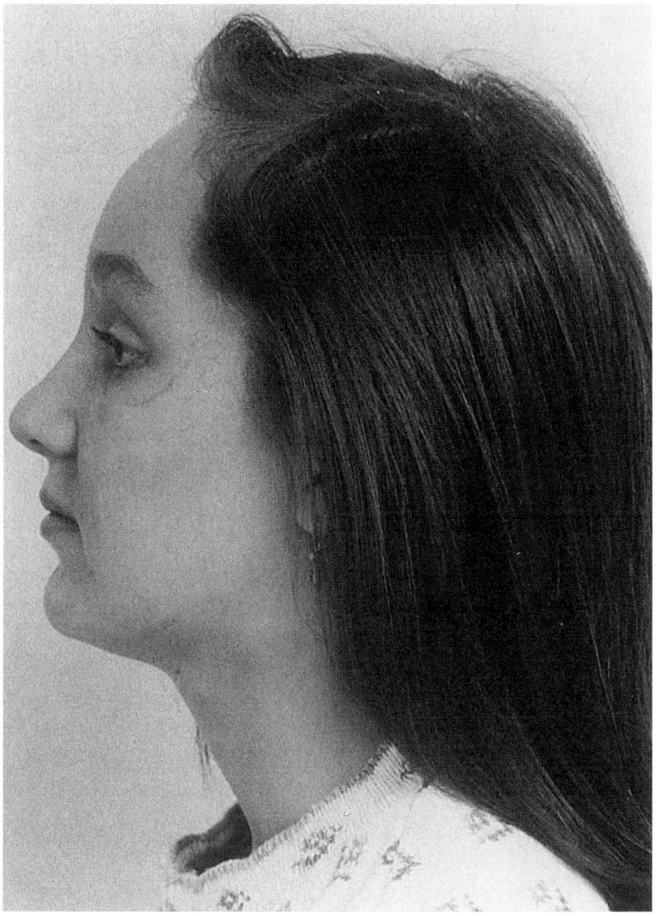

C

REFERENCES

1. Nordström REA. Scalp, hair, baldness and surgery. *Fac Plast Surg* 1985;2(3):173–177.
2. Nordström REA. Hair transplantation: The use of hair-bearing compound grafts for correction of alopecia due to chronic discoid lupus erythematosus, traumatic alopecia, and male pattern baldness. *Scand J Plast Reconstr Surg* 1976; suppl. 14 (Thesis and academic dissertation).
3. Nordström REA. Synchronous balding of scalp and hair-bearing grafts of scalp transplanted to the skin of the arm in male pattern baldness. *Acta Derm Venereol* 1979;59: 266–268.
4. Nordström REA. Antibiotics into the tissue expander to decrease the rate of infection. *Plast Reconstr Surg* 1988;81: 137–138.
5. Nordström REA, Devine JW. Scalp stretching with a tissue expander for closure of scalp defects. *Plast Reconstr Surg* 1985;75:578–581.
6. Nordström REA. Tissue expanders and flaps for surgical correction of male pattern baldness. *Br J Plast Surg* 1988;41: 154–159.
7. Norwood OT. Male pattern baldness: Classification and incidence. *South Med J* 1975;68:1359.
8. Nordström REA. The Nordström System: A specially designed tissue expander combined with the Nordström variety of the temporo-parieto-occipital flap and advancement flaps. In Unger WP, Nordström REA (eds.), *Hair transplantation*, 2nd ed. New York: Marcel Dekker, Inc., 1987, pp. 551–561.
9. Nordström REA. The Nordström variety of a temporo-parieto-occipital flap. In Unger WP, Nordström REA (eds.), *Hair transplantation*, 2nd ed. New York: Marcel Dekker, Inc., 1987, pp. 645–655.
10. Nordström REA. One variety of a temporo-parieto-occipital flap. *J Derm Surg Oncol* 1988;14:755–761.
11. Mayer TG, Fleming RW. Flaps in hair replacement. *J Derm Surg Oncol* 1986;12:782.
12. Juri J. The use of temporo-parieto-occipital flaps in the surgical treatment of baldness. *Plast Reconstr Surg* 1975;4: 456–460.
13. Nordström REA. "Micrografts" for improvement of the frontal hairline after hair transplantation. *Aesth Plast Surg* 1981;5:97.

14

Trunk

GUY MAGALON

The trunk can be divided into three different anatomical regions : abdomen, sides, and back. Each region displays particularities that must be taken into account for skin expansion.

1. Abdomen. The abdomen has two anatomical features that can impair tissue expansion, namely: (1) the umbilicus creates a fixed median line that must not be displaced and limits upward expansion, and (2) there is no underlying bone support for the expanders. Because of these features, expansion is very slow and patients should be properly informed (Figure 14-1).

2. Sides. The sides lend themselves better to tissue expansion because the bottom of the rib cage and the crests of the pelvis provide crucial support (Figure 14-2).

3. Back. Although the back is the largest skin surface of the body, expansion is not without problems. Undermining along the spinal column is difficult and can be hemorrhagic. Large volume expanders can be used but, as filling progresses, their weight pulls the skin downward. The periprosthetic capsule can thicken to such an extent that expansion is prevented after only a few weeks.

On the other hand, expansion of the latissimus dorsi muscle and microsurgery offer a new alternative for coverage of large areas. The expanded latissimus dorsi can be used as an isolated flap, a free musculocutaneous flap, or a local flap. Closure at the donor site is always easy. Section of the motor nerve may be useful to facilitate placement of the muscle and lessen retraction after section of the pedicle (Figure 14-3).

INDICATIONS

The routine indications for skin expansion anywhere on the trunk are removal of scars and nevi. However, this technique is not a panacea for giant hairy nevi. Even with serial expansion complete excision is often impossible, especially if the nevus is irregularly shaped and involves more than one region of the trunk. It is necessary to wait for 6 to 12 months between each course of expansion.

In addition to these routine applications, there are several specific applications for skin expansion in each region of the trunk, namely, (1) on the abdomen, it can be useful to reconstruct the abdominal wall after congenital or acquired evisceration; (2) on the sides, expansion allows revision of scars at the donor site of thoracic flaps used to cover defects on the upper extremities; and (3) on the back, expansion can be used to cover defects resulting from myelomeningoceles.

With regard to indications, it should be emphasized that skin expansion can be used for secondary reconstruction. For example, in a patient with melanoma, the lesion should be excised immediately and the defect covered with a mesh graft. If there is no recurrence, skin expansion can be used to replace the mesh graft. Similarly, in patients with open wounds, the defect can be closed with a thin skin graft and then skin expansion can be carried out later to correct the cosmetic result.

CONCLUSION

Tissue expansion opens new possibilities for management of defects on the trunk and deserves a prominent place in the therapeutic arsenal of plastic surgery.

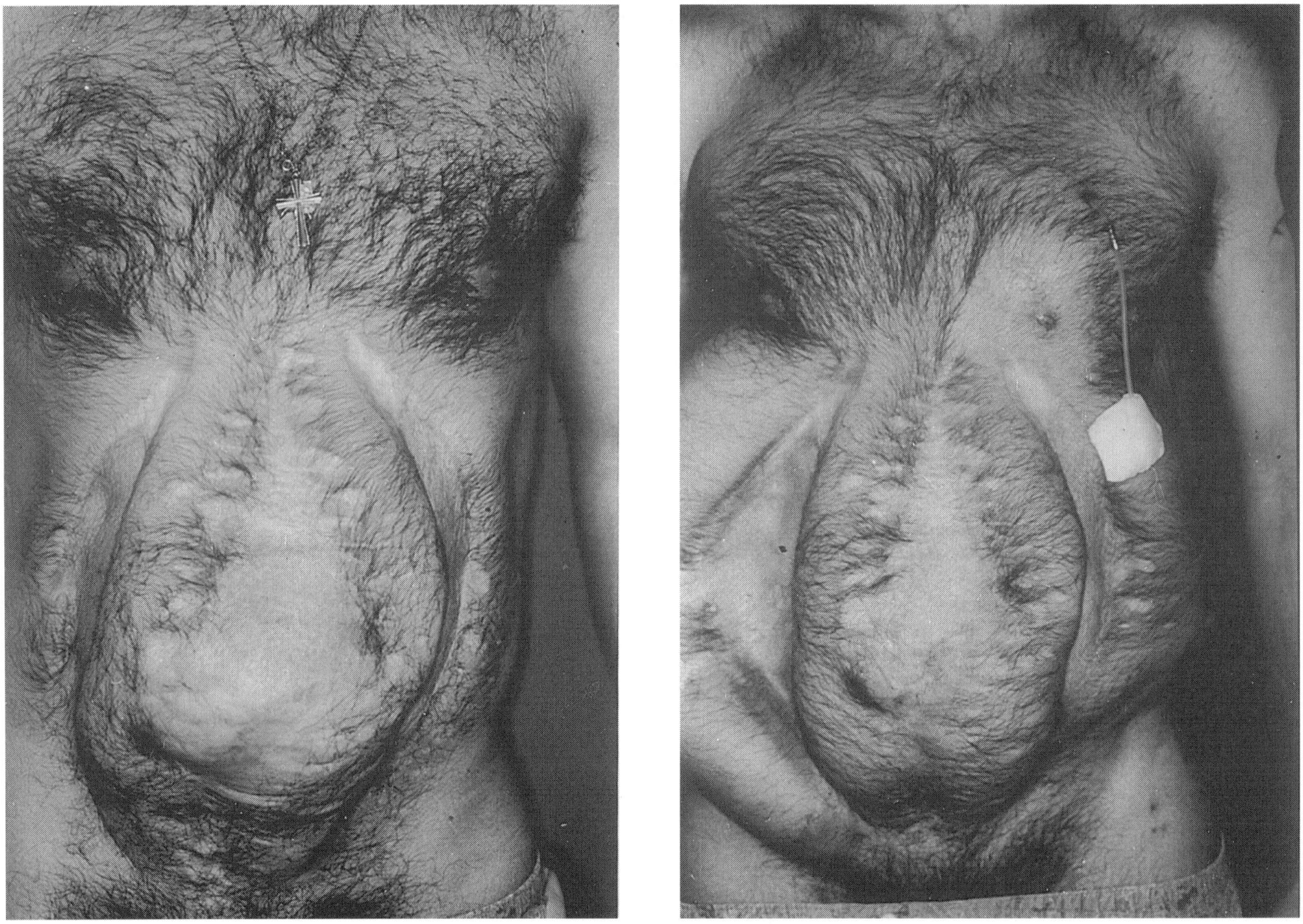

A B

FIGURE 14-1
Skin defect on abdomen. (A) Frontal view. (B) View during filling phase. (C) Late result, frontal view.

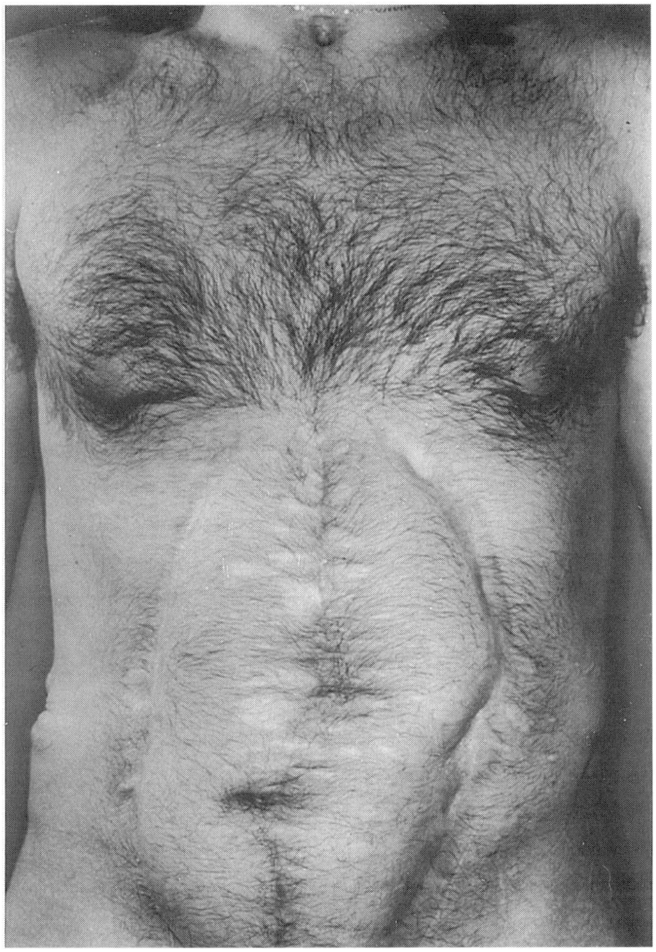

C

FIGURE 14-1 *(Continued)*

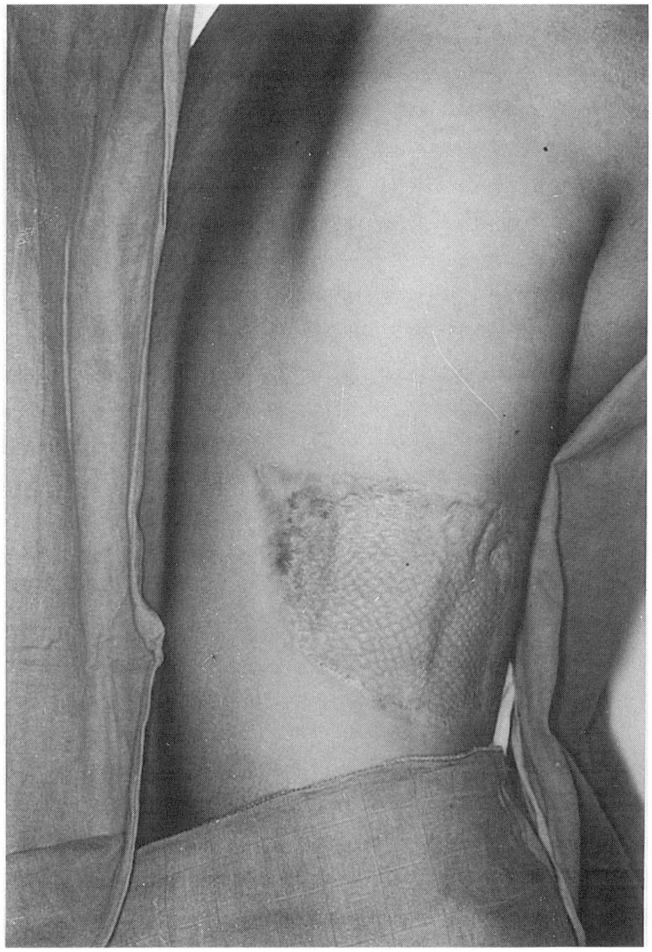

A

FIGURE 14-2
(A) Skin graft on right side of chest. (B) View at the end of filling phase. (C) Late result.

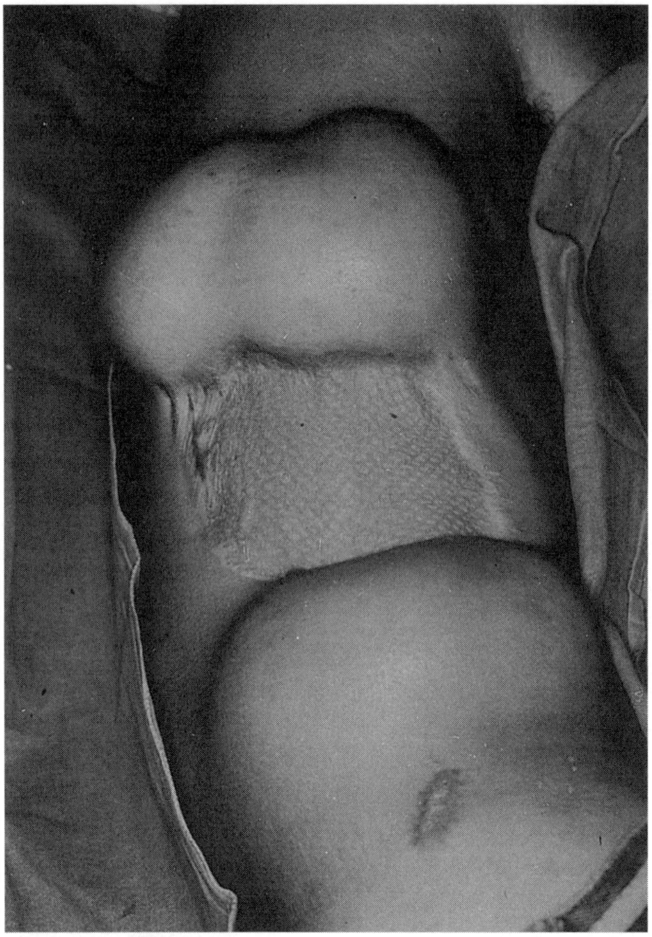

B

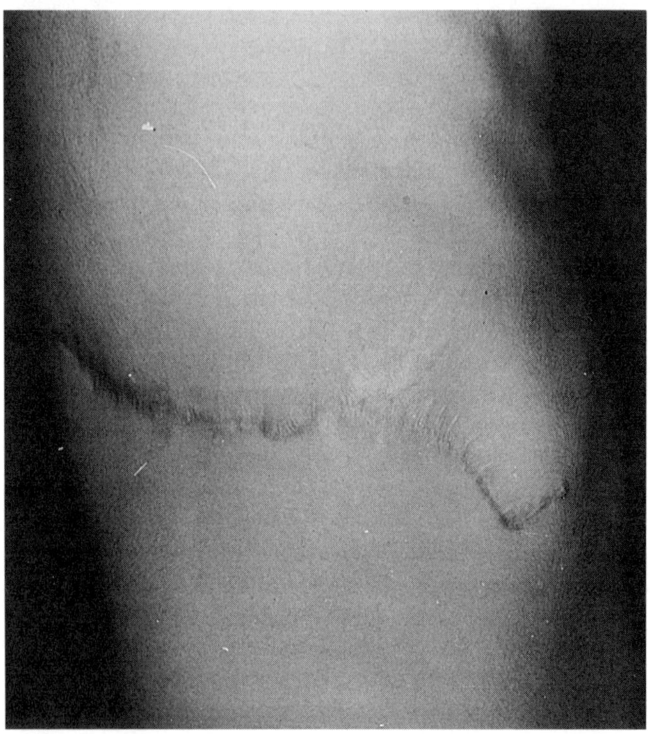

C

FIGURE 14-2 *(Continued)*

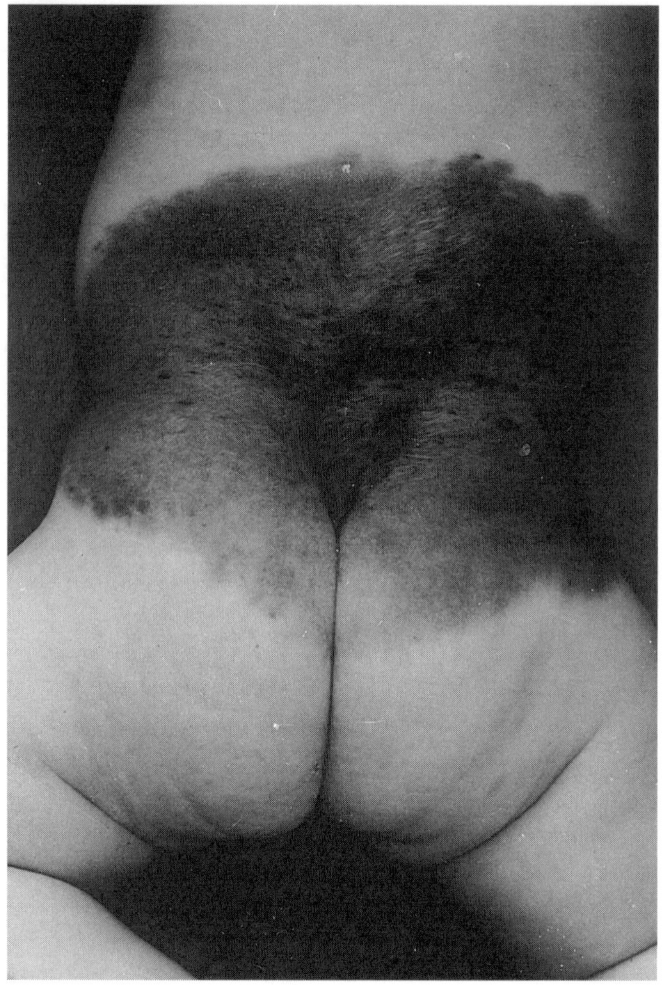

A

FIGURE 14-3
Pigmented nevus in lumbosacral region. (A) Initial view. (B) View at the end of filling phase. (C) Late result.

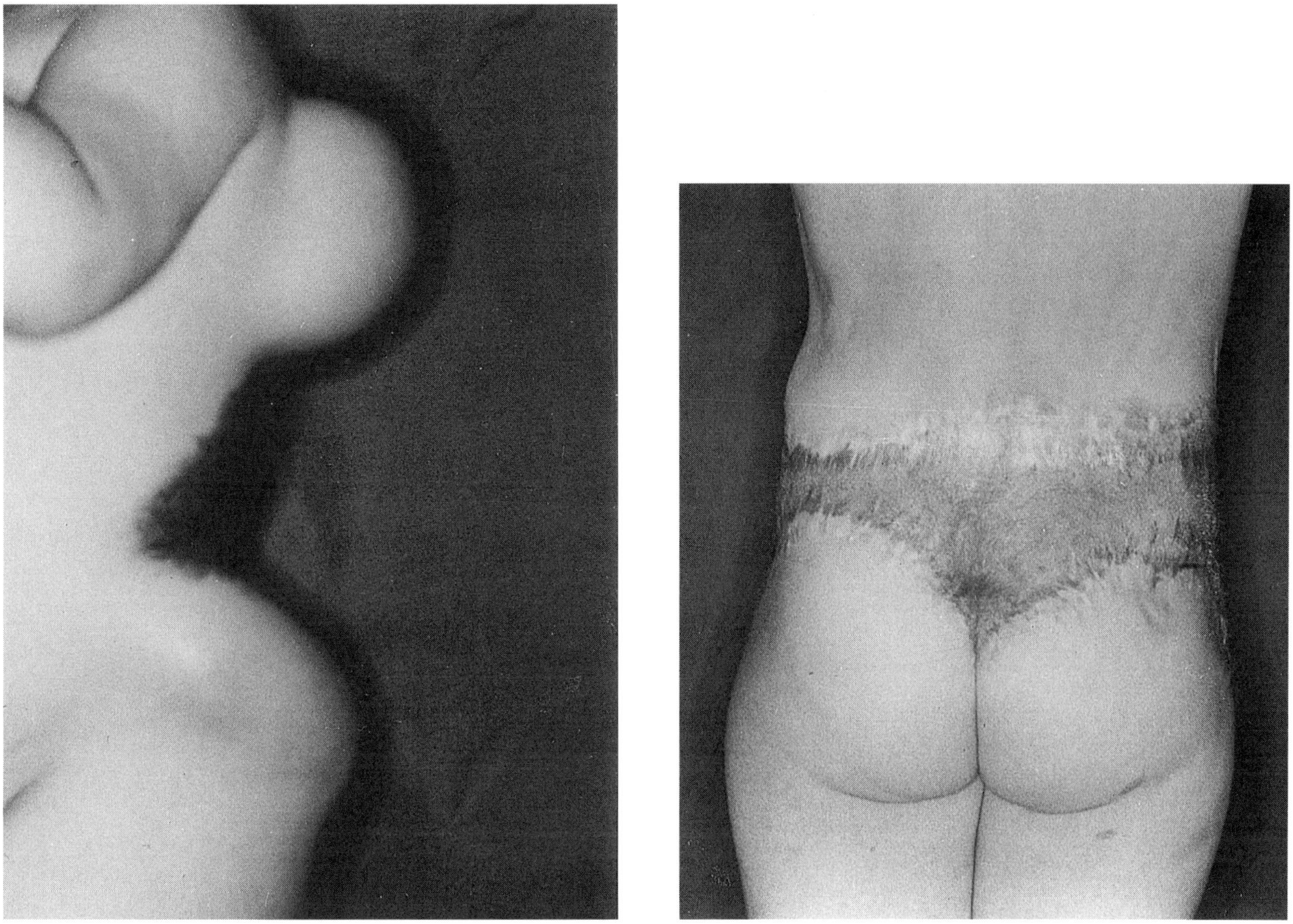

B

C

FIGURE 14-3 *(Continued)*

15

Extremities

GUY MAGALON

Management of skin defects on the extremities is always difficult. A wide range of reconstructive techniques is applicable—including skin grafts, local flaps, remote flaps, and free flaps. A recent addition to this therapeutic arsenal is skin expansion, which not only provides color, texture, sensory, and hair-bearing qualities but also reduces donor site scarring. Successful reconstruction and excellent results have been achieved on the extremities despite the limited surface available for expansion.

In using expansion on the extremities, three basic guidelines should be borne in mind: (1) the amount of useful skin decreases proximally to distally, (2) the skin dividend is always greater crosswise than lengthwise. Gains are easier to achieve circumferentially than longitudinally, and (3) in some cases it is necessary to work in both directions by using multiple expanders.

TECHNIQUE

The surgical technique used for skin expansion of the extremities is basically the same as in other locations. However, a few points should be emphasized. External valves are often easier to use than internal valves. Filling is usually slower, often requiring more than 3 months. Reconstruction should be achieved by simple advancement because complex flaps increase the risk of postoperative complications. The extremity should be kept in a cast until healing is complete.

Some areas of the extremities are not suitable for skin expansion, for example, the hand and foot, especially if the whole surface of one side is involved. An alternative technique that can be used in these cases consists of total replacement by expanded full-thickness skin grafts from the groin fold. Expansion of the inguinal region requires about 3 weeks. The graft is cut to fit the total defect, prepared according to standard procedures, and placed on the hand with a pledget, which should be left in place for 2 weeks. The wound at the donor site is easily closed with the remaining expanded skin.

Concerning the lower extremities, a noteworthy technique involves the expansion on the medial side of the leg to allow a cross leg flap and the closure by direct suture of the donor site.

INDICATIONS

Skin expansion is now used to treat a variety of indications. Similarly, a wide range of applications have been reported in the literature.

Scars

Two situations can be encountered in association with scars. If the problem is simply cosmetic, tissue expansion may be performed directly to remove the defect. However, if there is a functional problem, it may be necessary to perform tenolysis. The expanded skin flap lined with the collagenic capsule is an excellent sliding plane. If there is a neurologic problem, neurolysis should be performed when the expander is removed. Nerve grafts should be performed a few weeks after reconstruction with expanded skin.

Tumors
Expansion can only be used for benign tumors (Figure 15-1). The most common indications are pigmented nevi, which can often be removed from a segment of the arm in a single course of expansion. Giant hairy nevi are more difficult and often necessitate repetitive expansion and the use of other plastic surgery techniques.

Tattoos
Tattoos can be removed using two expanders, one on each side of the lesion, or a single large expander as described by Fenton.[1] The goal is to leave a straight scar (Figure 15-2).

Syndactylism
Since conventional treatment of syndactylism is simple and skin expansion is complex, the latter method will probably not be routinely adopted. However, skin expansion can be of adjuvant value in cases involving bone fusion, such as Apert's syndrome.

Stumps
After amputation, stumps may present defects due to the poor quality of the skin or the use of a mesh skin graft. Fitting patients with a prosthesis is difficult or impossible. In some cases, the prosthesis may even aggravate the problem. In some cases, skin expansion can be successful in resurfacing the stump with sensate skin, thus eliminating the need for shortening the stump to allow prosthetic fitting (Figure 15-3).

CONCLUSION

Our experience shows that if the usual guidelines for plastic and reconstructive surgery are followed, tissue expansion on the extremities is not only useful but, in some cases, is the only alternative. The surgeon should always consider this technique in his or her decision making.

REFERENCES

1. Fenton O. Expanding possibilities. *Br Med J* 1987;295: 684–693.

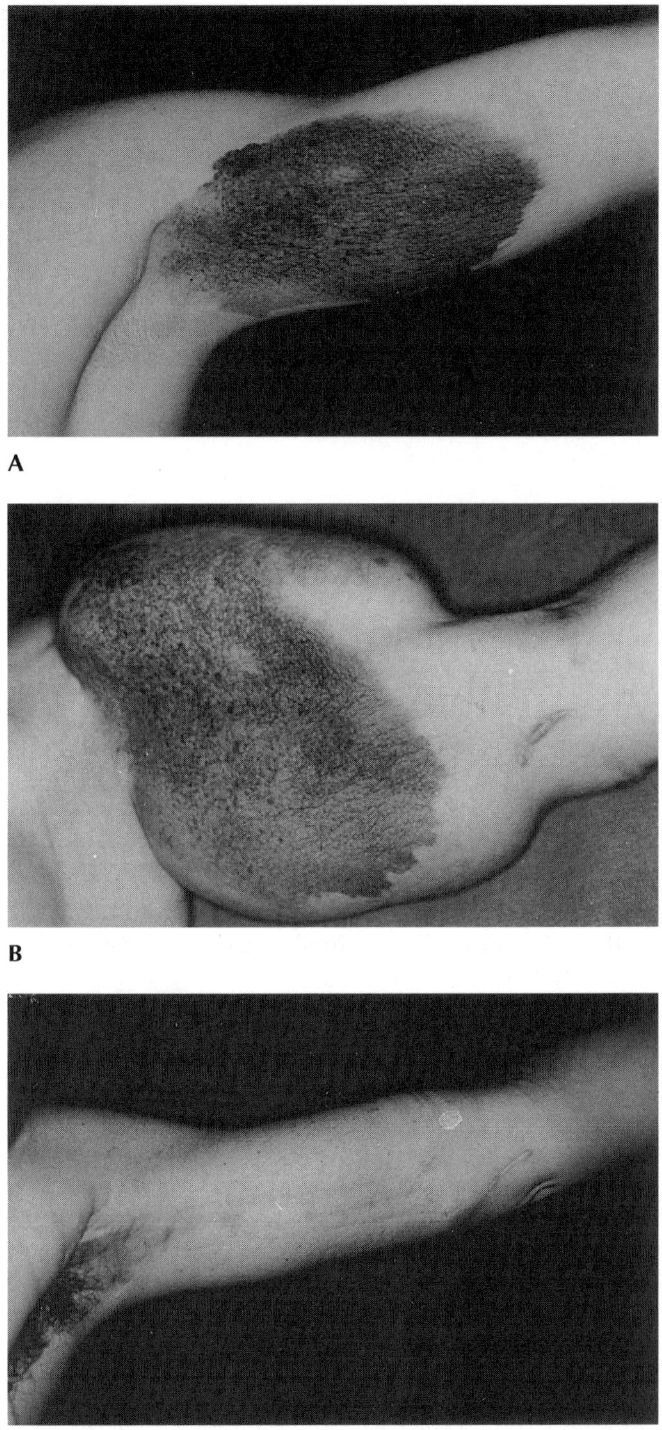

A

B

C

FIGURE 15-1
(A) Nevus on medial side of left arm. (B) View at the end of filling phase. (C) Late result.

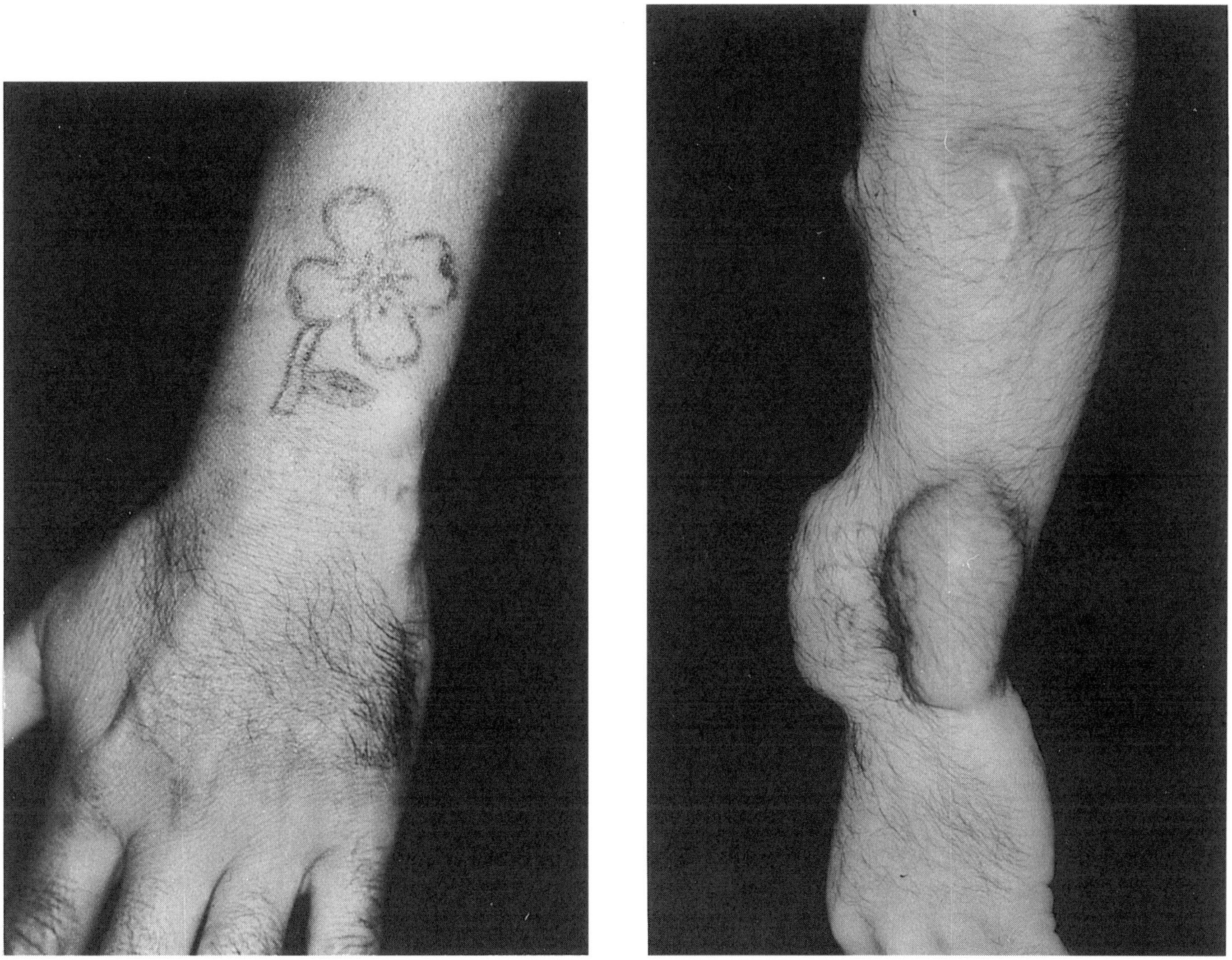

A B

FIGURE 15-2
(A) Tattoo on dorsal side of left wrist. (B) View at the end of filling phase. (C) Late result.

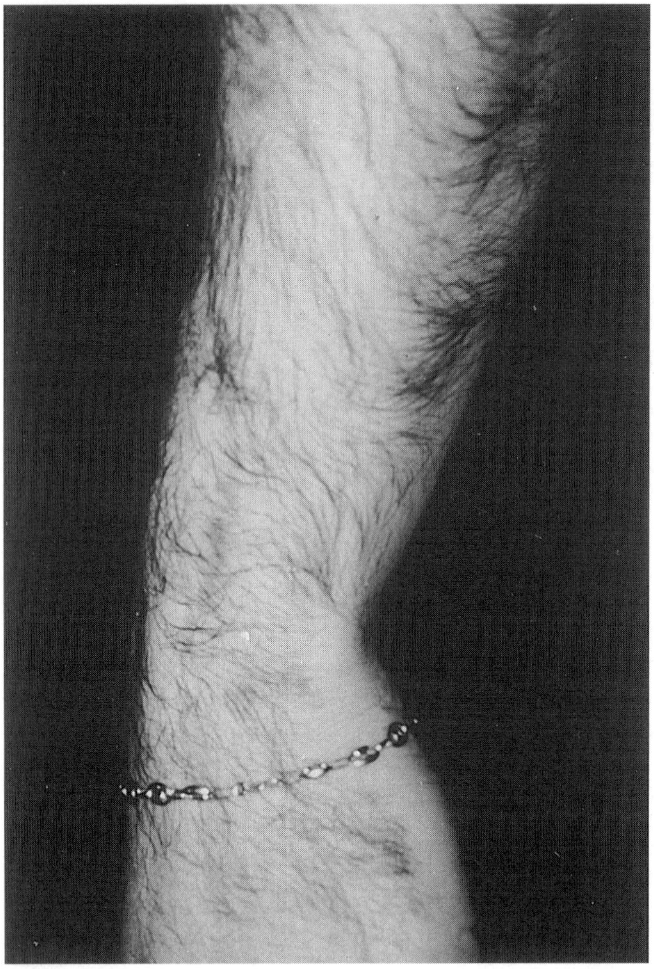

C

FIGURE 15-2 *(Continued)*

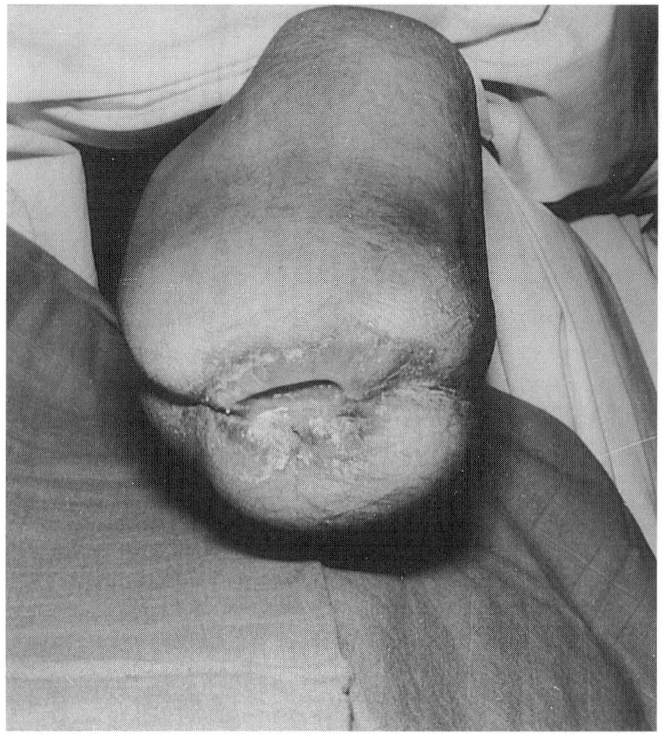

A

FIGURE 15-3
(A) Unstable scar after amputation of left lower leg, frontal view of stump. (B) View during filling phase. (C) Immediate result. (D) Late result.

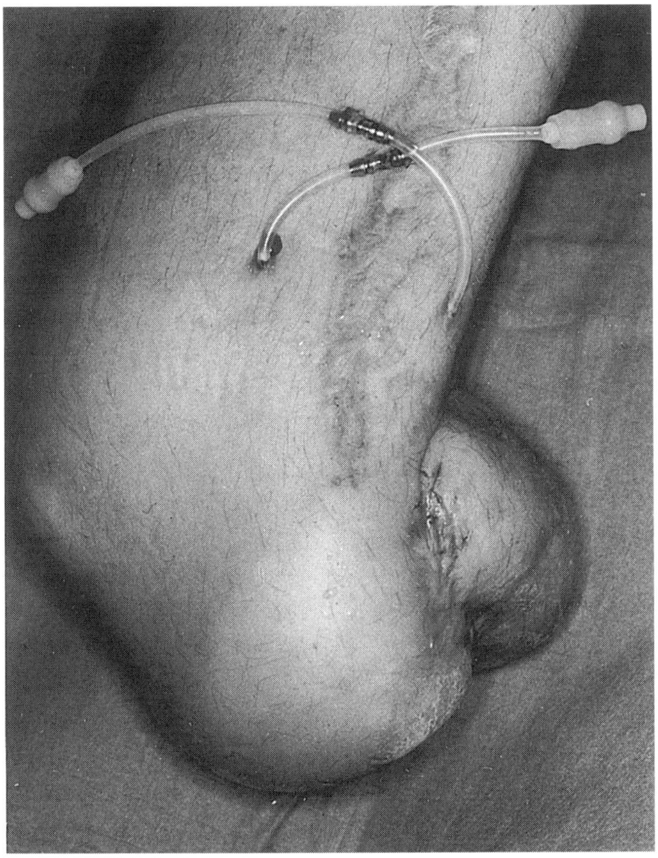

B

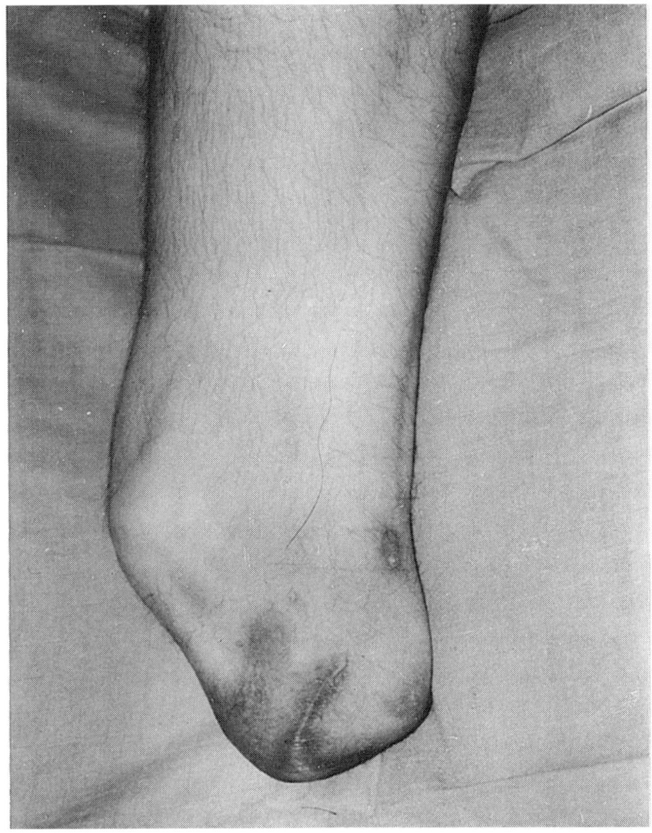

C

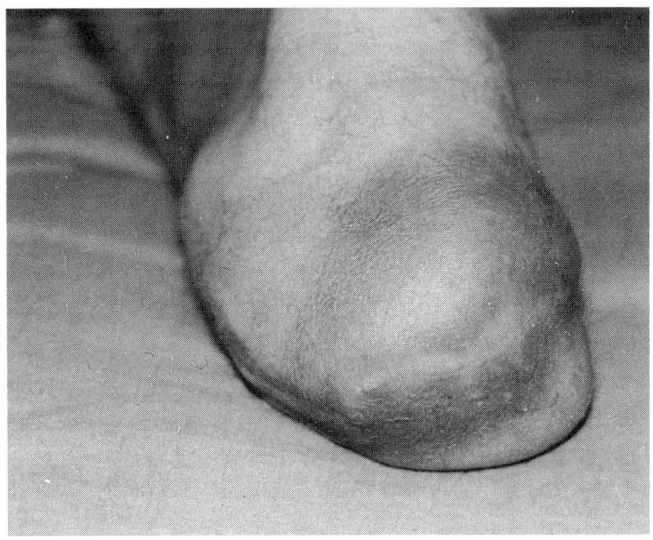

D

FIGURE 15-3 *(Continued)*

16

Genitalia

ANDREW G. BATCHELOR

Tissue expansion is at its best when used to resurface a defined defect, the geometry of which is maintained by surrounding structures. It is particularly useful for expanding a limited donor site to appropriately reconstruct a specialized surface with suitable tissue while minimizing donor morbidity.

Tissue expansion appears to have little direct application to the reconstruction of the male genitalia. In part, this is due to the absence of a rigid skeleton and the imperative for distensibility in the penis, together with its covering of skin. The unsatisfactory aspects of penile reconstruction that currently exist relate to erectile and sensory function, and it is unlikely that tissue expansion will make a significant contribution toward solving these difficulties.

On the other hand, tissue expansion is useful for reconstruction of a neovagina in case of congenital absence or for replacement of the vagina following surgical ablation or other trauma. The vagina is lined by specialized epithelium; furthermore, it is moist, hairless, very elastic, and sensitive to physiological hormonal changes. The cavity that the vagina occupies is enclosed by the other structures of the pelvis; thus, any reconstruction or new construction can be secured to remnants of the mullerian duct or surrounding structures.

The natural donor site for vaginal lining is clearly remaining vaginal lining or, in its absence, vulval skin from between the hair-bearing portion of the labia majora. The availability of this donor site is limited. The women of some South African tribes habitually stretch their labia minora by secret and unknown means to attain labia several centimeters longer than usual. This naturally suggests the application of tissue expansion of this area to provide a donor site sufficient to achieve complete vaginal lining.

VAGINAL RECONSTRUCTION

Multiple methods of vaginal reconstruction have been described in the past. The simplest, and most interesting from the point of view of tissue expansion, is the nonsurgical method in which a vaginal remnant or introital dimple is progressively deepened by the repeated forceful use of external dilators[1]. This is a variant of tissue expansion, but practical successes are rare if more than apocryphal.

The Williams operation[2] is designed to create a pseudovaginal pouch from vulval skin by uniting the margins of the posterior portions of the labia minora in the midline. Although the pouch is neither sufficiently large nor sufficiently supported to accurately mimic the vagina, the result is acceptable to some patients.

All the other techniques are variations on a theme, the theme being the creation of a tunnel between the rectum and bladder by blunt dissection and the lining of this cavity to maintain it. A wide variety of lining methods—each with its advantages and drawbacks—has been employed. Free grafts using split skin,[3] homograft amnion,[4] and bowel have been described. Local flaps of various designs that use the vulva and distant pedicle flaps from the thigh (gracilis),[5] groin (posterior labial artery),[6] abdomen (rectus abdominis),[7] and gut (colon or jejunum),[8] have also been described.

Even free vascularized tissue transfer from the scapular region has been used. None of these procedures is ideal.

By far the most widely used method for vaginal reconstruction has been skin grafting. The major problem with this is that the grafts take poorly and this is either associated with or followed by graft contracture.[9] Following reconstruction, the patient must maintain the new vagina by regular self-dilatation with an appropriate prosthesis several times a day. This is a demanding process, even for the highly motivated patient.

Because of the difficulty in maintaining graft vaginas, interest has increased in flap reconstructions. Ideally a flap reconstruction should reliably provide, in one stage, a stable vagina that does not shrink or require complex maneuvers to maintain it. The vagina should be elastic and moist. Flaps that use the gracilis are unreliable. A problem is that all skin flaps are dry and potentially hair-bearing. Furthermore, the donor defect should be invisible or minimal. Skin-flap donor sites are all visible—the least visible being the transverse rectus abdominis. Gut transfers overcome these problems but require laparotomy and gut resection with its attendant potential major morbidity. Vaginas that have been lined with gut are moist but produce profuse mucous discharge at all times—an unacceptable result for most patients. Although there is no visible donor site in this case, the laparotomy scar is visible.

By contrast, local flaps from the vulva have the advantage of being of quality similar to vaginal tissue—moist and elastic—and the resulting donor site is concealed. However, it is difficult to obtain enough tissue for total vaginal construction except in some patients with the adrenogenital syndrome. This deficiency can be overcome using tissue expansion of the labia. This technique provides the most satisfactory solution to vaginal absence currently available.

Patient Selection

Thus far, all the patients treated by this technique had some degree of congenital mullerian duct agenesis, ranging from vaginal atresia to Rokitansky-Küster-Hauser syndrome with complete absence of vagina and uterus. These patients are karyotypically and phenotypically female. However, it is feasible and indeed planned to construct vaginas in individuals with testicular feminization because the technique is applicable to all candidates who are phenotypically female. The technique could also be applied to transsexuals, although they lack the vaginal-like tissues.

Reconstruction following surgical excision is also feasible if the vulva remains, which is rare. On the other hand reconstruction following radiotherapy is commonly performed and has the advantage that the vulva is usually spared following intracavity radiation, which is routinely used for carcinoma of the cervix. Tissue expansion is relatively contraindicated when the donor site has received significant radiation.

These are all practical considerations related to surgical techniques. Selection of patients from a psychological point of view is more difficult.

Patient Counseling

Almost all our patients have been teenage girls. Diagnosis and referral are often drawn out and complicated because this condition is rare. The patient and family often will have received complicated, conflicting, and inaccurate advice prior to being seen. Because tissue expansion has a significant complication rate and is not a simple, one-stage procedure, it is very important to have full informed consent and cooperation from the patient. The procedure generally requires at least 5 weeks of hospitalization, and the patient experiences considerable discomfort and limitation of mobility, particularly toward the end of expansion. This occurs on a background of psychosexual difficulties, doubts about female identity, and frequently, knowledge that child birth will not be possible. It is essential to repeat full explanations and to involve the patient in decision making. Also, it is helpful to introduce prospective patients to those who have completed treatment.

Investigation

Patients who lack a vagina frequently have coexisting renal tract anomalies.[10] This should be investigated by ultrasound and intravenous urogram. Patients with vaginal atresia and a functioning uterus may present with hematocolpos. In this situation, ultrasound examination of the pelvis will accurately define the proximal anatomy. Patients who present before menarche or who lack functioning endometrium should undergo laparoscopy. If tubo-ovarian structures are not identified, then chromosomal and hormonal studies will define those patients who are karyotypical male despite their female phenotype.

Operative Management

This technique is accomplished in three stages, as is usual for tissue expansion. The first stage is the insertion of the expanders. The second stage is the progressive filling of the expanders. In the third stage, a second operation is performed to construct the vagina with the flaps developed by expansion.

Expander Insertion

This procedure is performed under general anesthesia on an inpatient basis. We recommend using broad spectrum antibiotics as a prophylactic measure. For this procedure, the patient is positioned in the Lloyd-Davies position.

Bilateral incisions are made in the groin crease just medial to the femoral vessels. The incision is deepened through the superficial fascia and a tunnel is made medially with a finger until the fingertip underlies the labia on that side. The finger tip then disrupts the superficial fascia in that area so that a pocket can be developed under the medial surface of the labia majora and the labia minora. Care should be taken that the pocket is not too lateral because the expander will then expand hair-bearing skin. It is also important not to tear the labia minora since this tissue is thin and delicate. Bleeding is not a problem because this is a blunt dissection and is very straightforward to perform.

Two 80 to 100 mL round or rectangular expanders are used. Care must be taken when inserting the expanders to fold them neatly such that they will unfold smoothly with few wrinkles. Folding is essential because the cavity is small and little fluid can be introduced at this stage. We use distant external valves, but internal injection domes would do as well now that effective topical local anesthetic is available. If using buried valves, some methylene blue in the fluid helps to confirm the dome puncture in case a seroma develops.[11] If external valves are used, then a long track to distant emergence sites is necessary.

Expansion

Expansion can begin as soon as the area to be expanded is comfortable, usually 3 to 5 days postoperatively. Initially, expansion may be performed on an outpatient basis and, with external valves, patients can perform the expansions themselves. Some patients prefer to have this control and can achieve rapid expansion in this manner. Small volumes, starting at 3 to 5 mL, injected on a daily or alternate day basis, seem to produce the most efficient expansion. Although expansion is not notably painful in general, one of our patients experienced problems with pain.

The patient is readmitted to the hospital when mechanical difficulties caused by chafing of the two expanding labia occur. Regular monitoring and limited mobility are required to avoid pressure necrosis between the expanders. Lubrication of the area with vaseline and the interposition of vaseline gauze provide symptomatic help. At this point in the procedure, the volumes being introduced can be increased to 10 to 15 mL. In cases of slow or painful expansion, an infusion pump has helped. Expansion is completed when sufficient tissue has been produced—usually when expander volumes reach 100 to 120 mL. If one of the expanders is lost due to infection or exposure, then unilateral expansion can continue to provide enough tissue from the one side; however, this is not ideal.

Those patients who have undrained hematocolpos will probably have had further menstruation suppressed with drugs. As soon as this type of reconstruction is planned, drug suppression should be discontinued so that "auto-expansion" of the proximal vaginal cavity will proceed and tissue will develop at the cervix that will be useful in the second operation.

Vaginal Construction

As in the first stage, this procedure is performed under general anesthesia with antibiotic prophylaxis and with the patient in the Lloyd-Davies position. As soon as the patient is anesthetized, the expanders are forcefully expanded rapidly to two to three times their preoperative volume. This maneuver disrupts the expander capsule; it also appears to decrease the elastic recoil of the flaps and perhaps achieves a little more expansion, as is consistent with intraoperative expansion (see Chapter 7).

Bilateral flaps are designed to overlie the expanders within the nonhair-bearing area (Figure 16-1). These are posteriorly based and indeed lie adjacent at their bases. The flaps are cut full thickness into the expander cavity. The flaps are then reflected posteriorly exposing the introital area such that a tunnel can be created by blunt dissection between the urethra and bladder anteriorly and anal sphincter and rectum posteriorly (Figure 16-2). Some sharp dissection may be needed, particularly in the midline, but this should be used judiciously because perforation of a pelvic organ or brisk hemorrhage from pelvic veins is a risk. In the patient with uterine absence, this tunnel is blind ended but should extend to the pelvic peritoneum. Where a hematocolpos exists, it should be exposed and an appropriate flap designed on it so that its lining can contribute to the reconstruction (Figure 16-2).

The two flaps are then sutured together with an absorbable stitch to form a tube prior to their vertical transposition into the neovaginal tunnel. In vaginal atresia patients, it is easier to unite what will become the posterior vaginal suture line only prior to suturing the flaps to the vaginal vault or cervix. A little imagination in this inset will avoid directly circular anastomosis (Figure 16-3). When there is no uterus, the vaginal tube can be completed before it is inserted into the cavity. However, it has to be suspended in position to maintain length and avoid prolapse. This is accomplished by suturing the new vaginal vault to the tubo-ovarian remnants. To manage this safely, we feel a small Pfannenstiel's incision is necessary to avoid damage to an abdominal viscus.

Once the vagina is in place, the remaining wounds, including the donor defects, can be closed directly over

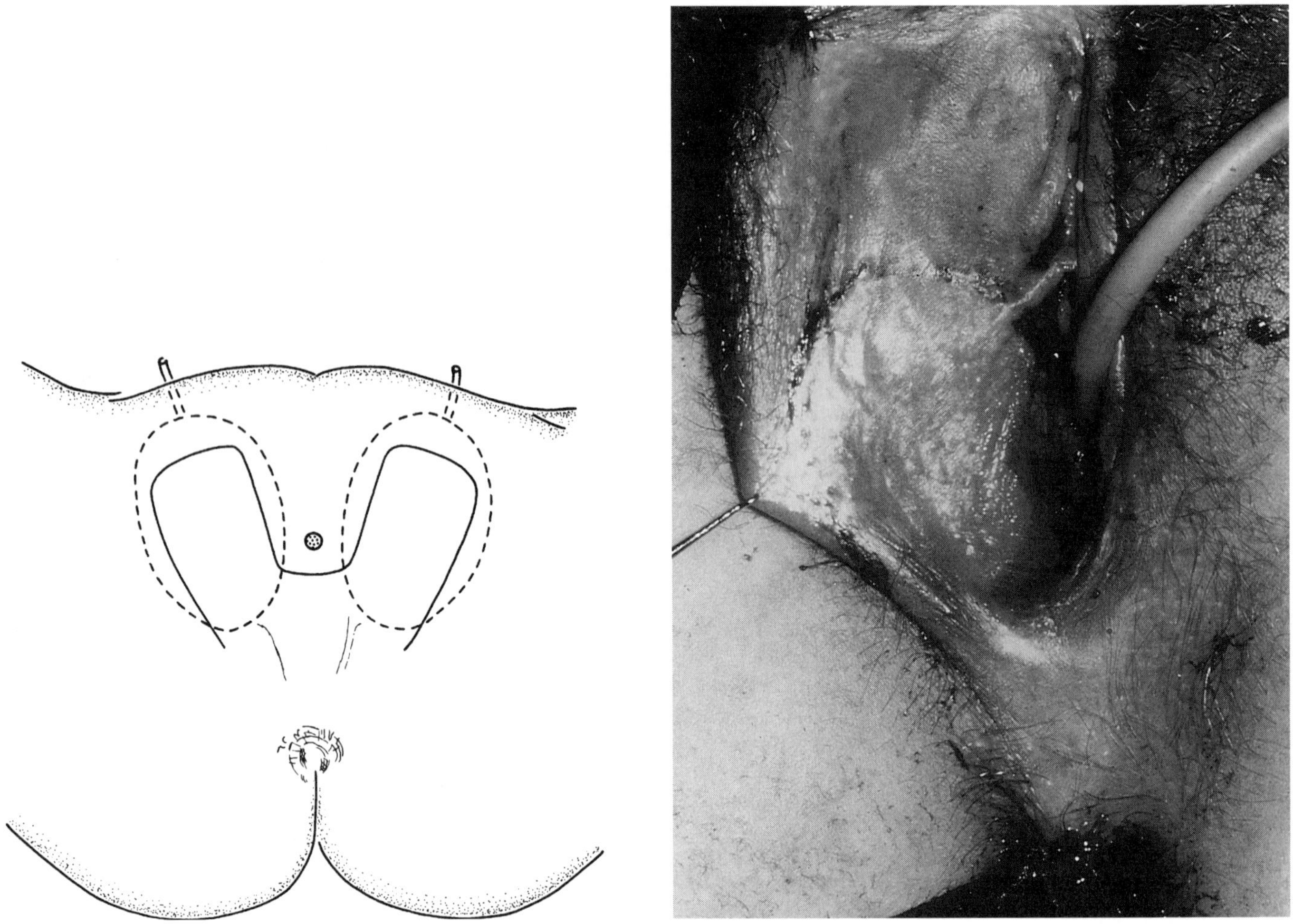

A B

FIGURE 16-1
(A) Diagram showing the position of the full expanders and the design of the flaps overlying them with reference to urethra and anus. (B) The available tissue in the hairless area following the removal of the expander.

bilateral suction drains. Urinary diversion via suprapubic catheter has been routine. No special postoperative measures are necessary, the drains and catheter being removed when appropriate. Postoperative stay is usually 4 to 7 days. Initially, we performed an examination under anesthesia at 7 days to ascertain complete flap survival. As no necrosis was observed, we have abandoned this practice.

Patients
We have commenced this procedure in 15 patients, all females aged 14 to 36 years, and all but one being teenagers. Eleven of these patients had no vagina or uterus but all had significant tubo-ovarian structures. Two patients presented with hematocolpos and had both uterus and cervix; one patient presented with hematometra, vaginal vault, and a dilated cervical canal that ended blindly without an external os, the last patient (age 36) presented with a long-segment lower-vaginal stenosis that ended in a pinhole closely associated with the urethra that was only visible under anesthesia. There were no renal tract anomalies although two patients had minor anomalies of the lumbosacral spine.

Complications
Two patients did not complete expansion because of exposure of the expanders early in the expansion

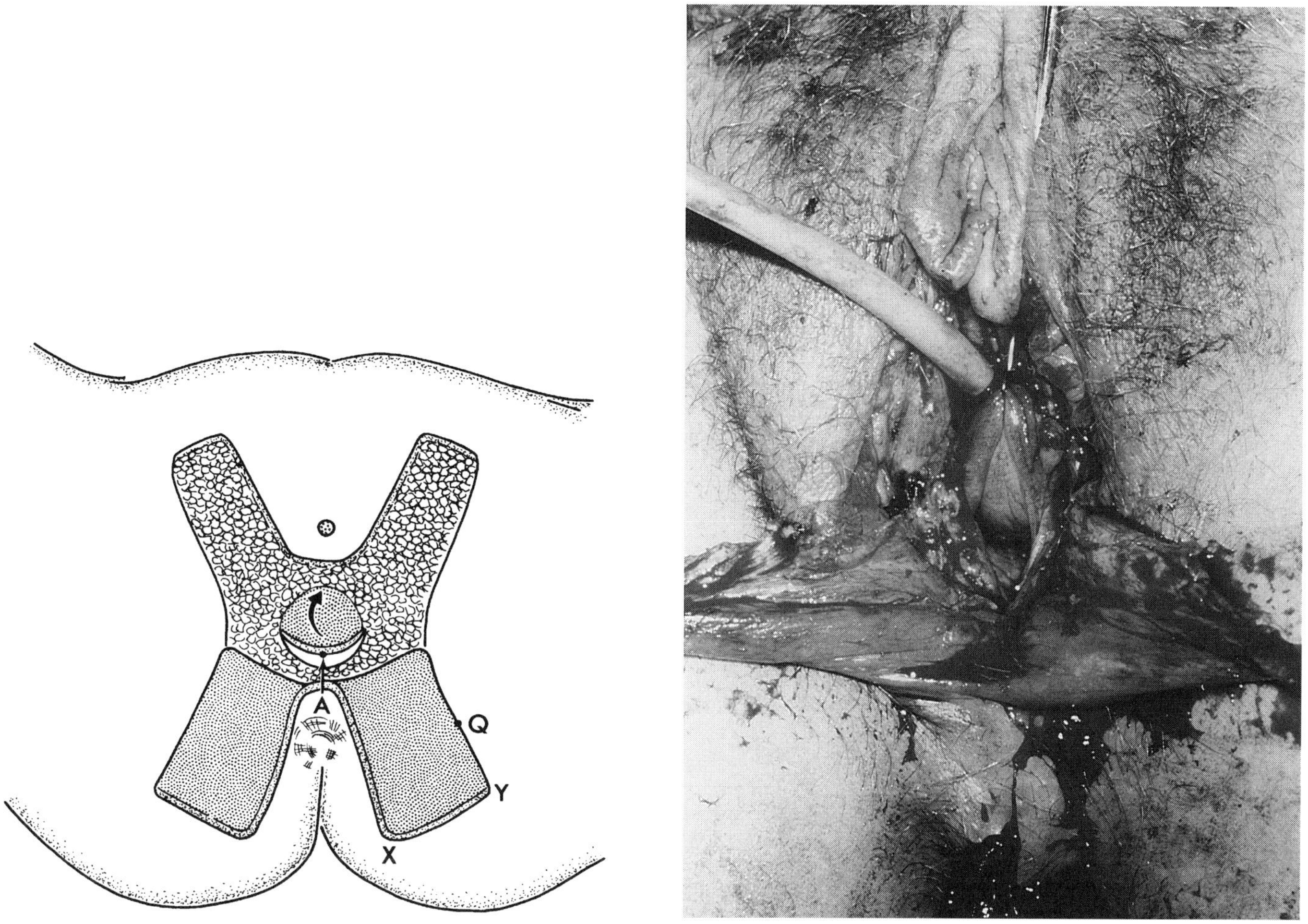

A B

FIGURE 16-2

(A) Schematic drawing showing reflected flaps and neovaginal tunnel in a patient with hematocolpos. A third flap (tip at A point A) has been cut from the vaginal vault and is transposed downward and forward (arrow) to interdigitate with the expanded flaps. (B). A photograph showing the flap inset having commenced.

period. This occurred early in the series when we were gaining experience both in placement of the expanders and management. At this time, expansion was performed entirely on an outpatient basis, and pressure necrosis occurred between the expanders at home. This has become less of a problem as we have readmitted the patients earlier and limited their mobility. We have found it more useful to keep patients in the hospital and perform the expansion more rapidly. One patient experienced expander exposure near the end of the expansion and this necessitated urgent completion of the reconstruction, which was carried out successfully.

The first patient reconstructed had presented with hematocolpos, which recurred after the second operation because the neovagina had not been attached successfully to the vaginal vault. At a subsequent operation, this was achieved without difficulty or further expansion.

One of the younger patients found that pain during expansion was a problem. This was overcome by hospital admission, analgesia, and the substitution of continuous expansion via infusion pump for intermittent expansion. Therefore, the patient completed reconstruction smoothly.

There were no specific perioperative complications of the expander insertion. However, after the reconstruction procedure, the patient developed a hematoma requiring a return to the operating room. Three

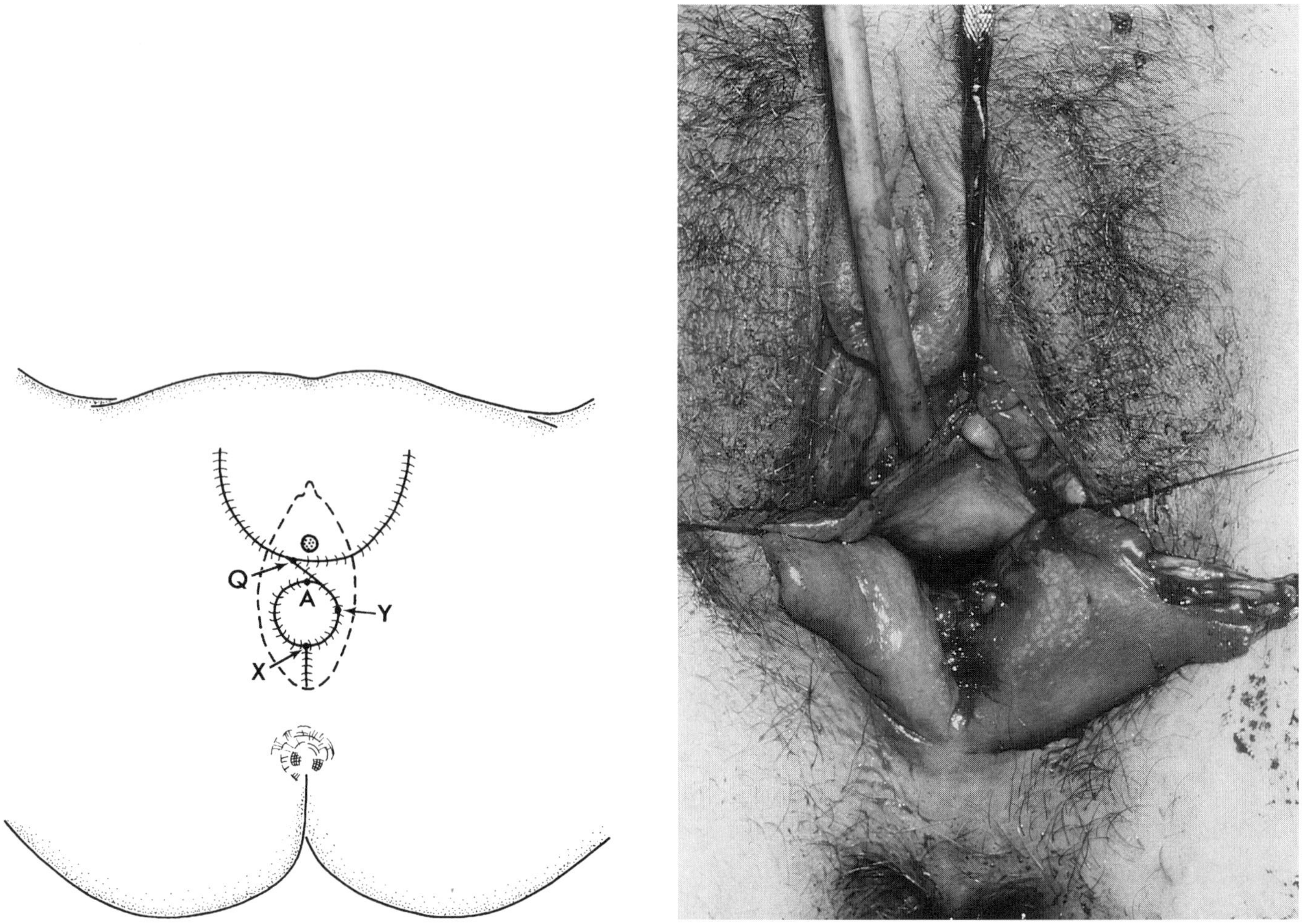

A B

FIGURE 16-3

(A) Schematic drawing showing the scheme of closure in a patient with hematocolpos, points A, Q, X, Y correspond to those in Figure 16–2a. (B). Near completion of the inset showing how flap A can be used to avoid a directly circular anastomosis.

patients developed minor lower urinary tract infections following the second operation.

Early in the series, one patient had a significant amount of hair transposed into the vagina because of poor flap design. This was successfully ablated using electrolysis.

Results

Of the 15 patients starting, 13 completed treatment. In all cases, vaginas of adequate dimensions were constructed. When healing is complete and the scars have settled, then two-finger, full-length vaginal examination is easily performed. No stents or molds have been used and the patients have not used dilatation. Several of the patients have become sexually active since completion of this procedure and report no difficulties. Most interestingly, some patients were inactive for more than a year following this procedure and then commenced or resumed intercourse without difficulty suggesting that shrinkage of the reconstructed vagina during periods without dilatation is not a problem.

The long-term appearance of the vulva is astonishingly normal—all dog-ears and irregularities simply disappear within weeks, leaving an unnoticeable donor site.

Patient satisfaction is high. They have all been counseled to use additional lubrication during intercourse, but some report that this is not necessary. We have counseled the patient with the neovagina sutured directly to the incompetent cervical canal that her partner should wear a condom because the prostaglandins present in semen would cause painful uterine contraction if the semen entered the uterus. Although pregnancy is technically possible in those patients who presented with hematocolpos, it has not yet happened, but previous reports after using other methods of reconstruction suggest that it may occur.

The two patients who experienced failure of the technique have been offered reexpansion, but neither has accepted. Both have chosen an alternative technique, and one of the patients has successfully completed reconstruction with posterior labial artery island flaps.

CONCLUSION

Preferred methods of vaginal reconstruction do not require complex maintenance by dilatation. The old concept of delaying surgery until marriage is imminent no longer applies. Surgery is appropriate earlier because sexual activity is not necessarily delayed until marriage. The expansion vaginoplasty appears to provide a solution to many of the problems associated with reconstruction and has the added advantage that appropriate tissue is being provided while leaving an inconspicuous donor site. It is the operation of choice.

Although there are complications, they are decreasing with time and experience. Some patients refuse this kind of manipulation or do not wish to continue the procedure in the event of difficulties. These patients have reconstructions performed with posterior labial artery island flaps if hair growth permits. If the patients have too much hair but are thin, then a rectus abdominis pedicle flap is used. If patients are too obese for this bulky flap to suffice, then a free tissue transfer from the thinner scapular area is employed. However, there is no doubt that in our experience the reconstructions of the best quality have been obtained by tissue expansion.

ACKNOWLEDGEMENTS

I should like to acknowledge the help and contribution of Professor Richard J. Lilford in cooperation with whom this technique has been developed and on whose patients these operations have been performed. The photographs and illustrations have been prepared by the Medical Illustration Dept. of St. James's University Hospital.

REFERENCES

1. Frank RT. The formation of an artificial vagina without operation. *Am J Obstetr Gynecol* 1938;35:1053–1058.
2. Williams EA. Congenital absence of the vagina: A simple operation for its relief. *J Obstetr Gynaecol Br Com* 1964;71:511–517.
3. McIndoe AH, Bannister JB. Operation for cure of congenital absence of vagina. *J Obstetr Br Emp* 1938;45:490–495.
4. Dhall K. Amnion graft for the treatment of congenital absence of the vagina. *Br J Obstetr Gynaecol* 1984;91:279–282.
5. Mathes SJ, Nahai F. *Clinical applications for muscle and musculocutaneous flaps.* St. Louis: C.V. Mosby, 1982, pp. 411–412.
6. Wee JTK, Joseph, VT. A new technique of vaginal reconstruction using neurovascular pudendal-thigh flaps: A preliminary report. *Plast Reconstr Surg* 1989;83:701–708.
7. Lilford RJ, Johnson N, Batchelor AG. A new operation for vaginal agenesis: Construction of a neo-vagina from a rectus abdominis musculo-cutaneous flap. *Br J Obstetr Gynaecol* 1989;96:1089–1094.
8. Baldwin JF. The formation of an artificial vagina by intestinal transplantation. *Ann Surg* 1904;40:398–402.
9. Hueston JT, McDonald LA. Congenital absence of the vagina. In CE Horton (ed.), *Plastic and reconstructive surgery of the genital area.* Boston: Little Brown & Co., 1973, pp. 423–424.
10. Fore SR, Hammond CB, Partan RT, Anderson EE. Urologic and genital anomalies in patients with congenital absence of the vagina. *Obstetr Gynecol* 1975;46:410–416.
11. Dickson MG, and Batchelor AG. A technical aid to tissue expansion. *Br J Plast Surg* 1987;40:305–308.

17

Breast Expansion Augmentation

HILTON BECKER

The fixed volume breast implant has been used in breast augmentation for almost 30 years.[1,2]

In breast reconstruction, however, where the tissue expander is more commonly used, the advantages of expandability and postoperative volume adjustment have become well recognized.[3–5]

The use of the expander implant in breast augmentation has been shown to have many advantages over standard breast implants: The expander implant can be inserted through a smaller incision than that needed by a gel implant because it is inserted empty. Once in position, the implant can be overexpanded. This stretches the pocket, assisting in the identification of tethering bands and thus facilitating dissection. The volume is then reduced to the desired extent (Figure 17-1). As the total volume does not need to be placed in the implant at the time of surgery, there is less tension on the incision and less postoperative pain. The adjustable implant also avoids the fear of selecting the wrong size, especially in cases of asymmetry and chest-wall deformities. Thus, fewer sizer implants and a smaller inventory of assorted breast implant sizes[6] are needed.

In patients with tight chest-wall tissues, asymmetry, or Poland's syndrome, expansion has been most helpful in achieving the desired size and shape of the breast (Figure 17-2). Overexpansion for a period of several weeks enables the shape of the breast to be changed and abnormally high nipples to be lowered. Furthermore, overexpansion has been found to be helpful in preventing capsular contracture.[5]

VOLUME ADJUSTMENT

Patient dissatisfaction with size is one of the most common problems seen following breast augmentation.[7,8] Initially, most patients feel that their breasts are too large. However, as time passes and patients become accustomed to their new breast size, many patients express the desire for their breasts to be larger. The expander implant enables the patient to have the size of her breast increased or decreased postoperatively to her satisfaction.

Volume adjustment has also been found to be of great value in treating patients with ptotic breast (Figures 17-3 and 17-4). In these cases, it is extremely difficult to assess the optimal size before or even during surgery.

THE IMPLANT

The expander implant is manufactured with either a smooth surface or a textured surface. Three types of expander implants are available:

1. 50 percent gel, 50 percent saline
2. 20 percent gel, 80 percent saline
3. 100 percent saline

Two filling domes are available: a larger filling dome for reconstruction cases and a microdome for breast augmentation.

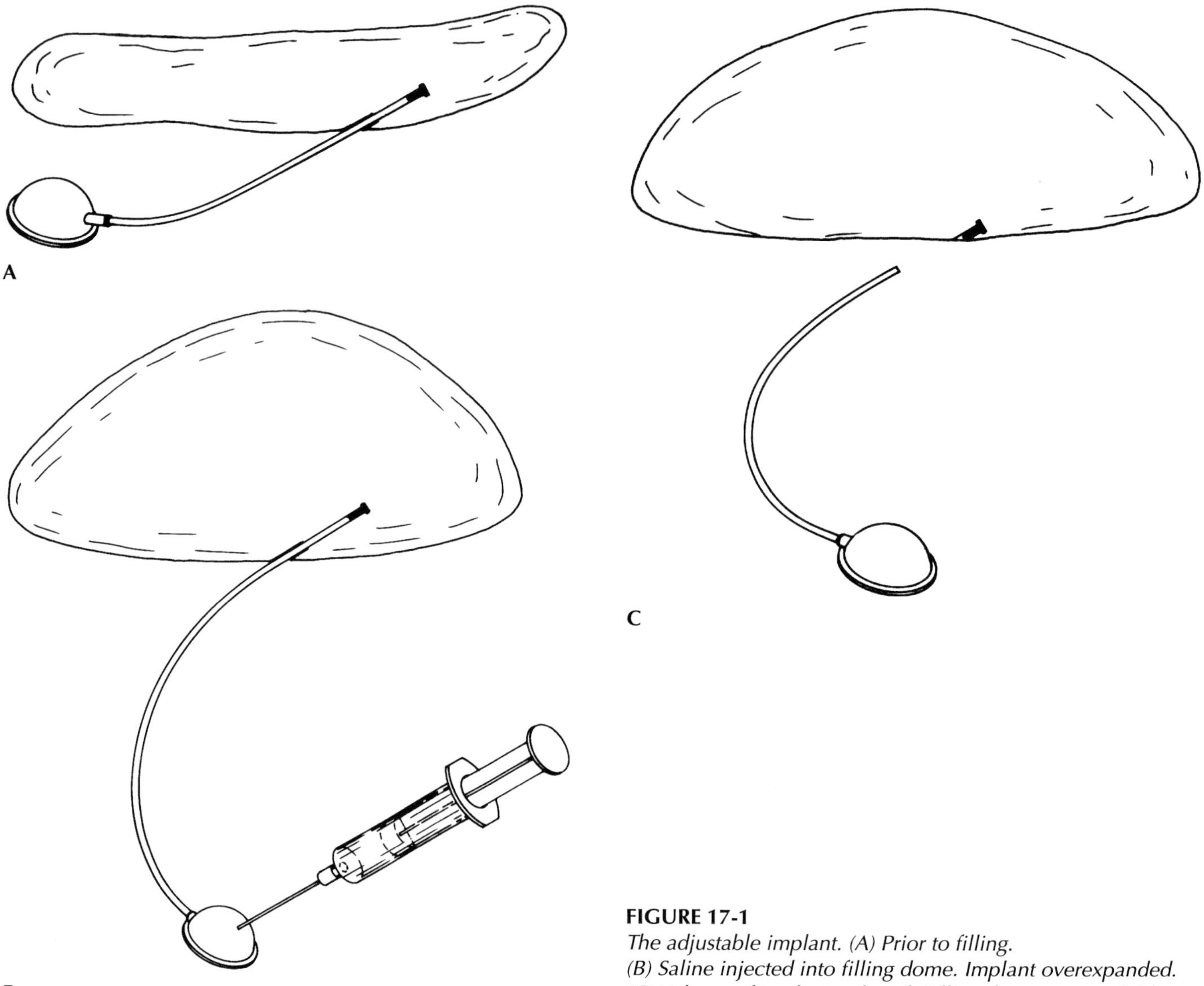

FIGURE 17-1
The adjustable implant. (A) Prior to filling.
(B) Saline injected into filling dome. Implant overexpanded.
(C) Volume of implant reduced. Filling dome removed. Plug valve seals implant.

OPERATIVE TECHNIQUE

The implant can be inserted through any of these three standard incisions: transaxillary, circumareolar, or inframammary. Placement of the implant can be either subglandular or submuscular. When using the transaxillary approach, the microinjection dome should be placed in a subcutaneous pocket close to the incision so that removal can be performed through the same incision.

Expansion is usually started 2 to 3 days after surgery when the pain has subsided. The filling dome is left in position for 3 to 6 months, at which time it is removed by partially opening the original incision or by creating a new small one. The incision is then closed with a few 5–0 catgut sutures.

COMPLICATIONS

Implant Related Complications

As with any other type of implant, rupture or leakage can occur. Should leakage occur, the saline will be absorbed resulting in a decrease in breast size.

Filling dome problems are rarely seen. These may include extrusion, infection, rotation, and leaking. If infection occurs around the filling dome, the dome should be removed.

Patient Related Complications

The usual complications reported with fixed-volume breast implants are seen with adjustable implants as well. These include hematoma, infection, capsular con-

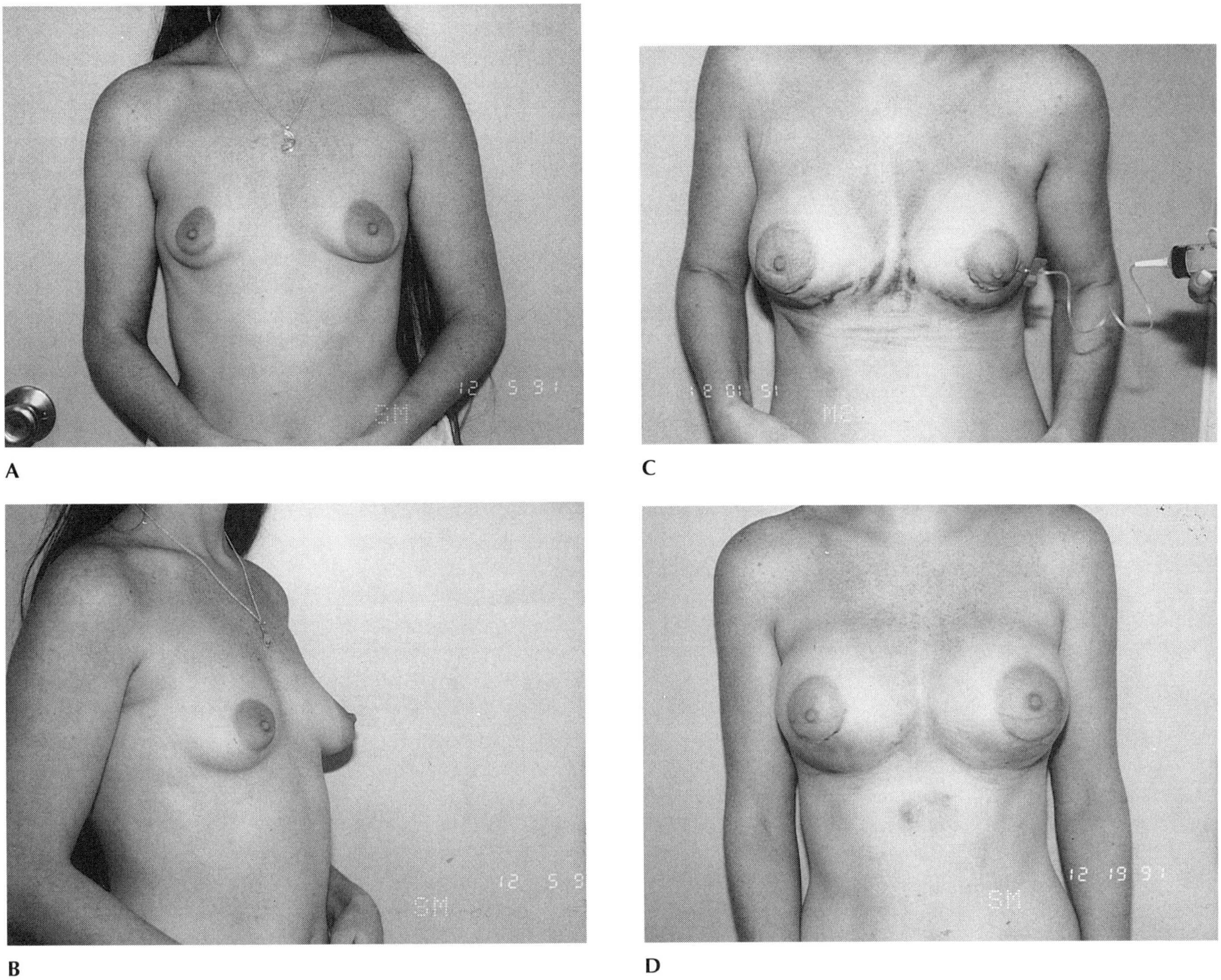

FIGURE 17-2

(A, B) A patient with mild asymmetry prior to surgery. (C) Early postoperative result. (250-mL implant filled to 180 mL, left and 200 mL, right.) Periareolar incision and subglandular implant. Saline being injected into filling dome. (D, E) Result at 2 weeks postsurgery. (F, G) Final result prior to filling dome removal (180 mL, left and 250 mL, right). (H, I) Filling dome being removed under local anesthetic.

tracture, and palpability, wrinkling, and malposition of the implant.[8,9]

SUMMARY

Postoperative volume adjustment allows the size and, to a certain degree, the shape of the breast to be altered. It appears that overexpansion and volume reduction have some effect on decreasing the incidence of capsular contracture.

Postoperative expansion and volume adjustment have particular benefit in treating patients with breast asymmetry, Poland's syndrome, and tubular breast deformities.

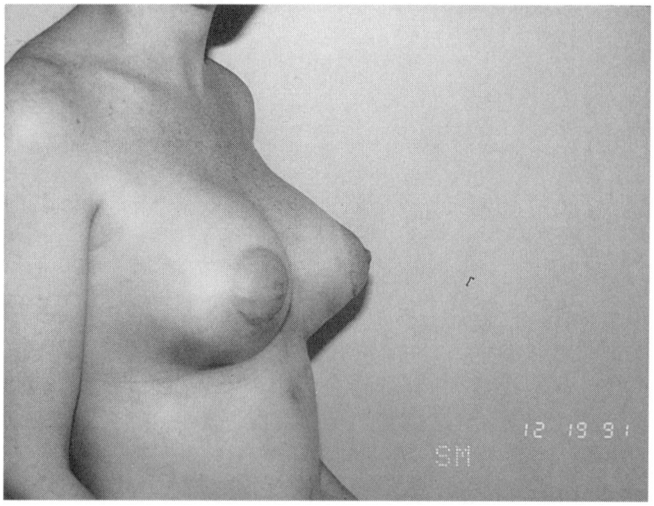

E

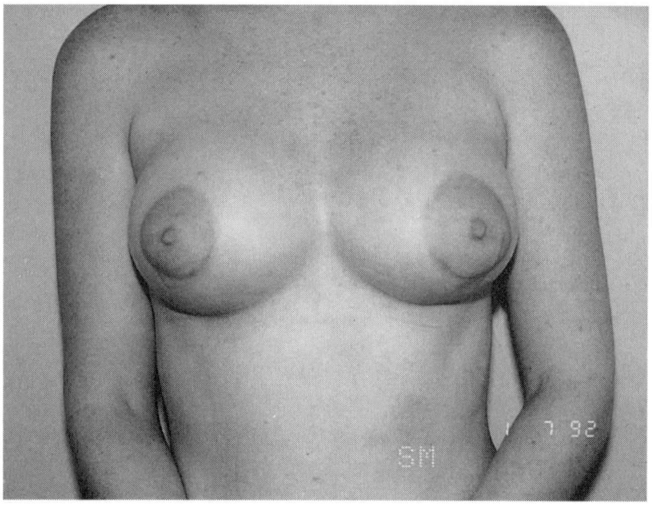

F

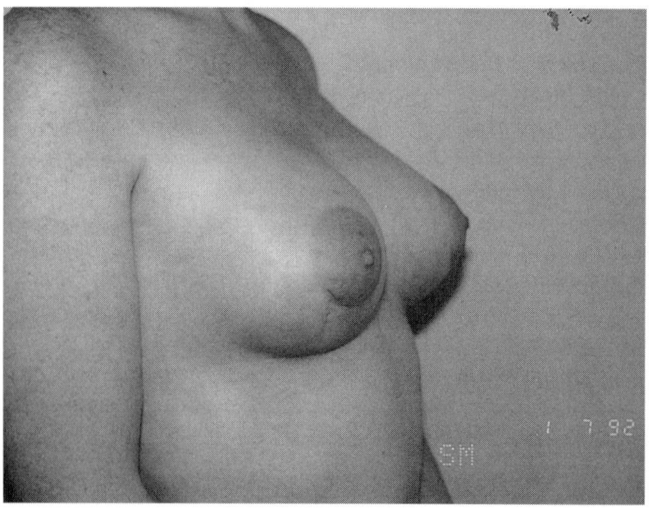

G

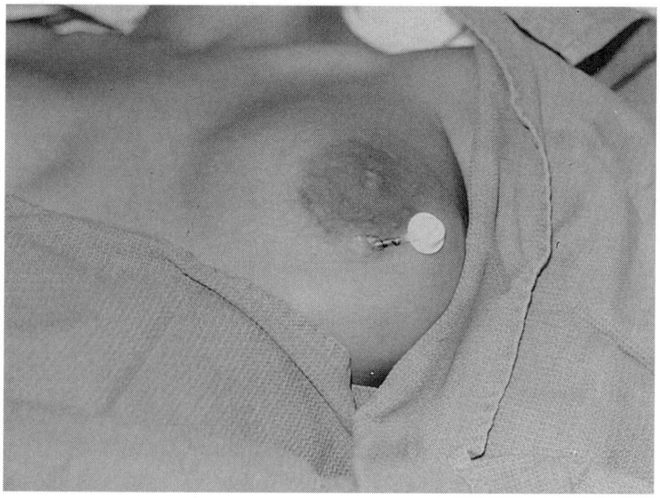

H

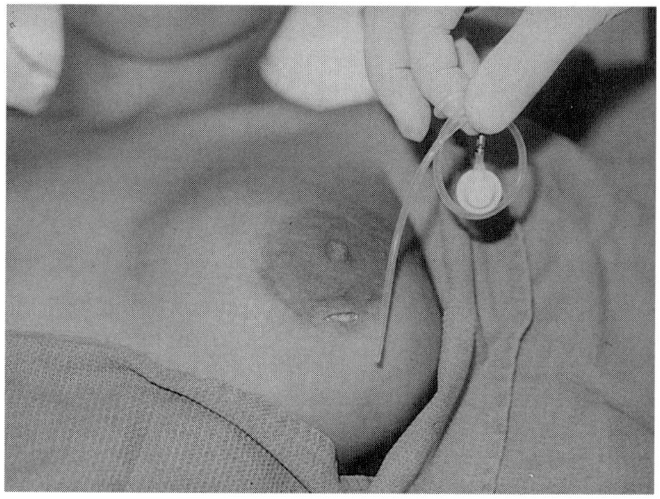

I

FIGURE 17-2 *(Continued)*

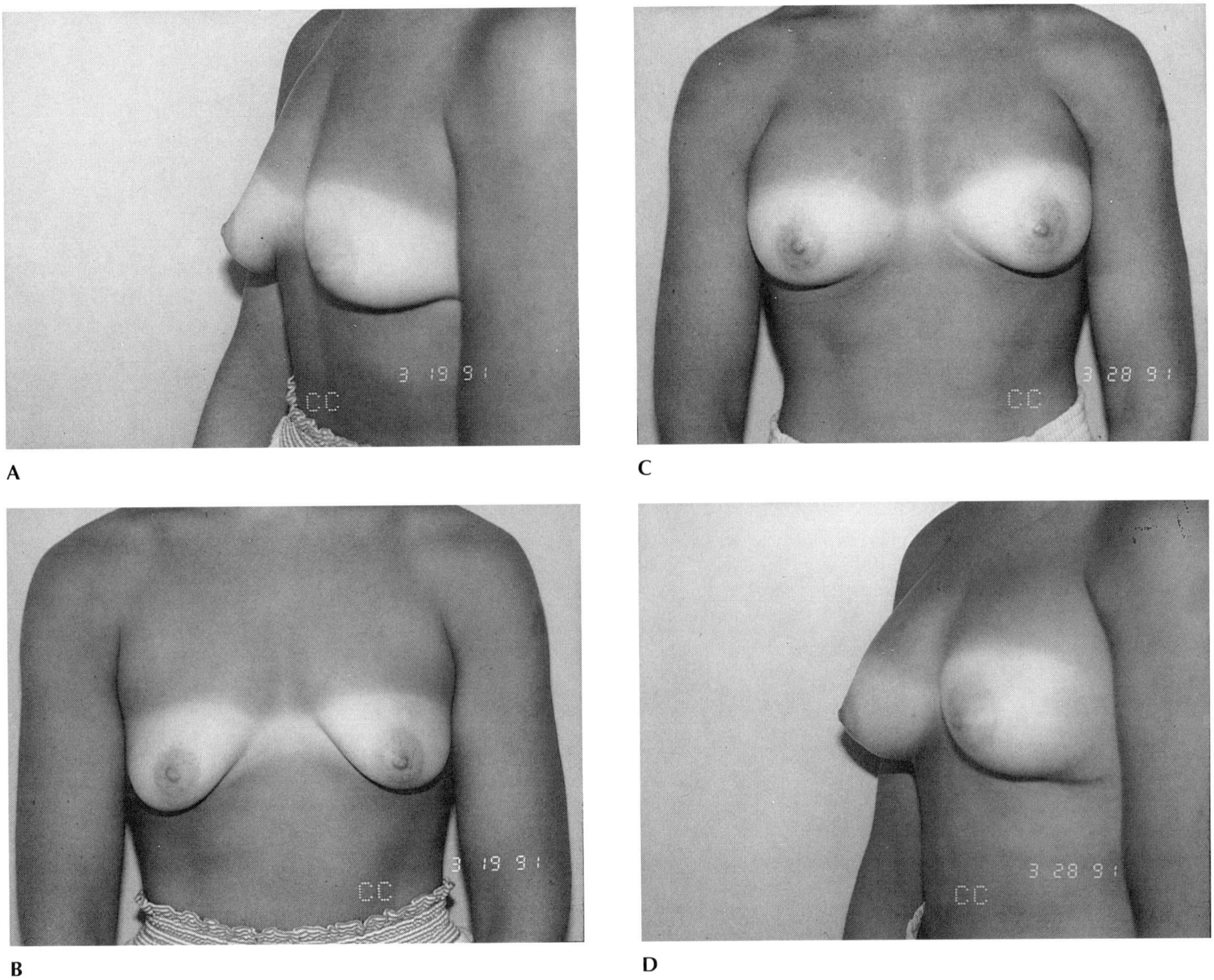

FIGURE 17-3

(A, B) Patient with moderate ptosis prior to surgery. (C, D) Early postoperative result (200 mL implants—150 mL added at surgery), 1 week after surgery. (E, F) Final result (4 months) (270 mL, left and 270 mL, right).

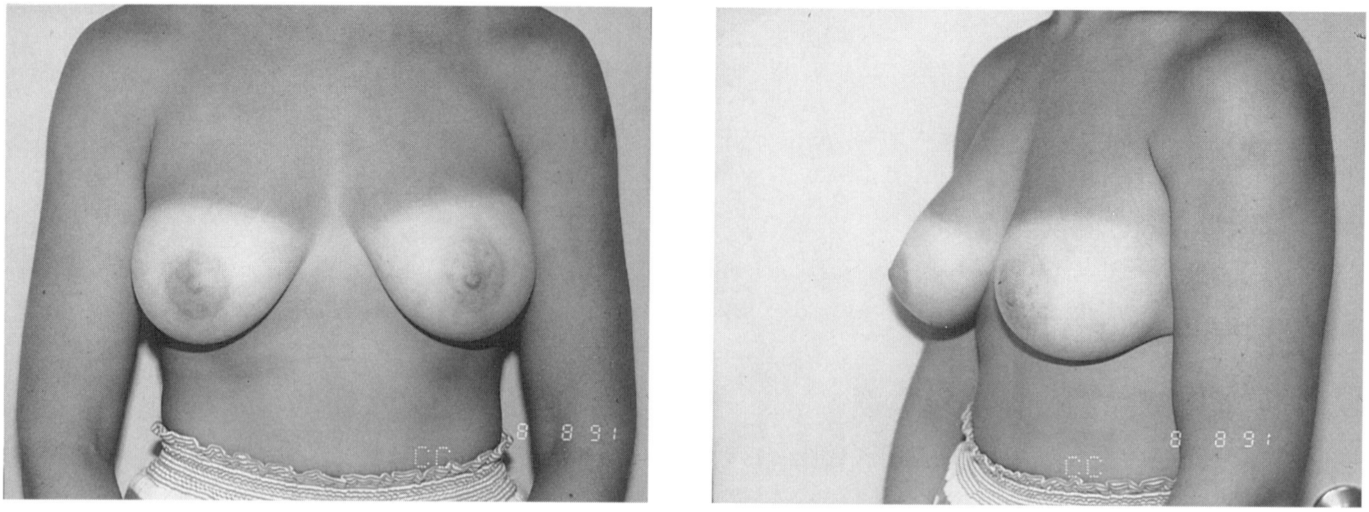

E F

FIGURE 17-3 *(Continued)*

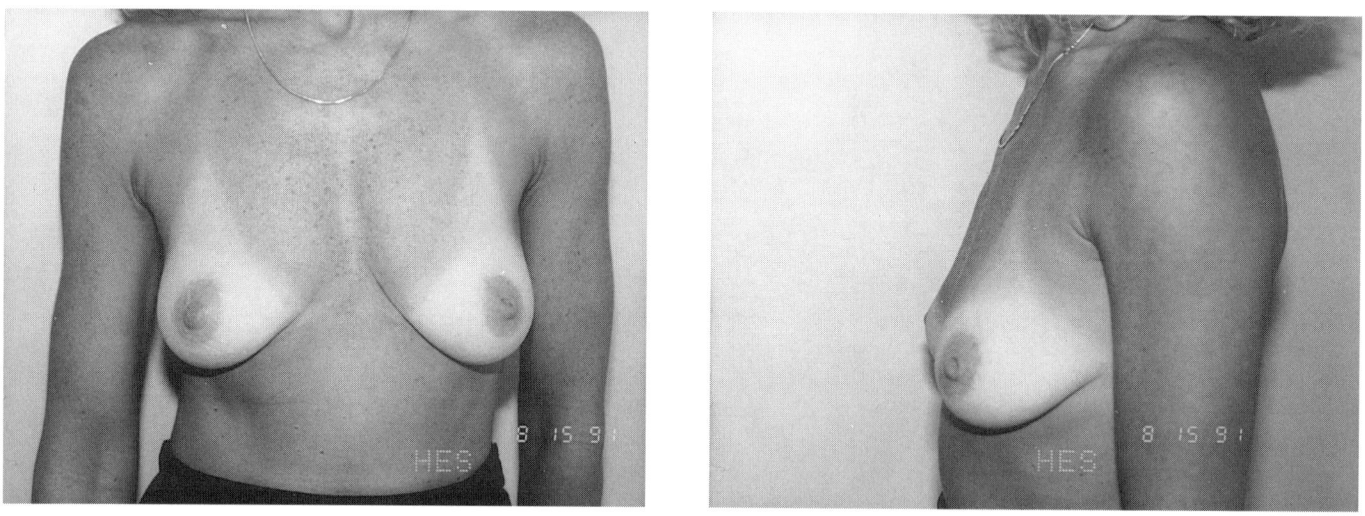

A B

FIGURE 17-4

(A, B) Patient with moderate ptosis prior to surgery. (C, D) Two weeks following transperiareolar subglandular augmentation (200 mL implants). (E, F) Final result at 3 months (160 mL, left and 220 mL, right).

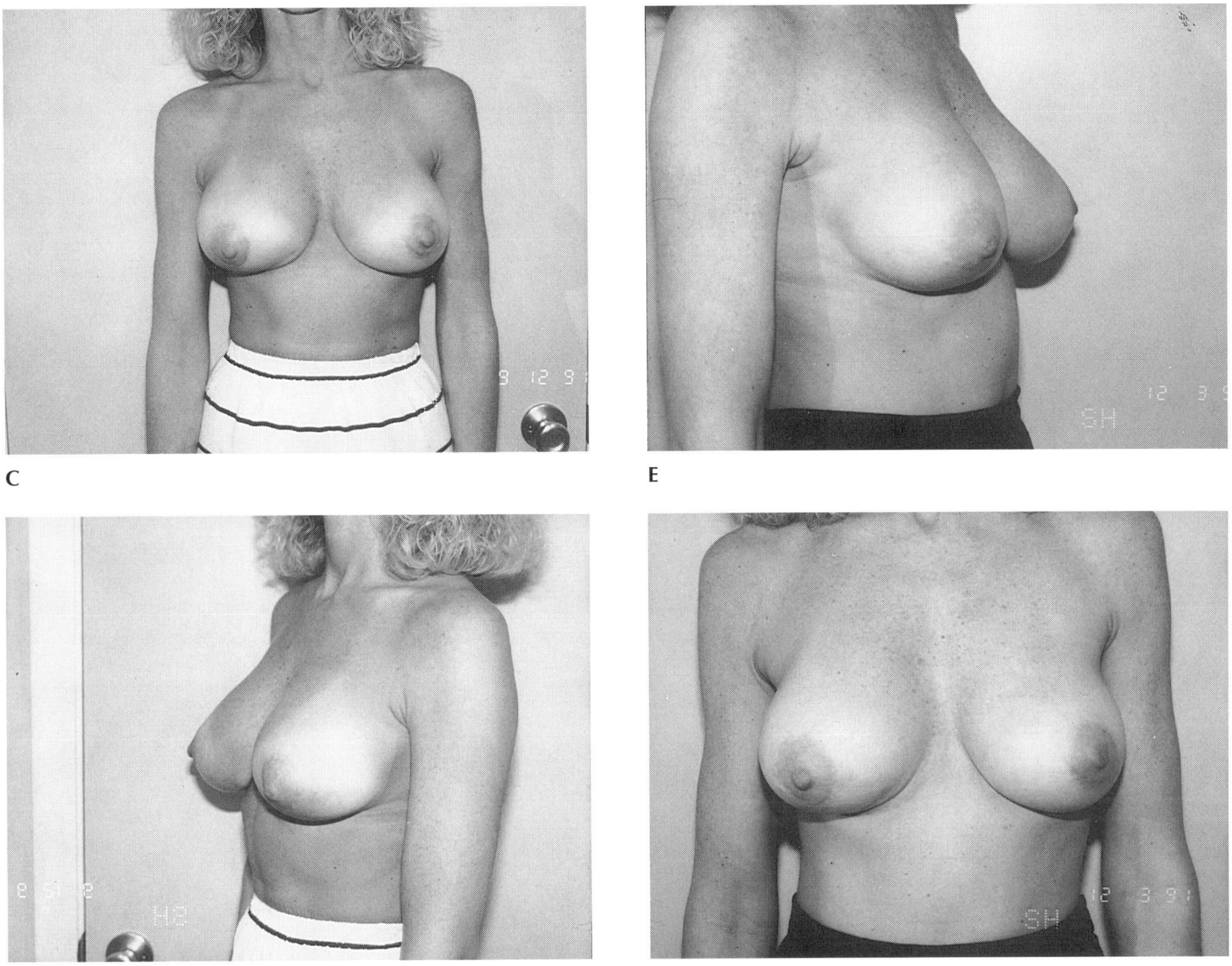

FIGURE 17-4 (*Continued*)

REFERENCES

1. McGrath MH, Burkhardt BR. The safety and efficacy of breast implants for augmentation mammaplasty. *Plast Recontr Surg* 1984;74:550.
2. Tebbetts JB. Transaxillary subpectoral augmentation mammaplasty: Long-term follow-up and refinements. *Plast Reconstr Surg* 1984;74:636.
3. Becker H. Augmentation mammoplasty—Expander use. In Vistnes LM (ed.), *Procedures in plastic and reconstructive surgery. How they do it.* Boston: Little, Brown, and Company, 1991, pp. 222–235.
4. Becker H. The permanent tissue expander. *Clin Plast Surg* 1987;vol.14.
5. Becker H. The expandable mammary implant: An update. *Perspect Plast Surg* 1989;vol.3.
6. Becker H. Breast augmentation using the expander mammary prosthesis. *Plastic Reconstr Surg* 1987;vol.79.
7. Hetter GP. Satisfactions and dissatisfactions of patients with augmentation mammaplasty. *Plast Reconstr Surg* 1979;64:151.
8. Cairns TS, de Villiers W. Capsular contracture after breast augmentation: A comparison between gel and saline filled prostheses. *S Afr Med J* 1980;57:951.
9. Regnault P, Baker TJ, Gleason MC, et al. Clinical trial and evaluation of proposed new inflatable mammary prosthesis. *Plast Reconstr Surg* 1972;50:220.

18

Breast Reconstruction

Stephen J. Mathes

RECONSTRUCTIVE REQUIREMENTS

The plastic surgeon has always recognized the value of skin and soft tissue adjacent to a defect. With similar skin tone, color, and subcutaneous tissue layers, a local advancement or rotation flap provides the ideal tissue for replacement at the defect site. Unlike most skin defects, the postmastectomy deformity lacks both skin for coverage and underlying breast mound. The skin envelope must have adequate laxity to allow the breast mound replacement to project and to remain soft in consistency. The traditional flap advancement, rotation, or transposition flap designed adjacent to the mastectomy site generally will not provide a suitable skin envelope to meet these reconstructive requirements.

With extensive defects associated with radical mastectomy, reconstructive surgeons select distant sites for a suitable flap. The latissimus dorsi musculocutaneous transposition flap is available for simultaneous replacement of missing skin envelope and the absent pectoralis major muscle. The skin island provides an adequate skin envelope to support an underlying silicone breast implant. The shift to a more conservative mastectomy defect with a modified radical mastectomy simultaneously reduces the reconstructive requirements. The surgeon has more local skin to use for the new breast envelope and does not have to replace missing chest-wall musculature.

With the introduction of the tissue expander, the surgeon can stretch the remaining local tissue to gain skin by both recruitment and epithelial growth.[1] With the decrease in the actual mastectomy defect, tissue expansion represents a useful technique to prepare the remaining mastectomy skin as the new skin envelope to cover the permanent breast implant. Trends in breast reconstruction over the past 13 years demonstrate an increase in tissue expansion with a corresponding decrease in latissimus dorsi musculocutaneous flap use for breast reconstruction.[2]

Indications

Selection of technique for breast reconstruction is based on three factors: (1) the mastectomy defect, (2) the contralateral breast, and (3) patient preference. For serial expansion to be effective, healthy mastectomy skin is important. If there is excessive scar due to prior partial mastectomy flap loss or prior radiation (adjuvant or associated with prior lumpectomy radiation therapy) the chest-wall skin will not stretch well. If the skin appears healthy and scar is limited to the mastectomy incision, tissue expansion is generally effective in enlarging the skin envelope.

The shape and size of the contralateral breast will also determine the usefulness of serial tissue expansion. Effective breast reconstruction should achieve symmetry. If the opposite breast is ptotic or very large, it may not be possible to establish a skin envelope of adequate size at the mastectomy site to reconstruct a breast with similar features. Frequently, the patient is interested in alteration of the opposite breast with either mastopexy or reduction mammoplasty. The opposite breast is modified at the time of the second stage procedure for permanent implant insertion at the mastectomy site. If the opposite breast is too small, augmentation may be performed at the second-stage procedure as an alternate approach.

The patient should actively participate in the decision regarding technique for breast reconstruction. Many patients remain concerned regarding safety of the silicone implant although current silicone-walled saline-filled implants are now exclusively used by most surgeons. This eliminates the risk of silicone gel leak, which is difficult to detect if implant rupture occurs. Therefore, in the patient who fears silicone implant use, autogenous reconstruction represents the best technique and is generally accomplished with the transverse rectus abdominis musculocutaneous transposition flap[3] or microvascular transplantation of the superior or inferior half of the gluteus maximus musculocutaneous flap.[4]

Although tissue expansion techniques require two operative procedures, each requires a minimal hospital stay and recovery period. The more complex flap techniques require both longer hospitalization and subsequent recovery. The postmastectomy patient with time constraints due to career or family obligations will frequently prefer the less complex technique of serial tissue expansion.

EXPANDER

The initial expander was designed to accept saline injections through a distant port (Figure 18-1A). The expander has a firm back and enlarges in the anterior and lateral direction as saline is added. Subsequently, four modifications have occurred in expander design: (1) The expander port was incorporated in the expander as an anterior diaphragm valve (Figure 18-1B). (2) An implant was designed with a distant port that is removable without manipulation of the implant.[5] This technique allows expansion with a permanent implant. (3) Anatomical shape was introduced to allow greater inflation at the inferior aspect of the expander. (4) Textured surfaces were introduced to avoid implant migration during the expansion process and to decrease the capsular contracture. Based on the success of the polyurethane-coated implants used to avoid symptomatic capsular contracture, the textured silicone is designed to simulate the polyurethane surface.[6,7]

Although personal preferences and prior experience determine expander preference, generally, the use of the incorporated port is preferred in expander selection (Figure 18-1B). This diaphragm, with its metal backing, is easily located with a magnetic probe or by direct palpation. The patient has less discomfort at the time of insertion of a 21- or 23-gauge butterfly needle due to partial denervation of mastectomy skin. This is in contrast to the well-innervated skin of the axilla or upper abdomen, where the patient notes pain during the passage of the needle into the distant port. Finally, the patient does not see or feel the bulge of the incorporated port compared to the distant port.

A

B

FIGURE 18-1
Expanders for breast reconstruction. (A) Smooth-walled silicone expander with distant port. (B) Textured expander with incorporated port.

The permanent implant-expander with a distant removable port will eliminate the necessity of a second operative procedure to replace the expander with a permanent implant and is advocated as a reliable treatment for breast reconstruction. However, the distant port is palpable to the patient and may cause discomfort. Of even greater significance, the expander design is based on a permanent implant and doesn't offer the specific advantages of an expander, which has the ability to stretch the overlying skin in multiple directions.

Variations in base width, height, and projection allow selection of an expander to fit the retained skin at the mastectomy site. Measurements of base width between the ipsilateral parasternal line and the anterior axillary line are used to select the expander size. Generally, the expander size is slightly larger than the estimated breast size to allow maximal recruitment of adjacent skin during the expansion process.

The textured expander surface allows precise placement of the inferior edge of the expander at the inframammary line. The expander will provide maximal stretch over the inferior anterior chest between the third through sixth intercostal spaces where the greatest expansion of the postmastectomy skin envelope is required. With the smooth expander, it is customary to place the inferior edge of the expander several centimeters below the determined inframammary line. In this position, superior migration of the expander would not result in an implant location in the infraclavicular region. However, the inframammary line is poorly defined and requires intraoperative adjustment at the second procedure for implant exchange.

TECHNIQUE

Stage I: Expander Insertion (Immediate reconstruction)

Recent studies show increasing interest in immediate reconstruction. Fortunately, tissue expander insertion at the time of mastectomy is readily accomplished for the majority of patients. The inframammary line is marked with the patient sitting prior to the mastectomy. The general surgeon will usually select appropriate incisions to accomplish the mastectomy. If possible, a transverse incision is made; it is more acceptable to most patients because it represents a less conspicuous scar on the reconstructive breast mound.

The expander may be placed directly under the mastectomy flaps or may be placed in a complete or partial submuscular space. With placement directly beneath the skin flaps, there is risk of implant exposure or migration because the mastectomy defect generally extends into the mid- or posterior axillary line. With complete submuscular coverage beneath the pectoralis major and serratus anterior muscle, the expander is located in a well-defined space. However, the muscle origin of the inferior pectoralis major muscle fibers results in malposition of the expander several centimeters above the sixth intercostal space or appropriate position for the inframammary line. Further muscular contraction will tend to push the implant in a subclavicular position and prevent expansion of inferior chest-wall skin envelope. For this reason, the pectoralis major muscle is released from its origin along the sternum (fourth through sixth intercostal space) and along the inferior costal margin (Figure 18-2A). This maneuver allows the inferior rim of the expander to rest exactly at the determined inframammary line. Furthermore, contraction of the superior intact pectoralis major muscle fibers will hold the expander in an inferior position. The lateral edge of the muscle is sutured to the chest wall to avoid migration of the implant laterally. This modification of submuscular implant position provides partial implant separation from the mastectomy skin but does not allow muscle to impede the serial expansion process.[3,4]

If the mastectomy skin has borderline circulation, complete submuscular implant placement is recommended. However, a technique has been developed to avoid excessive compression of the inferior half of the expander. A segmental rectus abdominis external oblique flap is utilized to cover the inferior part of the expander (Figure 18-2B).[8,9] Previous reports have used the rectus muscle based on the superior epigastric pedicle. However, this technique eliminates the future use of the superior-based ipsilateral rectus abdominis muscle or musculocutaneous flap for autogenous breast reconstruction. The preferred technique requires exposure of the ipsilateral rectus abdominis fibers of origin between the sixth rib through the costal margin. The rectus abdominis muscle is divided at the costal margin with the incision extending through the adjacent external oblique muscle. The rectus abdominis and the external oblique muscles are elevated as superior-based flaps to the sixth intercostal space. Here, segmental pedicles arise from intercostal artery and associated venae comitantes located at the inferior margin of the sixth rib. The flap is now rotated as a turnover flap in order to suture the distal edge of the segmental rectus abdominis external oblique flap to the previously released inferior fibers of origin of the pectoralis major muscle. Together, these muscle flaps provide complete muscular coverage of the inferior part of the expander. This technique preserves the implant position at the inframammary line and avoids muscular pressure on the inferior aspect of the expander. In the event of partial mastectomy flap loss, the expander is protected by vascularized muscle flap and is left in position while skin

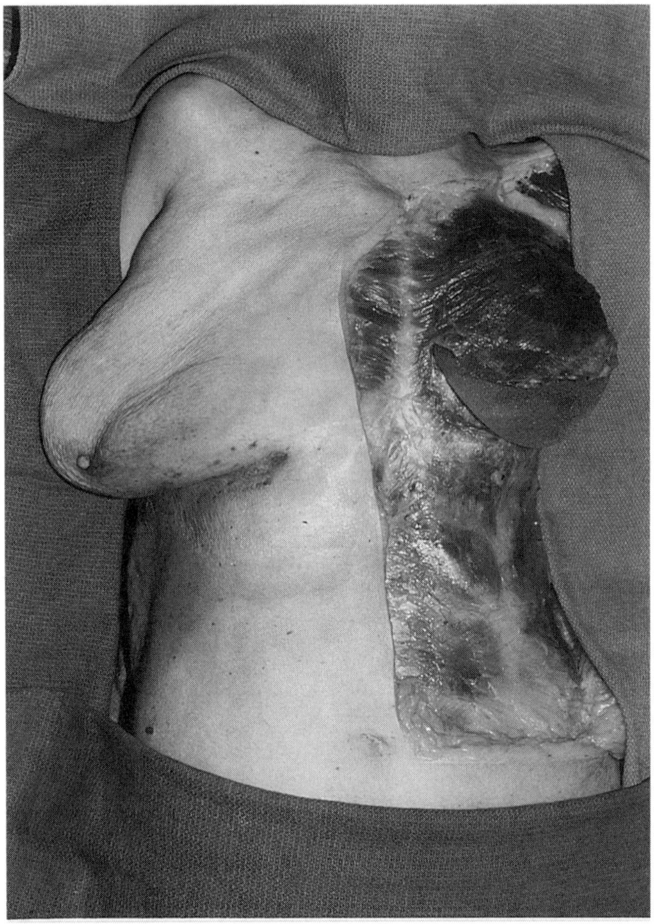

A

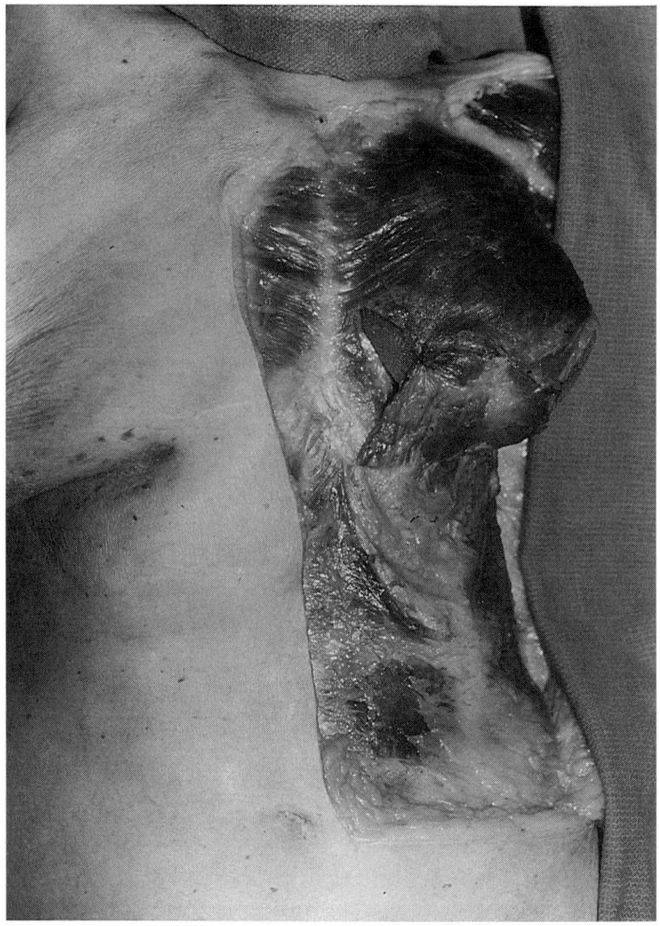

B

FIGURE 18-2

Expander position at mastectomy site. (A) Cadaver dissection with partial muscular coverage of expander implant. Release of pectoralis major muscle origin from costal margin and inferior sternum allows implant to maintain position of inframammary fold. (B) Cadaver dissection with complete muscular coverage of expander implant. Segmental rectus abdominis external oblique turnover flap approximated to inferior pectoralis major muscle.

healing occurs in conjunction with local wound care (Figures 18-3 and 18-4).

Stage 1: Expander Insertion (Delayed reconstruction)
Expander insertion for delayed reconstruction is essentially the same as immediate reconstruction. The lateral half of the mastectomy incision is incised. The lateral edge of the pectoralis major muscle is identified and a submuscular pocket is elevated from the previously determined inframammary line to the subclavicular space extending from the parasternal line to the anterior axillary line. Upon completion of the pocket, a fiber-optic lighted retractor is used to visualize the pectoralis major fibers of origin between the fourth and sixth intercostal space and along the costal margin. These fibers are divided to avoid muscle pressure against the inferior aspect of the expander. The selected tissue expander is placed in the pocket using care to position the inferior edge of the expander exactly at the inframammary line. After skin closure, the incorporated expander port is located with the magnetic probe through the overlying skin envelope and the implant is partially filled with 50 to 100 mL saline. The extent of volume inflation at the time of expander insertion depends on laxity of the mastectomy skin. It is preferable to avoid excessive skin compression, particularly in immediate reconstruction where skin circulation is more vulnerable to expander compression (Figure 18-5).

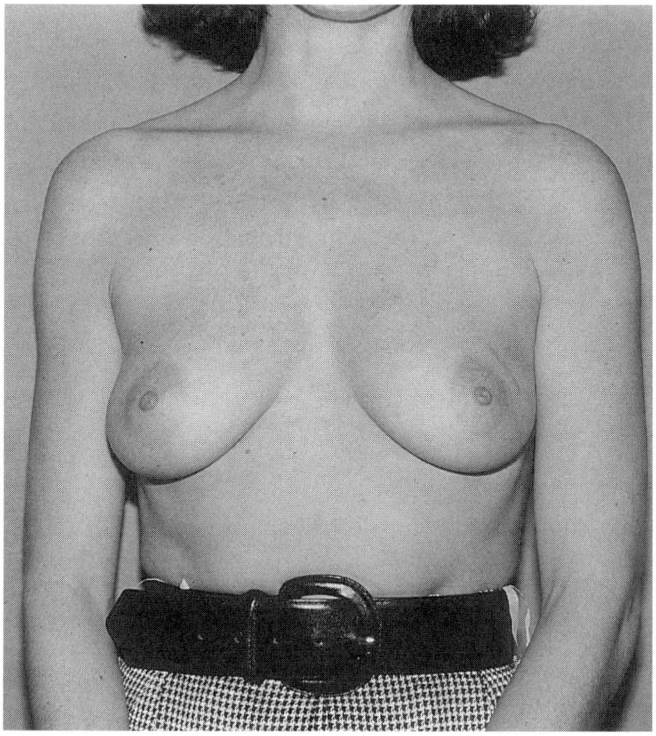

A

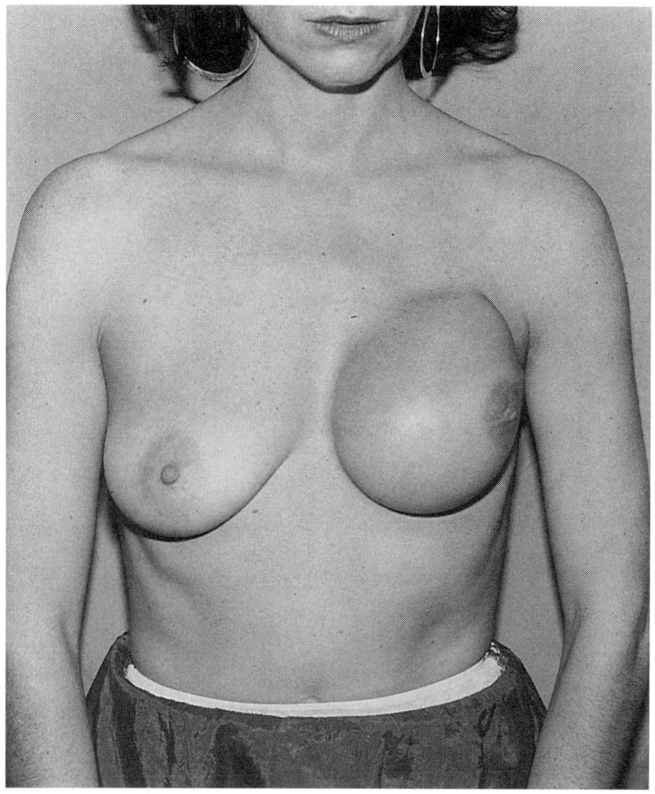

C

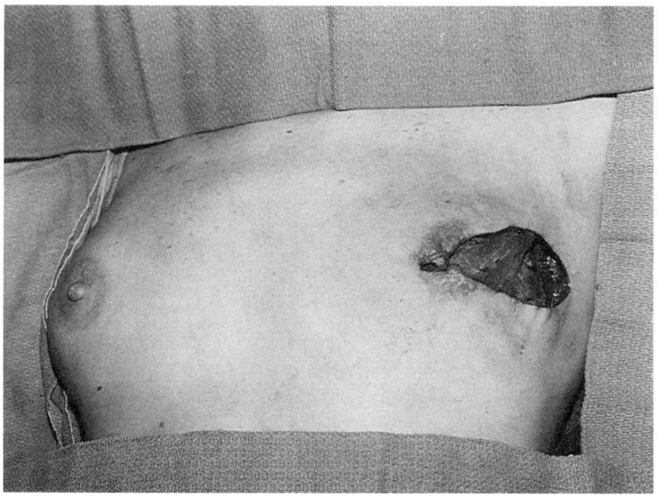

B

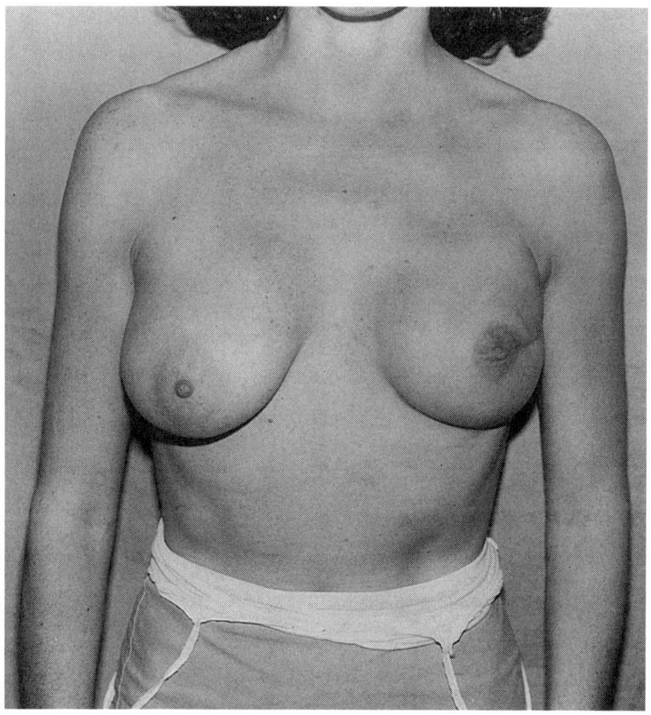

D

FIGURE 18-3
Expander implant. Immediate breast reconstruction after modified radical mastectomy. (A) Preoperative anterior view. Left modified radical mastectomy. (B) Surgical defect post-mastectomy. Areola is preserved with removal of nipple and complete ductal system with mastectomy. (C) Partial submuscular expander in place with completion of serial expansion. (D) Postoperative result. One year postoperative removal of expander and replacement with textured saline implant. (E) Lateral view postoperative. Note well-defined inframammary line.

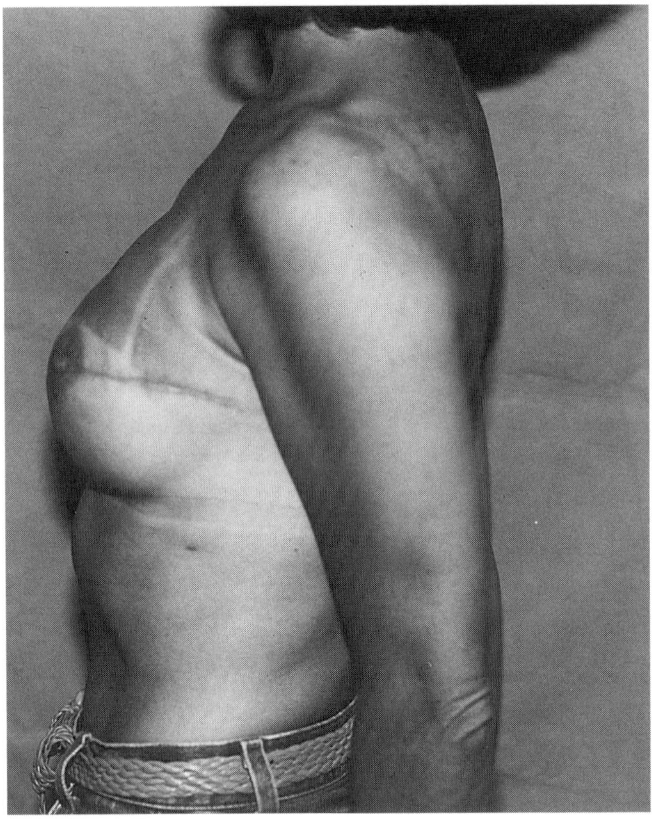

E

FIGURE 18-3 (*Continued*)

Closed suction drains are used for immediate reconstruction. One drain is placed in the inferior pocket and the second placed within the axilla if axillary nodal dissection has been performed. The expander drain is generally removed in 3 days and the axillary drain in 3 to 7 days, based on the extent of serous drainage.

Serial Expansion

The expander is usually inflated over a 3-month period. If the patient is undergoing adjuvant chemotherapy, the second stage is delayed until chemotherapy is completed. The patient is seen as an outpatient at 3-week intervals during which saline is placed in the expander using a 21- or 23-gauge butterfly needle. Eventually, the expanded side will have a volume 10% to 20% greater than the estimated volume of the contralateral breast.

Second Stage Permanent Implant

When expansion is completed, the patient is consulted regarding implant size. If the opposite breast demonstrates no ptosis, the implant size is generally based on the contralateral breast size. If the patient prefers the larger breast size of the expanded skin envelope, then an appropriate size implant will be inserted in the opposite breast to accomplish augmentation. If the opposite breast demonstrates ptosis or macromastia, then an appropriate surgical alteration of the opposite breast will be performed at the time of the implant exchange on the mastectomy side.

Implant selection is based on the opposite base, height, and projection. The biodimensional implant provided by McGhan Corporation offers implant selection based on transverse, vertical and projection base measurements. Preoperative measurements are made in preparation for implant selection (Figure 18-6).

The lateral half of the mastectomy incision is reopened to enter the expander space. The expander is removed. With the patient in a semisitting position, gel-filled silicone sizer implants are placed in the pocket to confirm appropriate implant position with regard to the inframammary line. Actual implant size and dimensions are not based on the gel-implant sizes. If the inframammary or lateral position are not symmetrical with the opposite breast, sutures are placed to plicate the capsule to adjust the implant position. Either permanent or absorbable sutures may be used for this purpose.

If surgical adjustments are planned for the opposite breast, these are performed at this time. Transverse, vertical, and projection measurements are made on the opposite breast. Based on this data, a permanent silicone-walled saline-filled implant is selected. It is inserted in the expander pocket and the breast again is visualized with the patient in a sitting position. If adequate symmetry is observed, the incision is closed and compressive dressing applied. No closed suction drainage is used unless internal adjustments or partial

FIGURE 18-4

Expander implant. Immediate breast reconstruction after bilateral total mastectomy. (A) Preoperative anterior view. (B) Preoperative lateral view. (C) Anterior view. One year postoperative bilateral total mastectomy reconstructed with initial partial submuscular expanders followed by permanent saline textured implants. (D) Lateral view. Mastectomy incision extends from medial edge of areola to anterior axillary line. Entire ductal system and breast have been resected.

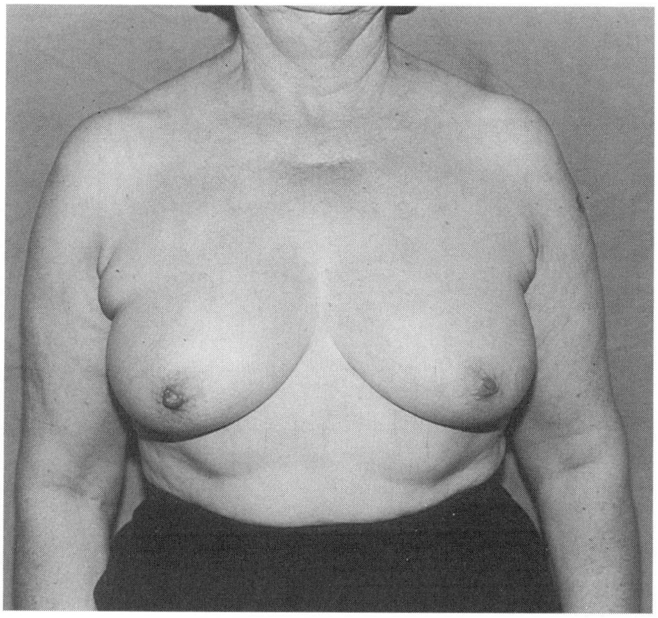

A

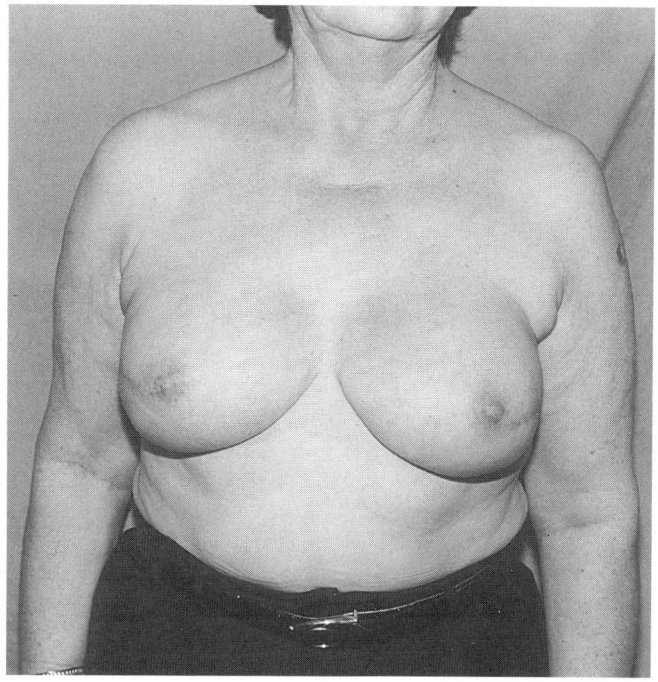

C

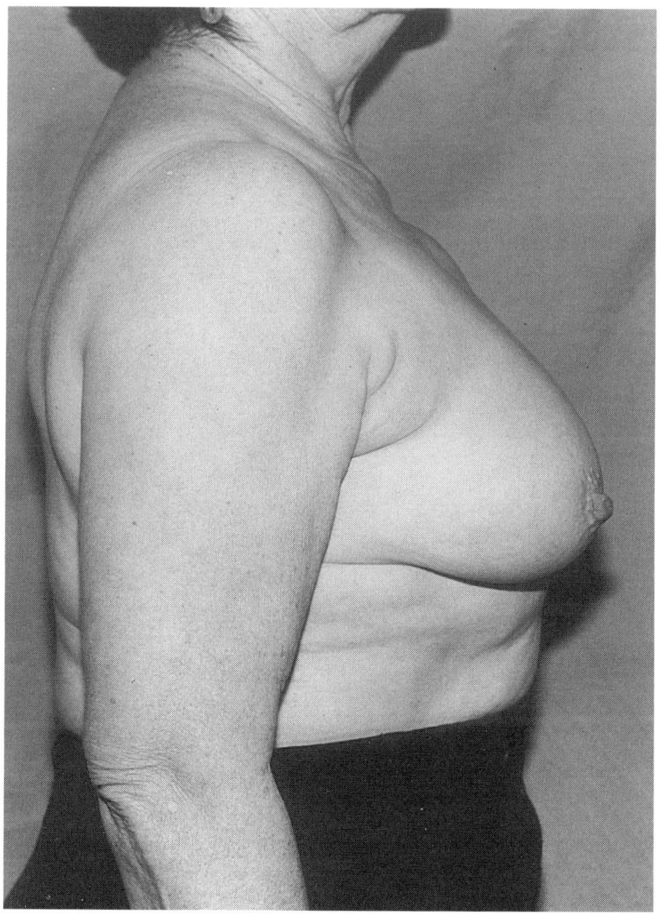

B

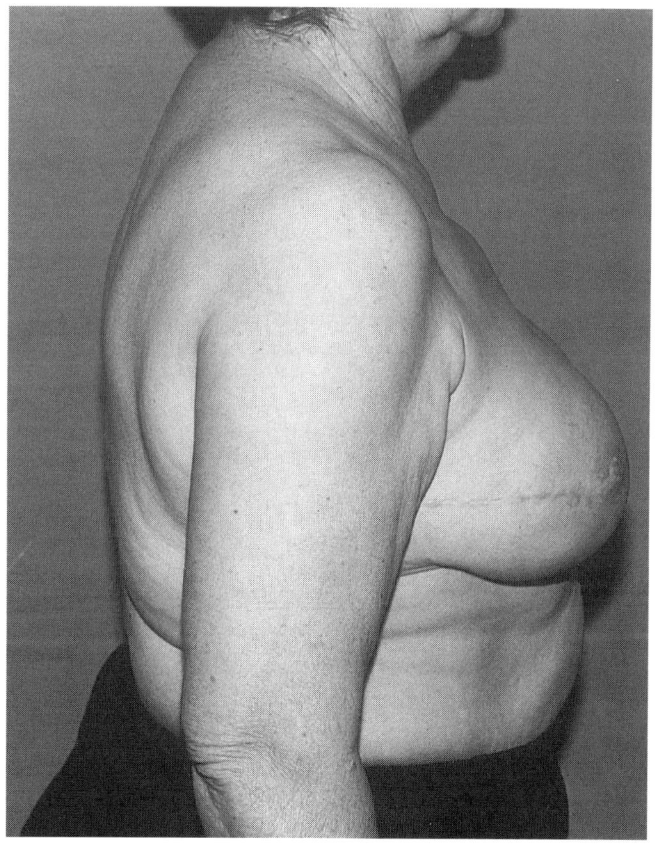

D

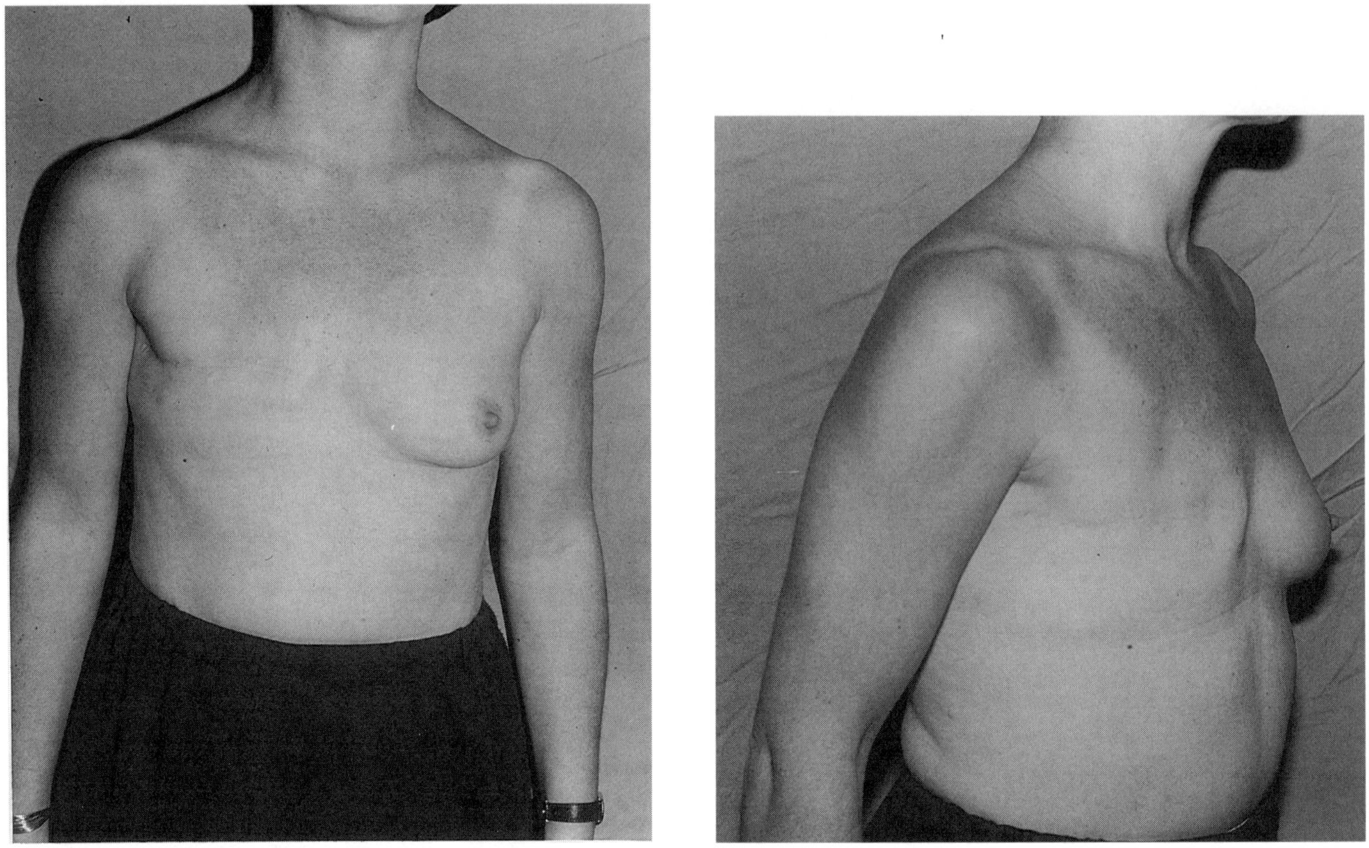

A B

FIGURE 18-5

Expander implant. Delayed breast reconstruction after modified radical mastectomy. (A) Preoperative anterior view 1 year postmastectomy. (B) Preoperative oblique view. (C) Three months postoperative submuscular expander insertion. Textured expander provides well-defined inframammary line. (D) Postoperative 1 year breast reconstruction with expander followed by insertion of permanent textured saline implant. (E) Oblique view. Patient has undergone nipple-areola reconstruction using full-thickness groin skin graft with local dermal fat flaps for nipple mound reconstruction.

capsulectomy is required. If used, the drain is generally removed in 1 to 3 days.

Patients may resume normal activities within 3 to 5 days postoperatively in both the first- and second-stage procedures. After the second stage, sporting activities are avoided for 2 to 3 weeks. No specific dressing or bra requirements are necessary after 3 to 5 days.

COMPLICATIONS

Complications of hematoma, seroma, and infection are rare. If cellulitis is observed, the patient will receive a 5- to 7-day course of antibiotic therapy. Development of abscess requires expander removal. Fortunately, this complication is rarely observed in immediate or delayed primary reconstruction.

Inappropriate patient selection or unrealistic expectations on the part of the patient present the most common complication in patients who present for revision or secondary reconstruction. If the patient has mild ptosis on the opposite breast, expansion alone will not provide ideal symmetry but will provide a result that is pleasing for the well-informed patient. For the patient who stresses optimal symmetry with a ptotic opposite breast, the surgeon should use a flap to reestablish an adequate skin envelope similar to the opposite breast. The breast mound can be established with an implant for the latissimus dorsi musculocutaneous flap or with

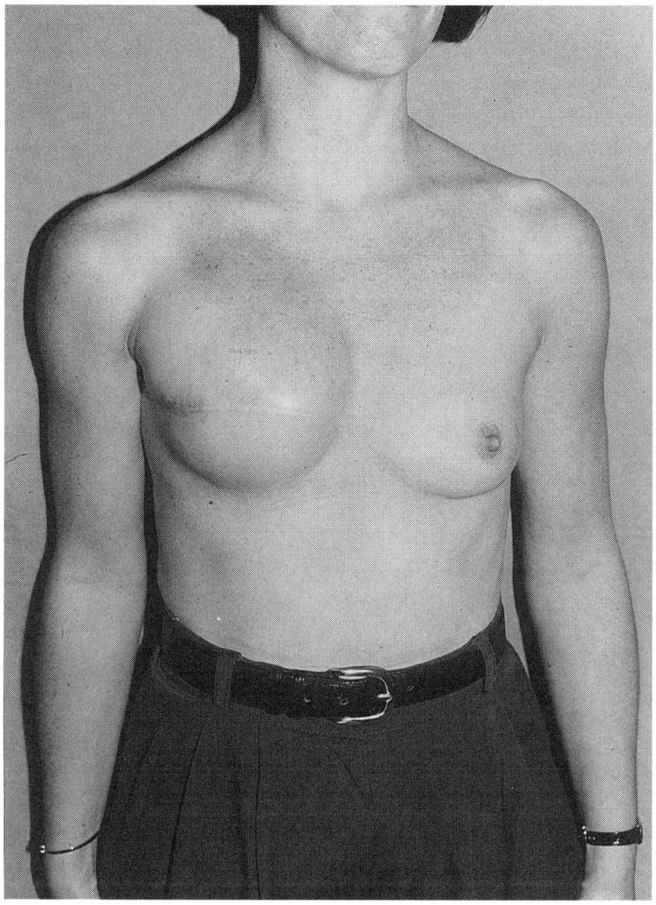

C

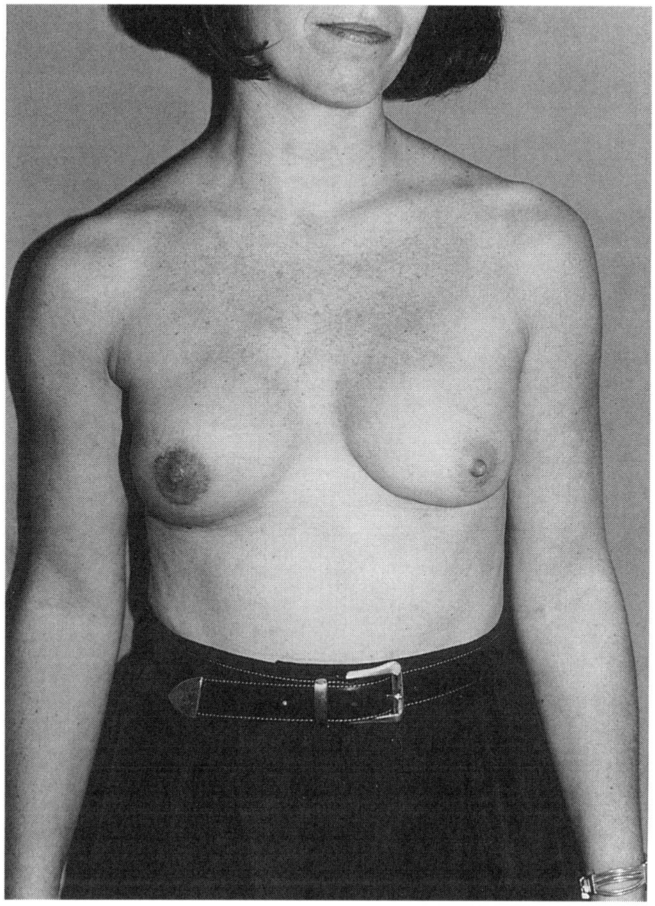

D

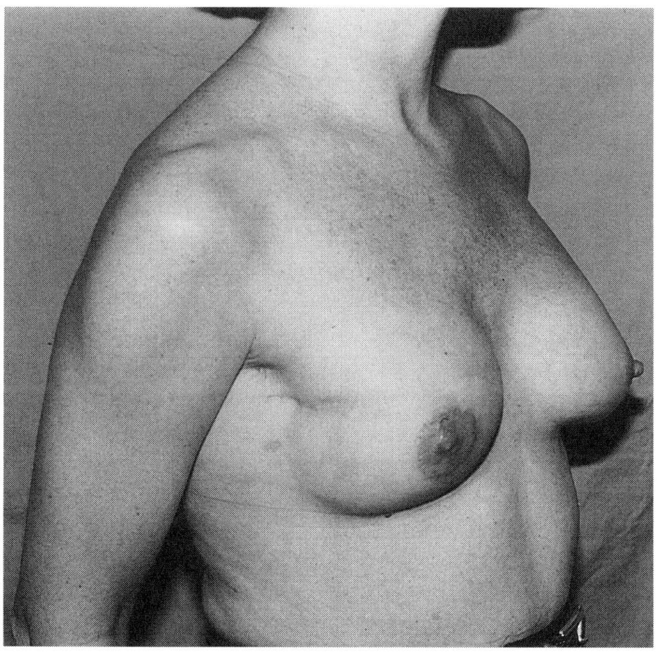

E

FIGURE 18-5 *(Continued)*

A

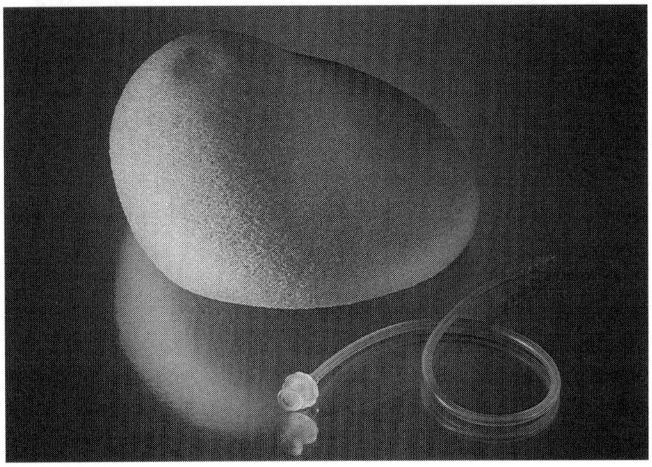

B

FIGURE 18-6
Textured implant. (A) Permanent saline implant with low profile. (B) Permanent saline implant with high profile.

autogenous tissue for the rectus abdominis musculocutaneous flap.

Late capsular contracture remains a risk for all patients reconstructed with an implant. The textured silicone-walled saline implant has performed well with a low acceptable incidence of capsular contracture. With aging of the opposite breast, either mastopexy on the contralateral breast or revision of the reconstructed breast may be required to maintain symmetry.

REFERENCES

1. Radovan C. Breast reconstruction after mastectomy using the temporary expander. *Plast Reconstr Surg* 1982;69: 195–206.
2. Trabulsy P, Anthony J, Mathes S. Changing trends in postmastectomy reconstruction: A 13-year experience. *Plast Reconstr Surg* 1994;93:1418–1425.
3. Bunkis J, Walton RL, Mathes SJ, et al. Experience with transverse lower rectus abdominis operation for breast reconstruction. *Plast Reconstr Surg* 1983;72:819–827.
4. Shaw WW. Breast reconstruction by superior gluteal microvascular free flap without silicone implants. *Plast Reconstr Surg* 1983;72:490–497.
5. Becker A. Breast reconstruction using an inflatable breast implant with attachable reservoir. *Plast Reconstr Surg* 1984;73:678–683.
6. Maxwell GP, Falcone PA. Eighty-four consecutive breast reconstructions using a textured silicone tissue expander. *Plast Reconstr Surg* 1992;89:1022–1028.
7. Barone FE, Perry L, Maxwell GP. The biomechanical pathological effects of surface texturing with silicone and polyurethane and tissue implantation and expansion. *Plast Reconstr Surg* 1992;90:77–83.
8. Bohmert H. Experience in breast reconstruction with thoracoepigastric and advancement flaps. *Acta Chir Belg* 1980;79:105–111.
9. Holle J, Pierini A. Breast reconstruction with an external oblique abdominis turnover flap and a bilateral abdominal skin flap. *Plast Reconstr Surg* 1984;73:469–473.

19

The Expander Mammary Implant for Breast Reconstruction

HILTON BECKER

The expander mammary implant was originally introduced in 1984. The implant is now available in two versions, a saline-filled single-lumen implant and a saline-filled double-lumen implant that has a thin layer of gel between the two membranes (Figure 19-1). The implant is also available with a Siltex® (Mentor Corp., Goleta, CA) textured surface.

The initial concept of the implant was to combine the temporary tissue expander and permanent implant into one device.[1] This was achieved by designing a detachable injection dome that enables the volume of the implant to be adjusted and expanded postoperatively and then to allow the expander to seal on removal of the injection dome.[2] Since its introduction, the expander implant has found wide application in breast reconstructive procedures.

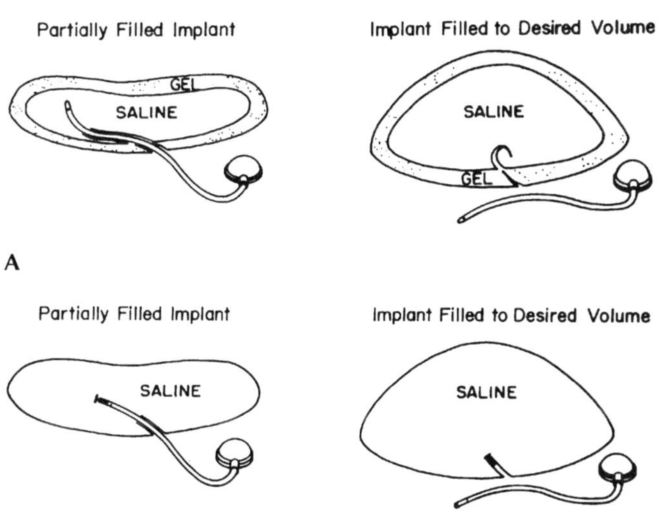

FIGURE 19-1
Cross section of expander implant. (A) Double-lumen implant. (B) Single-lumen implant.

SECONDARY RECONSTRUCTION

The most suitable patients for reconstruction by tissue expansion are those who have had a modified radical mastectomy and who have sufficient skin and muscle intact to allow for expansion procedures.

The procedure is basically very similar to that of breast augmentation. The reconstruction can be performed under general anesthesia or local anesthesia with sedation, often on an outpatient basis.

The implant is usually inserted through the original incision and placed in a submuscular plane. The inferior and inferomedial portion of the pectoralis major muscle is detached with a cautery at the time of surgery. Intraoperative overexpansion is performed to magnify any tethering bands or irregularities. It is then important to place the patient in a sitting position to make sure that the implant is in the correct position relative to the opposite breast.

The overexpansion is initially performed by injecting air into a temporary expander; once satisfied with position, the temorary expander is removed and replaced with the expander implant. Saline is injected into the implant. It is not necessary to place the full

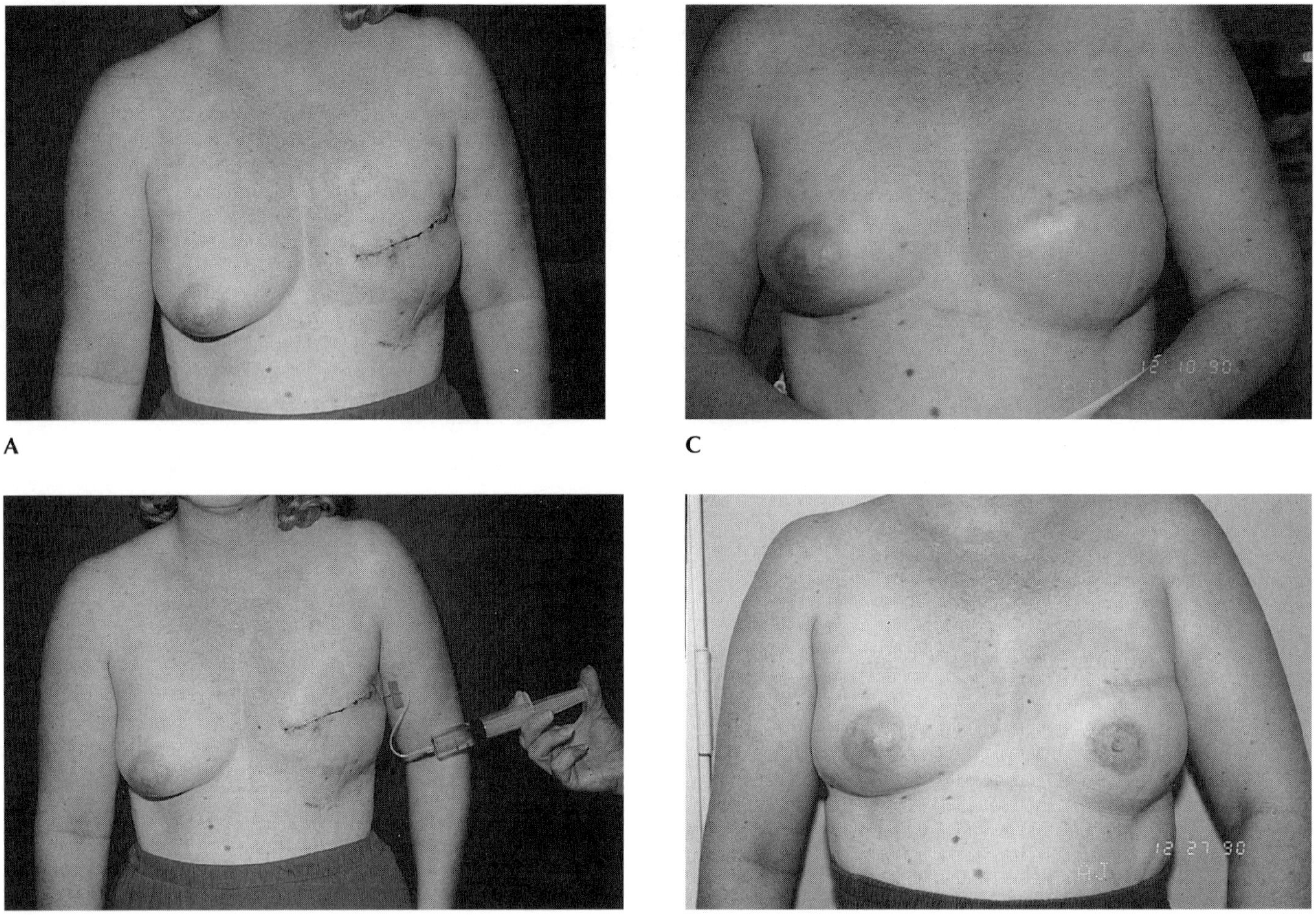

FIGURE 19-2

(A) Forty-two-year-old woman following immediate reconstruction with a 300-milliliter expander implant filled to 200 milliliters. (B) Saline being added to implant 3 days after surgery. (C) Implant expander filled to 500 milliliters (saline). (D) Final result. Saline volume 400 milliliters.

amount of saline into the implant because this can be done postoperatively without hurry. Usually about one-quarter to one-half of the anticipated volume is placed at the time of the initial surgery. This has been found to reduce the postoperative pain and also decrease the risk of implant extrusion.

Drains are usually placed in the pocket and left in position for 2 to 3 days.

Expansion is started early postoperatively using sterile technique. Overexpansion of approximately 50 percent is usually performed and maintained for several months, depending on the patient's tolerance. Injection domes are usually removed between 3 and 12 months.

PRIMARY RECONSTRUCTION

It is in primary reconstruction that the expander implant has found its greatest application. General surgeons are very receptive to the insertion of an expander at the time of mastectomy,[3] especially if it is placed in the submuscular plane and does not take a great amount of time to perform (Figures 19-2 through 19-4).

The possibility of a one-stage procedure with decreased incidence of wound problems, greater operative simplicity and enhanced results have won the favor of the general surgeon. Some surgeons are now adopting a more aggressive approach in the management of the younger patients with breast cancer. Bilateral mas-

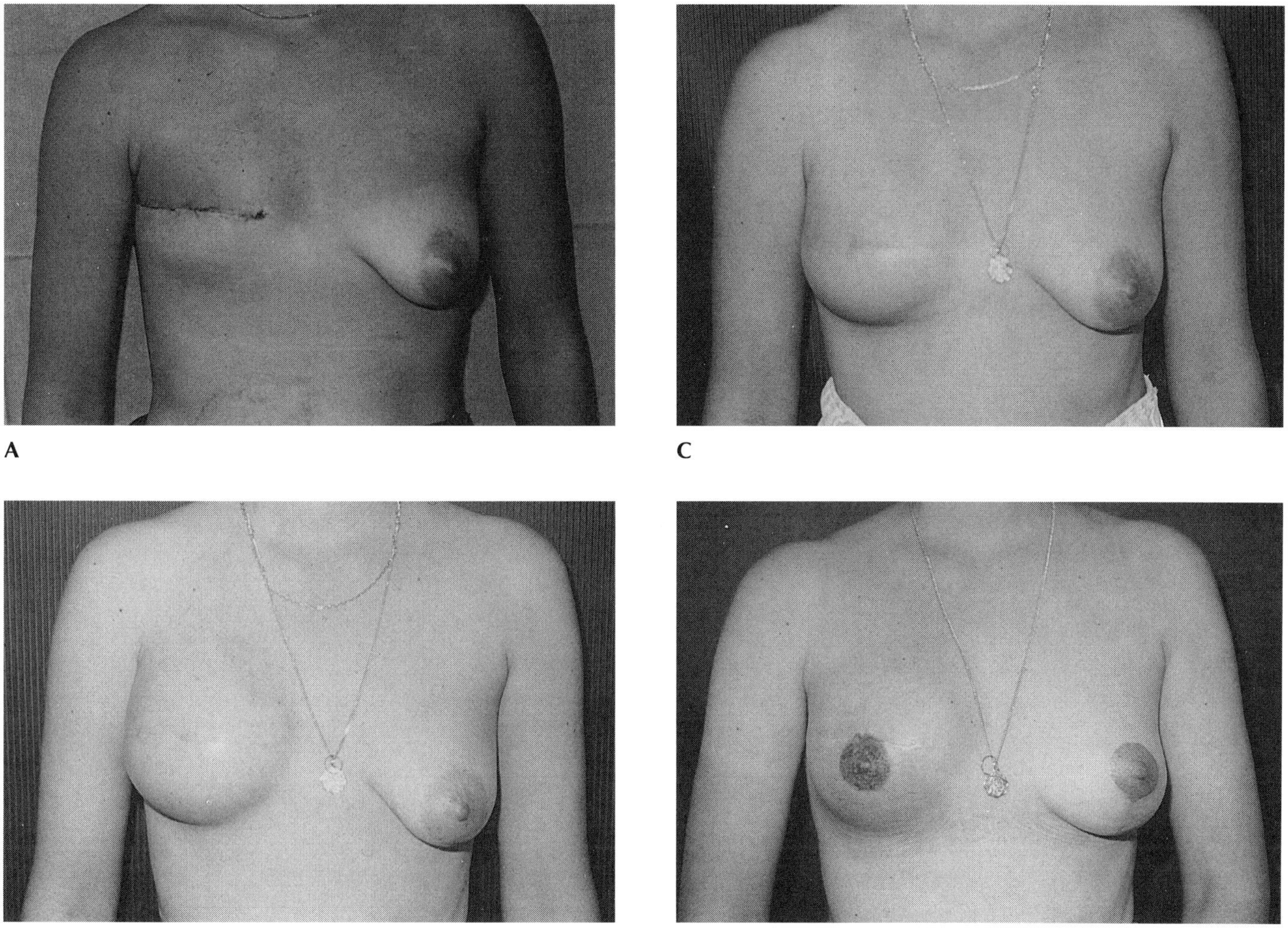

FIGURE 19-3
(A) Thirty-two-year-old woman following immediate reconstruction of the right breast with a 250 milliliter expander implant. (B) Implant overexpanded to 450 milliliters (saline). (C) Final volume 310 milliliters (saline). (D) Final result following nipple reconstruction and left mastopexy.

tectomy and immediate reconstruction is being offered to patients with unilateral breast cancer in whom the opposite breast is believed to be at higher risk.

The operative technique of immediate reconstruction is more complicated than that of delayed reconstruction. It is important to perform the mastectomy together with the general surgeon, planning favorable incisions and preserving as much skin as possible. It is also important to maintain the inframammary crease[4] (Figures 19-5 and 19-6).

The surgeon should not violate the pectoralis major muscle and should keep the pectoralis fascia intact. This will facilitate submuscular dissection.

At the completion of the mastectomy, the mastectomy site is irrigated with antibiotic solution and drapes are changed. The submuscular pocket can be approached through the pectoralis major muscle fibers if the muscle and inferior fascia are sufficient and if a moderate size breast is going to be reconstructed. If a larger breast is going to be reconstructed, it may be advisable to elevate the muscle-fascial layer below the inframammary crease area and elevate the muscle-fascial flap superiorly to allow unimpaired expansion of the pocket.

The key point is to detach the fascial band that runs between the pectoralis major muscle and the rectus

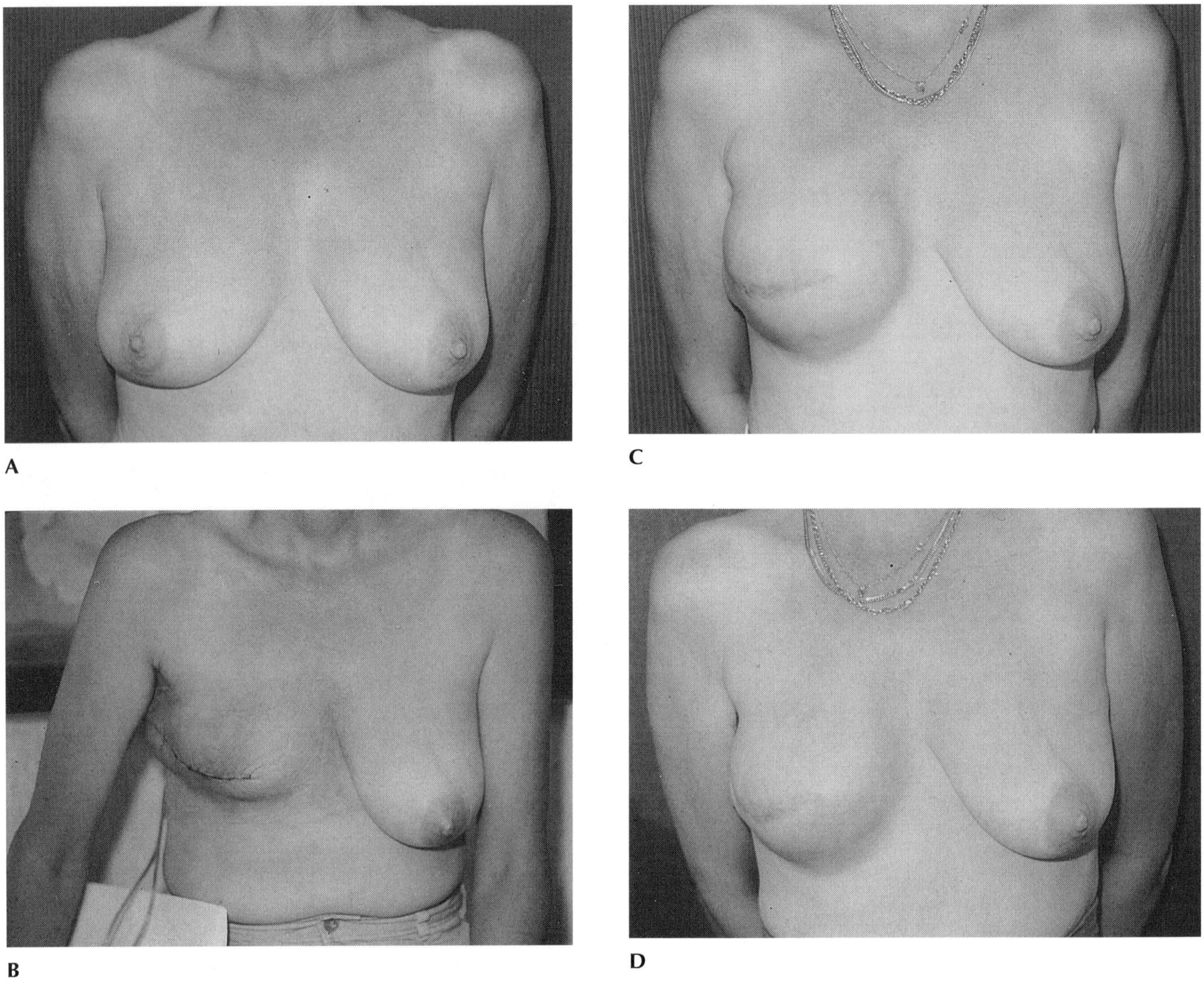

FIGURE 19-4
(A) Forty-seven-year-old woman prior to right mastectomy. (B) Immediate reconstruction using a 500 milliliter expander implant filled to 200 milliliters (saline). (C) Implant overexpanded to 500 milliliters (saline). (D) Final result. Implant volume 390 milliliters (saline).

abdominis muscle, because this will keep the implant in a superior position, if not detached. It is also important to maintain complete muscle coverage or at least to have a viable inferior flap onto which the muscle flap can be sutured.

Once again, drains are used. Care should be taken to place the drain through a fairly long subcutaneous pocket to avoid contamination of the injection dome.

Expansion following immediate reconstruction is delayed for several days to make sure that the skin flaps are viable. Expansion is then carried out in a similar fashion to that of delayed reconstruction.

COMPLICATIONS

Early postoperative complications include problematic wound healing, skin flap necrosis, and malposition of the implant. The most common complication is that of the high-riding implant.[5]

Most of these problems are related to surgical technique, emphasizing the importance of adequate plan-

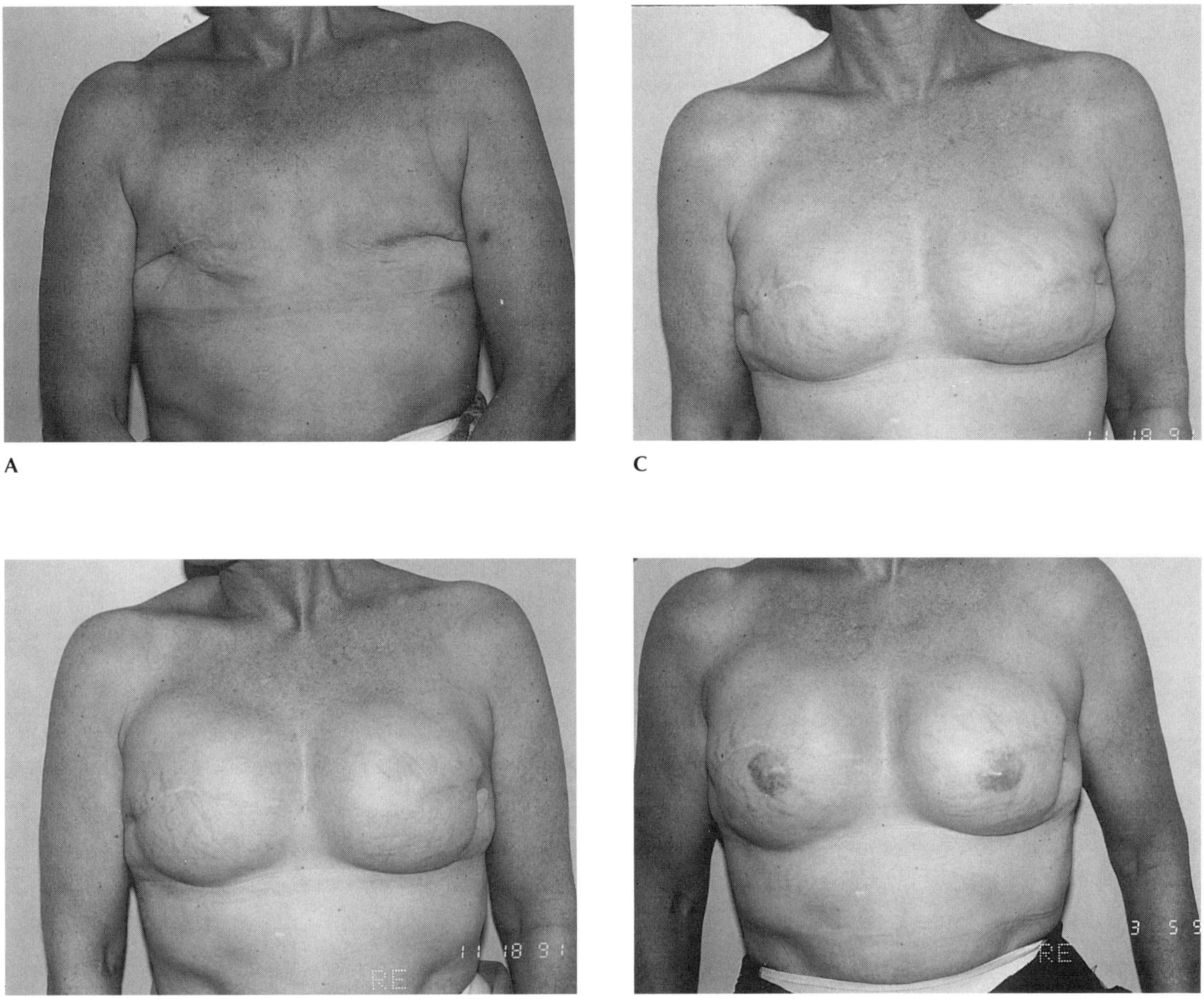

A C B D

FIGURE 19-5

(A) Forty-seven-year-old woman following bilateral mastectomies. (B) Bilateral reconstruction performed using 400 milliliter expander implants. Implants overexpanded to 600 milliliters (saline). (C) Volume reduced to 450 milliliters saline on the left and 500 milliliters saline on the right. (D) Final result following nipple reconstruction.

ning and technique. Other complications that have been seen include capsular contracture, which is seen in less than 5 percent of cases; infection, seen in less than 2 percent of cases; and mechanical problems relating to the implant itself, including rotation of the injection dome, kinks of the fill tube, the insertion of a needle into the breast implant, or spontaneous implant leakage. Should leakage occur, the saline will be absorbed, resulting in a rapid decrease in size. Replacement of the implant at this stage is easily performed through the original incision under local anesthesia.[5]

Patient satisfaction has generally been high, with approximately 80 percent of patients being adequately treated in a one-stage procedure. Approximately 20 percent of patients will require a second procedure, most often to lower the implant.[6,7]

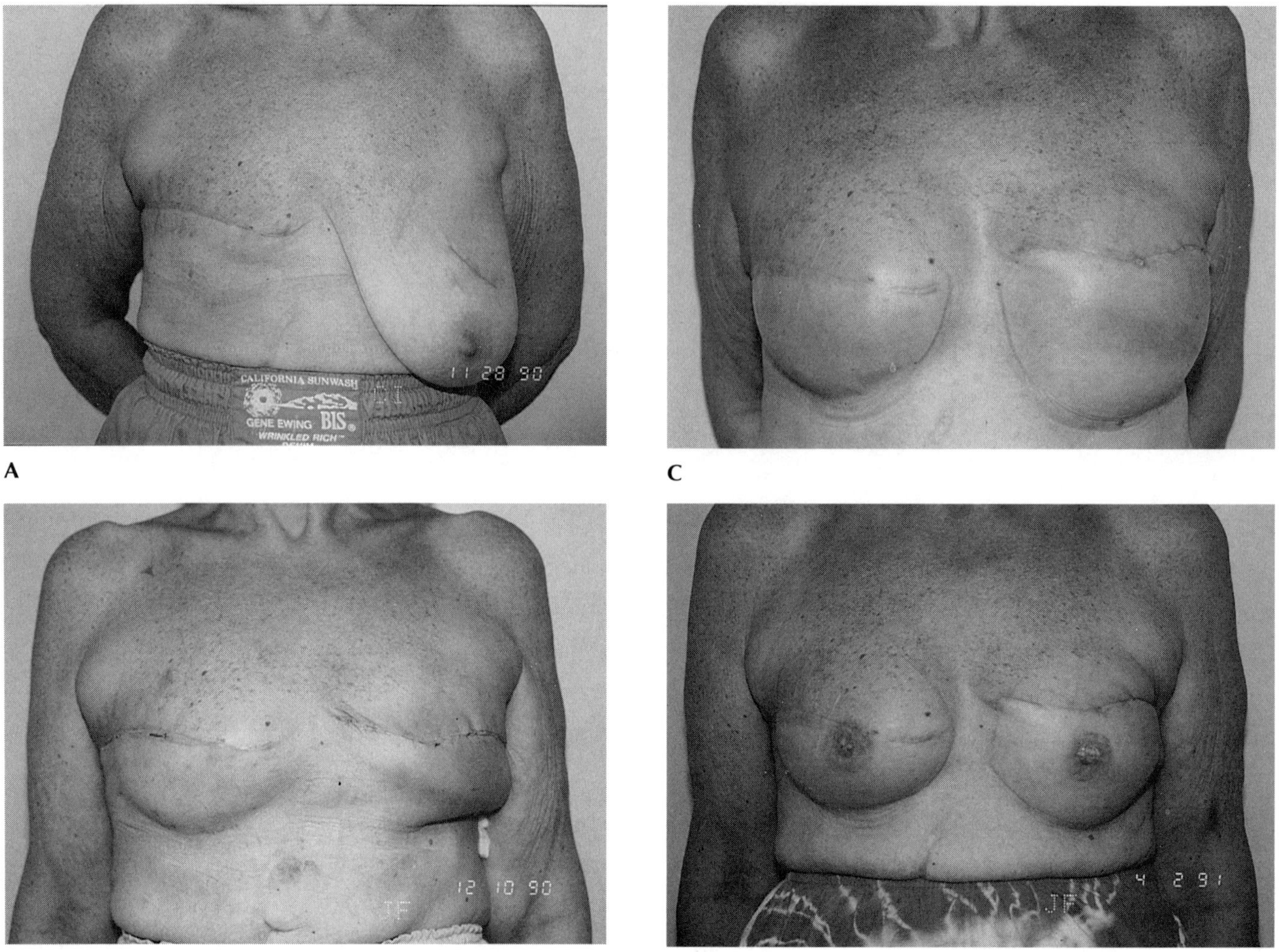

A B C D

FIGURE 19-6

(A) Sixty-two-year-old woman following right mastectomy with a biopsy positive lesion in the left breast. (B) Bilateral reconstruction with a 300 milliliter expander implant on the left and a 350 milliliter on the right. (C) Implants overexpanded to 500 milliliters on the left and 400 milliliters on the right. (D) Final result. Implant filled to 250 milliliters on the right and 180 milliliters on the left.

REFERENCES

1. Radovan C. Breast reconstruction after mastectomy using a temporary expander. *Plast Reconstr Surg* 1982;69:195–208.
2. Becker H. Breast reconstruction using an inflatable breast implant with detachable reservoir. *Plast Reconstr Surg* 1984;73:678–683.
3. Becker H, Maraist F. Immediate breast reconstruction after mastectomy using a permanent tissue expander. *South Med J* 1987;80:154–160.
4. Becker H. The permanent tissue expander. *Clin Plast Surg* 1987;14:519–527.
5. Van Heerden JA, Jackson IT, Martin JK Jr., Fisher J. Surgical technique and pitfalls of breast reconstruction immediately after mastectomy for carcinoma: Initial experience. *Mayo Clin Proc* 1987;62:185–191.
6. Becker H. The expandable mammary implant: An update. *Perspect Plast Surg* 1989;3(1):164–178.
7. Jackson IT. Immediate breast reconstruction with prolonged overexpansion using the Becker permanent expander prosthesis. *Eur J Plast Surg* 1992;15:79–84.

20

Correction of Anomalies of the Breast

ARMAND D. VERSACI

Anomalies of the breast are relatively uncommon; however, they are of serious consequence to the adolescent female. Fortunately, it has become possible to address and ameliorate the deformities in a relatively simple and satisfactory manner since the introduction of the silicone mammary prosthesis. Following the introduction of the tissue expander, further refinements in the field of breast reconstruction have occurred. Its use in breast reconstruction was initially directed toward postmastectomy deformities, but it was rapidly applied to the treatment of nonmalignant deformities, including the burned breast.[1–4]

It has always been difficult to estimate the volumetric need in any breast reconstruction. The problem of volume estimation in the reconstruction of an anomalous breast is even more difficult because the patient usually presents for treatment soon after the breast discrepancy is first noted. At that stage, it is impossible to accurately determine what the ultimate size of the uninvolved breast will be. Prior to the advent of tissue expanders,[5] it was not unusual to begin reconstruction early and then perform a second procedure after further asymmetry was noted as the normal breast enlarged. Expansion of the breast envelope was achieved by the serial exchange of the implant until full growth occurred. The method was actually a form of tissue expansion that relied on multiple operations, but it was deficient because it did not permit volumetric adjustments to be made intermittently to maintain symmetry.

With the introduction of the tissue expander, the surgeon had a device that removed the guess work from the determination of the volumetric needs of a patient.[5–8] It was possible to insert an expander when the deformity became apparent and then expand it incrementally over a period of years, if necessary, to maintain symmetry of the two breasts. The exchange of the expander for the implant was delayed until it appeared that full growth had occurred.

A further major change in the treatment of all breast deformities followed the introduction of the tissue expander/implant in its present form.[9,10] I am speaking of the Becker 50/50 (Mentor Corporation) expander-implant, which we have been using since 1988. The device has two lumens. The outer lumen is filled with silicone gel. The inner lumen can be filled by the percutaneous injection of saline into a valve that is left in the subcutaneous tissue close to the expander. The fill-tube passes through the outer, silicone gel chamber and into the inner chamber that is to be filled with saline. Two pigtail valves, one in the saline chamber and the other in the gel chamber, provide dual protection against leakage. With this device it is no longer necessary to exchange the tissue expander for a prosthesis. Only the fill-dome needs to be removed when the expansion has been completed.

A more recent technological development has been the application of textured silicone to the silicone shell of a prosthesis in an attempt to reduce capsular formation and contracture. It is still too early to know what the long-term effect of a textured silicone surface will be on capsule formation, and whether there will be a beneficial effect on capsular contraction.

Much of the anguish of growing up with asymmetrical breasts can be alleviated by beginning the reconstruction at an early age. An almost instant psychological change usually occurs once the device is implanted and mammary enlargement becomes apparent. We have not noted any adverse psychological responses in our series of patients with unilateral hypoplasia, Poland's syndrome, or tuberous and tubular breasts.

UNILATERAL HYPOPLASIA

In unilateral hypoplasia, the involved breast is easily noted to be smaller than the uninvolved breast. There is an insufficient amount of skin along with an insufficient amount of breast parenchyma. Usually the two deficiencies are proportional to one another and are centered around the nipple-areola. Often the areola is small. The breast and the areola may be more highly positioned on the chest wall than the more fully developed breast, which has undergone its normal descent.

In more severe cases, there may be an asymmetrical deficiency in both skin and breast parenchyma. The two deficiencies are usually present in the lower hemisphere of the breast. In these cases, the inframammary fold and the nipple-areola appear more highly displaced. Determining the proposed position of the new fold becomes more of a problem than in the less severe form of breast hypoplasia.

There are analogies and differences between treating these patients and patients desiring augmentation mammaplasty to enlarge small but well-shaped breasts. Indeed, some patients presenting with asymmetrical breasts may have unilateral macromastia that requires reduction mammaplasty, while others have minor discrepancies that can be handled as unilateral augmentations.

The preoperative analysis of the deformity is of vital importance in planning the reconstruction. When the breast is small but symmetrically shaped, the principal concern is that of accurately locating the inframammary fold. When a small breast is determined to have asymmetrical deficiencies in both the skin cover and glandular volume, the planning becomes more difficult. One must be concerned about the position of the inframammary fold, but also the base diameter of the reconstruction and what portion of the breast is to be selectively enlarged (shape of the reconstructed breast). The principal concern is that of determining the base diameter of the segment to be reconstructed. The question of volumetric requirement assumes a lesser role than that of base diameter size and positioning of the device because the volume can be adjusted over a wide range, irrespective of the diameter of the device used. The question of ultimate areolar size is deferred until the expansion is completed, since the areola does expand rapidly as the mound grows.

After placement and expansion of the device, we have noted the gradual descent and enlargement of the nipple-areola complex and what appears to be the stimulation of growth of the skin envelope beyond that which might be attributable to the direct expansion of the skin. In these young females, the presence of an expanding mass in the breast area appears to trigger enlargement and downward movement of the breast. It is for this reason that we recommend beginning the reconstruction early and proceeding slowly to take advantage of unknown forces in normal development that are unleashed by the reconstruction (Figure 20-1).

The size of the device (expander or expander-implant) is chosen by measuring the base diameter of the anticipated size of the breast parenchyma of the normal breast. The skin overlying the breast can be compressed inward to assist in determining the width of the glandular elements. The device will be excessively wide if the measurement of the base of the skin envelope is used in selecting the expansion device.

Initially we felt that the device should only be placed in the subglandular plane so that the normal downward movement of the mound would occur. In cases where there is a marked discrepancy in the position of the two breasts, we tend to use this placement position. In most of the other cases we place the device in the submuscular plane. We determine the position of the inframammary fold on the normal breast and transpose it to the opposite side. The measurement serves as a guide in choosing the position of the fold on the breast to be reconstructed. It is preferable to err by placing the incision about 1 centimeter below the measured position than placing it too high.

The access incision is within the proposed inframammary fold.[11] It is 5 to 6 cm in length and is carried through the skin and subcutaneous tissue to the pectoral fascia underlying the fold. The fascia and the underlying muscles are divided down to the anterior surface of the rib. By employing a combination of sharp and blunt dissection, a pocket of appropriate size to accommodate the device is made beneath the pectoralis muscle and the serratus anterior muscle. The pocket is made only large enough to accommodate the expander to minimize its movement. The device is inserted into the pocket; the tear in the muscle and fascia is not sutured, thus leaving the lower portion of the device exposed to the subcutaneous layer. The valve is left in the subcutaneous tissues, and the wound is sutured. We always leave a suction drain in the submuscular pocket and bring it out through a stab-wound in the midaxillary region. When using a conventional expander, it is

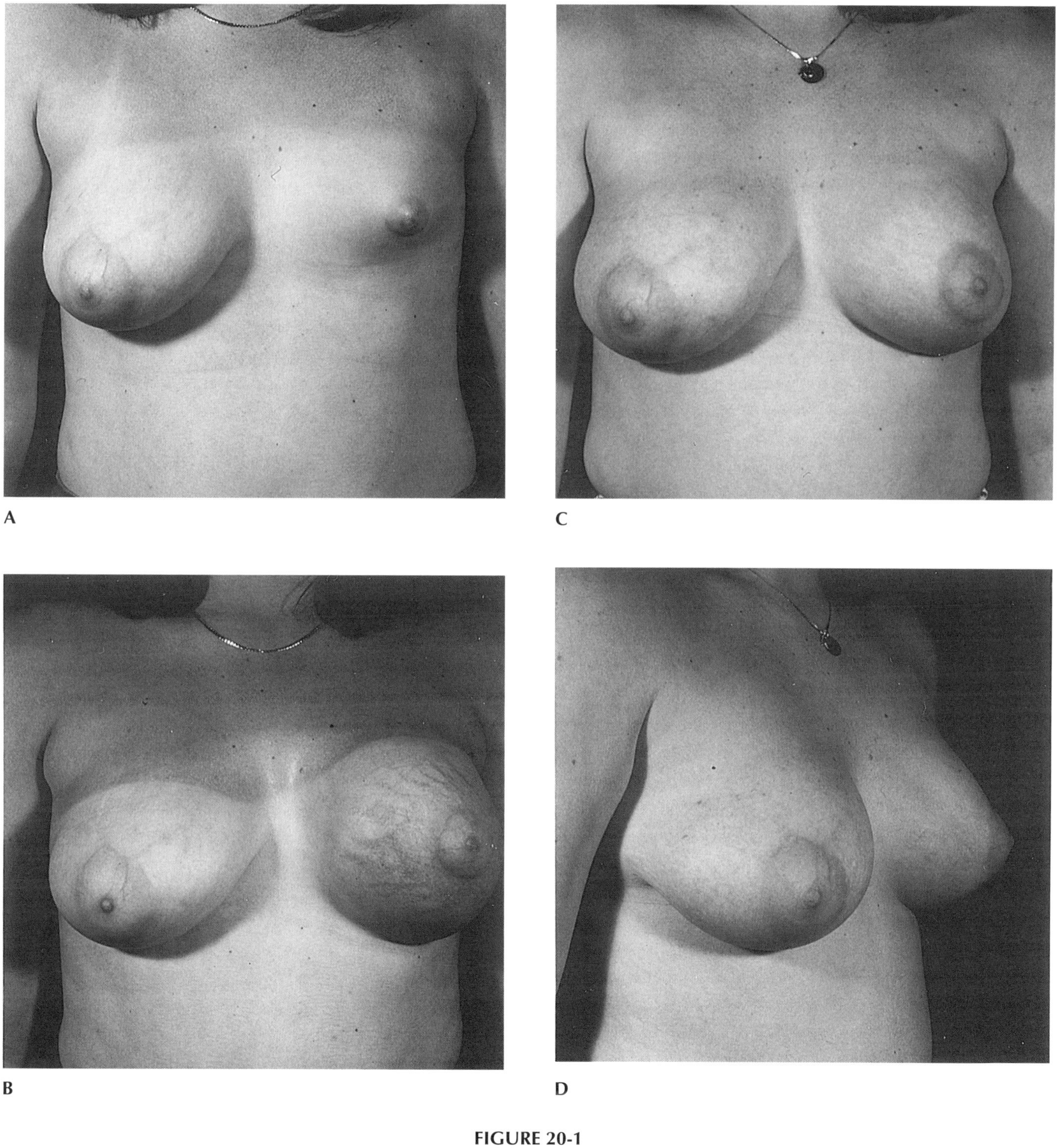

FIGURE 20-1

Unilateral hypoplasia of breast. (A) Preoperative photo. (B) Expander inflated to 200 milliliters above the desired volume. (C) Final view. Expander exchanged for gel-silicone implant. (D) Final view.

filled with saline to a comfortable volume to avoid any undue pressure of the skin. When using a Becker 50/50 expander-implant, we do not initially instill saline into the device. Antibiotics are given in the perioperative period and continued until the drain is removed.

Because we try to take full advantage of the unknown normal factors governing normal growth and development, we refrain from being too hasty in filling the device. We are content to fill either device in 25-milliliter aliquots of saline first injected at 3 to 4 weeks postoperatively and then as infrequently as possible to follow the natural changes that may be stimulated by the presence of the device.

We have patients apply a snug fitting strap around the chest that crosses the upper hemisphere of the breast to apply downward pressure on the expanding breast mound. The pressure assists in driving the expander downward, thus placing the fullness where it is most beneficial in producing a pleasing breast shape.

The expander is not replaced by an implant, nor is the valve of the expander-implant removed until full development of the normal breast is observed and been matched by the reconstruction. It may be a year or more before further surgery is undertaken. The final surgical step may include exchange of the expander for an implant; removal of the fill-port of an expander-implant; enlargement of the areola; modification of the opposite breast by mastopexy; or a combination of procedures.

POLAND'S SYNDROME

A complex type of unilateral hypoplasia is found in Poland's Syndrome.[12–14] This syndrome represents degrees of aplasia of the breast, thorax, and upper extremity. Unlike Poland's original description, the vast majority of affected individuals have minimal or no extremity deformities. The most common deformity is the absence of the sternal head of the pectoralis major muscle. There may be a deficiency of breast or nipple-areola; deficiency of subcutaneous fat and axillary hair; abnormalities of the costal cartilages and anterior ends of ribs; various abnormalities of the upper extremities including shortening and syndactyly; and abnormalities of the latissimus dorsi, deltoid and supra- and infraspinatus muscles. Because of the potential complexity of this syndrome, a precise assessment of the deficient tissues is essential before embarking on a plan for reconstruction. The goal of reconstruction is to restore a more natural contour to an area that is distorted at many levels. In women, the aims are to provide a symmetric breast mound, a natural-appearing nipple-areolar complex, and an adequate correction of the infraclavicular defect that results from the absence of the sternal head of the pectoralis major muscle.

Construction of the breast mound is similar to that for the unilateral hypoplastic breast described earlier. The skin expander is enormously useful in developing the breast envelope and makes volumetric equality possible.

Similarly the nipple-areola deformity is treated as it is in the case of unilateral hypoplasia. This unit tends to descend to a relatively normal position as the breast mound is expanded. Additional procedures to enlarge the areola or alter its shape may also become necessary.

The chest-wall deformities are exaggerated by the absence of the breast mound. Their presence is not associated with disability in my experience. A deformity usually does not require reconstruction and is generally obscured by the creation of a normal breast mound that places it clearly between the two breasts, where it is not very conspicuous. Sometimes it will appear as though the patient has a pectus excavatum (Figure 20-2).

When an aplasia or hypoplasia of the pectoral musculature exists, the reconstruction must also include the restoration of fullness in the infraclavicular area. This is a most difficult goal that is seldom if ever fully achieved. A shaped silicone prosthesis can be attached to or fabricated with the mammary prosthesis and inserted into the expanded pocket; or a shaped infraclavicular prosthesis can be inserted into a separate pocket above the mammary implant to assist in obscuring the defect (Figure 20-3).[15] Neither of these has proven to be totally satisfactory because evidence of the defect usually persists. The absence of muscle renders it impossible to camouflage the prosthesis adequately. The shape or volume of the implant may be such that a hollow or a ridge persists between the mound and the infraclavicular defect. This will occur even if the infraclavicular implant is incorporated into the design of the breast prosthesis.[15]

The transposition of a muscular pedicle flap offers the best hope for the successful reconstruction of the infraclavicular defect, but at the price of creating a similar defect in the donor area.[12–14] The latissimus dorsi muscle flap transposition has been used in two ways for this purpose (Figure 20-4). It can be transposed without dividing its insertion and sutured into the lower border of the clavicular head of the pectoralis muscle. If a sufficient portion of this muscle remains, a satisfactory reconstruction will occur as shown independently by Hester, Argenta, and Seyfer.[13,14] What remains is the loss of definition of the anterior axillary fold. The fold can be incompletely reconstructed by detaching the insertion of the latissimus dorsi muscle, rerouting it to the anterior thoracic region and reinserting it into the humerus.

The operation is done through either a vertical incision along the anterior border of the latissimus or a horizontal incison in the bra-strap line. A second incision

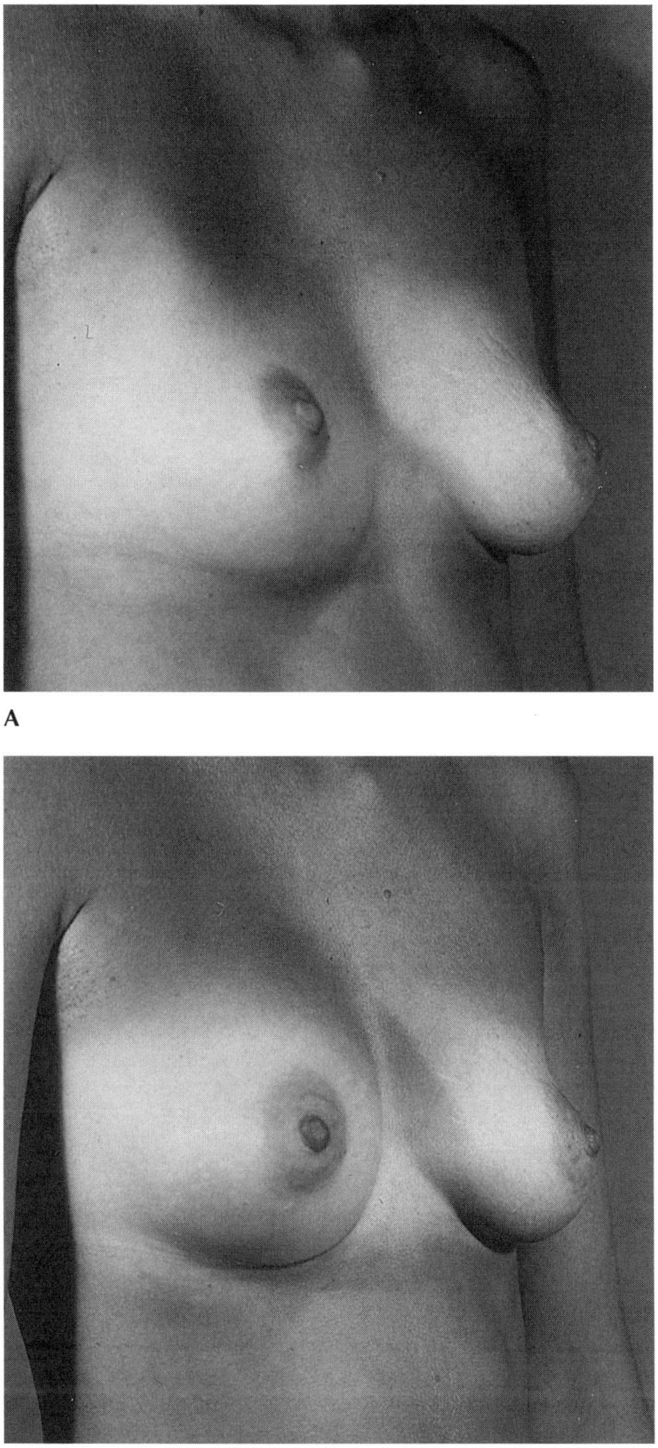

A

B

FIGURE 20-2
Unilateral Poland's syndrome. (A) Preoperative. Note the depression due to the absent sternal head of the pectoralis major muscle and the depression of the thoracic wall. (B) Final view. A Becker 50/50 implant was inserted through an inframammary incision and expanded to the desired volume. Note that the hollow in the thoracic region has been obscured by the implant.

is made in the apex of the axilla to detach the insertion of the latissimus. The skin is dissected from the latissimus and the muscle is detached from the thoracic wall. Anteriorly, it is detached from the teres major. The thoracodorsal vessels are identified and the branch to the serratus is divided. The humeral insertion of the latissimus is identified and severed to free the muscle completely from its neurovascular pedicle.

A tunnel is created to deliver the muscle to the anterior chest pocket. The anterior incison is made along the inframammary fold. If a tissue expander has been used to create a breast mound, it is removed at this time and exchanged for the permanent implant. The latissimus is sutured medially to the fascia overlying the lateral margin of the sternum, inferiorly to the fascia of the chest wall, superiorly to the clavicular portion of the pectoralis. The anterior axillary fold is created by suturing the latissimus tendon to the periosteum of the humerus below the insertion of the clavicular head to the pectoralis major muscle. To prevent the lateral dislocation of the implant, sutures are placed through the muscle and the chest wall along the lateral border of the breast mound. The tendency for implant displacement has diminished since the introduction of textured implants. If a tissue expander has been used, the permanent implant is placed within the capsule, eliminating the concern about displacement.

Our preference is to sequence the reconstruction by first creating the breast mound utilizing a tissue expander or, more recently, an expander-implant (Becker 50/50, Mentor). The expansion is done slowly in the hope that a normal pendulous breast shape will be realized, as described under the section on unilateral reconstruction. Once the mound is created, attention is turned to the chest wall defect. If desired, the latissimus transfer is performed as described. If the patient is fully developed when first seen and if the pectoralis defect is significant, transposing the latissimus muscle and placing the expander or the expander-implant beneath the muscle is the preferred method.

In cases where the infraclavicular defect is small and discrete, the teres major muscle can be used to fill out the infraclavicular hollow, as described by Dr. Philip Godfrey and colleagues (Figure 20-5).[16] A vertically oriented incision is made just posterior to the free border of the latissimus muscle. The teres major muscle is identified after identifying and mobilizing the latissimus dorsi and its neurovascular bundle, the subscapular, circumflex scapular, teres major vessels, as well as the lower subscapular nerve. Superficial dissection posteriorly over the teres major and latissimus allows circumferential control over the tendon. The muscle is separated from the scapula, its tendon is divided, and unwanted vascular tethering is freed. The

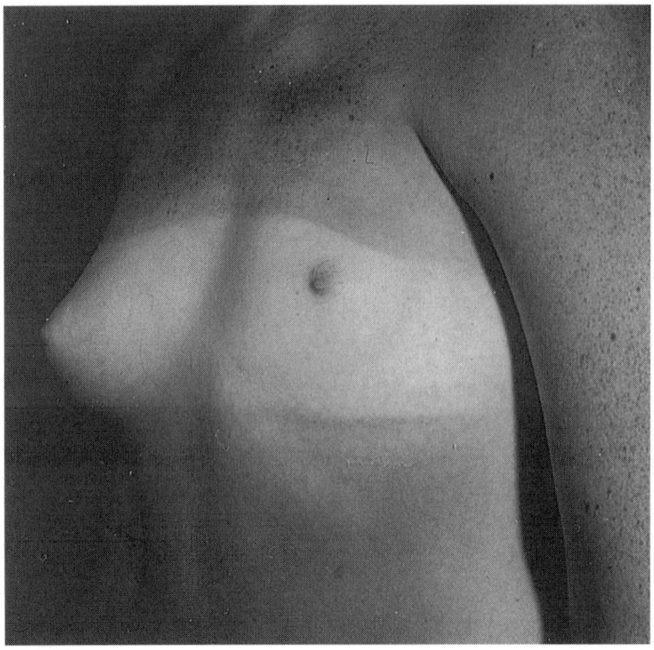

A

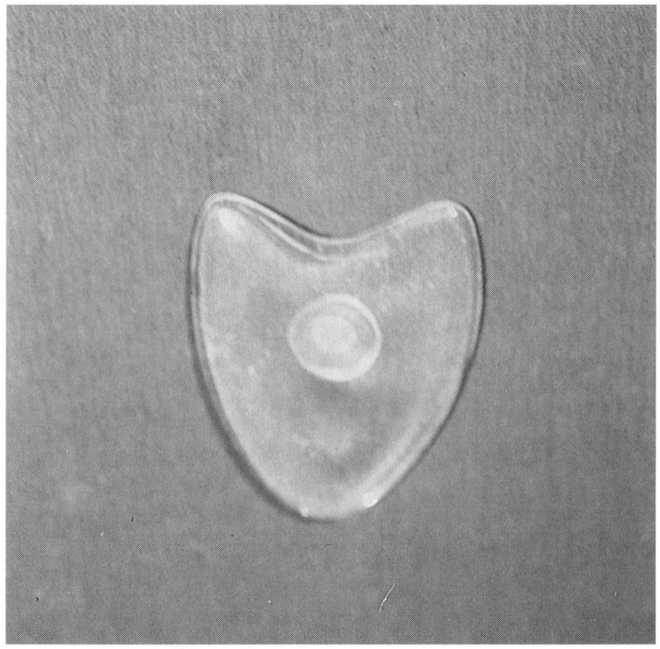

C

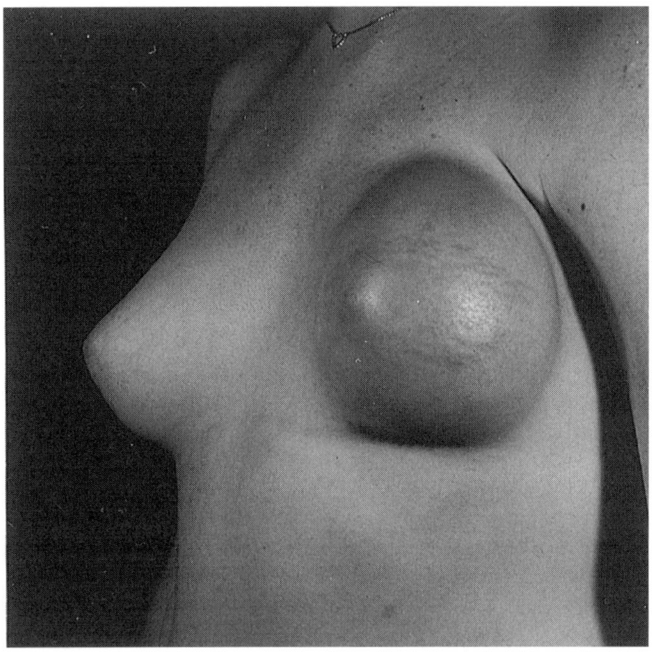

B

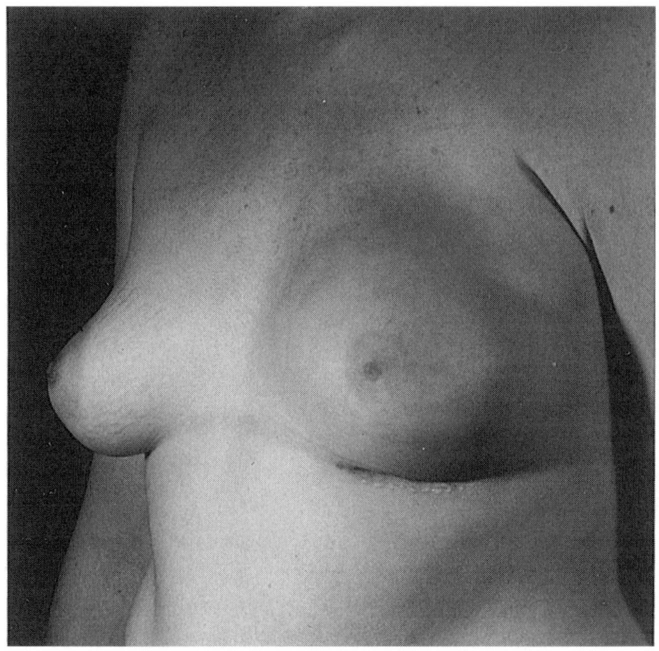

D

FIGURE 20-3

Unilateral Poland's syndrome. (A) Preoperative view. Note the significant deformity due to the absence of the sternal head of the pectoralis muscle. (B) Status of the patient at the completion of expansion, which was 200 milliliters above the anticipated volume of the permanent implant. (C) An infraclavicular implant made of a silicone shell filled with gel. (D) Final view following the exchange of expander for a silicone-gel implant and the insertion of the infraclavicular prosthesis through a second incision made within the axilla. Note that the infraclavicular defect has been incompletely corrected by the use of the infraclavicular implant.

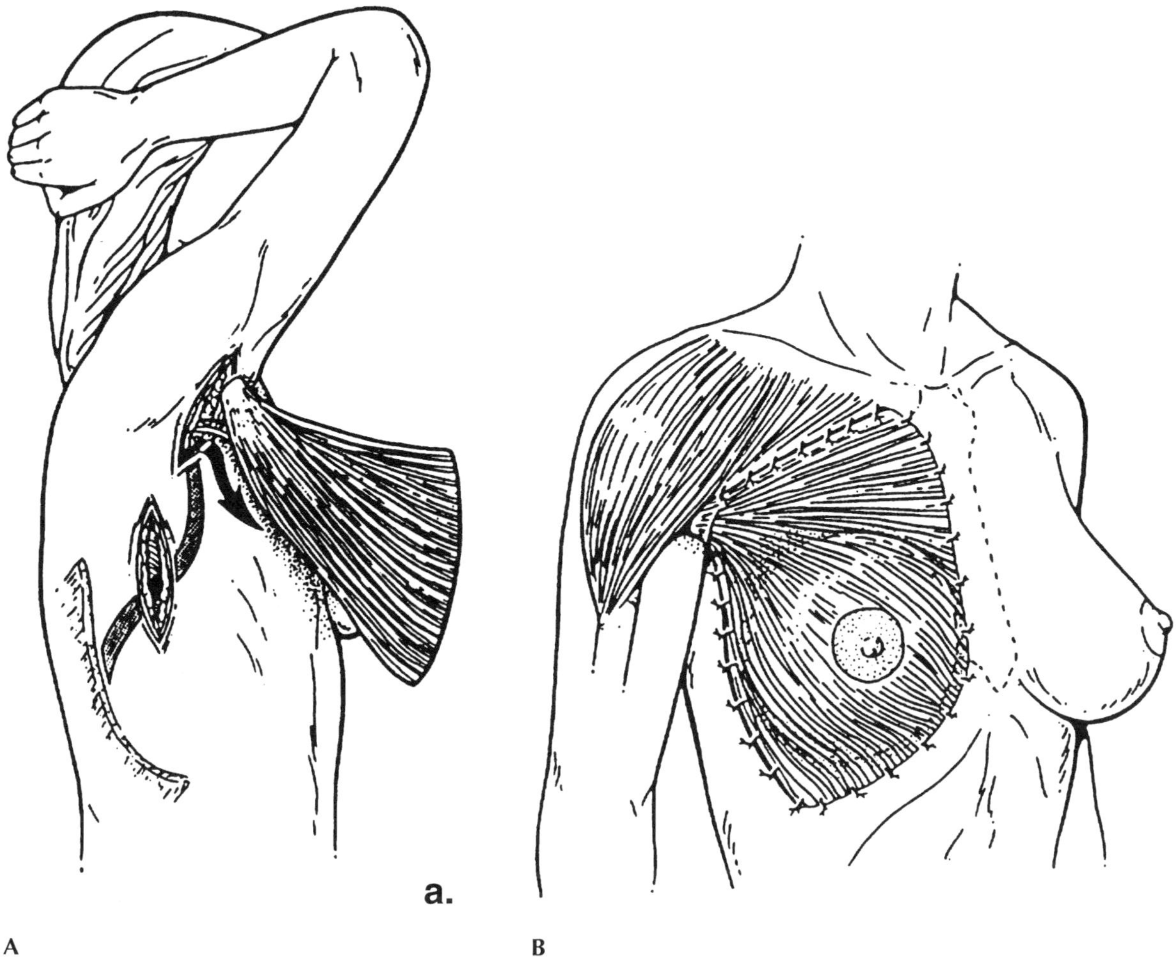

FIGURE 20-4
Latissimus dorsi flap. Transposition of the latissimus dorsi flap to the anterior chest wall and the reinsertion of its tendon into the tendon of the clavicular head of the pectoralis major muscle. The position of the implant centered beneath the nipple-areola is shown. (Reproduced with permission. Hester TR, Bostwick J III. Poland's syndrome: Correction with latissimus muscle transposition. Plast Reconstr Surg *1982;69(2):226–233.*

muscle-tendinous unit is then transposed with its neurovascular pedicle intact. The muscle is sutured to the periosteum along the ribs and its tendon is sutured to the humeral insertion of the clavicular head of the pectoralis major. An excellent description of the technical considerations can be found in the original paper.[16] It is essential that the muscle be innervated and placed under proper tension to maintain its contractility and, therefore, to preserve the anterior axillary fold.

Dr. David Sharpe of Bradford, England has also performed two teres major reconstructions to correct the infraclavicular deformity of Poland's syndrome.[17] His first patient initially had difficulty with shoulder motion that cleared within several months. His second patient did not encounter difficulty with shoulder motion.

Correction of the infraclavicular defect by the transfer of the teres major muscle is a significant alternative to using the latissimus dorsi transfer. Apart from the healed incision, there is no contoural donor site deformity as seen when the latissimus is transferred.

In the more severe forms of the deformity, where the skin is atrophic, the latissimus transfer is an excellent method of reconstruction because the muscle bulk is sufficient to cover all, or at least a significant portion, of the implant and still fill the infraclavicular defect. Fortunately, most cases are not that severe. In such

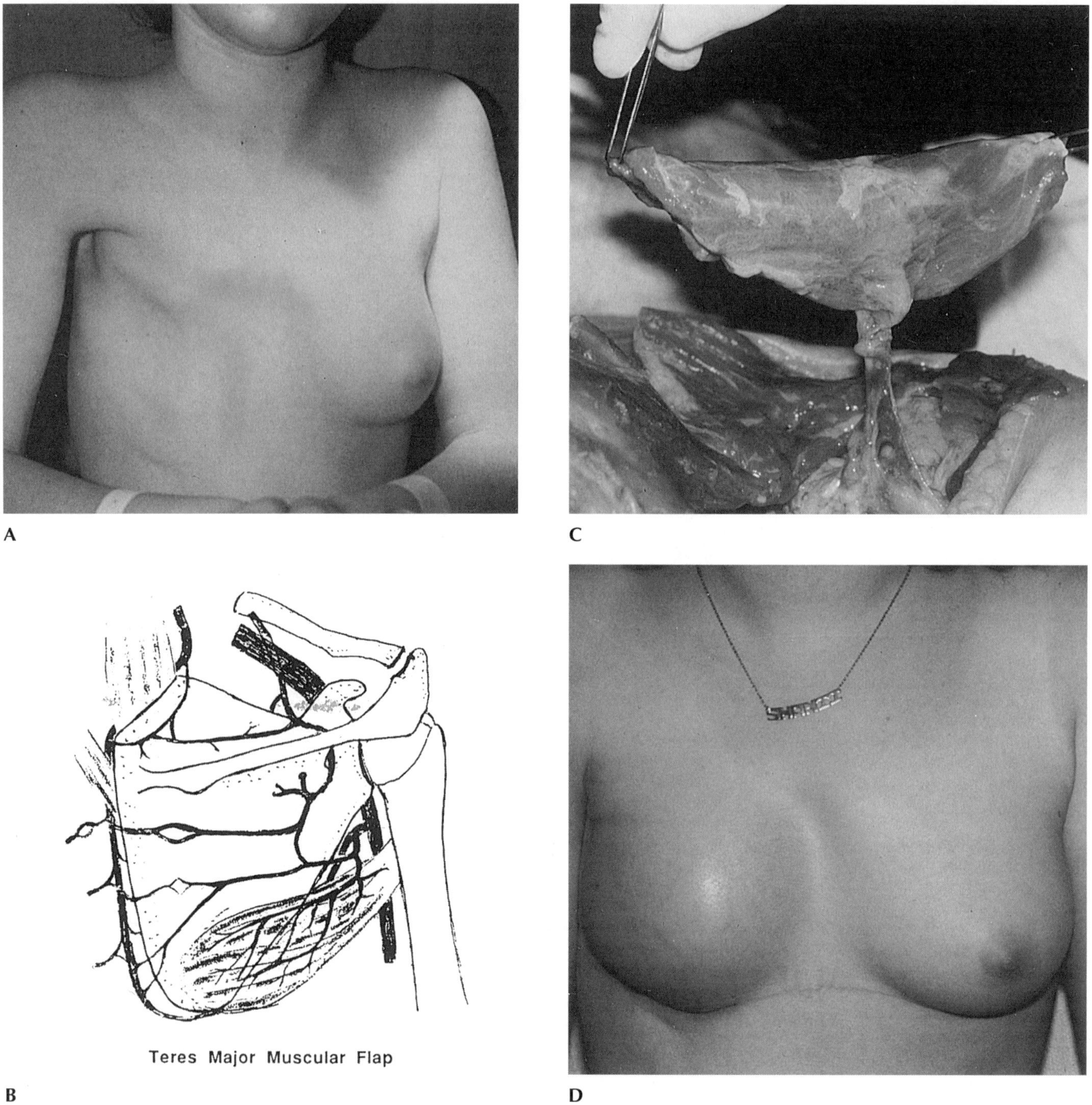

FIGURE 20-5

Unilateral Poland's syndrome. (A) Front view showing the preoperative appearance of the breast. Note the absence of the nipple-areola. (B) A diagram showing the teres major muscular flap. (C) The teres major muscular flap elevated and showing the intact neurovascular bundle. (D) The final appearance of the case with the teres major flap transposed to replace the sternal portion of the pectoralis major muscle. (Courtesy of Dr. Philip Godfrey) cases, many patients are content with the reconstruction of the mound and do not actively pursue the correction of the infraclavicular defect if it requires the transposition of a muscle pedicle flap. When patients weigh the possible gains against the certainty of having a new deformity at the donor site, they frequently reject the procedure.

TUBEROUS OR TUBULAR BREAST

The term tuberous breast is used to describe a breast deformity that is characterized by a circumferential deficiency of skin envelope at the base of the breast, and a deficiency of parenchyma. The deficiencies of both structures are usually asymmetric and concentrated in the lower hemisphere of the breast. The nipple-areola is deformed by a ring of tissue at the periphery of the areola through which breast tissue "herniates" into the areola, causing it to protrude. The level of the inframammary fold is usually raised. The condition may be unilateral or bilateral (Type 1 or "true tuberous breast").[18]

Another related defect is the tubular breast. It is characterized as having a circumferential deficiency of skin envelope along with generally symmetrical deficiency of skin. Although the nipple is usually large, herniation of breast tissue into the areola does not usually occur. Such a breast is usually ptotic, tubular in shape, and the nipple-areola points downward (Type II or "tubular breast")(Figure 20-6).[18] Many consider the two defects to be variants of the same deformity.

The treatment of tuberous breast deformities is a highly individualized endeavor. The deformity is so varied that a critical evaluation of each case is necessary. Although the deformity falls into two broad categories (tuberous and tubular), there are many possible variations even if the condition is bilateral. The most important thing we have learned is that the tissue expansion technique will allow the breast mound to be created without introducing pedicle tissue from the thoracic region.

A review of the literature reveals that many surgical procedures have been used with varying degrees of success in the treatment of the tuberous breast. The treatment objectives of the procedures reviewed are these: expand the base circumference of the breast; expand the skin of the lower hemisphere; release the skin tightness at the skin-areola junction; lower the position of the inframammary fold; increase the volume of the breast; reduce the size of the areola and correct its deformity; and perform a mastopexy to correct ptosis, if necessary.[18–26]

Our method of treatment differs from the others that rely heavily on extensive surgical procedures to achieve the above objectives. We correct the circumferential skin envelope deficiency with the use of a tissue expander.[27]

The treatment varies in accordance with the deformity. In general, we feel that there are two broad types of deformity as suggested and described by Rees and Aston—tuberous and tubular.[18] Since the two types have been seen in the same patient, one may speculate that the two deformities may be variants of the same developmental anomaly.

The tuberous breast is small but the deformity is accentuated by the asymmetrical deficiencies of both skin and parenchyma that are localized in the lower hemisphere of the affected breast. It is further accentuated by the skin tightness at the areolar rim, which creates the impression that a herniation of parenchyma into the areola exists. The inframammary fold is raised. A device with a base diameter that corresponds to the diameter of the specific region to be expanded is selected. The tissue expander is placed in an asymmetric position, either beneath the gland or in the subpectoral muscle plane, and is centered beneath the deficient skin and parenchyma. It is introduced through an inframammary incision located in a position that corresponds to the normal inframammary sulcus of the uninvolved breast (Figure 20-7).

Intraoperative expansion is limited to approximately 100 to 150 mL of saline when the traditional round expander is used (Dow Corning's Lapin design). When the Becker 50/50 expander-implant (Mentor) is used, saline is not added at the time of insertion. Subsequent percutaneous injections of saline commence once the wounds heal. Injections of 25 to 50 mL of saline are carried out at monthly or longer intervals to permit expansion to occur slowly, especially in the young patient who has not achieved full breast development. The slower rate of expansion theoretically favors true growth of skin as opposed to the "borrowing" or mobilization of skin from the periphery of the wound.

It is a good idea to overexpand the skin to minimize the skin tightness at the rim of the areola. Selective expansion in this particular area, if it occurs, helps correct the nipple-areola deformity and at times obviates surgical correction (Figure 20-7). The excess saline is removed either before or at the time of the exchange of the expander for the permanent implant or when the valve is removed from the expander-implant.

There is no clear cause for the areolar deformity seen in the tuberous breast. It is clear that through tissue expansion, the apparent paucity of skin at its areolar junction can be addressed without rotating a pedicle flap. However, in this group of patients, we have not been able to expand the skin sufficiently to completely

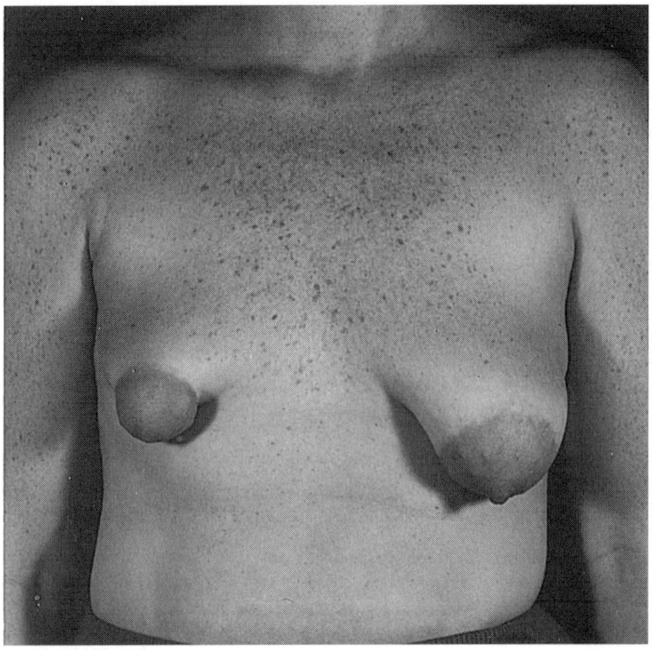

A

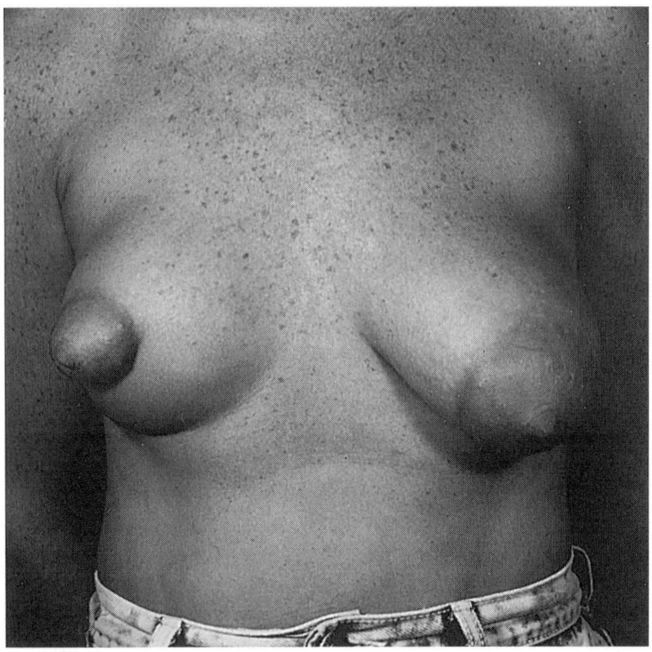

B

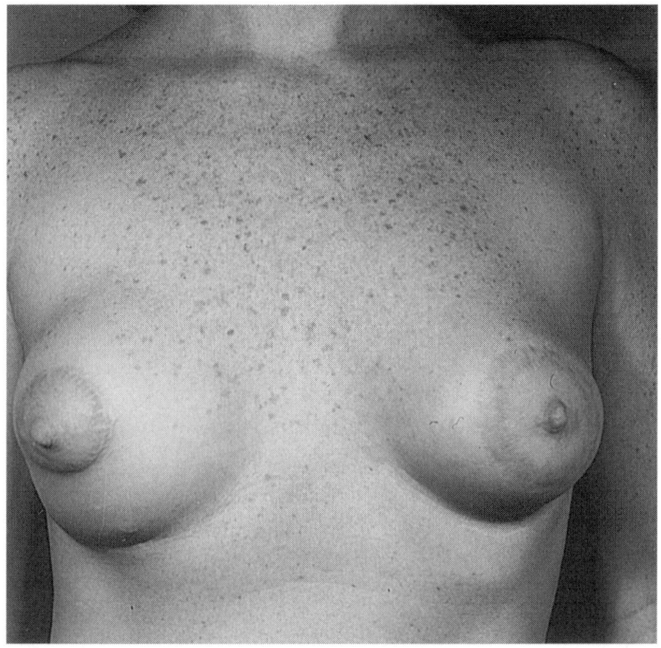

C

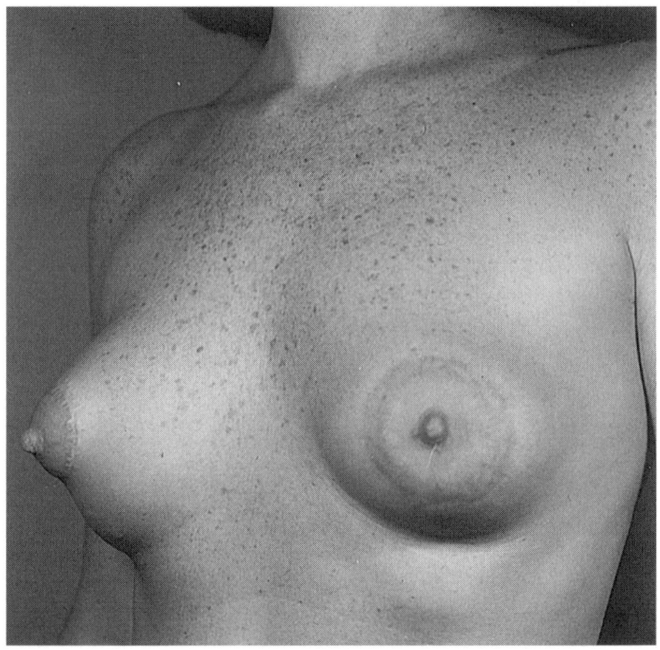

D

FIGURE 20-6

Tuberous breast and tubular breast. (A) Front view, preoperatively. (B) Front view following the use of Becker 50/50 expander-implants to expand the tuberous breast and the tubular breast. The right device was placed asymmetrically to expand the lower hemisphere of the breast. The left implant was placed more centrally. Note that the tubular breast projects well away from the thoracic wall, thereby exaggerating the deformity. (C) Front view following the removal of the Becker 50/50 expander-implant on the right side. No attempt was made to alter the skin-areola junction on that side. On the left side, a circular mastopexy was performed to suture the base of the parenchyma to the thoracic wall. The diameter of the areola was reduced by performing a doughnut resection. (Reprinted with permission from Versaci AD. Treatment of tuberous breasts utilizing tissue expansion. Aesth Plast Surg *1991;15:307–312.)*[27]

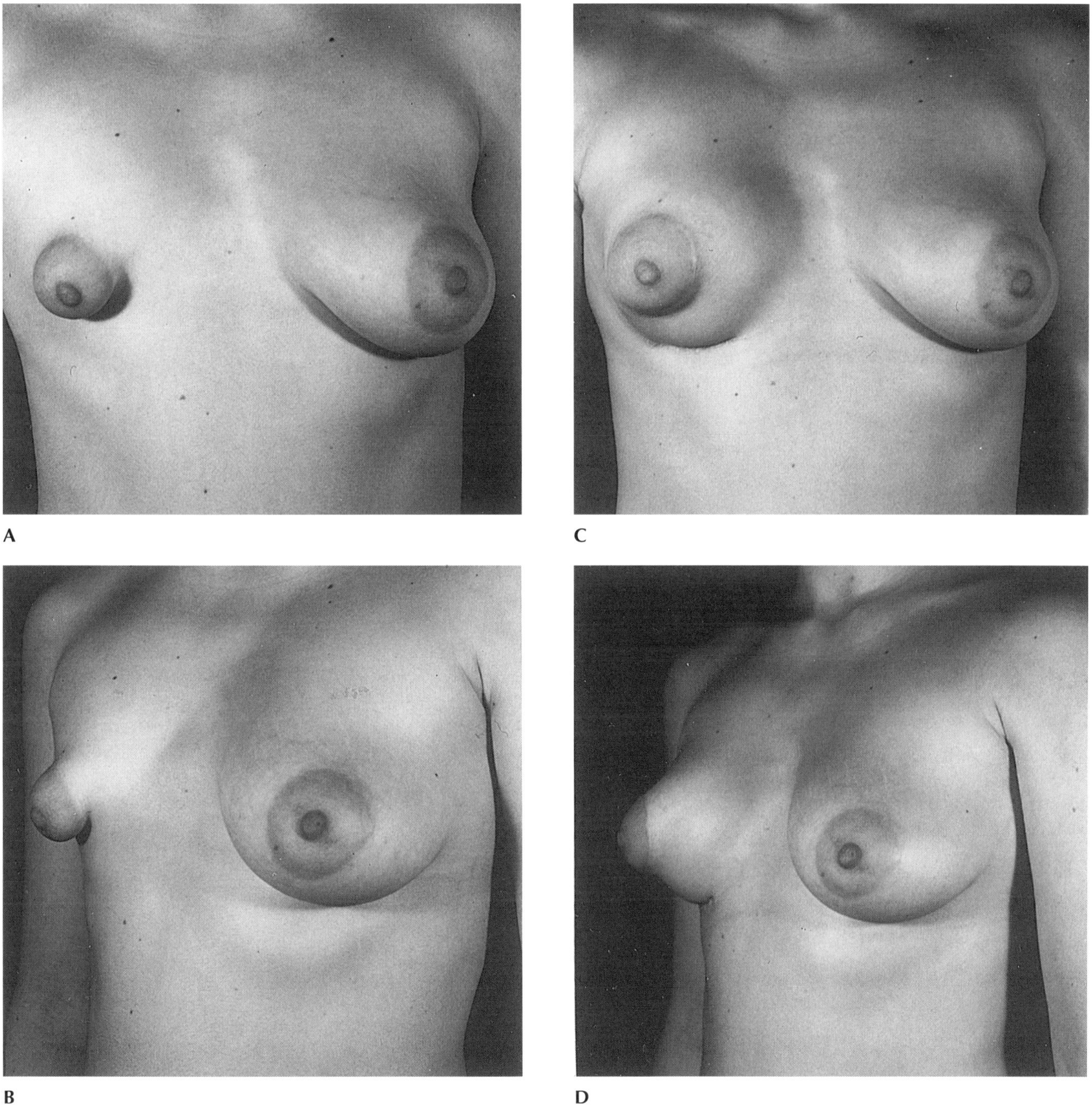

FIGURE 20-7

Unilateral tuberous breast. (A) Front view of the deformity. (B) Three-quarter view of deformity. (C) Final view after patient had the breast tissue expanded through the use of a conventional expander that was exchanged for a silicone-gel implant. (D) Three-quarter view showing that the defect at the skin-areola junction was not completely corrected by expansion alone.

eliminate the deformity, even after performing a circumareolar mastopexy.

Surgery to correct the nipple-areola deformity and to telescope the herniated parenchyma is performed as part of the operation to either exchange the expander or remove the fill-port. We prefer to use the Bass modification of the Aston-Rees procedure to reduce the size of the areola.[18,19] This eliminates the need to remove medial and lateral wedges of skin when suturing the undermined skin to the areola. Although we have not performed the operation described by Teimourian,[21] we feel that it may positively affect the final appearance of the reconstruction if used in conjunction with the treatment plan that we have been following.

The tubular breast is symmetrically small because both the skin and parenchyma deficiencies appear centralized on the breast. The base circumference is decreased, the breast has a cylindrical shape, its nipple-areola is enlarged, and the breast may be ptotic.

The expander is centered beneath the deformed breast to expand the skin and increase breast volume symmetrically. It is introduced through an inframammary incision that is planned at the level of the normal fold. Its position and size will determine the base diameter of the reconstructed breast. The expander is placed in either the subglandular or the subpectoral plane. Expansion proceeds as for the tuberous breast. However, during expansion, ptosis becomes accentuated because the breast tissue is simply pushed ahead of the expander. Very little overexpansion becomes necessary. Reduction in size of the areola, along with the introduction of an implant or the removal of the fill-port, are done in conjunction with the mastopexy to correct ptosis.

We have encountered our greatest difficulty in treating those patients with a bilateral deformity in whom one breast is tuberous and the other is tubular. Throughout the entire expansion phase of the therapy, it is difficult to get a clear sense that the two breasts are expanding as desired. In one breast there exists an asymmetric deformity of the mound, while in the other there is a concentric lack of breast mound development. As the expansion proceeds, the tubular breast continues to project forward and therefore appears to be more elongated in shape than the tuberous breast, whose base is being expanded into a breast mound. The problem here is trying to estimate how an effective mastopexy can be planned and executed without resecting ptotic parenchyma tissue. We prefer to use the traditional circumareolar mastopexy, but in one case it was necessary to suture the breast tissue to the pectoral fascia. We initially felt that the breast mound had been abnormally raised when sutured to the pectoral fascia, but within a short time it descended to its normal position. We now feel that it is a safe procedure because the breast will descend naturally from the initially elevated position, as seen when performing a mastopexy for other conditions.

COMPLICATIONS

We have noted the appearance of an erythematous reaction of the skin in some patients undergoing expansion. It occurs during the expansion phase and was initially misconstrued to be an inflammatory reaction associated with a deep infection. Instead, it proved to be the development of a rich vascular network arranged in a radial pattern surrounding the areola. It subsides slowly. Initially it appears as though skin striae will result, but in fact the vascularity clears without leaving a blemish, although the process may take a year or more to complete.

Capsule formation accompanies the introduction of an expander irrespective of the position of its placement or the chemical composition of the expander. Fortunately, in most instances the capsule that forms is relatively thin and pliable and, therefore, does not cause the expander to alter its position. In some cases, the body's reaction to the implant is such that the capsule may become hypertrophic and distorted. The capsular contracture may be reflected as simple firmness of the reconstruction; it may cause the position of expander to be altered; or it may be so severe as to compromise the quality of the reconstruction. The latter condition invariably dictates that further surgery is required to salvage the reconstruction. By placing the implant in the subpectoral-subserratus space, capsular contraction becomes less of a problem than a contracture that occurs in the subglandular space. It is too early to know if the textured implants will minimize its occurrence.

Whenever the tissue expansion technique is used, the expander may become exposed either through pressure or infection. Exposure of the expander or the implant may occur if the pocket in which it is placed is too tight. There is a tendency to develop too small a pocket in an attempt to minimize implant migration. This tendency must be resisted because tightness does increase the chance for exposure.

Antibiotics are given in the perioperative period theoretically to minimize the incidence of infection and subsequent exposure of the implant or expander. If a drain is to be used, antibiotics are continued until it has been removed.

We have not had a rupture of an implant in our patients nor have we had a patient who developed signs of a connective tissue disorder or arthritis.

SUMMARY

Anomalies of the female breast are relatively uncommon; however, they can be treated satisfactorily and should be treated as early as is practical—preferably during the adolescent years. Following the introduction of the tissue expander and more recently the combined tissue expander-implant (Becker 50/50, Mentor), the expectation for attractive breast reconstructions in this group of patients has been enhanced.

REFERENCES

1. Radovan C. Development of adjacent flaps using a temporary expander. *Plast Surg Forum* 1982;69:69.
2. Austad ED, Rose GL. A self-inflating tissue expander. *Plast Reconstr Surg* 1982;70:588–593.
3. Lapin R, Daniel D, Hutchinson H, et al. Primary breast reconstruction following mastectomy using a skin expander-prosthesis. *Breast* 1980;6:97.
4. Radovan C. Adjacent flap development using expandable silastic implants. Annual Meeting of American Society of Plastic and Reconstructive Surgeons, Boston, MA, September 30, 1976.
5. Argenta LC, Marks MW, Grabb WC. Selective use of serial expansion in breast reconstruction. *Ann Plast Surg* 1983;11:188–195.
6. Versaci A. Soft tissue expansion in reconstructive surgery. University of Michigan Medical School, Post-graduate course, November 19–20, 1982.
7. Versaci AD, Balkovich ME. Tissue expansion. In Habal M (ed.), *Advances in plastic and reconstructive surgery*. Chicago: Year Book Medical Publishers, Inc., 1984, p. 129.
8. Versaci AD, Balkovich ME, Goldstein SA. Breast reconstruction by tissue expansion for congenital and burn deformities. *Ann Plast Surg* 1986;16:20–31.
9. Gibney J. The long term results of tissue expansion for breast reconstruction. *Clin Plast Surg* 1987;14:509–518.
10. Becker H. Breast reconstruction using an inflatable breast implant with detachable reservoir. *Plast Reconstr Surg* 1984;73:678–683.
11. Versaci AD. Reconstruction of a pendulous breast utilizing a shaped expander. *Ann Plast Surg* 1989;23:469–478.
12. Hester TR, Jr., Bostwick J III. Poland's syndrome: Correction with latissimus muscle transposition. *Plast Reconstr Surg* 1982;69:226–233.
13. Argenta LC, VanderKolk C, Friedman RJ, et al. Refinements in reconstruction of congenital breast deformities. *Plast Reconstr Surg* 1985;76:73–80.
14. Seyfer AE, Icochea R, Graeber GM. Natural history and long-term results of chest wall reconstruction in 33 patients. *Ann Plast Surg* 1988;22:776–782.
15. Marks M, Argenta L, Izenberg P, et al. Discussion by Alan E. Seyfer, M.D.; Management of the chest-wall deformity in male patients with Poland's syndrome. *Plast Reconstr Surg* 1991;87:679–681.
16. Godfrey PM, Godfrey NV, Romita MC, Guthrie RH. The teres major muscle flap in breast reconstruction. *Ann Plast Surg* 1990;25:402–408.
17. Sharpe, D. Personal communication, 1991.
18. Rees TD, Aston S. The tuberous breast. *Clin Plast Surg* 1976;3:339–347.
19. Bass CB. Herniated areolar complex. *Ann Plast Surg* 1978;1:402–406.
20. Longacre JJ. Correction of the hypoplastic breast with special reference to reconstruction of the "nipple type breast" with local dermo-fat pedicle flap. *Plast Reconstr Surg* 1957;14:431–437.
21. Teimourian B, Adham NM. Surgical correction of the tuberous breast. *Ann Plast Surg* 1983;10:190–193.
22. Toranto IR. Two-stage correction of tuberous breasts. *Plast Reconstr Surg* 1981;67:642–646.
23. Vecchione TR. A method of recontouring the domed nipple. *Plast Reconstr Surg* 1976;57:30–32.
24. Dinner MJ, Dowden RV. The tubular/tuberous breast syndrome. *Ann Plast Surg* 1987;19:414–420.
25. Williams JE, Hoffman S. Mammaplasty for tubular breasts. *Aesth Plast Surg* 1981;5:51–56.
26. Elliott MP. A musculocutaneous transposition flap mammaplasty for correction of the tuberous breast. *Ann Plast Surg* 1988;20:153–157.
27. Versaci, A. Treatment of tuberous breasts utilizing tissue expansion. *Aesth Plast Surg* 1991;15:307–312.

21

Soft Tissue Expansion in Children

GUY MAGALON

Soft tissue expansion is widely used for reconstructive surgery in children and has greatly modified the treatment of many congenital anomalies. In addition to providing expanded skin grafts to cover large defects, this technique improves cosmetic results by providing tissue with matching characteristics and color. In this chapter, I discuss the methods, indications, and results of tissue expansion in children.

METHODOLOGY

Standard Techniques

Preoperative Planning. Like all candidates for skin expansion, children must undergo a thorough examination to note the size and shape of the lesion, the condition of skin, the area available for expansion, and the presence of risk factors such as flexion folds, scars, chronic wounds, and infection. If necessary, the surgeon can use a template to simulate the prosthesis and draw a sketch to show the position of the expanders, valves, and placement incisions. At the end of the examination, the surgeon should be able to estimate the final dividend of expanded skin and determine whether the lesion can be completely excised in a single course as well as where the residual scars will be.

A crucial part of the planning phase is the selection of the expander. It is important not to leave any unexpanded skin around the lesion. Off-the-shelf expanders are available in several volumes and shapes. Rectangular prostheses are the most widely used. To meet special requirements, custom-made expanders are available. To order a custom-made expander, the surgeon must send the manufacturer a transparent sketch showing the shape of the expander along with the desired specifications (volume, length, width, and projection).

The choice of filling valve is also an important preoperative decision. The conventional technique calls for the use of remote internal valves, which are available in diameters ranging from 1 to 3 cm. Large valves are more reliable but they require a longer placement incision (3 centimeters) and more undermining for placement and removal.

In some cases, external valves offer an interesting alternative. They can be placed without further undermining using a suction drain probe and are easy to remove during reconstruction. The skin opening can be used for suction drain placement. External valves also avoid the need for painful transcutaneous injection; this is a point to be taken into account when treating children. However, external valves require greater attention to hygiene by the patient and the use of an occlusive dressing between fillings. The metallic adaptor used to connect the external valve to the filling tube should be placed externally, away from the skin opening.

Before beginning expansion, the surgeon must explain the procedure to the family. It is especially important for the family to understand that this is a long procedure that requires both hospitalization (for placement and reconstruction) and weekly visits (for filling). Rather than giving specific dates, the surgeon should prepare the child and the family to wait as long as necessary to achieve the planned goals.

Placement

APPROACH. The location of the placement incision is very important. An intralesional incision should only be used if the skin is normal. However, if the skin is of poor quality, (for example, adherent scar or old skin graft), the incision should be made in a normal zone outside the limits of the lesion. To avoid tension and allow filling to be started as early as possible, we make placement incisions radially with respect to the lesion and at a good distance away from the zone of expansion.

The length of the incision depends on the size and type of valve used; 2 centimeters is sufficient for an external valve as opposed to 3 centimeters for a standard *internal remote* valve. We recommend placement of only one expander through each incision. Placement of several expanders through the same incision reduces residual scarring but, in case of infection, it results in contamination of all the expanders. With separate pockets, it is possible to treat infection in one pocket while continuing expansion in the uncontaminated pockets.

UNDERMINING. The zone to be undermined should be outlined on the skin before the beginning of the operation. The pocket should be slightly larger than the expander. After making the incision, undermining is initiated with scissors and completed with blunt dissectors of various lengths and shapes. We use a blunt hooked dissector to break any remaining adherences. Placing a long expander on the abdomen or extremities may require a short radial counterincision in front of the opposite margin of the planned pocket in order to easily perform the desired amount of undermining.

Undermining is usually carried out just above the fascia. Resistance is weak at that level but it is necessary to avoid damaging the overlying fat layer. Submuscular undermining (for example, under the frontalis muscle or platysma) is performed to obtain expansion of muscular or myocutaneous flaps. Subfascial undermining is useful on the extremities, for example, for lower leg reconstruction, but the fascia must be fully opened to minimize the risk of compartmental syndrome.

An easy method of checking the pocket is to inject a large quantity of saline with a wide-nozzle syringe. If the pocket is large enough and no adherences are seen, the expander can be inserted. Injection of saline also rinses out any fat debris that may be left inside the pocket. If further undermining is required, the pocket should be rinsed again.

DRAINAGE. Although undermining is usually performed blindly, bleeding is minimal when the procedure is performed skillfully with blunt dissectors. Thus in most cases there is no need for drainage. However, if there is any doubt, a suction drain should be inserted. It should always be placed under the expander to avoid a "knuckle" under the skin that could lead to erosion and exposure. The tube should be brought out away from the zone of undermining. Except in hairy zones, the drain tube should be secured with Steristrip (3M, St. Paul, MN) to avoid stitch marks.

INSERTION. Because skin expanders are very electrostatic, exposure to air as well as handling must be kept to a minimum. Gloves should be rinsed. The first step in placement is to fully expel any air trapped inside the balloon. Next, the prosthesis is rolled around a blunt dissector and pushed into the pocket. Some expanders have a small pouch that allows them to be attached to the end of the dissector. This feature makes it easier to push the device into the pocket. As soon as the expander is in the pocket, it should be temporarily filled with saline. This allows the surgeon to recheck the quality of the pocket, locate leaks, and ensure that the connecting tube and filling valve are functioning.

Remote internal valves should be placed in a pocket of exactly corresponding size to prevent capsizing. Their distance from the expander should be at least 5 centimeters. External valves should be located about 8 centimeters from the prosthesis, and the metallic adaptor should be placed as far away from the skin opening as possible.

When all the components are in place, the system should be retested by filling from 10% to 20% of capacity. There should be no wrinkles or creases because they can lead to erosion and necrosis. There should be no sign of skin ischemia in the expanded area. After this final check, the placement incision can be closed using inverted stitches at the deep level and a subcuticular running suture at the surface level. Suction is then applied to the drain. The dressing should not cover the expanded zones to allow postoperative inspection.

FILLING. Filling can be started after removal of the suction drain. Several guidelines must be observed. Filling must be performed under sterile conditions using individual doses of sterile saline. Filling must not cause discomfort. Injection must be stopped as soon as the child complains of tightness. If the expanded skin demonstrates blanching (pale discoloration), the operator should not hesitate to withdraw saline.

If a remote internal valve is used, the overlying skin must be thoroughly disinfected. For standard sizes, a 23-gauge butterfly needle is suitable, but for smaller models a 25-gauge butterfly needle should be used.

If an external valve is used, the valve must be carefully disinfected. There is no need for a needle because the syringe is connected directly to the one-way valve. The dressing over the skin opening should be changed at each filling session. The valve should be wrapped inside the dressing without twisting or kinking.

For outpatients, filling sessions are usually scheduled weekly and the duration of filling varies from 3 weeks to

4 months, depending on the amount of skin needed for reconstruction. However, expansion can be achieved more quickly using a biweekly schedule. For hospitalized children, even more rapid expansion can be accomplished with an automatic pressure-controlled infusion pump. To expedite expansion, we advise massaging the expanding skin twice a day with hydrating ointments.

Noting the amount of saline injected is not an accurate method of estimating the skin gain. The proper technique consists of measuring the expanded zone at each filling and, if possible, comparing it with the unexpanded opposite side. The amount of skin obtained by expansion must be greater than the amount needed for reconstruction. An excess of 1 to 2 cm provides a good safety margin.

RECONSTRUCTION. After administering anesthesia, the expander should be vigorously overfilled. In this way, one can obtain several centimeters of additional skin almost instantly. Blanching is a normal consequence of overfilling.

The expander should be removed first and then the valve. After the initial incision with a knife, electric cautery should be used to open the capsule, taking care not to puncture the expander. The expander is usually extracted full and the connecting tube is cut. In some cases, internal remote valves must be removed through a counterincision over the valve. External valves can simply be pulled out of the encapsulated tunnel.

Some surgeons advise capsulectomy to limit retraction, especially around facial orificia. However, since capsulectomy is a hemorrhagic procedure that greatly reduces the blood supply, we prefer the conservative approach, although it may be necessary to incise the sides and back of the capsule to facilitate advancement of the expanded flap. In some cases, the edges of the capsule may form a thick fold that should be resected if it diminishes the length of the flap. We do perform capsulectomies for some indications, such as nose reconstruction with a forehead flap. Some practitioners have advocated tacking the capsule to the recipient bed to reduce the risk of retraction.

Although expanded skin can be used to fashion all types of conventional flaps, the simple advancement flap allowing linear suture is the best technique. However, it is feasible only if the expanded skin is located at the edge of the lesion and if advancement is not impaired by unexpanded skin. Before resecting the lesion, it is necessary to be sure that there is enough expanded skin to cover the defect without excessive tension on the suture line. Incomplete reconstruction is always preferable to wound dehiscence or necrosis of the leading edge of the flap.

Careful hemostasis is required. One or more suction drains should be inserted, depending on the size of the residual pocket. If filling was accomplished through an external valve, the encapsulated tunnel can be used for drain placement. This avoids further scarring.

Closure is achieved in layers. The suture technique used for the superficial layer depends on local conditions, but the goal is always to minimize stitch marks. Whenever possible, we perform a subcuticular running suture with steel wire. If the dermis is thin, we use thin interrupted nylon sutures. If a second course of expansion is planned, for example, for serial resection of a large hairy nevus, special attention is required not to injure the expanded flap. Stitches should involve only the dermis on the flap side and dermo-epidermal tissue on the lesion side.

Postoperative Care. During the immediate postoperative period, a noncompressive dressing should be used to enable inspection of the flap and suture lines. On the extremities, a cast may be applied to reduce pain and prevent damage from inconsiderate movements after recovery from anesthesia. The drain should be left in place for several days. The subcuticular running suture can be removed between 14 and 21 days after reconstruction. We recommend use of Steristrip to limit scar widening as well as use of compressive garments and gel sheeting to prevent hypertrophy.

To avoid lengthening of the scar, dog-ears should not be resected immediately. If necessary, they can be corrected 6 months later with little additional scarring. In most pediatric cases, dog-ears disappear as the child grows.

Serial Expansion

Serial skin expansion is often necessary in children to treat large lesions, especially giant hairy nevi.

Expanded Skin Graft

Skin expansion of the supraclavicular or inguinal region allows creation of single-piece, full-thickness skin grafts that can be used for coverage of large defects anywhere on the body (see Chapter 22). Direct closure of the donor site is always possible.

INDICATIONS

Head and Neck

Lesions on the head and neck are frequent indications for skin expansion in children. These include alopecia, benign tumors, scars, and reconstruction.

Alopecia. Alopecia can be congenital or acquired. This distinction is important because greater precaution is necessary for acquired alopecia because the skin is usually damaged. This is especially true for burn-related lesions, which require strictly sterile conditions and expander placement under healthy skin (Figure 21-1).

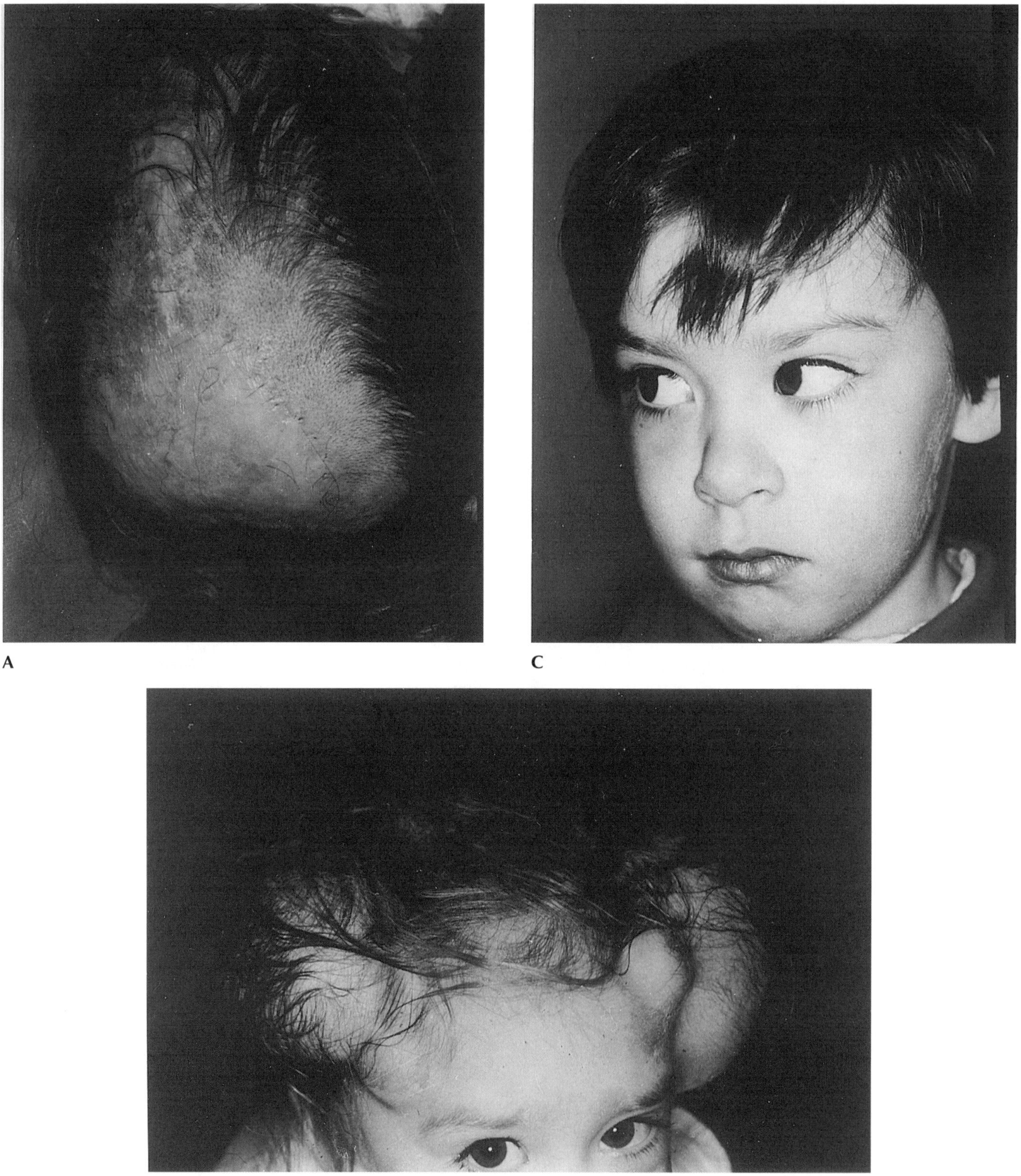

A C B

FIGURE 21-1
(A) Congenital alopecia (width: 8 cm; height: 14 cm), top view. (B) View at the end of expansion phase. (C) Late result.

Benign Tumors. The most common benign tumors of the head and neck are the nevi. Removal is necessary for cosmetic reasons but in some cases excision is also needed to prevent malignant transformation. For example, Jadassohn nevi frequently degenerate to basal cell epithelioma in adulthood. Before the advent of tissue expansion, a large full-thickness skin graft was often required to replace nevi from the face (Figure 21-2).

Scars. Tissue expansion greatly facilitates scar excision. As previously noted, strict attention to sterility is necessary when dealing with burn-related scars. To avoid recurrence due to wound compression, a neck brace should be worn during the postoperative period.

Reconstruction. For post-traumatic nasal defects, tissue expansion adds several advantages to conventional nose reconstruction. Expanded flaps from the forehead provide matching tissue that is more pliable because it is thinner. Furthermore, the tissue is more reliable because it has a better blood supply. In addition, scarring of the donor site is minimal because it can be closed directly (Figure 21-3). Tissue expansion is useful for nose reconstruction in patients with Binder's syndrome, a mandibulofacial dysostosis characterized by hypoplasia of the nasal bones. In this case, it is very difficult to insert a bone graft to reconstruct the bridge of the nose because of the paucity of soft tissue available. An expander can solve this problem, creating a pocket for the graft.

Tissue expansion can also be used for reconstruction in children with congenital auricle malformation (see Chapter 11).

Trunk

In children, the most frequent indications for tissue expansion on the trunk are giant hairy nevi. In young girls, expanders should not be placed near the breasts because of the risk of subsequent asymmetrical gland development. Similarly, it is important to take great care expanding regions where pubic hair will grow to avoid unsightly results of misplaced pubic hair.

Tissue expansion constitutes a considerable advance in the treatment of Poland's syndrome (see Chapter 20). The expander is inserted under the aplastic breast at the age of 12 or 13 and left in place for months or years. In this way the areola of the affected side can grow concomitantly with the contralateral side. At the end of development, the expander can be replaced by a permanent prosthesis matching the normal contralateral breast mound.

Extremities

In children, the limited area available on the extremities, especially the upper extremities, is a restricting factor for tissue expansion (see Chapter 15). As in other regions, the main indications for tissue expansion on the upper and lower extremities are benign tumors (especially nevi) and scars. In addition to these lesions, we have also treated lymphangiomas on the lower extremities.

Tissue expansion may also have a promising future for the treatment of syndactylism. For this application, tubular expanders measuring 0.5 to 1 cm in diameter and 2 to 3 cm in length are placed on the dorsum of the hand, while the injection ports are placed on the wrist. This technique can avoid skin grafts that leave scars at the donor site and tend both to retract and change color.

Thus, tissue expansion gives new perspectives in the treatment of congenital hand malformations. However, because this technique is long and difficult, indications must be carefully considered. Complex syndactyly with bone fusion and Apert's syndrome are indications of choice. Expanded skin-grafts can be useful on the hands and feet.

Skin expansion can begin in infants very early, but most practitioners wait until the age of 6 months to expand the trunk and extremities. On the head, Bauer et al.[1] have reported expansion in very young infants. However, we prefer to wait until 18 months when the bones have sufficiently hardened.

RESULTS

The outcomes of tissue expansion depend mainly on preventing a number of major and minor complications (see Chapter 8). Major complications are defined as those resulting in complete failure of the procedure. They can occur during the placement of the expander or during the course of the expansion. The main major intraoperative complication is damage to the overlying skin during undermining. The main major postoperative complications are infection, necrosis, and exposure.

Minor complications are defined as those that hamper expansion. They can occur during or after expansion. The main minor complications during expansion are exposure of the injection port and valve migration. The main minor complications after expansion are dehiscence, seroma, and scarring.

In growing children, scar widening is a problem that almost always deteriorates the final result. Because of this complication, it is important, when possible, to place the scar vertically rather than horizontally.

REFERENCES

1. Bauer BS, Johnson PG, Lovato G. Applications of soft tissue expansion in children. *Ped Dermatol* 1986;3:281–292.

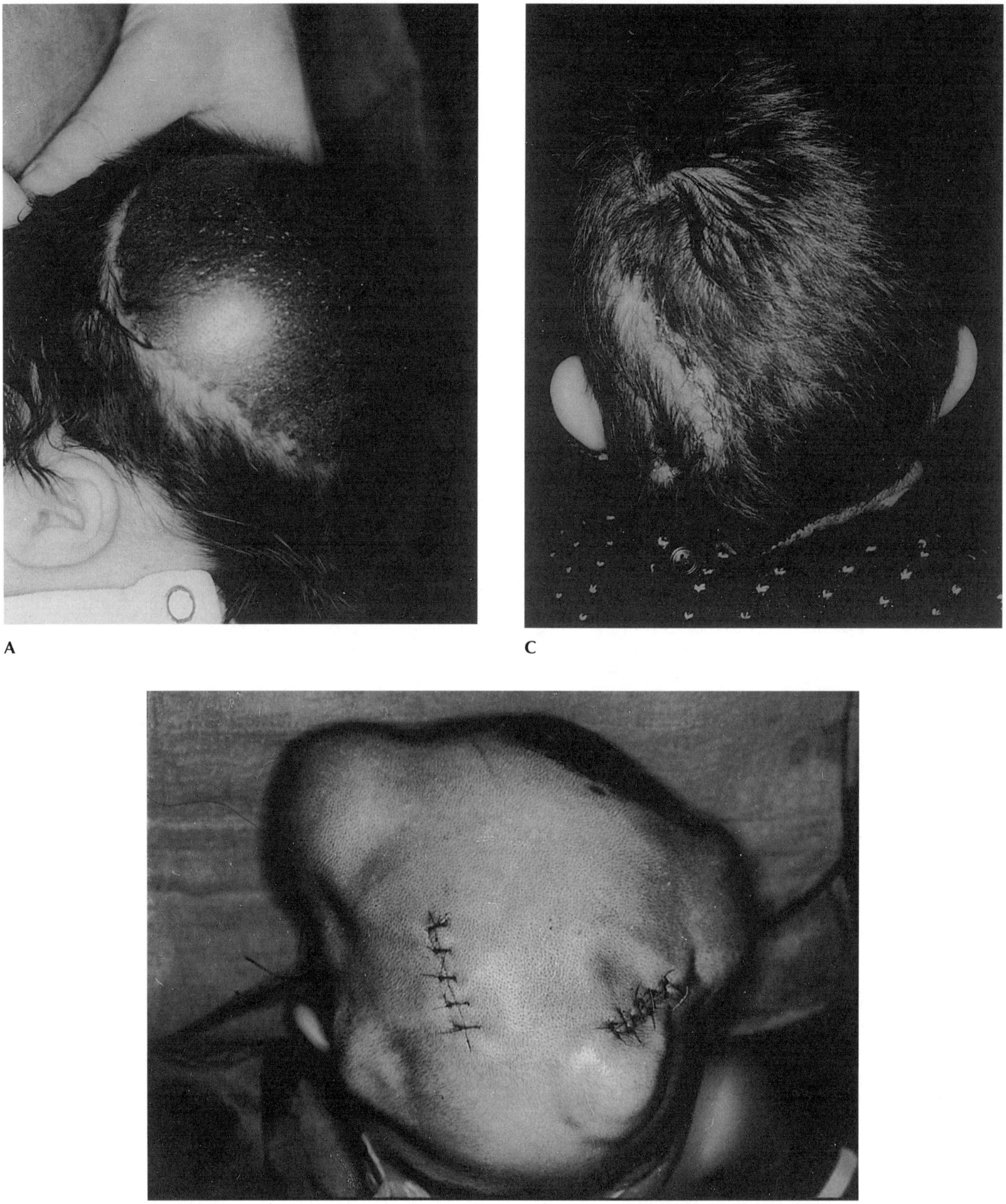

A

C

B

FIGURE 21-2

(A) Giant oval-shaped blue nevus in occipital region (length: 10 cm; width: 9 cm), profile view. (B) View after placement. (C) Late result.

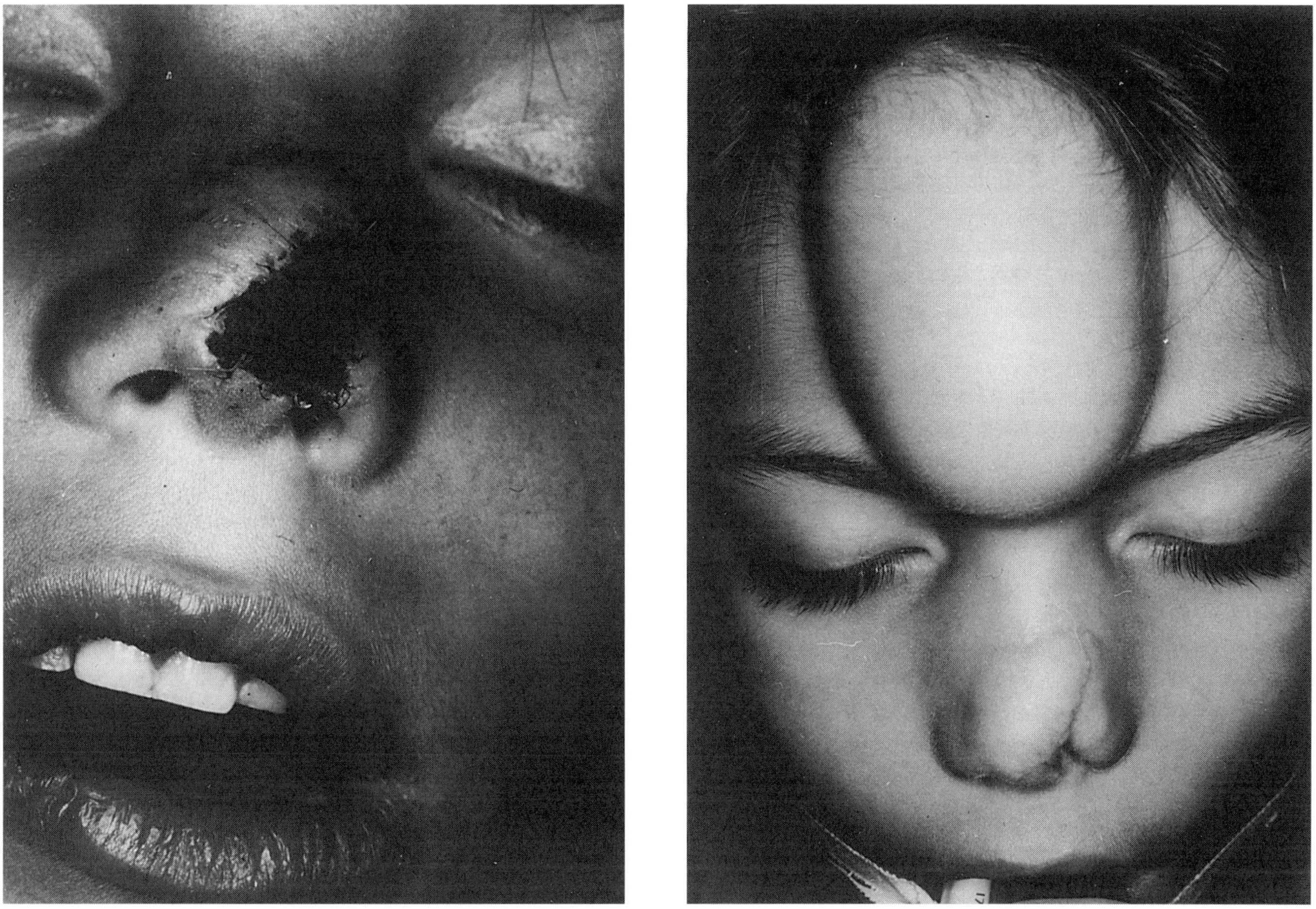

A B

FIGURE 21-3

(A) Defect in the left ala of the nose following dog bite. (B) Forehead flap at the end of filling phase. (C) After transposition of the forehead flap. (D) Late result.

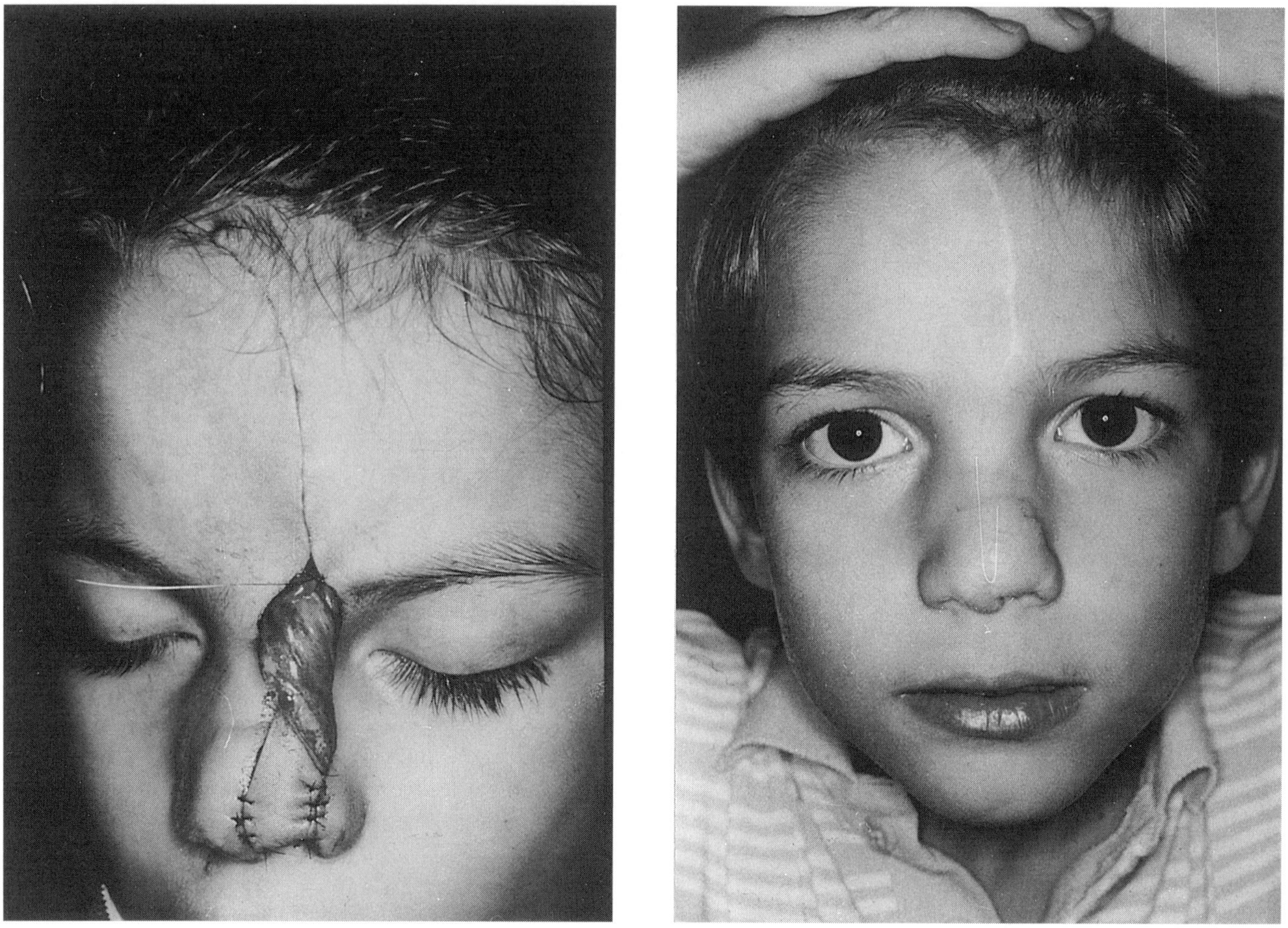

C D

FIGURE 21-3 *(continued)*

22

Expanded Full-Thickness Skin Grafts

BRUCE S. BAUER AND FRANK A. VICARI

Tissue expansion is now recognized as a means of providing additional tissue for the reconstruction of large skin defects and soft tissue defects. The limitations and complications of tissue expansion (TE) are well documented.

The concept of donor site expansion arose out of a desire to make use of the benefits of TE in cases where regional flap tissue was either too limited to expand or the risk of regional expander complications was too great to warrant application of this technique. Despite the inability to provide expanded flap coverage in these cases, reconstruction with full-thickness vs. split-thickness skin would in many cases provide more durable coverage and better aesthetic results.

The application of TE to the full-thickness skin graft (FTSG) donor site seemed the logical answer to some of these selected problems, while at the same time minimizing risks of donor site complication and improving aesthetics of the full-thickness skin graft donor site.

Since the initial presentation of our experience with tissue expansion as a means of harvesting large FTSGs in children (at the First International Tissue Expansion Symposium in San Francisco, 1987), we have continued to widen our applications of expanded full-thickness grafts (EFTG). We have also tried to define the benefits of larger EFTGs in cases where our primary treatment modality previously would have been a split-thickness skin graft (STSG).

This chapter will discuss our experience with EFTGs with particular emphasis on case selection, planning, and management, with some specific comments relating to management in the pediatric population. Although our patient population is entirely pediatric, the technique and patient management is readily applicable to patients of all ages.

CHOICE OF GRAFT AND DONOR SITE EXPANSION

Skin grafting is a well-established technique for providing new tissue for the reconstruction of skin and soft tissue defects where available regional tissue is inadequate or unavailable. Large defects have commonly been grafted with STSGs rather than FTSGs due to the limitation in available full-thickness donor sites or to unacceptable donor site deformity. However, for many reconstructive procedures, full-thickness grafts would clearly be preferable to STSGs due to the improved durability, minimal graft contraction, retained skin adnexa, and aesthetics of FTSGs. Tissue expansion provides a means of virtually eliminating size as a factor in the decision to use FTSGs or STSGs.

Of the common full-thickness donor sites, we have used the clavicular area (supra- and infraclavicular), groin, and bilateral groin/lower abdomen for the majority of cases. Postauricular skin could also be expanded as an additional choice. Although the groin and lower abdominal skin may not be the ideal choice for single aesthetic unit reconstruction in the face, the uniformity of color that is achieved when it is used as a single sheet in resurfacing the face (or the majority of the face) may outweigh the pigmentation differences (which have been minimal in our experience).

The size and type of expander is chosen partly on the basis of the surgeon's preference and partly on the location of the donor site and size of the patient. We have used rectangular remote injection port expanders in all cases, placing the largest expander possible given the anatomic confines of the donor area chosen. When expanding the bilateral groin/lower abdomen in a child, one may need to start with 325 to 500 milliliter rectangular expanders on either side and then replace these with an even larger single expander. (On one occasion, two expanders were used.) We have used up to a 3200-milliliter expander for our largest donor site expansion in a 13-year-old girl.

The expander is placed through an incision several centimeters beyond the proposed position for the final donor site scar (for example, approximately 2 to 3 cm below the groin crease) so that following donor graft harvest and closure of the donor site, the scar advances to the appropriate position. The expander is placed above the deep fascia.

At the time of expander placement, the expander is filled comfortably to eliminate dead space in combination with the placement of one or more 19-gauge butterfly IV drains (connected postoperatively to vacutainer tubes). Expansion begins approximately 10 days after placement of the expander and follows our standard expansion protocol of one expansion session each week.

The length of expansion will vary based on the size of the graft to be harvested. One must take into account both the size of graft needed plus the additional amount of skin necessary to close the donor site. We have found that the most reliable way of determining adequate expansion is by transferring a cloth or plastic pattern (or pattern made from a surgical glove) of the defect to the dome of the expanded donor site. If correctly planned, not only will there be adequate FTSG, but the donor site closure will be accomplished with little or no undermining.

When a larger graft is needed than can be obtained with the expander fully expanded (due to patient size), the expander can be replaced with a larger expander and the expansion can proceed further until the desired donor site expansion is obtained. In such cases of "in situ" serial expansion, we connect the new expander(s) to the already securely positioned injection port(s) of the previous expander(s). "In situ" serial expansion is most likely to be used in a small child where a single large full-thickness graft without seams is desired (for example, a full facial graft).

GRAFT HARVEST AND PLACEMENT

At the time of graft harvest, a pattern is taken of the defect to be grafted and this is transferred to the dome of the expanded donor site. The pattern is positioned with one edge along the incision used for expander placement. The graft is then harvested from the subcutaneous fat and capsule in what is usually a readily entered and relatively bloodless plane. Typically the graft can be defatted as it is harvested, but it should be inspected in the event that additional defatting is needed.

Once the graft is harvested, the exposed layer of subcutaneous fat and capsule is excised with a "hot knife" and discarded. If there has been adequate expansion, little if any undermining will be required to close the donor site. The closure is accomplished in layers and drains are placed that are similar to those used at the time of expander placement. For very large donor sites, such as bilateral lower groin/abdomen, a larger bore suction drain is placed in addition to the butterfly drains for the first few postoperative days.

The basic handling of the EFTG is no different from a nonexpanded graft in terms of preparation of the recipient site and fixation of the graft. Hemostasis must be meticulous prior to placement of the graft. In cases with very large recipient defects and potential for excessive oozing following scar or lesion excision, consideration should be given to delaying the graft placement for 24 hours.

We typically secure the graft with interrupted and running absorbable sutures and dress the graft with a layer of greasy gauze. This is followed by cotton soaked in mineral oil and saline and a final layer of Reston foam ® (Reston Products, Hinsdale, IL). The foam is secured with adhesive after careful defatting of the skin surface surrounding the graft and it is further secured with sutures through the foam and into the skin a centimeter beyond the edge of the defect.

The donor area is further immobilized with whatever technique is appropriate for the given area augmented in the pediatric patient with a regimen of postoperative sedation previously reported.[1] The dressing is left in place for 1 week before inspecting the graft.

CHARACTERISTICS AND MATURATION OF THE EXPANDED FULL-THICKNESS SKIN GRAFT

Expanded full-thickness grafts, as with nonexpanded full-thickness grafts, go through a period of gradual maturation from the point of graft take until return of normal skin characteristics. Small areas of epidermolysis may take longer to blend with the surrounding graft color but by two to three years postoperatively, uniformity of color and texture have been excellent. It is our impression that this time course is a bit slower than with nonexpanded grafts, but this has not been quantified. This is particularly true with regard to pigmentation of the graft, which appears to change for the better as much as 3 to 4 years after grafting in some cases.

Expanded full-thickness grafts appear to behave no differently than their nonexpanded counterparts in regard to all other physical properties. There appears to be no difference in graft contracture or retention of skin adnexa. Although some surgeons anecdotally have mentioned problems with greater contraction of expanded grafts, these cases uniformly appear to have had the donor sites rapidly expanded over the course of only a few days to a week rather than the gradual weekly expansion that we have described. This rapid expansion may account for the difference due to the greater component of stretch in rapid expansion.

EXPERIENCE AND COMPLICATIONS

Experience in our first 16 cases has clearly supported the benefits and wide application of the technique. As mentioned, our experience has been in the pediatric population with patients ranging in age from 6 months to 15 years. Follow-up has ranged from 9 months to 7 years. Grafts have ranged in size from approximately 60 cm^2 to greater than 700 cm^2. Grafts have been used for single and multiple aesthetic unit reconstruction of the face and a single full facial graft as well as for coverage of defects of the hand, extremities, and a full plantar resurfacing (Figures 22-1 and 22-2). The technique has been used for reconstruction in patients with both congenital and acquired defects.

Graft loss has been confined to a 3 percent loss on one facial graft and 4.5 percent loss of the plantar graft (on the nonweight-bearing portion of the foot). Donor site complications have been limited to a 1×3 cm^2 area of separation of the lateral extent of a bilateral groin/lower abdomen EFTG that healed uneventfully. No secondary procedures have been required for treatment of the donor areas. Revisional procedures on the grafted areas, following initial coverage of the two areas of loss, have been confined to scar revision with Z-plasty along the graft/normal skin junction in the nasolabial area and periorbital area in two of the facial EFTGs.

SUMMARY

Expanded full-thickness grafts share the characteristics of nonexpanded full-thickness grafts including minimal graft contraction, retained skin adnexa, growth, durability, and improved aesthetics compared to split-thickness skin grafts. In addition, donor site expansion results in minimal graft size limitation, ease of donor site closure, and improved donor site appearance. These considerations may define the major contribution of this technique.

While our experience is confined to the pediatric population, the technique is readily applicable to pa-

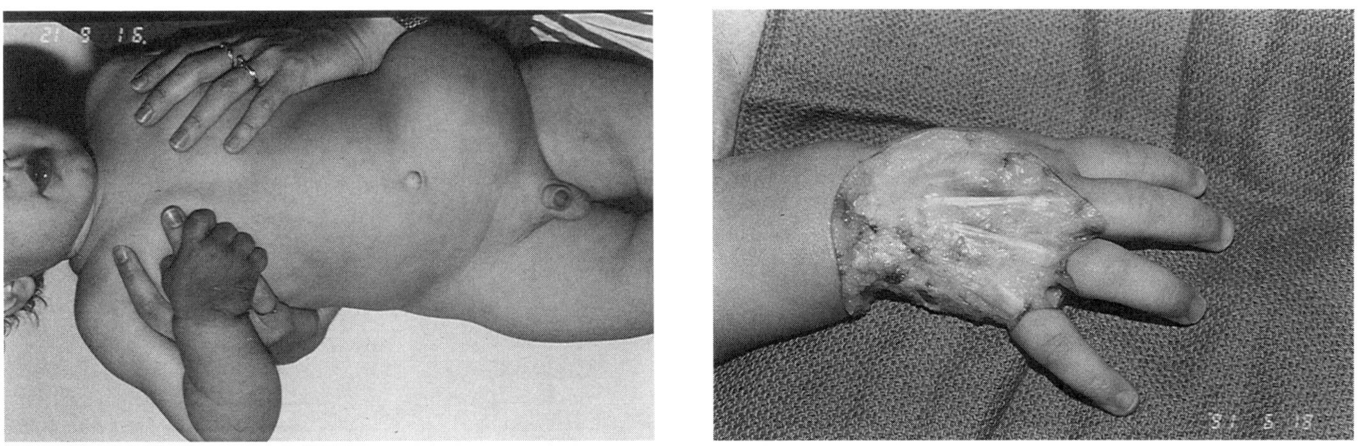

A B

FIGURE 22-1

An 18-month old child with congenital pigmented nevus of hand involving most of the dorsum of the right hand and extending over hypothenar area volar surface. (A) The nevus is visible on the right hand as is the expanded left groin donor site. The defect is shown (B) dorsal and (C) volar views prior to grafting. (D) The expanded FTG is being harvested off the dome of the expander (donor site of different patient). The subcutaneous fat and capsule are excised following graft harvest and the donor site closed. (E,F) The graft is shown in place. (G,H,I) The result is shown 13 months post-operatively.

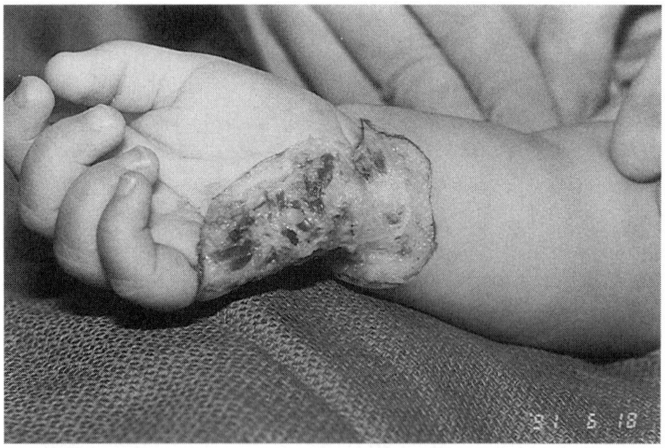

C

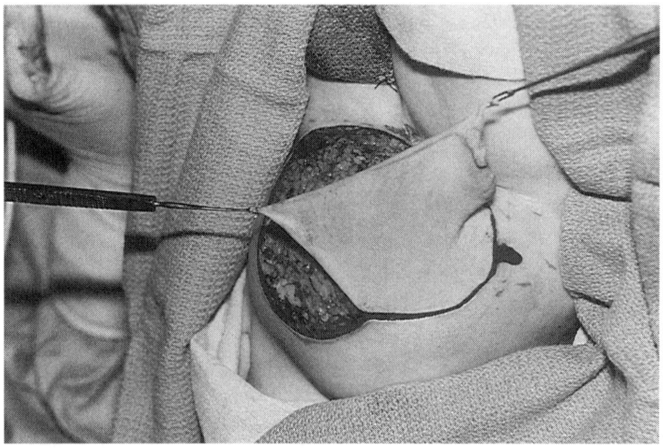

D

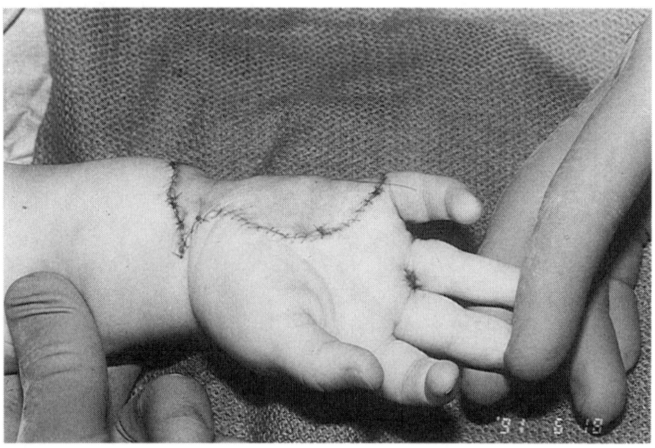

E

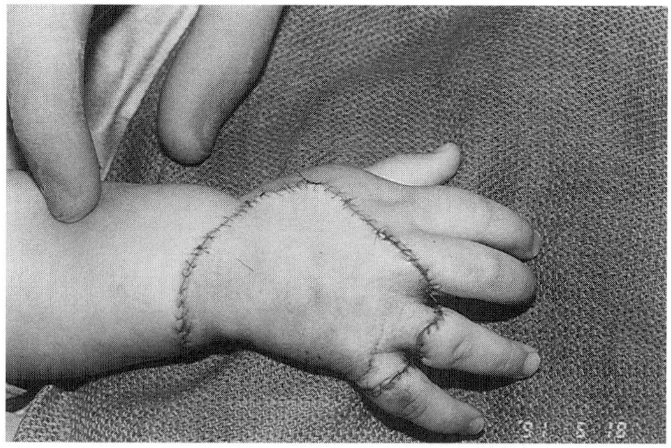

F

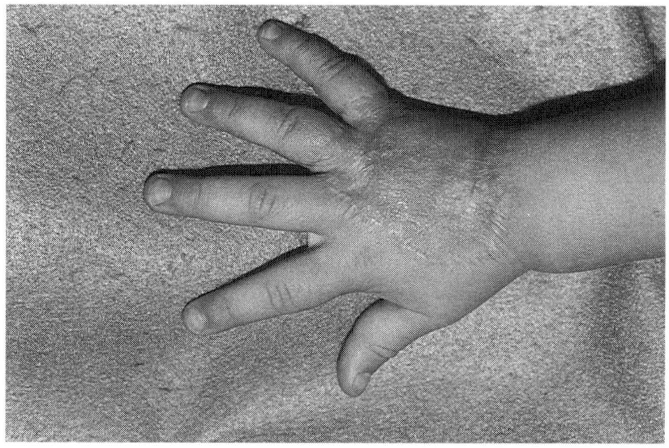

G

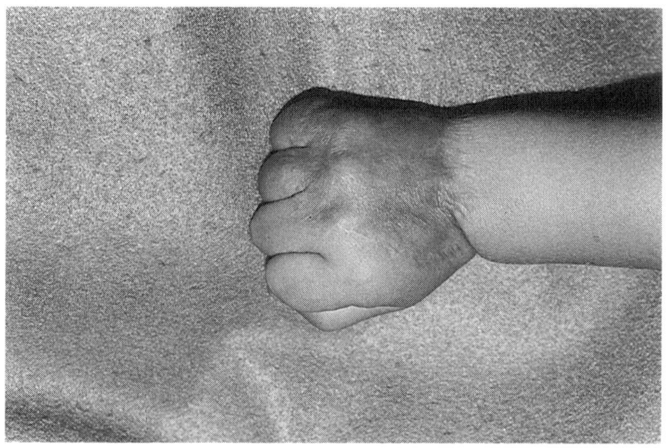

H

FIGURE 22-1 *(Continued)*

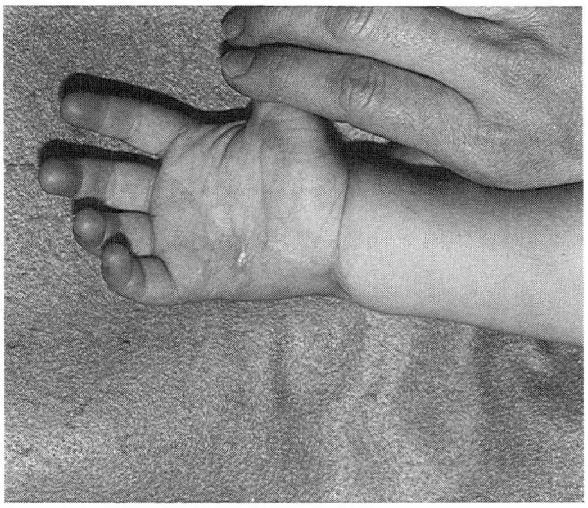

I

FIGURE 22-1 *(Continued)*

tients of all ages, although some of the need for "in situ" serial expansion in the child would not arise in the adult. In addition, postoperative immobilization and patient cooperation would in most cases be easier in the adult population.

With careful patient and case selection and strict adherence to preoperative, intraoperative, and postoperative protocols, expanded full-thickness skin grafting can be performed safely and effectively with minimal risk of complications and optimal reconstructive results.

REFERENCES

1. Bauer BS, Vicari FA. An approach to excision of congenital giant pigmented nevi in infancy and early childhood. *Plast Reconstr Surg* 1988;82:1012–1021.

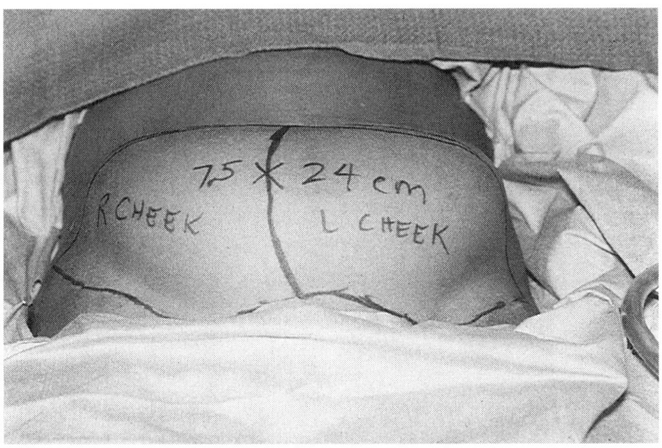

A

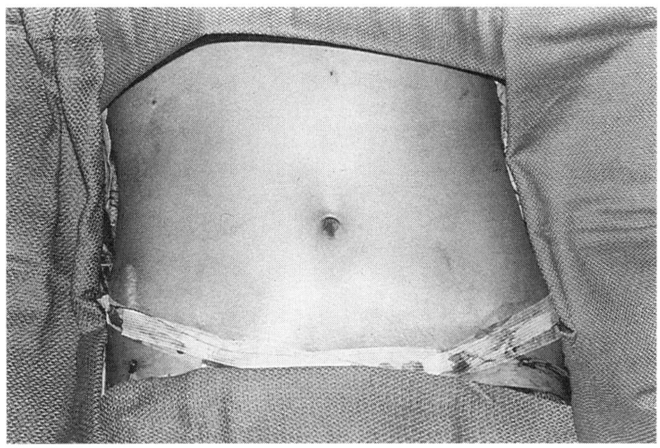

C

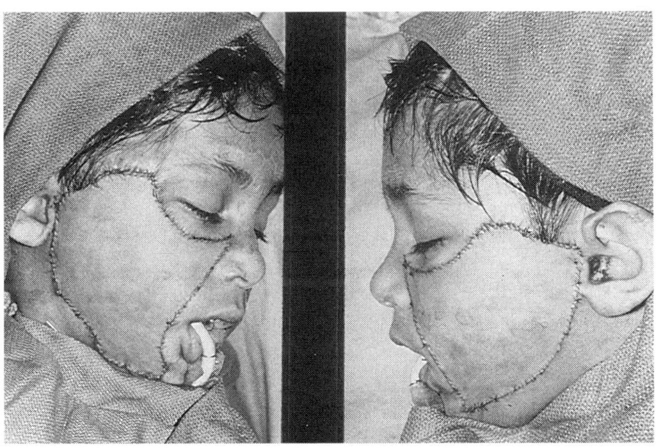

B

FIGURE 22-2
(A) The expanded donor site used in a burn reconstruction is shown with the pattern for each cheek aesthetic unit illustrated. (B) The expanded FTGs are shown in place in the composite view of both cheeks. (C) The donor site closure was readily accomplished without need for undermining.

23

Burn Scar Expansion

JAIME ANGER

The current knowledge of the histopathological aspects of expanded tissue and the large clinical experience achieved with the use of the tissue expansion technique have resulted in a more effective control of the expansion process. The incidence of complications has thus decreased.

Scar tissue expansion has usually been avoided. The risk of complications when expanding scars is emphasized in the literature.[1]

In 1987, Marks et al. reported good results in the expansion of mammary tissue affected by postburn scar, followed by the successful placement of a silicone gel prosthesis.[2] In 1987, Versaci reported keloidal scar sequelae in the thorax where the expansion of the breast region improved the appearance of the scars as a result of the pressure caused by the balloon. This action was defined as "a Jobst effect (pressure garment effect) from inside to the outside."[3] It was also possible to insert silicone gel mammary prostheses without complications, achieving the desired results. Nelligan, in 1989, reported scar tissue present in the donor areas that sustained the expansion and contributed to the success of the procedure.[4]

We routinely see patients who present with large scar sequelae resulting from burns and with scar contractions in flexion zones without normal skin areas in the surrounding area and thus without a possibility of correction with nearby flaps. These patients have usually undergone correction attempts with the use of skin grafts.

The new option of using microsurgical free flaps has brought many benefits to some of these cases, as have the musculocutaneous and fasciocutaneous flaps. But the aesthetic outcomes and the difficulties in the techniques can be limiting factors in certain patients.

In 1989, we achieved a successful expansion of a scar in a part of an expanded flap. In this case, we observed an expansion of the nearby skin without modifications of its integrity and with an apparent improvement of the aesthetic appearance. Morever, it is important to point out that we observed no dehiscence or necrosis of the expanded scar edges.[5]

Based on this experience and also on the experimental works of Sasaki and Pang,[6] who observed an increase of vascularization and an increase in the survival rates of the expanded flaps, we tried to use controlled expansion of scar tissue in the flexion zones of the limbs. These presented functional impairment and we tried to obtain a sufficient tissue gain from the expansion to allow free movement, while avoiding contraction relapse.[7]

PATIENT SELECTION

Patients with scar sequelae resulting from burns in limbs or flexion zones and who have functional limitations can be treated. The scar areas are evaluated especially to look for those that have some mobility of the deep planes and for the presence of subcutaneous tissue. The presence of previous grafts or small ulcerations in the zone to be expanded does not necessarily contraindicate the technique. Patients in whom follow-up is difficult and who, for any reason, cannot carry out the physiotherapy treatment following the final surgical stage should be excluded.

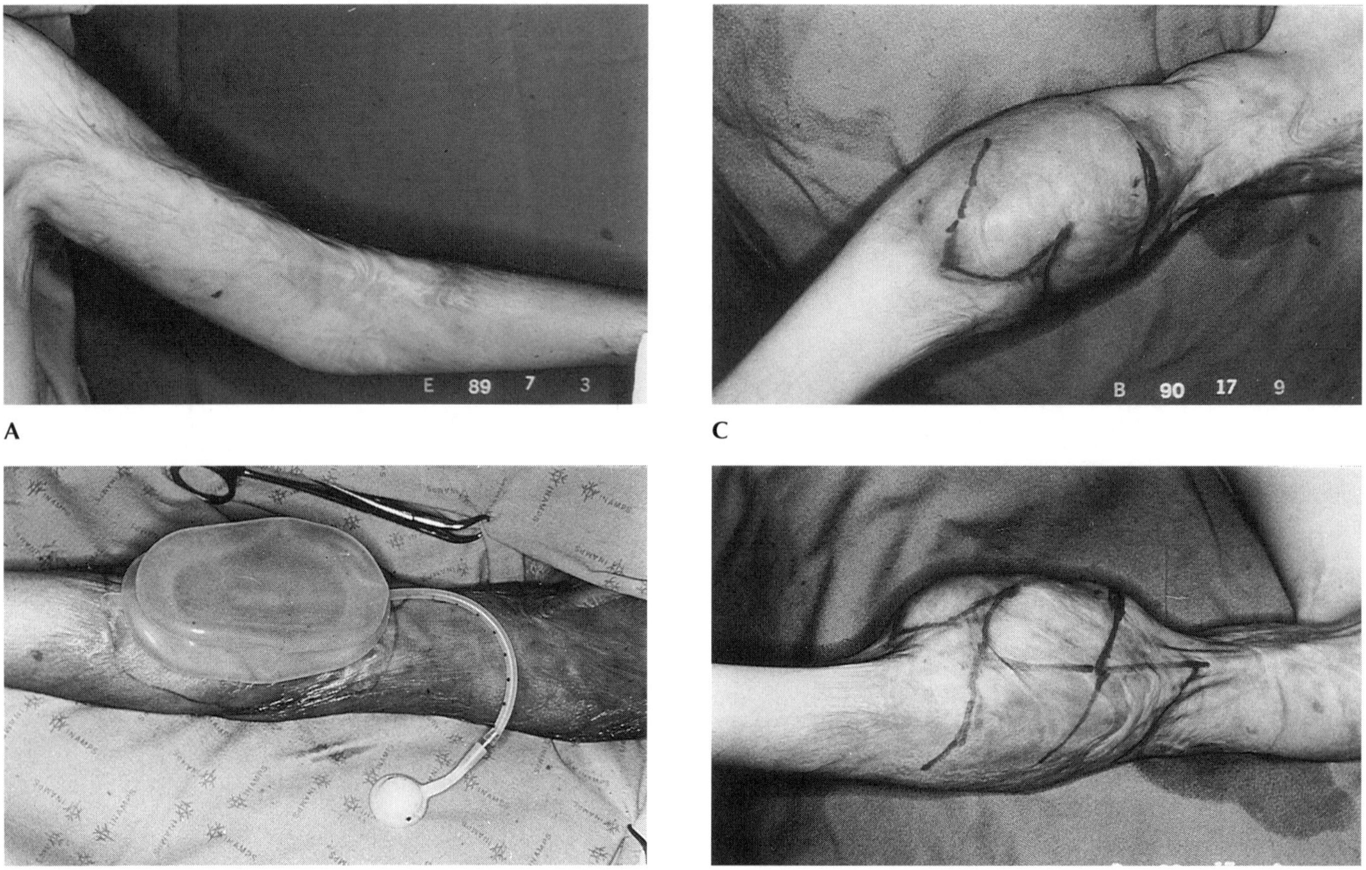

A C B D

FIGURE 23-1

(A) Five-year-old boy with postburn sequelae and arm contraction. (B) A rectangular double chamber with a small valve was used. (C) Full expansion and (D) design of the flaps. Mixture of Z-plasty and Converse flap. (E) Flap incised through all the layers including the capsule. (F) Immediate postoperative result. (G) One year later.

THE TECHNIQUE AND EVALUATION OF THE RESULTS

The objective is to expand the scar tissues in the contraction zones and adjacent areas in order to obtain an adequate amount of tissue to provide sufficient mobility in the retracted regions. The simple expansion of the scar areas in the flexion zones and adjacent tissues, whether they contain normal skin or not, is performed. The aim is to obtain an area gain that will be used to create flaps that, when interposed, increase the length of the flexion area. The resulting scars should be located so that they avoid future contraction.

The number and dimensions of the expanders to be used depend upon the size, shape, and condition of the area to be expanded. Expanders are inserted through incisions made in the scar areas. The length of the incisions should be as small as possible and, if needed, multiple incisions should be made. The direction of such incisions should be perpendicular to the flexion line. The dissection is initially done with blunt-tipped scissors followed by digital undermining to achieve a correct level of the dissected plane.

The valves used are the smallest possible, especially in terms of height. They are positioned as far as possible from the flexion area under a sufficient amount of subcutaneous tissue, even if it contains scar tissue. A vacuum drain is placed under the balloon to avoid hematoma.

Saline is injected in the balloon to fill up the dead space. The expansion starts 10 to 14 days postopera-

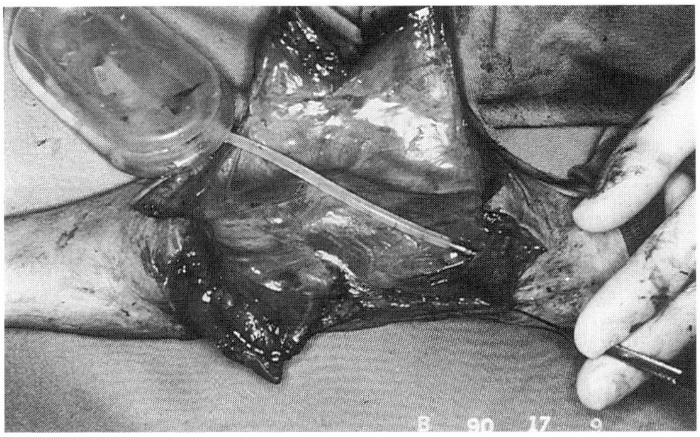

E

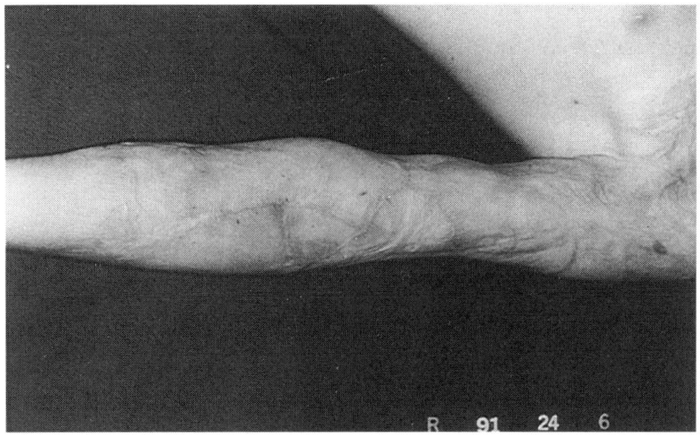

G

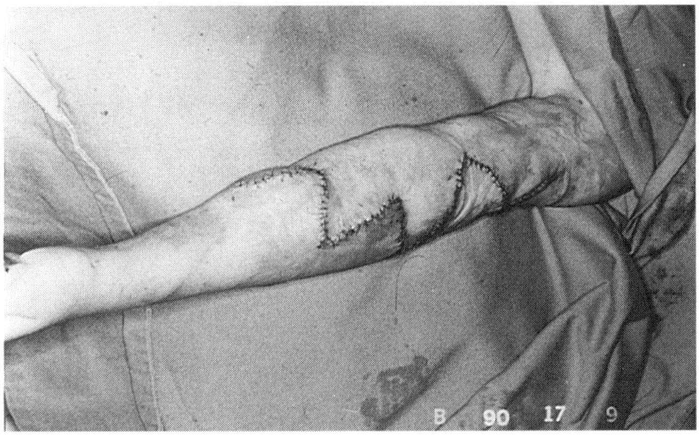

F

FIGURE 23-1 *(Continued)*

tively with a volume of about 5 percent of the nominal final volume of the balloon. The expansion is repeated at 7-day intervals with a gradual increase in the injected volume, depending on the tissue condition.

To provide better vascularity, the flaps obtained at the end of the expansion process should include the capsule. Once the flaps are in place, absorbable sutures are used in the subcutaneous layer and nonresorbable sutures are used in the superficial layer (Figure 23-1).

The use of elastic garments is indicated in the expansion period as well as in the postoperative phase. Continuous physiotherapy is started 15 days postoperatively until the scar has matured. The use of hydrating lotions and massage is recommended.

RESULTS

At present, our clinical experience with cases having more than a 2 year follow-up includes 13 patients aged

TABLE 23-1
Patients Submitted to Scar Expansion (1989 to 1991)

Case	*Body Area*	*Age (Years)*	*Previous Procedure*	*Functional Extension Limits (Degrees)*
1	antecubital	5	skin graft	160
2	antecubital	9	skin graft and skin flap	150
3	popliteal	6	—	165
4	axillar	14	skin graft	130
5	antecubital	6	—	160
6	axillar	19	—	145
7	popliteal	8	—	150
8	popliteal	9	skin graft	160
9	antecubital	19	—	145
10	foot	7	skin graft	100
11	antecubital	5	—	150
12	antecubital	9	—	160
13	axillar	7	skin graft	110

TABLE 23-2
Nominal Volume, Type of Expander, Days of Expansion, and Final Volume of the Expander

Case	*Expander Volume (mL) and Shape*	*Days of Expansion*	*Final Volume (mL)*
1	200 round	65	220
2	160 rectangular	73	180
3	160 rectangular	64	167
4	400 round	94	515
5	160 rectangular	71	170
6	200 round	67	233
7	100 round	72	120
8	200 round	32	hematoma and extrusion
9	200 round	70	215
10	100 round	75	95
11	100 rectangular	67	115
12	160 rectangular	73	187
13	320 rectangular	75	330

5 to 19 years. The postburn period varies from 8 months to 2 years. In 11 patients, the results obtained were good. In one patient (case 5) there was an improvement, but the elbow extension achieved only 170 degrees due to previous articulatory problems. One major complication was noted, which led to the removal of the expander and interruption of the procedure. Concerning minor complications, there was one hematoma (case 4) and one small displacement of the expander, neither of which affected the final result (Table 23-1 and 23-2). Cases number 5 and 6 presented ulcerations in the scar that healed as the expansion progressed, giving no further complication (Figure 23-2).

The interval between the initial and the definitive surgery for the removal of the expander was between 64 and 94 days, with an average of 72 days.

In all cases, the final volume exceeded the nominal value of the expander; the maximum was in case 4,

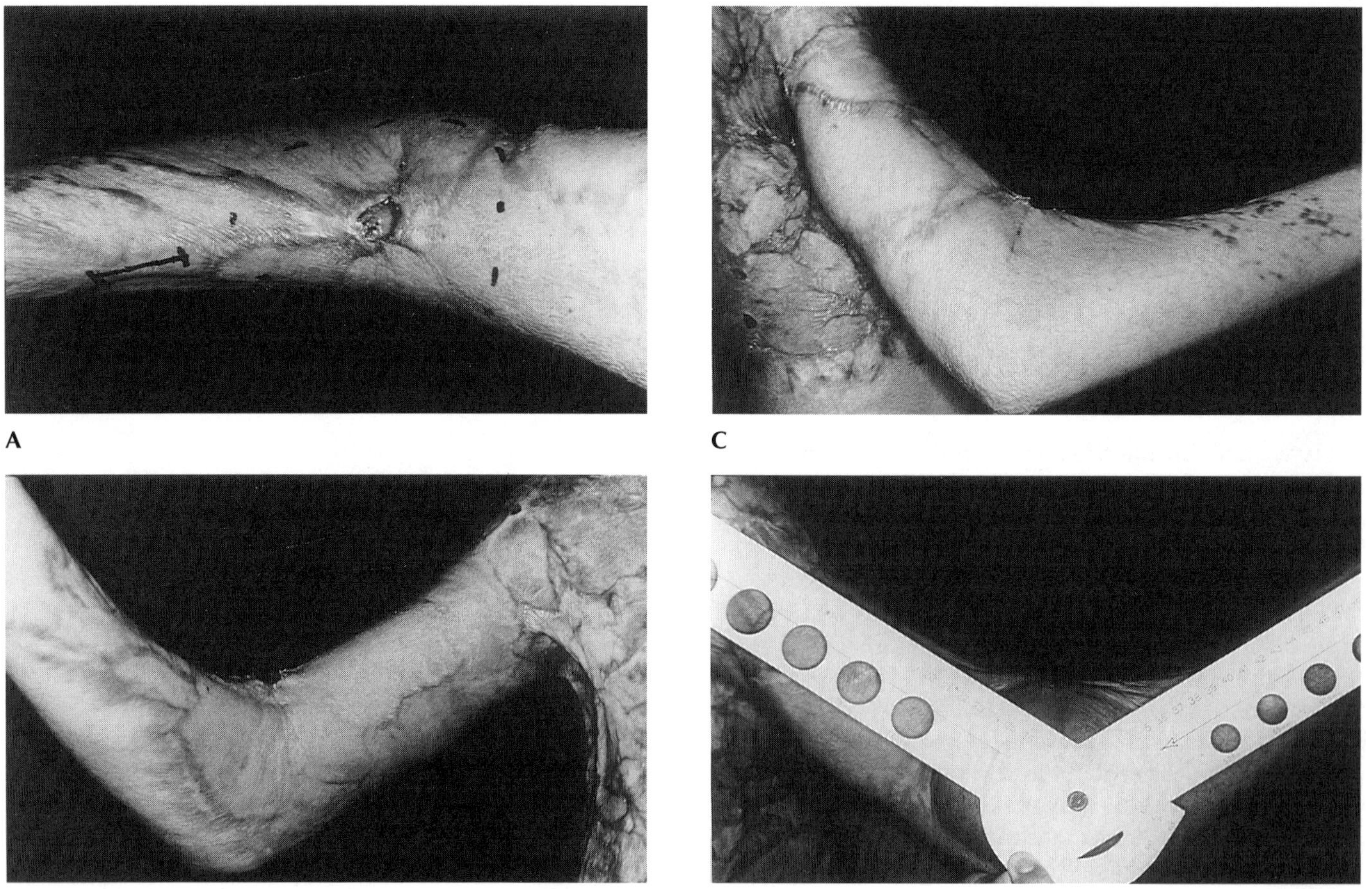

A C B D

FIGURE 23-2
(A–F) Nineteen-year-old boy with 145 degree elbow flexion limitation. There was an ulcer present. (G) Total healing at the end of the expansion.

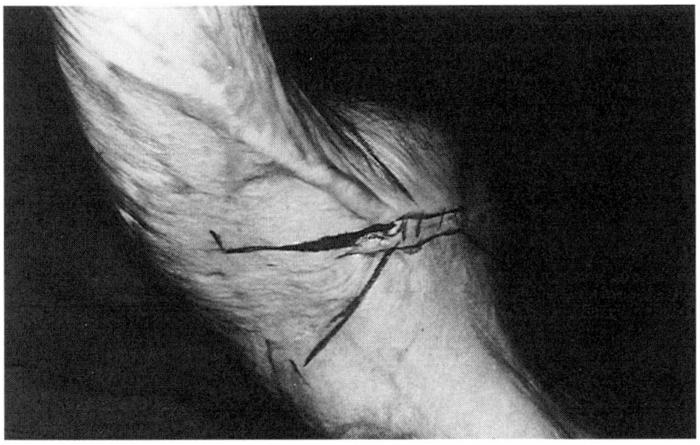

E

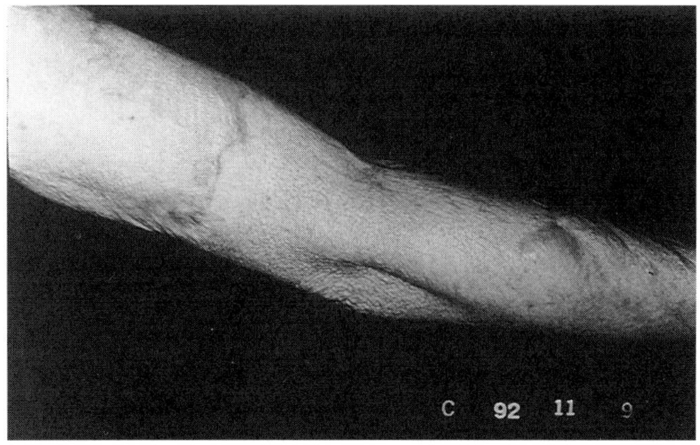

G

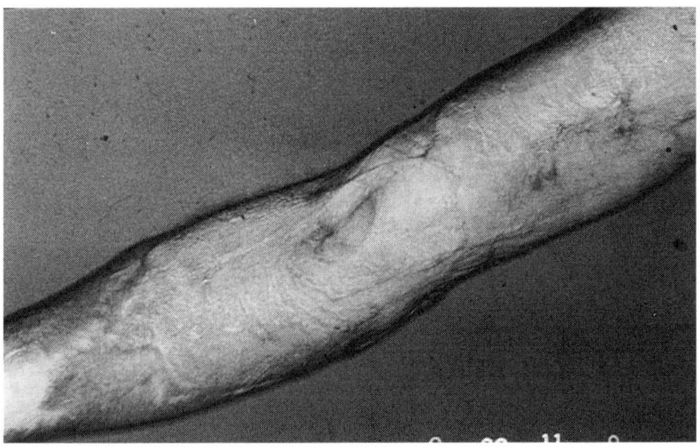

F

FIGURE 23-2 *(Continued)*

which achieved 29 percent above the volume indicated by the manufacturer. There were no postoperative complications of the final surgery. The postoperative follow-up was at least 2 years and at most 6 years following the final surgery. There was no case of contraction relapse.

DISCUSSION

Postburn scars are generally irregular, and frequently, in our clinical experience, we use an amount of scar tissue as part of a donor area without an increase in the rate of complications. This requires more care in the surgical procedures and in the expansion sessions. The results obtained have shown that expansion of scar tissues is feasible.

The current use of videoscopy for the placement of expanders has made the dissection under the scar areas easier and safer[8] and has also improved the accuracy of the hemostasis. Even when there is no evident bleeding during the dissection, large amounts of drained blood are found until the second or third postoperative day. Therefore, the use of drains is mandatory.

To prevent major complications, the joints must be immobilized, and bed rest during the first week is recommended after leg surgery. This implies that the patient or the relatives must strictly follow these guidelines.

An important result is the improvement of the aesthetic appearance of the scars. This is the result of the reduction of the traction forces upon these tissues and of the mobility gained.

This procedure is still limited in our clinical experience, but it is sometimes the only way to improve the clinical condition of these patients. In most cases, the surgical alternative would be skin grafting, but some of them had already submitted to this procedure before without success. Finally, it is important to keep in mind the possibility of using this procedure and to learn to choose carefully the indication (Figure 23-3).

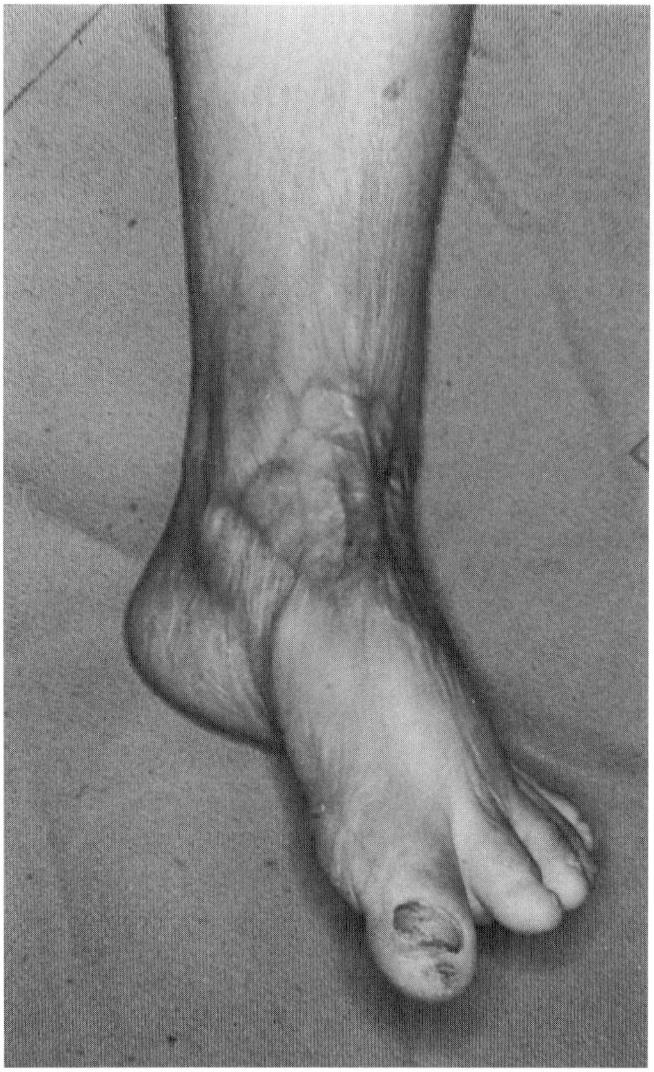

A

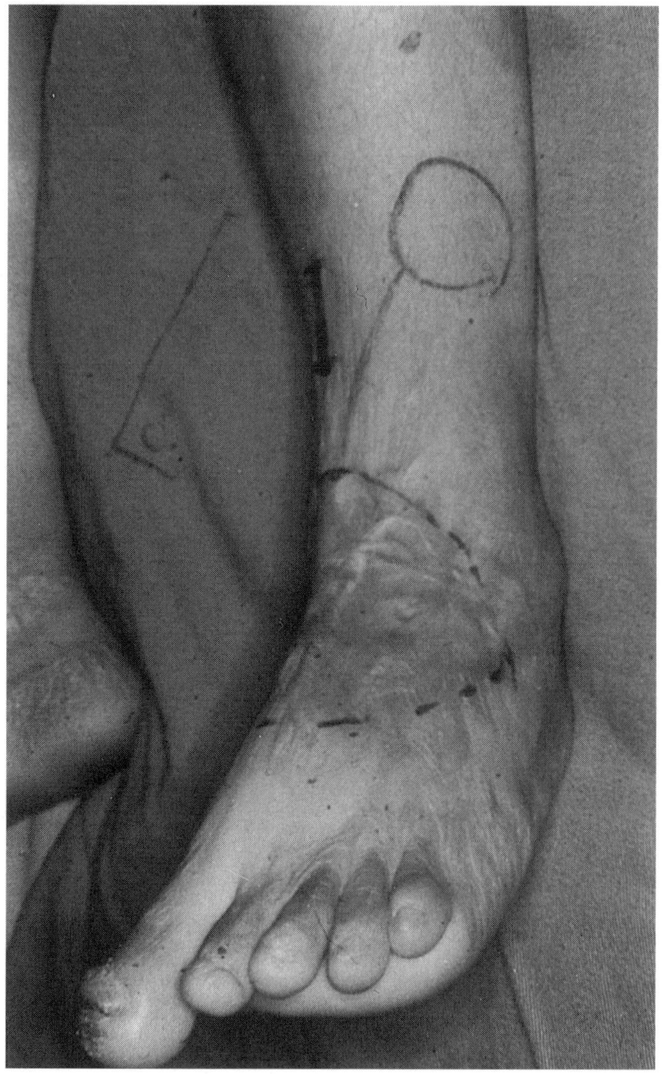

B

FIGURE 23-3

(A,B). A 7-year-old girl with difficulty in wearing shoes and a limitation of movements. (C,D) A round 100-milliliter expander was introduced. (E,F) The long-term postoperative result. In this patient, a Z-plasty was performed, including all the expanded scar to achieve an elongation of the skin on the dorsum of the foot.

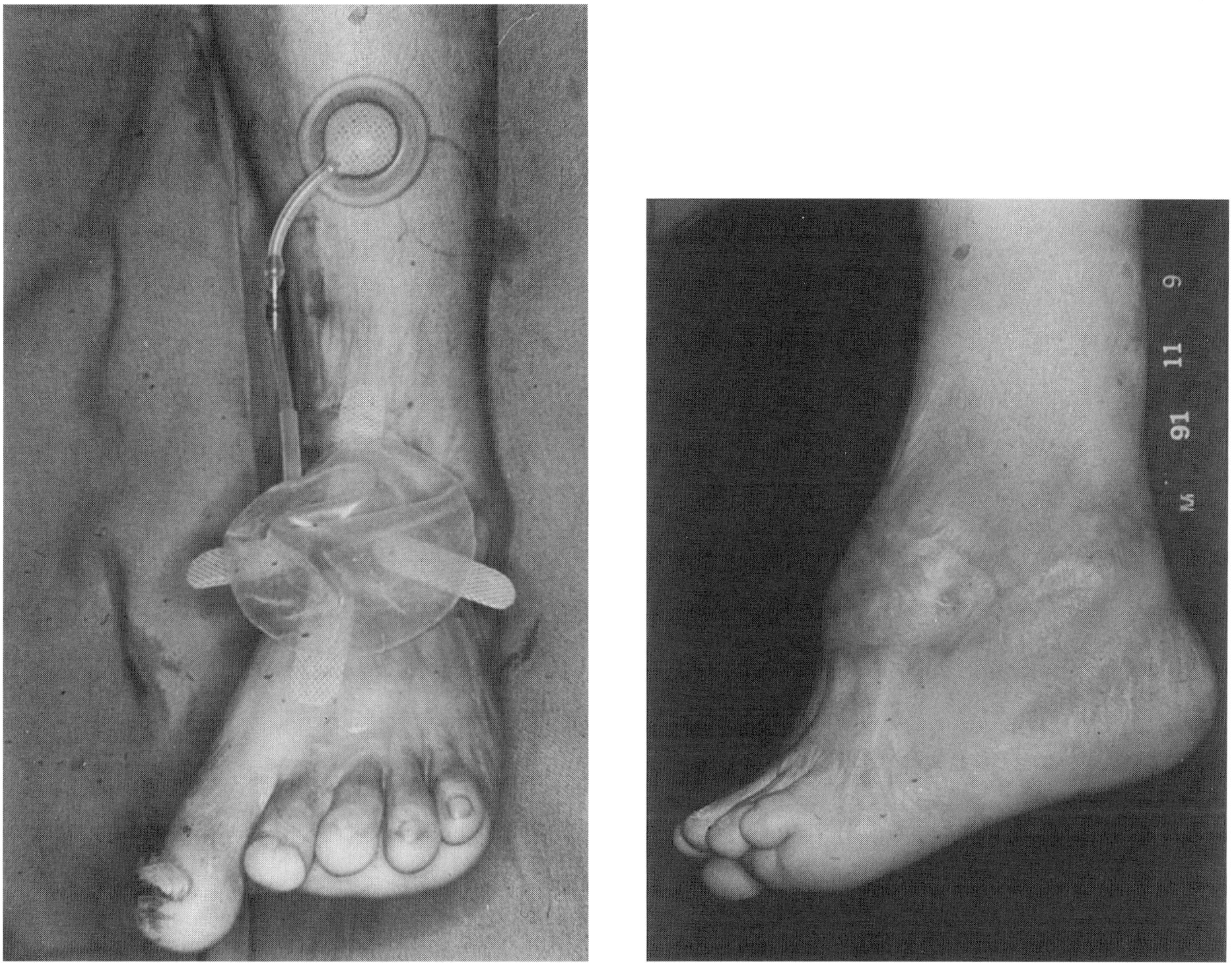

C D

FIGURE 23-3 *(Continued)*

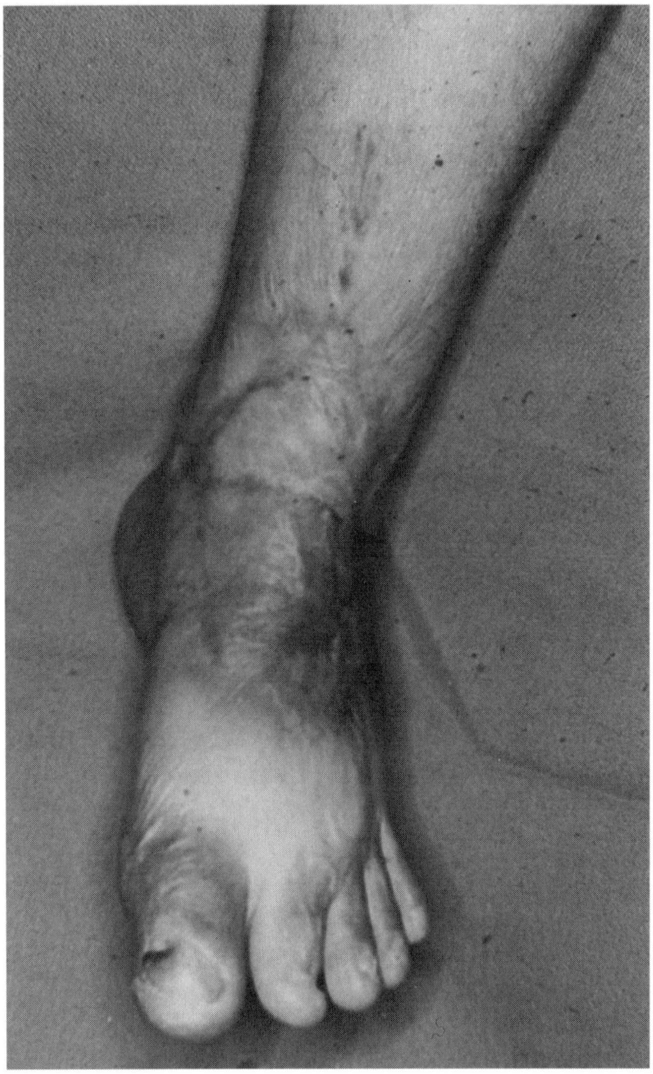
E

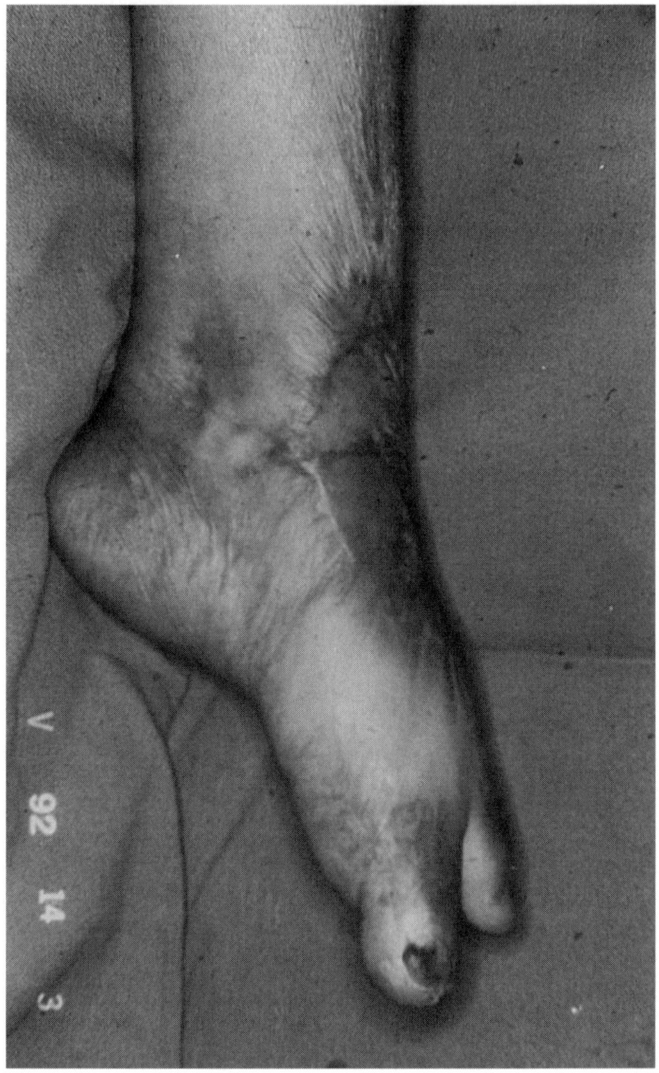
F

FIGURE 23-3 *(Continued)*

REFERENCES

1. Austad ED. Complications in tissue expansion. *Clin Plast Surg* 1987;14:549–550.
2. Marks MW, Argenta LC, Thornton W. Burn management: The role of tissue expansion. *Clin Plast Surg* 1987;14:543–548.
3. Versaci A. Breast reconstruction. First International Tissue Expansion Symposium, San Francisco, CA, 1987.
4. Neligan PC, Peters WJ. Advances in burn scar reconstruction: The use of tissue expansion. *Ann Plast Surg* 1989;22: 203–210.
5. Anger J. Principios basicos da espansio controlada de tecidos na infancia. In Avelar J. (ed.), Cirurgia plastica na infancia. Sao Paulo: Ed. Hippocrates, 1989, p. 601D4.
6. Sasaki GH, Pang CY. Pathophysiology of skin flaps raised on expanded skin. *Plast Reconstr Surg* 1984;74:59–65.
7. Anger, J. Burn scar expansion. In Ohura T. (ed.) *Tissue expansion symposium.* Sapporo: Sanco Inc., 1991, pp. 325–329.
8. Anger J, Szego T. Use of videoscopy as an aid in the placement of tissue expanders. *Rev Pau Med* 1993;111:363–366.

24

Videoscopy in Controlled Tissue Expansion

Jaime Anger

Today, the technique of tissue expansion is well known and widely used, thanks to the experience gained by various practitioners. Although complications are still reported,[1,2] their occurrence has become less common. Most of these complications are directly related to the first surgical step, that is, the insertion of the expander. Subcutaneous pocket insufficiently dissected, inadequate hemostasis or lack of drainage, dehiscence of the incisions, and technical errors of dissection are common causes of such complications.[1,2]

An analysis of these complications shows that most are due to lack of direct and perfect visualization.

Extrusion of prostheses through access incisions has been described by several authors.[2–5] To reduce such complications, in 1985 Sasaki suggested choosing small incisions placed radially in relation to the expander's main axis to avoid traction on the sutures after expansion begins, with a distance of at least 2 centimeters from the edge of the balloon.[3] Others recommend tunneling between the incisions and the place where the balloon will remain.

Still other authors suggest choosing access routes of larger dimensions on the edge of the lesion near the expander to allow for easier visualization during the operation. But this means a 2-week delay for expansion to begin, which is not desirable because over time the fibrotic capsule that normally forms around the balloon gradually grows, thus making progression of expansion more difficult.[4]

Videoscopy, as a surgical technique, adheres to the principles idealized by plastic surgery, namely, tiny incisions and easy execution. At present, there is no doubt that this new procedure is a solid contribution to the technique of controlled tissue expansion.

INDICATIONS OF THE TECHNIQUE

Placement of Expanders

The use of videoscopy for expander placement allows for better visualization of the tissues and a direct control of hemostasis, even with a small incision. These incisions can be made in areas located some distance from the region where the tissues will be under the traction produced by expansion. The small size of the incisions and their position far from and perpendicular to the main axis of the expander prevent the dehiscence of the wounds, thus permitting an early start of the filling procedure.

The possibility of direct vision allows for exact dissection of the desired plane. This prevents imperfections in the contours of the undermined pocket. If the balloon is placed in a space insufficient for its size, it becomes wrinkled, and as it expands, sharp points are formed that can provoke erosion and possibly perforate the tissues.

One basic principle in the surgical planning of tissue expansion is to benefit as much as possible from the donor area. Better dissection of the desired area allows previously inaccessible zones to be approached, especially in situations where important vessels or nerves are present. One example is the preparation of a frontal flap in nasal reconstruction, in which identification of the supraorbital vessels may enlarge the donor area available to be expanded (Figure 24-1).

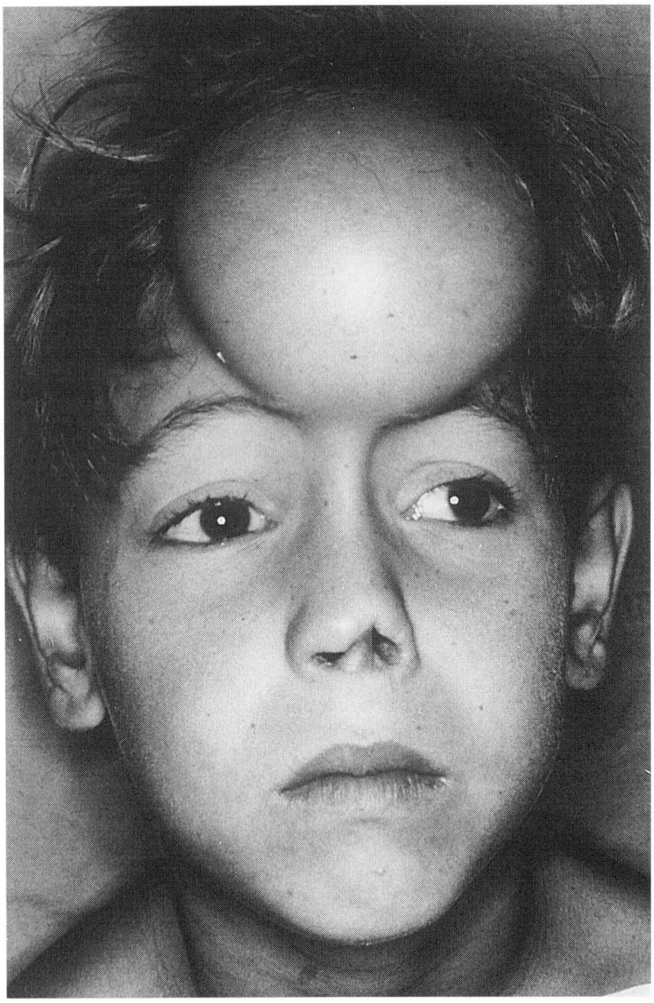

A

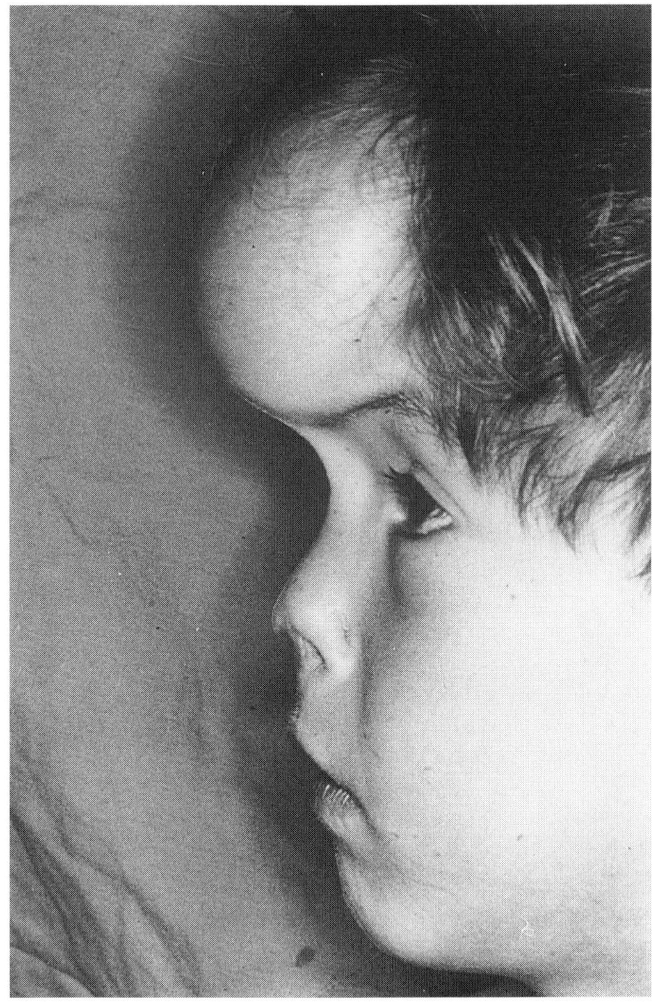

B

FIGURE 24-1
Seven-year-old child with nose tip loss. A round shape expander of 200 milliliters was introduced through a 3-centimeter incision inside the scalp area.

Access to regions with irregular contours, such as the thigh or thorax, is made easier, even with a small incision.

Another important region suitable for videoscopy is the anterior cervical region. Here it is preferable to place the expander in a supraplatysmal plane. Using a conventional approach, the required dissection is risky and difficult without wide incisions, which in most cases are contraindicated because of the aesthetic importance of the region (Figure 24-2).

Surgical Revision

Surgical revision may be necessary in certain complications during the expansion phase. Common complications in this phase are the rotation or damage of the valve or accidental perforation of the balloon.[6] The position of the valve can be corrected by endoscopy with the help of Kelly-type forceps.

The use of extensive incisions near the expander may provoke an inflammatory reaction resulting in extrusion. In these cases, the use of small and distant means of access is indicated.

Final Surgical Stage

In the final stage (removal of the expander), the endoscopic option may occasionally be used to perform capsulotomies to increase the gain achieved by the final intraoperative expansion.

EQUIPMENT AND INSTRUMENTS

The equipment necessary to perform videoscopy in tissue expansion can be divided into two groups: the

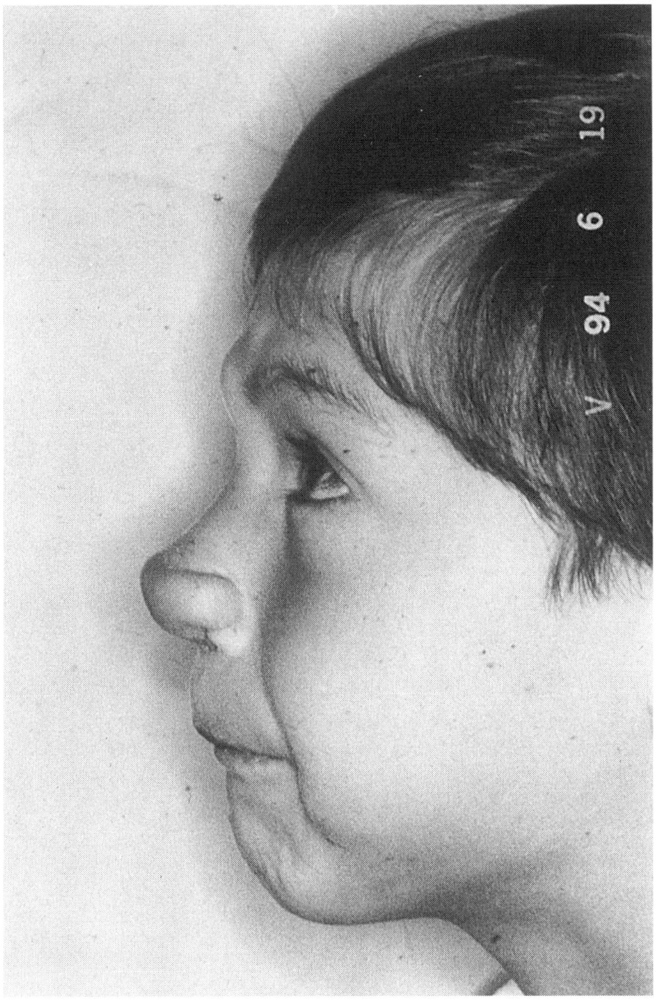

C

FIGURE 24-1 *(Continued)*

equipment for capturing, processing, and monitoring the image; and surgical instruments.

The basic optical equipment consists of the following:

- two video cameras, one 10 millimeter in diameter with 0-degree angle of vision and one 4 millimeter in diameter, 16 to 18 cm in length with 30-degree angle of vision
- image processor
- source of cold light with potency of 250 to 350 w
- fiberoptic cable
- pneumostat with pressure control and measurements of volume used (Figure 24-3)
- CO_2

The surgical instruments vary according to the clinical indications and the regions of the body, as follows:

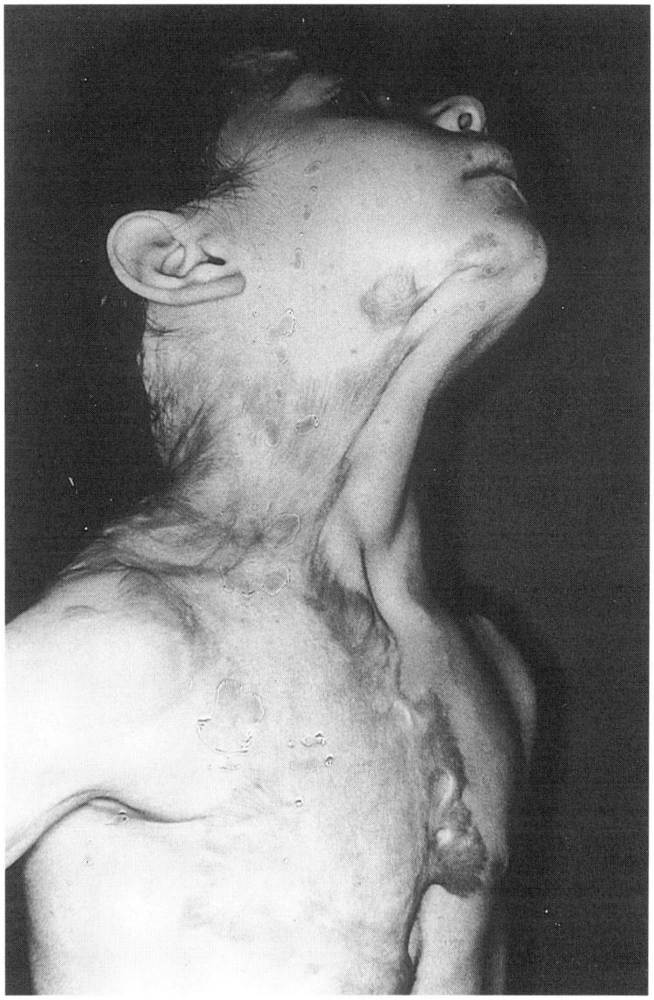

A

FIGURE 24-2
Five-year-old girl, with scar burn contraction in the cervical area. Two small incisions inside the scar, radial to the expander axis, are shown.

- 5-millimeter trocars with retaining valve
- scissors with curved, blunt tips, 25-centimeters in length, covered to allow connection to an electric scalpel
- forceps for hemostasis

The specific equipment and surgical instruments for the technique of controlled tissue expansion are more variable compared to those used for other procedures in plastic surgery, and their proper selection depends on the clinical indications, according to the need to use CO_2 and, in some cases, the visibility and dimensions of the area to be examined.

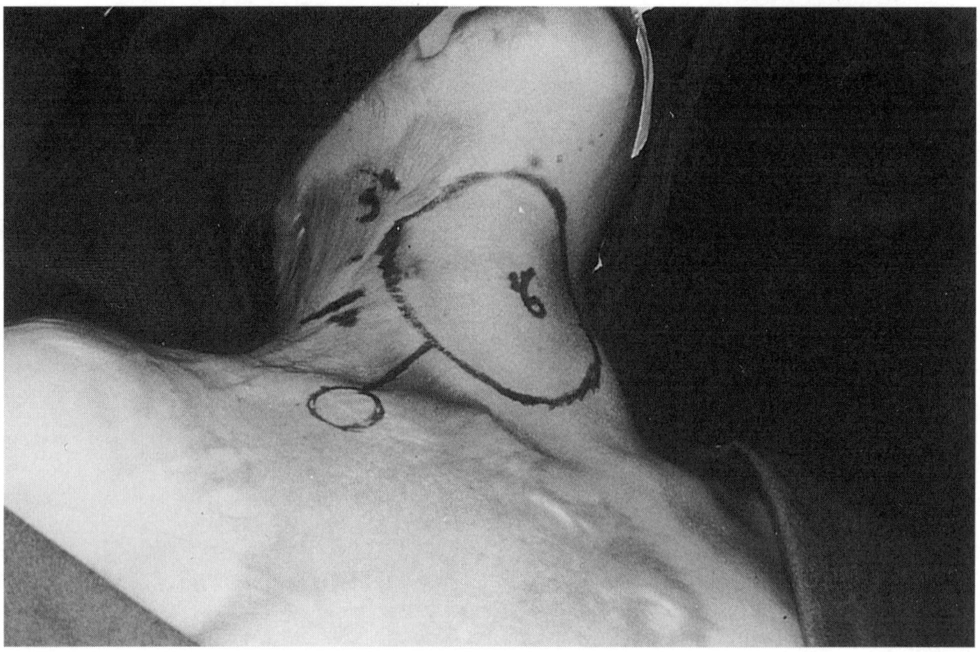

B

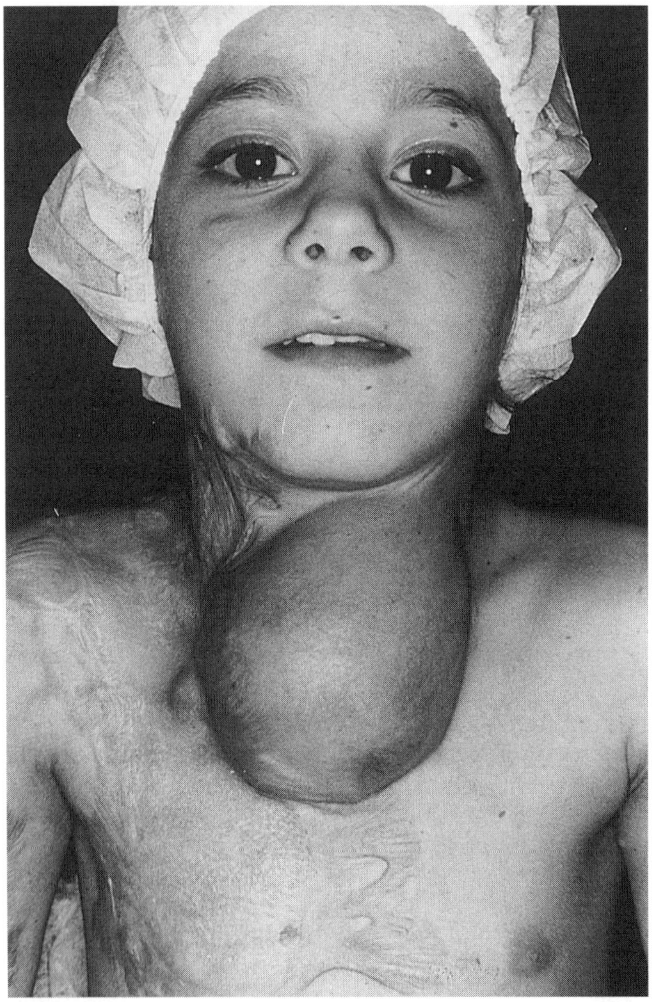

C

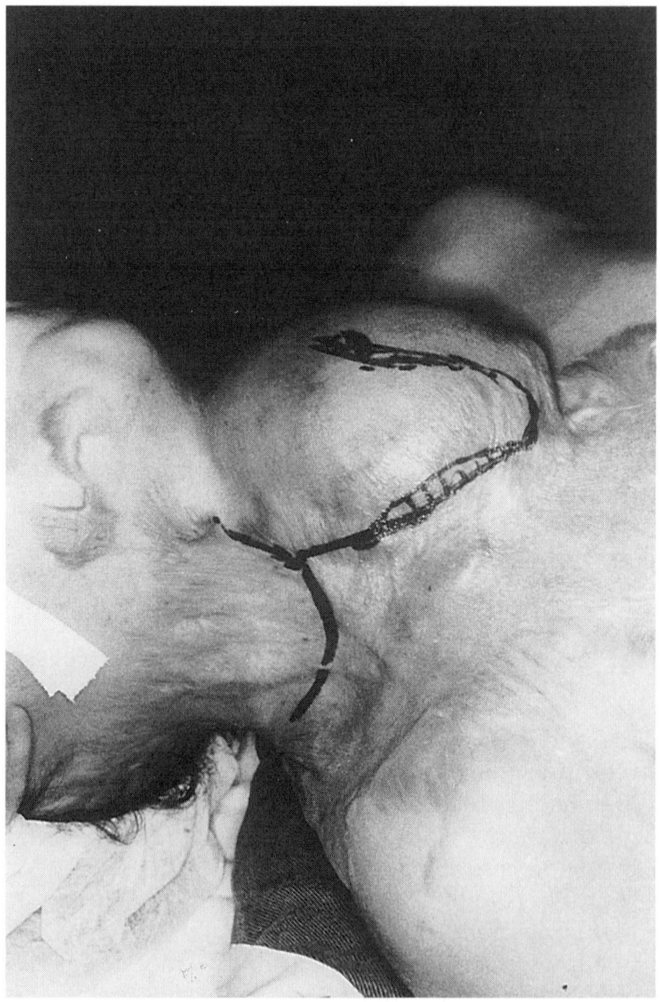

D

FIGURE 24-2 *(Continued)*

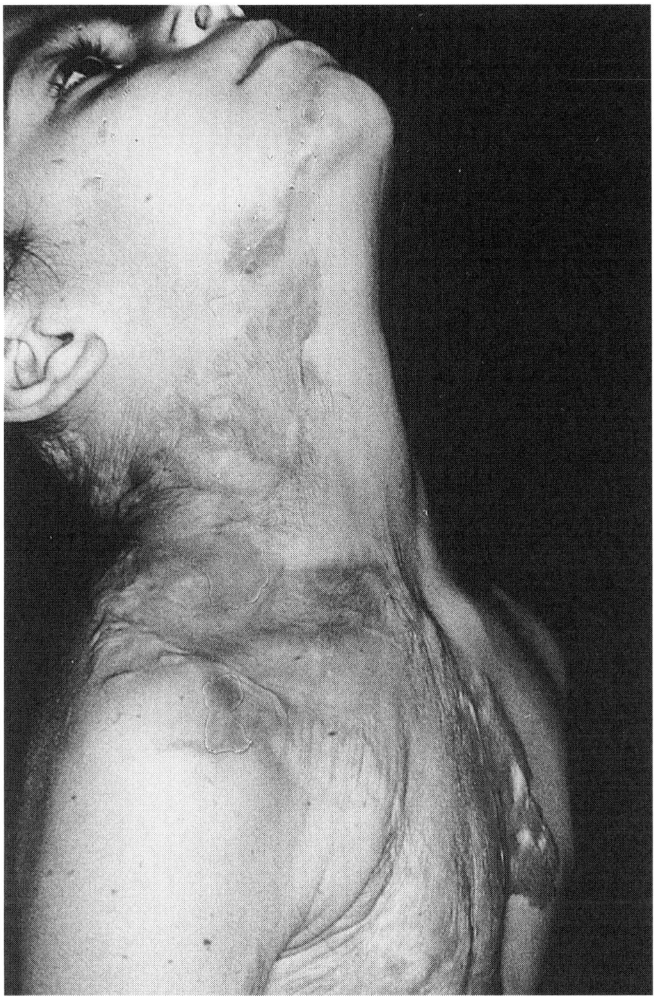

E

FIGURE 24-2 *(Continued)*

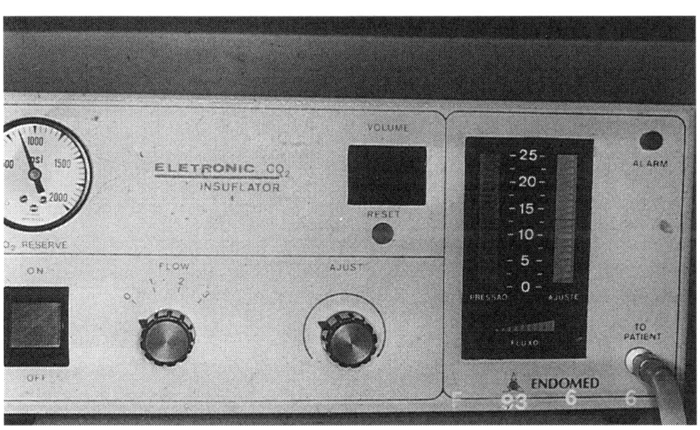

FIGURE 24-3
A pneumostat with volume and pressure control.

SURGICAL TECHNIQUE

Only two incisions for access are necessary; one of 10 millimeters for positioning the trocar and video camera and a second of 5 millimeters as a service channel for scissors or forceps for hemostasis. One of the incisions should be enlarged at the end of the procedure for placement of the remote valve and the expander; the final size depends on the dimensions of the type of valve (between 2 and 3.2 cm). The incisions are positioned at least 3 centimeters from the edge of the area undermined for the expander (Figure 24-4). The incisions should be at least 5 centimeters apart. When the space required for the balloon exceeds 20 centimeters, a third incision may be necessary.

After the incision is made in the skin and subcutaneous area, the tissues are identified and the desired level to place the expander is chosen. Careful undermining of the pocket is then performed, using a common pair of blunt-tipped scissors. As much space as possible is freed, exerting minimum pressure and avoiding damage to vessels and nerves. Obstacles that can prevent the progression of dissection are essentially fibrotic areas, vessels, and nerves. Anyhow, in this phase, most of the necessary space is generally freed.

After the scissors are removed, a 10-millimeter trocar is inserted for the video camera. Carbon dioxide is insufflated to a maximum pressure of 8 mmHg. At this level of pressure, the gas that penetrates the dissected space expands the area and visually projects the limits of dissection. This allows the operator to identify what has already been dissected and what still needs to be undermined (Figure 24-5). CO_2 does not provoke subcutaneous emphysema at these levels of pressure.[5]

The microcamera is then introduced and the dissection is revised. Vision is broad and retractors are unnecessary because the space is kept expanded by the insufflated gas. Even if CO_2 does escape, the visual field is maintained by the pneumostat that insufflates CO_2 as soon as the pressure goes down. As a result, the tissues never stick together.

A second incision is made for the second trocar. The scissors are used for dissection until the necessary space is achieved. Hemostasis is carried out with the same scissors connected to an electric cautery unit. The freeing of the tissues in the deep planes is accompanied by the gas-induced expansion, which identifies precisely the limits already dissected.

Once the area for expander placement has been prepared, the video system is disconnected and the trocars removed. The balloon is inserted through the larger incision, most easily if it is rolled lengthwise into a tube. This procedure is made easier if the expander has a reinforced base. Once the balloon and suction drain are

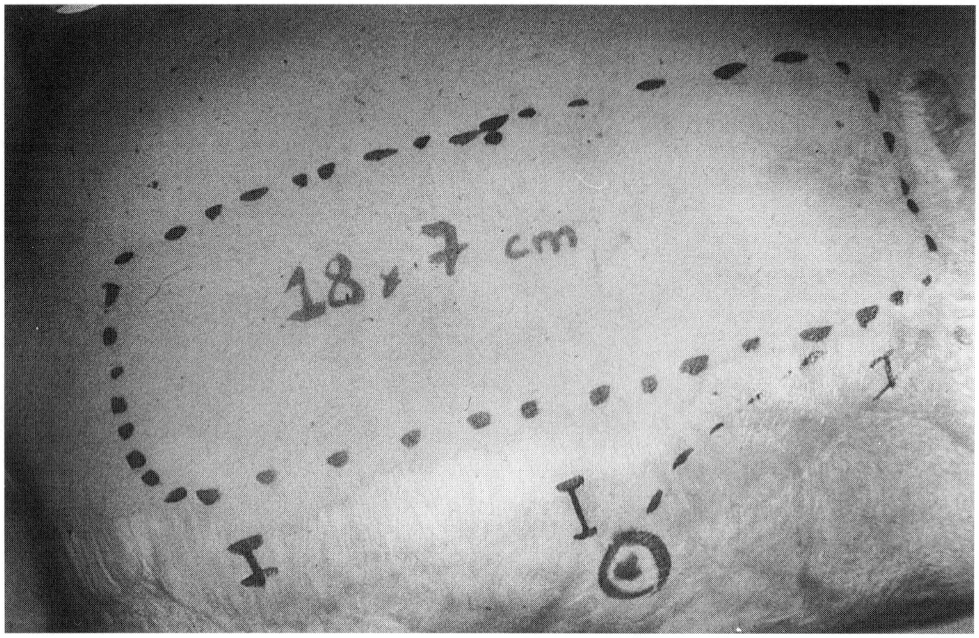

FIGURE 24-4
Posterior thorax. A 480-milliliter rectangular expander was used (17 × 6 cm). Two small incisions were used. The round mark represents the position of the filling valve.

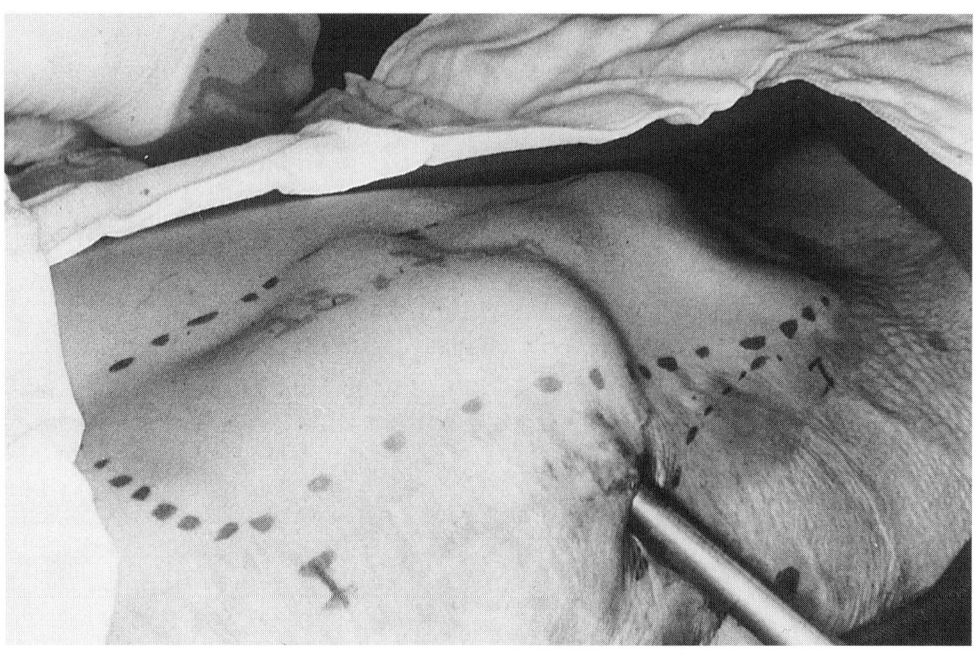

FIGURE 24-5
The area was inflated with CO_2, showing the distention.

in place, the valve is positioned and the desired amount of physiologic saline injected.

In areas where the dissection plane is located over bone, the use of gas may not be necessary. In these areas the video camera system may serve as a retractor, as happens in the frontal region when subperiostal rhytidectomy is performed. The equipment and the operating technique in these cases are the same as for rhytidectomies, using videocameras of 4 millimeters and 30-degree angle of vision.

GENERAL COMMENTS

Although videoscopy is a technique that started only recently,[5] the experience gathered previously in the technique of tissue expansion quickly enabled us to define the clinical indications for its use. We can classify these indications into the following groups:

(1) Patients in whom the position and routes of access are of the utmost importance because of the risk of dehiscence or the difficulty of finding suitable areas for incisions for expander insertion. The most characteristic examples are patients with scar sequelae from burns.
(2) Regions that offer few options for access, principally for aesthetic reasons, and where direct vision for dissection is fundamental. This group includes the anterior cervical region, the frontal region, and the face.
(3) Regions where the plane of dissection is curved, thus making blind dissection difficult. The lateral and medial regions of the thigh and the edges of the thorax and abdomen are two examples.
(4) Patients in whom the lesion to be removed may itself be used as a way of access; for example, a giant congenital nevus (Figure 24-6).

The main objections to the use of videoscopy are essentially the cost, the length of the operation, and, in cases when CO_2 is used, possible absorption and hypercapnia during the operation. Cost is related to equipment and instruments. The most expensive piece of equipment, the video system, is already installed in most hospitals all over the world, thanks to the large number of videolaparoscopic procedures that are performed. Some of the instruments are commonly used in frontal rhytidoplasties or placement of breast implants with a transaxillary approach. This means that the highest cost is related to the use of the material or, when the material belongs to the doctor, the cost for wear and tear. This wear on the equipment and instruments is nonetheless directly linked to the quality of the result and especially to the risk of complications.

The operation is longer because more time is spent installing and preparing the equipment and the dissection requires prolonged movements and programming. Indeed, what has been learned from the experience of over 50 operations performed is that the area dissected with the videoscopic system is never greater than 15% of the total undermined area. This dissection may be quicker and more effective than that achieved by a more conventional approach. In the first 25 patients (using videoscopy) with burn sequelae, where it was necessary to use only one expander, the length of the operation decreased from 100 to 120 minutes in the first operations to an average of 50 minutes in the last five operations.

The most controversial topic is related to the use of CO_2 in the subcutaneous area. Initial work on animals revealed that it would not provoke subcutaneous emphysema at relatively low pressures (around 8 to 12 mmHg).

Published reports of complications from the use of CO_2 are related to gynecological and gastroenterological operations. The intercurrences may be summarized as hypercapnia, subcutaneous emphysema, pneumothorax, and gas embolism. The total number of these complications is small compared to the total number of operations performed. Moreover, they are exclusively related to the insufflation of large volumes of CO_2 under high pressures.

Some of the most serious intercurrences are the result of errors in the position of the Verres needle in gynecological laparoscopy, causing subcutaneous emphysema.

The descriptions for subcutaneous emphysema in videolaparoscopy for gastroenterological emphysema refer to a trajectory of gas that, at a very high pressure, passes from the peritoneum cavity to the mediastinum near the esophagus, provoking pneumomediastinum, pneumothorax, and subcutaneous emphysema.[7]

Some descriptions of subcutaneous emphysema without other alterations being present reveal a possible relation with the escape of gas around the trocar. Generally, this gas escapes because of the poor positioning of the trocar or by increasing the pressure too much as a result of an error in the pneumostat.[7]

Hypercapnia is the most frequent complication associated with the use of CO_2. This generally occurs during longer operations (over 2 hours) because of the gas absorbed. Since CO_2 has a very great capacity for diffusion, its blood level can become high. However, for this to happen, there must be associated errors in a patient's ventilation. To avoid any complications in using CO_2, a capnometer should be used in all procedures

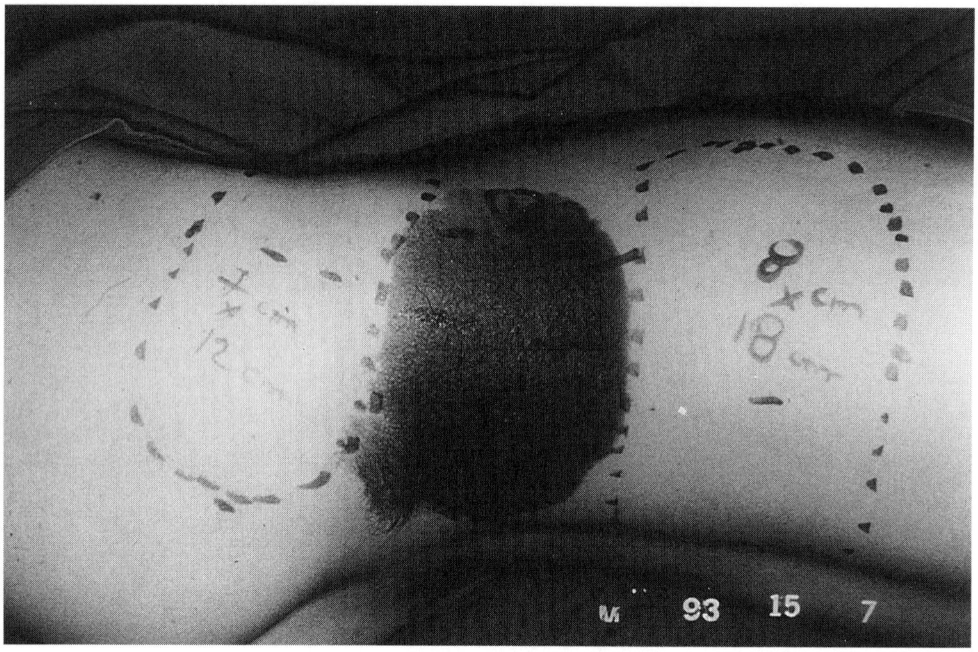

A

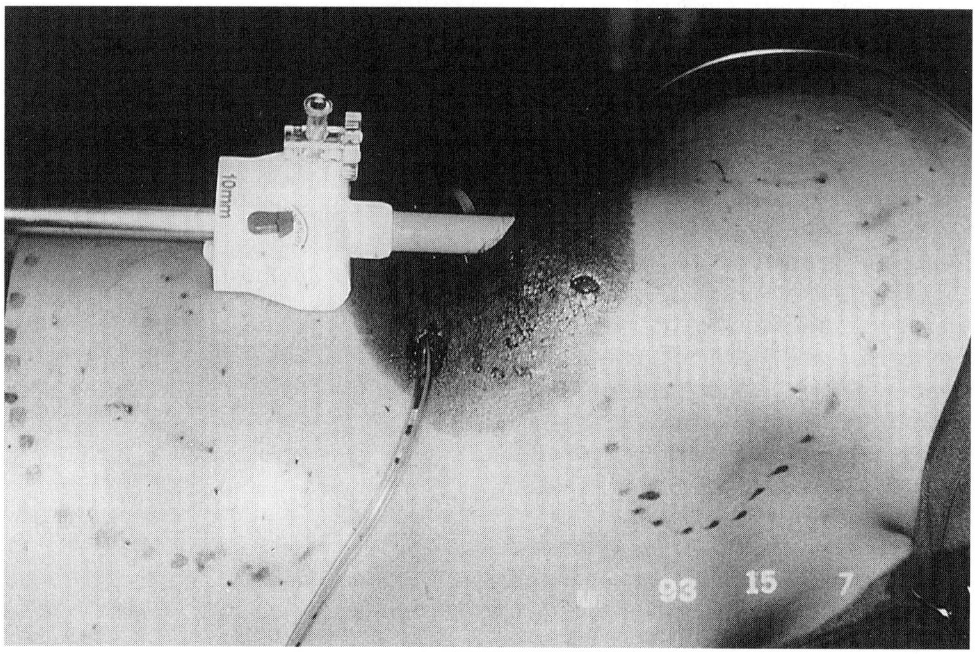

B

FIGURE 24-6

(A–D) A giant congenital nevus. Only one access was used. (C) Shows the patient after 3 weeks with saline injected.

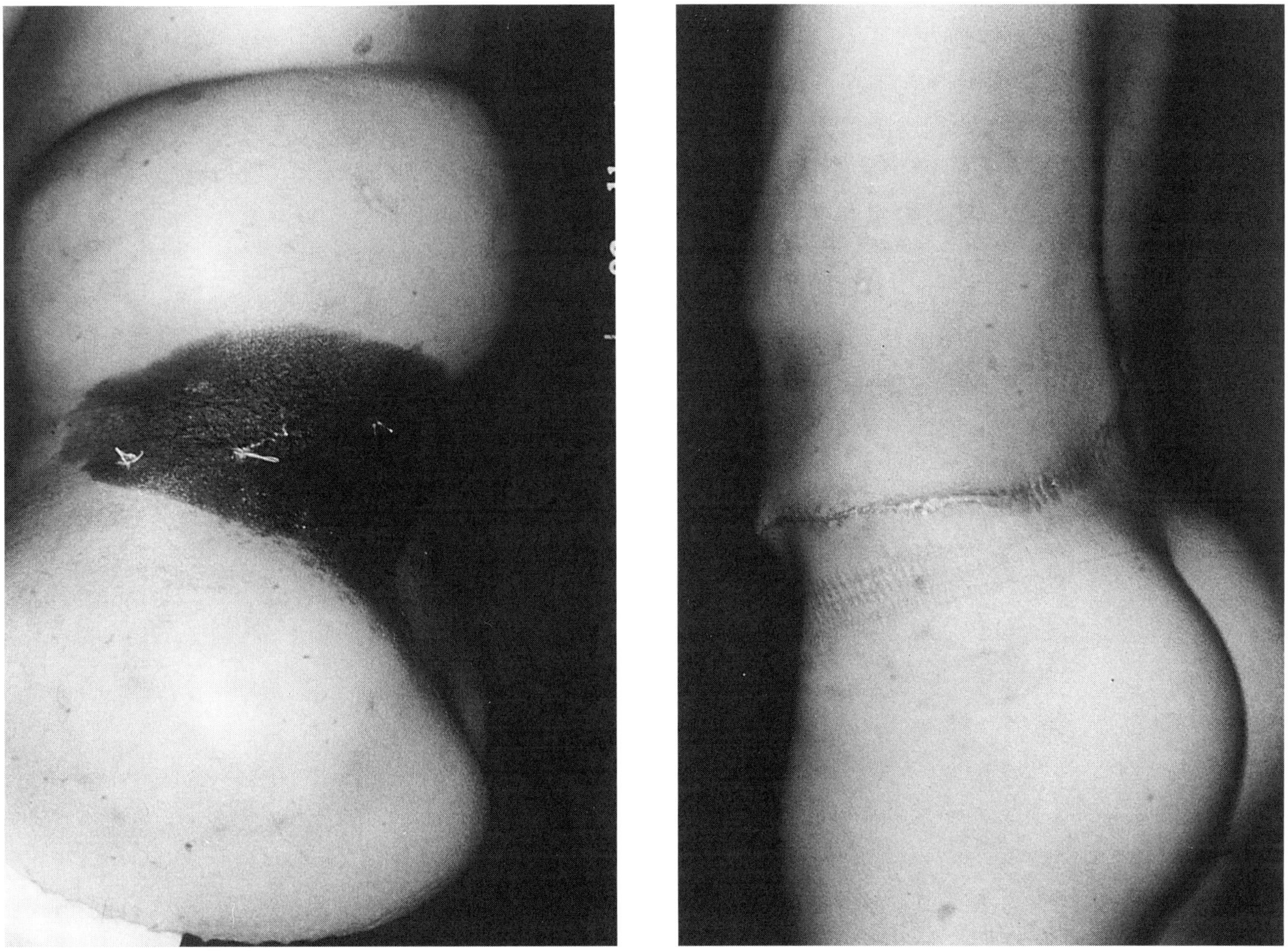

C D

FIGURE 24-6 *(Continued)*

where CO_2 is insufflated, as this quickly detects any increase in the level of CO_2 caused by excessive absorption. This can easily be corrected by diminishing the volume of insufflated gas and lowering its pressure.

To study possible alterations in the subcutaneous area, we used a capnometer in 25 patients. In the first ten patients, we carried out arterial blood gas tests every 30 minutes during the operation. The longest operation lasted 145 minutes, with the placement of two expanders, and the shortest lasted 30 minutes. No change was detected.

In the first 47 expanders placed with videoscopy, the incidence of major complications was 0%. Five of the 47 expanders were placed without CO_2. We consider these first results excellent and we think that this new modality will give more security to the technique of controlled tissue expansion.

REFERENCES

1. Anger J. Principios basicos da espansio controlada na infancia. In: Avelar, J (ed.), *Cirurgia plastica na infancia.* Sao Paulo: Ed. Hippocrates, 1989.
2. Austad ED. Complications in tissue expansion. *Clin Plast Surg* 1987;14:549–50.
3. Sasako GH. Tissue expansion. Guidelines and case analyses. Dow Corning Wright form L080-0204, Arlington, 1985.
4. Manders EK, Oaks TE, Wong RKM, Furrey JA, Davis TS, Graham WP. Soft tissue expansion in lower extremities. *Plast Reconstr Surg* 1988;81:208–217.
5. Anger J, Szego T. Use of videoscopy as an aid in the placement of tissue expanders. *Rev Pau Med* 1993;111: 363–366.
6. Anger J. The reuse of tissue expanders in developing countries (letter). *Plast Reconstr Surg* 1991;88:1114.
7. Bard PA, Chen L. Subcutaneous emphysema associated with laparoscopy (letter). *Anesth Analg* 1990;71:101–102.

25

Alternative Devices and Techniques

JINCAI FAN, EDOARDO RAPOSIO, AND ROLF E.A. NORDSTRÖM

Traditional tissue expansion, although successfully used (and improved) for over 15 years, still has limitations, basically related to the number of outpatient visits needed to complete the inflation of the implant, the long period required by the whole filling procedure, as well as the temporary cosmetic disfigurement produced in the expanded area.

Furthermore, traditional tissue expansion is not a useful method for providing coverage of acute soft tissue loss and infected wounds. Alternative techniques and devices have been developed that take advantage of the viscoelastic properties of the skin, producing extra tissue without the drawbacks mentioned. In this respect, it is worth noting two important viscoelastic phenomena of the skin: "creep" and "stress relaxation."[1] Creep occurs when a piece of skin is stretched and the stretching force is kept constant; the skin continues to extend, depending on the amount of tension applied. During this process, as the fibers in the stretched dermis become aligned, in a somewhat parallel rearrangement, tissue fluid and ground substances are progressively extruded from between the collagen fibers. Therefore, the more fluid in the dermis, the greater is the amount of creep. Stress relaxation occurs when a piece of skin is stretched to a given length and held constant; the force required to keep the skin stretched gradually decreases. The following devices and techniques have been designed in an attempt to take advantage of these viscoelastic properties of the skin.

PRESUTURING TECHNIQUE

In 1988, Liang and associates described a new technique, termed *presuturing*, for aiding in the closure of large skin defects that otherwise may have required a skin graft or flap.[2] The presuturing technique is performed under local anesthesia and consists of plicating intact skin, by means of several 2-0 nylon sutures, over the area of planned excision the night prior to operation. This serves to stretch the skin surrounding the area to be reconstructed in a linear fashion. The direction of stretch has to be chosen according to the anticipated direction of primary closure, for example, by the shape of the lesion (as in the case of scar revision) or, with anticipated circular defects, by lines of minimal tension. Presuturing is an easy and clinically useful technique that helps make bigger excisions possible with closure under acceptable tension. In a pig model, Liang et al. have found that the decrease in force required to close a standard-sized wound was 40.1 percent less than in the control in which there was no presuturing. These results were confirmed in several clinical reconstruction cases.[2] The presuturing technique has also been shown to increase the amount of skin excised in face-lift surgery.[3]

SUTURE TENSION ADJUSTMENT REEL

The suture tension adjustment reel (Miami Star; George Tiemann, Plainview, NY), as described by Cohen and Cosmetto in 1992,[4] is a device developed to tighten or

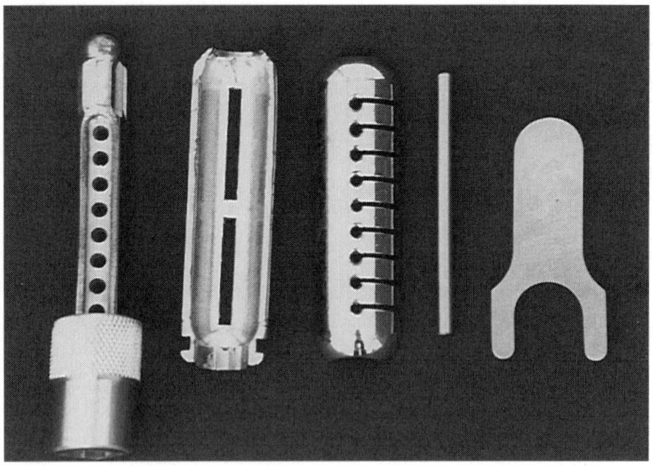

FIGURE 25-1
The suture tension adjustment reel (Miami Star Device).

loosen a suture ligature after the knot has been tied (Figure 25-1). This device is designed in such a way that by passing a continous loop of suture through it, tying the knot, and turning the reel portion of the device, the suture loops tighten in an incremental fashion. Thus, the inclusion of the STAR in the suture loop is equivalent to an adjustable knot that may be tightened or loosened without untying or cutting the suture material. The suture tension adjustment reel is useful for both preoperative and intraoperative tissue stretching of the area of an excision site, thus causing mechanical creep and allowing for subsequent closure with diminished tension. When used intraoperatively, it is advisable to load cycle the tension, to maximize efficiency, preserve perfusion, and minimize time.

The contoured design of the device mechanically bolsters the suture materials' contact with the skin, thus diminishing the likelihood of skin deformation or laceration. However, prolonged attachment of this device to the skin surface as well as overzealous tightening will end, obviously, in the same complications one would expect from overtightening of a material in the skin.[4]

FRECHET'S SCALP EXTENDER

Frechet has described the use of two thin strips of silicone elastomer with a row of titanium hooks on the two opposite ends (MXM-Les Mimosas, Quartier Croix Rouge, 0600, Antibes, France) to enhance the outcome of scalp reduction procedures (Figure 25-2). The extender strips are less than 1-millimeter thick and can vary in shape, the simplest form being a rectangle of varying length (average 5 centimeters). This sheet

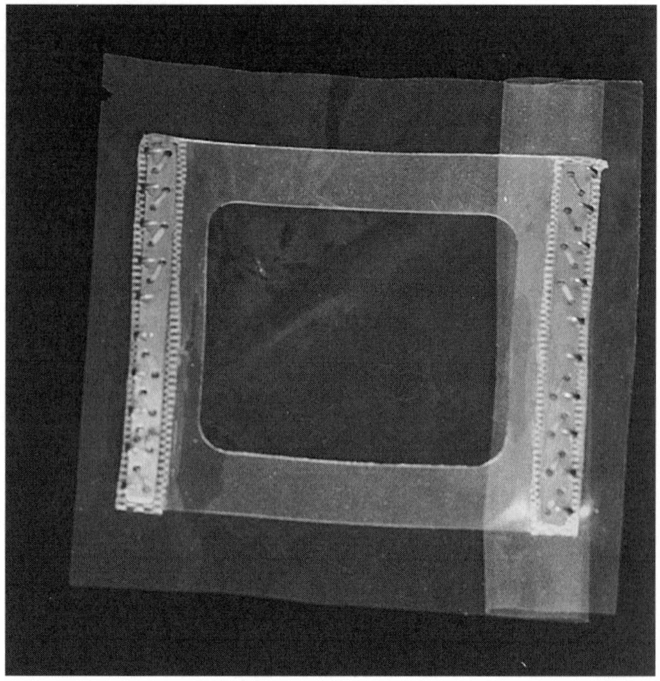

FIGURE 25-2
Frechet's scalp extender.

has elastic properties that enable it to stretch up to 100 percent with a natural tendency (memory) to return to its original length; an additional silicone sheet protects from overstretching more than 100 percent. The extender is designed to be inserted during a conventional scalp reduction procedure. After the excision of the designated area of bald scalp and before closure of the galea under moderate tension, the appropriately stretched extender is positioned approximately at the midline of the incision and perpendicular to it. Its two rows of hooks are attached to the underside of the galea using a specially designed instrument. The extender is then left in place for about 30 days. The use of the scalp extender has been reported to significantly enhance conventional scalp reduction by means of an increase in the amount of bald area excised and a shortening of the total treatment time.[5]

In our experience, the scalp extender has the ability to resist "stretch-back," which usually follows scalp reduction procedures.[6]

CUTANEOUS EXPANSION USING ENFORCED POSITION

Chretien-Marquet et al.[7] have reported the successful closure of large skin defects in children, taking advantage of both the flexibility of the joints and the extensibility of the skin, using an enforced position. To obtain the enforced position, flexion, extension, inclination, or rotation of the joints close to the defect were used. The joints were thus positioned in such a way to allow suturing without any tension. In most of the patients, no undermining was performed. This method protected the scar from stretching (for 6 weeks) in order to achieve sufficient strength. Thereafter, free movements across the joints were permitted to allow gradual tension in the skin surrounding the scar. This tension, due to the natural tendency of the joints to return to a normal position from the enforced position, elicited a gradual and progressive tissue expansion. The reported final result was a permanent gain of tissue replacing the excised lesion.[7] In our opinion, this method cannot be generalized; nevertheless, it allows a natural, painless, and progressive expansion that avoids the need for undermining and for multistep procedures involving tissue expanders. This method also avoids the problems related to an eventual expansion of pathologic tissue.

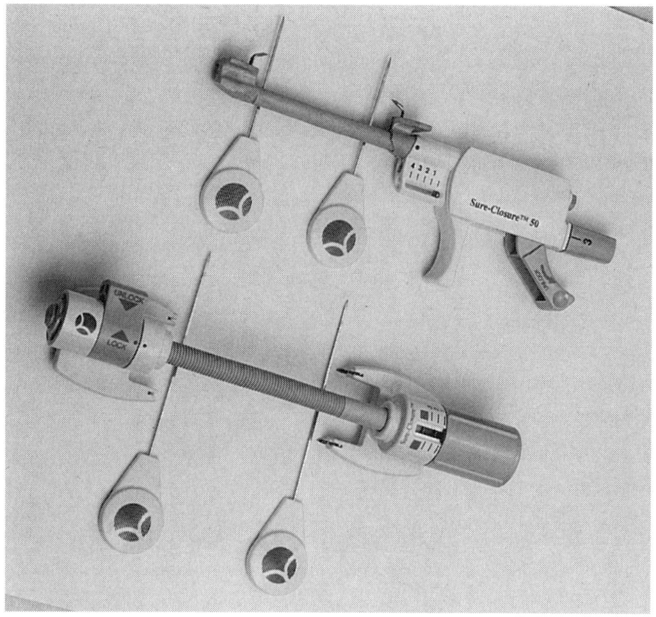

FIGURE 25-3
"Sure-Closure" skin stretching system. (Reprinted with permission from MedChem Products, Inc., Woburn, MA.)

"SURE-CLOSURE" SKIN STRETCHING SYSTEM

In 1993, Hirshowitz et al. developed an external skin stretching device designed to harness the viscoelastic properties of the skin using incremental traction (Figure 25-3).[8] It consisted of two long pins that had to be threaded through the dermis on either side of the wound/lesion, two U-shaped arms with inward-facing cutting hooks on the undersurface that engaged the pins through the overlying skin, and a threaded screw that passed through the center of the arms. The distal arm, which was loose, rode over the screw and was pulled closer when the screw was turned with a tension knob at its free end. In reported clinical experiences,[8] the skin-stretching device was tested over three different periods of time: preoperatively (1 to 2 days), intraoperatively (20 to 30 minutes), and postoperatively (several hours to 3 days). However, the stretching force was always applied in increments without periods of relaxation as in load cycling, while the device allowed the force to spread over a wide area, thus preventing mechanical damage of the stretched skin. No undermining was required. After suturing, disengagement of the skin-stretching device was easily accomplished by turning screws counterclockwise and pulling out of both intradermal pins. This device, now known commercially as the "Sure-Closure" skin stretching system (MedChem Products, Inc., Woburn, MA) has been proven to be both simple in its application and valuable in helping to close problematic areas of skin shortage, which would otherwise have required more complicated procedures.[8]

KIRSCHNER WIRES

A simple, speedy, and effective technique of wound closure by skin traction for use with multiple mass casualties as, for example, in a wartime situation, was described by Bashir in 1987.[9] With this technique, several Kirschner wires, 2 millimeters in diameter, are inserted into the skin, passed through the dermis for some distance and then out again. Two fine silver wires are then looped around the cross points of the Kirschner wires at the narrower diameter of the wound. The wires pass under the skin edges to meet and are twisted together at the center of the wound. The twisting of the wires will bring the wound edges together. Twisting is continued until the skin blanches, while the wound is left open for daily inspection. The wires are twisted tightly each day, advancing the wound edges by 3 to 4 mm. At the end of the procedure, the Kirschner wires lie almost parallel and the wound edges are approximated. The Kirschner wires are left in place for a few days until healing is completed and are then removed. Bashir reported that all the cases treated with this method had a

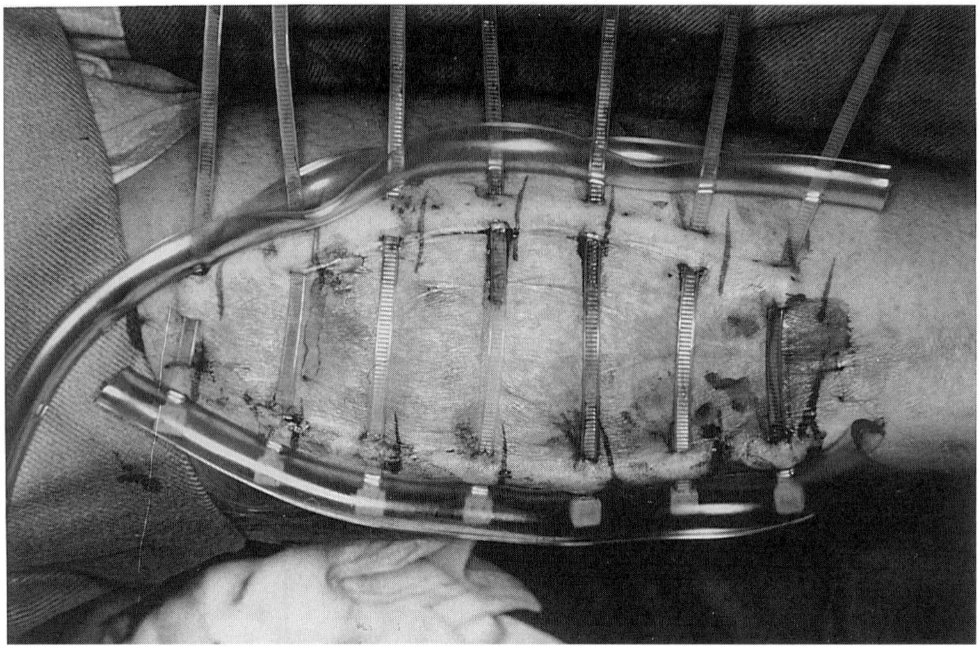

FIGURE 25-4
Blomqvist-Steenfos' external device.

satisfactory outcome; furthermore, the final scar was always aesthetically acceptable.

BLOMQVIST-STEENFOS' EXTERNAL DEVICE

In 1993, Blomqvist and Steenfos developed and clinically tested a skin extension device that produced rapid and effective skin extension prior to excision of a skin defect (Figure 25-4).[10] This device, made of polyamide or silicone rubber, consists of two holding bars with several straps between them. Each strap has a stopper at one end and a one-way locking device at the other end. After infiltration of local anesthesia, each strap is placed subcutaneously under the skin defect through small incisions on each side of the lesion to be excised; the straps are parallel to each other, 2 to 3 cm apart. On each side of the area to be covered, the holding bars are attached to the straps on both sides; a locking device is then attached. After tension has been applied, the patient is instructed to increase the tension slightly every day by pushing the one-way locking device. After 1 to 3 weeks, the extension is completed, the whole device is removed, and the defect may be excised, to allow a primary closure of the wound. Our clinical experience confirms the successful results reported by Blomqvist and Steenfos. With such a device, it is possible to expand skin before excision of skin defects in an efficient, simple, and inexpensive way.

REFERENCES

1. Gibson T. Physical properties of skin. In McCarthy JG (ed.), *Plastic surgery*. Philadelphia: W. B. Saunders Company, 1990, pp. 207–220.
2. Liang MD, Briggs P, Heckler FR, Futrell JW. Presuturing—A new technique for closing large skin defects: Clinical and experimental studies. *Plast Reconstr Surg* 1988;81: 694–702.
3. Heden P. Presuturing in rhytidectomy: A case report. *Aesth Plast Surg* 1991;15:161–165.
4. Cohen BH, Cosmetto AJ. The suture tension adjustment reel. A new device for the management of skin closure. *J Dermatol Surg Oncol* 1992;18:112–123.
5. Frechet P. Scalp extension. *J Dermatol Surg Oncol* 1993;19: 616–621.
6. Nordström REA. "Stretch-back" in scalp reductions for male pattern baldness. *Plast Reconstr Surg* 1984;73: 422–426.
7. Chretien-Marquet B, Benaceur S, Carceau M. Cutaneous expansion using enforced position in the treatment of large skin defects. *Plast Reconstr Surg* 1994;93:337–340.
8. Hirshowitz B, Lindenbaum E, Har-Shai Y. A skin-stretching device for the harnessing of the viscoelastic properties of the skin. *Plast Reconstr Surg* 1993;92:260–270.
9. Bashir AH. Wound closure by skin traction: An application of tissue expansion. *Br J Plast Surg* 1987;40:582–587.
10. Blomqvist G, Steenfos H. A new partly external device for extension of skin before excision of skin defects. *Scand J Plast Reconstr Surg Hand Surg* 1993;27:179–182.

26

Topical Agents and Drugs to Enhance Tissue Expansion

EDOARDO RAPOSIO AND ROLF E.A. NORDSTRÖM

In recent years, many efforts have been made to evaluate a wide range of substances with the hope of discovering a topical and/or systemic drug able to reduce one of the major drawbacks associated with tissue expansion technique, namely, the length of the procedure. There are two important morphological and functional changes in the expanded skin that are partly responsible for the length of the procedure. The first change is the formation of a fibrous capsule surrounding the implant.[1] The capsule increases the mechanical resistance of the expanded tissue. The second change is the alteration of local blood flow of the expanded skin. This is due to the compressive force applied by the underlying expander to dermal vessels, eventually leading to skin necrosis.

Regarding the mechanical resistance of the expanded skin, histologic studies have shown that the expander quickly becomes surrounded by a fibrous capsule composed mainly of collagen and of both active and modified fibroblasts. The modified fibroblasts, named myofibroblasts by Gabbiani et al. in 1971,[2] are believed to arise locally from fibroblasts or from primitive mesenchymal cells. It is also possible that they are derived from vascular smooth muscle cells. The myofibroblasts are thought to be responsible for the contraction of the capsule surrounding the expander as well as for the contraction of granulation tissue in healing wounds.[3,4] The rate of synthesis, the orientation, and maintenance of many connective tissues are, in part, stress dependent, and it has been postulated that contractile myofibroblasts arise in the capsular tissue in response to mechanical tension and that they tend to oppose the applied force.[5] These unique cells combine ultrastructural features of both the fibroblast and the smooth muscle cell. Myofibroblasts are found free in the connective tissue space without a plasmalemma and have a well-developed system of rough endoplasmic reticulum. Both these features are characteristic of fibroblasts, but the cytoplasm of these cells contains thick bundles of microfilaments resembling those of smooth muscle as well as high levels of chemically extractable actomyosin. Furthermore, their cytoplasm shows a selective localization of antismooth muscle serum. Ultrastructurally, there are multiple indentations present in the nuclei, including folds that mimic the nuclear changes observed in contracted smooth muscle, venular endothelium, and myocardium. A basal lamina-like material surrounds these cells, and there are cell-to-cell and cell-to-stroma connections through which myofibroblast contraction can be transmitted to the tissue as a whole. When tested pharmacologically, strips of granulation tissue contract or relax similarly to smooth muscle.[6,7]

Once the filling procedures are carried out, the capsule becomes less cellular and progressively more collagenized. Pasyk et al.[8] have found that the expanded dermis, especially the reticular layer, contains large bundles of compacted collagen fibers. These fibers are interpreted as being secreted by the active, elongated, and flattened fibroblasts that are found embedded in the capsular tissue. This is a response to the continuous stretching effect induced by the implant. According to the data, the capsular tissue may be seen as a contractile organ that consists of a contractile component, the myofibroblasts, and a stromal component, the collagen fibers, which act as a mesh to support the function of the myofibroblasts.

Regarding the alterations in dermal blood flow that occur during tissue expansion, it is of the greatest importance to maintain an adequate perfusion of the expanded skin during all the expansion procedure. Little is still known about both morphological and functional changes in this field. Obviously, an excessive intraluminal expander pressure that compresses the dermal microvessels will result in ischemia of the overlying skin. This jeopardizes the completion of the filling procedure as well as the future viability of the expanded flap. When care is taken to avoid this complication, the time required to achieve the desired amount of expanded tissue is often long. If capsular contraction is the result of myofibroblastic activity and if myofibroblasts respond to pharmacological agents as do smooth muscle cells, then smooth muscle antagonists can be used to diminish capsular contraction significantly. Moreover, the use of drugs affecting collagen synthesis and cross-linkage, both in the capsule and in the surrounding dermis, may be of some value in decreasing the mechanical resistance of the expanded tissue. Finally, the administration of topical vasodilating drugs may increase the filling safety margin, thus reducing the risks of ischemic complications during expansion. In this way, the expansion can proceed in a more rapid, safe, and efficient manner, both allieviating the patients from unnecessary discomfort and lowering the total cost of the procedure.

ANTICONTRACTILE AGENTS

Theophylline

Theophylline, an alkaloid that occurs in plants that are widely distributed geographically, is a methylated xanthine. The most important pharmacological action of theophylline, with regard to tissue expansion, is its ability to relax smooth muscle cells, probably by inhibiting both cyclic nucleotide phosphodiesterases and receptor-mediated actions of adenosine. *In vitro*, it has been found that theophylline (about 0.2 millimolar or above) causes relaxation of vascular smooth muscle in the presence of various stimulators of contraction (for example, norepinephrine and angiotensin.)[9]

Although relaxation probably results from a reduction of the cytosolic concentration of Ca^{++}, it is still not clear to what extent this methylxanthine-altered Ca^{++} binding and transport directly or indirectly influences these smooth muscle functions by means of changes in cyclic nucleotide metabolism. In a recent study, Matt et al.[10] have shown that local treatment with theophylline (40 mg/kg) delivered by a modified bilumen expander (a tissue expander enclosed by a perforated concentric envelope that provides a uniform distribution of the anticontractile agent into the tissue surrounding the implant) significantly enhances tissue expansion in guinea pigs, presumably by means of an inhibition of myofibroblastic contraction into the capsule.

Papaverine

Papaverine and its analogues dilate peripheral arteries in animals and humans when administered parenterally, but only small increases in blood flow have been found. Lee et al.[11] have studied the effects of papaverine (delivered by a catheter surrounding the expander) on the rate of tissue expansion in guinea pigs. There was a statistically significant increase in the extent of expansion for the experimental group as compared with the control group. In a more recent study, Canady et al.[4] have shown that a topical pretreatment with papaverine hydrochloride (30 milligrams) in guinea pigs enhances the postexpansion blood flow of the expanded skin. This is probably due to both the relaxation of the contractile fibroblasts and the local drug effects on the vasculature, leading to vasodilation and an increase in perfusion.

Ethaverine, a synthetic derivative of papaverine, seems to have at least twice as much smooth-muscle relaxant activity as papaverine, although it has not yet been extensively studied.

Cytochalasin D

In 1985, Lee et al. reported that cytochalasin D, (administered in guinea pigs by means of a modified catheter into the tissue surrounding the expander) significantly increases the rate and extent of tissue expansion in rats.[11] Histologic examination of the expanded skin underlined the role of relaxation or inactivation of myofibroblasts in the fibrous capsule surrounding the expander. Moreover, the authors found that the capsule was less fibrous, and the fibroblasts and the collagen fibers were less well oriented. On the other hand, although cytochalasin D is reported to be an antiactin agent, Miranda et al.[12] found that cells in culture that are exposed to cytochalasin D rapidly undergo a long-sustained tonic contraction. Coincident with this contracture, the thin microfilaments of the cortex become compacted into feltlike masses. The raveled filaments of these masses remain actinlike and bind heavy meromyosin. On continued exposure to cytochalasin D, 'myoid' bundles (containing thick, dense filaments) and larger fusiform or ribbonlike—aggregates putatively myosinoid—may appear. These modifications have been interpreted as consequences of a state of hypercontraction without relaxation induced by cytochalasin D.[12] The modifications do not occur in cytochalasin D-treated cells that are prevented from contracting by inhibitors of energy metabolism, and the effects are readily reversible on withdrawal of cytochalasin D.

Trocinate
Beta-diethylaminoethyldiphenylthioacetate (trocinate) is one of the most promising drugs that can improve tissue expansion through a decrease in capsular contraction, although its functions are yet to be further evaluated. Trocinate inhibits myofibroblast function by affecting smooth muscle cells directly rather than influencing parasympathetic or sympathetic nerve function. In 1974, Madden et al. confirmed this hypothesis, studying the effects of topical trocinate on contraction of experimental wounds. They found that trocinate prevents wound contraction by specific inhibition of myofibroblast activity. Neither the rate of collagen synthesis nor morphologic patterns of collagen fiber aggregation has been affected by trocinate applications. In the same study, when trocinate was removed (after 10 days), wound contraction resumed immediately without a latent interval.[5] These data suggest that trocinate paralyzes rather than destroys the contraction mechanism.

DRUGS AFFECTING DERMAL AND CAPSULAR FIBROUS TISSUE

Colchicine
Colchicine, an anti-inflammatory alkaloid, is an antimitotic agent and inhibits the release of histamine-containing granules from mast cells. Moreover, as stated by Chvapil et al.,[13] administration of colchicine appears to increase synthesis of collagen but ultimately results in reduced total collagen content of newly synthesized connective tissue, such as the capsule surrounding the expander. A possible explanation of this seeming paradox is that colchicine-induced collagen synthesis results in production of collagen molecules that are not utilized in the assembly of connective tissue. Another possibility is an increase of collagenolytic enzyme activity following colchicine treatment. After administration of high doses of colchicine (0.10 to 0.15 mg/kg intramuscularly) in rats, Chvapil et al. evaluated the impact on healing wounds. Based on biochemical and ultrastructural data, they speculated that colchicine could be topically cytotoxic to granulocytes and macrophages, both of which contain collagenase. The stimulation of fibroblast activity may be explained by the release of factors from activated macrophages. More recently, in a rodent model, Netscher et al.[14] studied the effects of colchicine (0.125 mg/kg, administered both topically and intramuscularly) on the relationship between inflated volume and mean intraluminal expander pressure. They found that the rate of expansion was significantly greater in the colchicine group compared with the control group. Colchicine, currently used to treat gouty arthritis (0.5 milligrams four times a week), is well tolerated in moderate dosage. The most common side effects (nausea, vomiting, diarrhea, abdominal pain) reflect the action of the drug on the rapidly proliferating epithelial cells in the gastrointestinal tract.

Dimethyl Sulfoxide
Dimethyl sulfoxide (DMSO), an industrial solvent and by-product of paper manufacture, was first presented as a therapeutic agent in 1963 by Jacob et al.[15] Early human clinical trials with DMSO indicated beneficial anti-inflammatory and analgetic effects in musculoskeletal injuries, arthritis, and bursitis.[16] Berliner and Ruhmann[17] found that DMSO inhibited fibroblast proliferation *in vitro*, while Gorog and Kovacs,[18] studying the anti-inflammatory potential of DMSO in adjuvant-induced polyarthritis of rats, found that it showed potent anti-inflammatory properties in this model.

Furthermore, because lysosomes can be stabilized against a variety of harmful agents by cortisone, the concentration of cortisone necessary to stabilize lysosomes is reduced 10- to 1000-fold by DMSO. The possibility suggested by Weissman et al.[19] is that DMSO may render steroids more available to their targets within tissues (membranes of cells or their organelles). In 1983, Goldblum et al. reported that topical 95 percent DMSO does not affect epidermal wound healing in pigs.[16] DMSO is both an effective scavenger of hydroxyl radicals and a drug that produces analgesia (with a longer duration of action than morphine) by acting both locally and systemically.[20]

Numerous animal studies have documented decreased tensile strength in tissues treated with DMSO, while light and electron microscopic investigation of DMSO-exposed tissues has revealed the presence of characteristic collagen fiber swelling.[21] The suggested etiology of this phenomenon is that collagen fiber intra- and intermolecular cross-links are ruptured. Gries et al.[22] studied rabbit skin before and after 24 hours of *in vitro* exposure to 100 percent DMSO. After immersion in DMSO, the collagen fraction extractable with neutral salt solution was significantly decreased. Jacob and Herschler[15] have reported that topical DMSO in humans exerts a significant effect on the pathological deposition of collagen in postirradiation subcutaneous fibrosis. This did not appear to change the equilibrium of collagen metabolism in normal tissue. Urinary hydroxyproline levels are increased in scleroderma patients treated with topical DMSO,[23] while keloids biopsied in humans before and after DMSO therapy show histological improvement toward normality.[24]

Wong et al.[25] studied the absorption, excretion, and biotransformation of topical 80 percent DMSO gel in humans. They found that its absorption rate ranged from 25% to 40% of the total dose, and that it is partly transformed into dimethyl-sulfone ($DMSO_2$) and dimethyl-sulfide (DMS). DMSO and $DMSO_2$ are excreted in the urine, whereas DMS is eliminated in the expired air. Skin contact may result in primary irritation with redness, itching, and, sometimes, scaling; urticarial wheals are not uncommon. In 1967, Adamson et al. markedly enhanced the circulation of pedicle flaps in an animal model by daily local treatment with 70 percent DMSO solution.[26] Because this improvement disappeared if the animals were treated with an antihistamine at the same time, the authors postulate that the action of DMSO is similar to histamine.

Arturson and Khanna[27] improved the circulation in partially necrotic skin flaps in rats, thereby increasing the amount of surviving tissue, with 70 percent DMSO topical treatment. These results have been confirmed by Haller et al.,[28] who significantly enhanced blood flow and survival of rat abdominal island flaps with the use of systemic (1.5 g/kg) DMSO. On the other hand, Koehnlein and Lemperle[29] applied topical (89 percent) DMSO gel to rat dorsal flaps and reported more extensive sloughing than in the control (untreated) group. In 1987, Liang et associates tested the use of topical (99 percent) DMSO in rats as an enhancer of tissue expansion, and they obtained a significant reduction in expansion pressures.[30] In 1991, Vinnik and Jacob reported the clinical use of both topical (70 percent) and intravenous DMSO to greatly enhance tissue expansion during breast reconstruction, large lesion excision, and abdominoplasty.[31] Recent studies from our department (Raposio et al., unpublished data, 1994, Department of Plastic and Reconstructive Surgery, National Institute for Cancer Research, Genoa, Italy) have confirmed this data, showing that prefilling topical treatment with 60 percent DMSO significantly enhances tissue expansion rate in breast reconstruction.

Hyaluronidase

Hyaluronidase catalyzes the hydrolyses of links between 2-acetamido-2-deoxy-D-glucose and D-glucuronate residues in hyaluronate. In addition, the enzyme attacks chondroitin and mucoitin sulphate; it also exhibits transglycosylation activity. In 1988, Netscher et al. noticed that hyaluronidase (150 units, instilled via an irrigation catheter mounted on the expander) remarkably enhances the rate of expansion (as determined by the volume instilled, the surface area increased, and the arc increased across the expander dome in response to inflation). Furthermore, they also found that there is no difference in histology or capsule thickness between the treated and the untreated expanded groups.[14]

VASODILATOR DRUGS

Whether or not vasodilators can enhance the safety of tissue expansion procedure is still unknown. An ideal drug for the prevention and/or the treatment of ischemic expanded skin would dilate blood vessels and increase blood flow only in the areas of deficient blood supply. Unfortunately, such agents do not currently exist. Even local administration of vasodilators by the intra-arterial route may be detrimental if redistribution effects an increase in muscle blood flow at the expense of skin blood flow.[32] Because vasodilators that directly relax blood-vessel walls increase both skin and muscle blood flow, and because agents that interfere with the sympathetic nervous system primarily increase cutaneous blood flow (little sympathetic vasoconstrictor tone exists in muscle), to enhance expanded skin perfusion it would be better to use the latter than the former. Furthermore, because the vascular sympathetic receptors of skin are mainly of the alpha-type, alpha-adrenergic blocking compounds are more suitable to achieve maximal vasodilation in the expanded skin than beta-adrenergic antagonists.

Isoxsuprine

Isoxsuprine produces vasodilation (predominantly on the precapillary arteriolar side of the vascular system) probably by long-acting beta-adrenergic receptor stimulation, thus generating prolonged relaxation of the smooth muscle in vessel walls. It has also been shown that Isoxsuprine can diminish necrosis both in experimental neurovascular island muscle flaps and in neurovascular island skin flaps in rats.[33] Otherwise, in studies on normal animals and human subjects, muscle blood flow increases, but skin blood flow is usually unaffected by isoxsuprine.[34]

Propranolol

Propranolol is a nonselective beta-adrenergic antagonist. It lacks intrinsic sympathomimetic activity, and does not block alpha-adrenergic receptors. Jonsson et al.[35] enhanced the survival of experimental skin flaps with propranolol administration. Voorhees et al.[36] reported that the most important pharmacological action of propranolol seems to be its capacity to block the norepinephrine-stimulated accumulation of cyclic AMP, thus augmenting epithelial cell proliferation rates. This peculiar pharmacological effect could be used to significantly stimulate the mitotic index of keratinocytes during tissue expansion, thus shortening the time required for the completion of the filling procedure.

Prostaglandin E_2

In most species (including man) and in most vascular beds, the prostaglandins of the E type (PGEs) are potent vasodilators. The dilatation appears to involve arterioles, precapillaries, sphincters, and postcapillary venules; large veins are not affected by PGEs. PGE_2 gel is used in obstetric practice and is provided prepackaged in a concentration of 0.5 mg PGE_2/3 g. Sasaki and Pang[37] showed that subcutaneous administration of PGE_2 (2mg/kg/day) significantly enhances the extent of skin-flap survival in rats. More recently, Netscher et al. have applied prostaglandin E_2 twice daily to expanded skin in rats; they found significant differences in transcutaneous oxygen tension level between the control and PGE_2 groups, as measured immediately before and immediately after each inflation.[14] The authors thought that current methods of expansion might be made safer by means of topical PGE_2 administration.

Phenoxybenzamine

Phenoxybenzamine, a haloalkylamine that blocks alpha-1- and alpha-2-adrenergic receptors in smooth muscle, causes a progressive decrease in peripheral vascular resistance. It also causes an increase in cardiac output that is due, in part, to reflex sympathetic nerve stimulation. Phenoxybenzamine has proved successful in increasing flap survival when administered preoperatively and postoperatively in rats by both the systemic route and topical injection.[38,39]

Phentolamine

Phentolamine, an imidazoline, is a competitive alpha-adrenergic antagonist that has similar affinity for alpha-1 and alpha-2 receptors. It can also block receptors for 5-hydroxytryptamine (5-HT), and it releases histamine from mast cells. The pharmacokinetic properties of phentolamine are not known, although the drug is metabolized extensively. Treatment with phentolamine (5–10 mg/kg/24 hours) has proved to be successful in lowering the peripheral vascular resistance, thus increasing the survival area of skin flaps in rats.[35,40]

Because ischemic complications during tissue expansion are attributed mainly to the mechanical pressure applied by the implant rather than to peripheral vasoconstriction, we feel that the administration of phentolamine, (similar to the administration of isoxsuprine, prostaglandin E_2, and phenoxybenzamine) to enhance skin blood flow during the filling procedure is of little value.

Pentoxifylline

Pentoxifylline is a vasoactive agent used for cerebral vascular insufficiency. Although it produces slight vasodilation, its main pharmacological effect, particularly regarding the prevention of ischemic complications during tissue expansion, is the enhancement of capillary blood flow by both increasing the ATP content of erythrocytes (thus augmenting their deformability) and decreasing blood viscosity and platelet aggregation.

In 1975, N. Olivari and B. Olivari demonstrated that pentoxifylline augmented flap survival in rats,[41] while, in 1980, Takayanagi and Ogawa reported its successful clinical application (200 milligrams orally three times a day) in the treatment of ischemic skin flaps.[42] These results are reasonably attributed to alterations in the rheologic properties of blood that was observed after pentoxifylline treatment.

The former considerations suggest that pharmacological control of both capsular contraction and ischemic complications during the course of tissue expansion may be possible. Although the reported data indicate several clinical applications of numerous pharmacological agents, more experimental work is needed both to test their safety and to accurately evaluate the timing of treatment, dosages, and routes of administration.

REFERENCES

1. Austad ED, Pasyk KA, McClatchey KD, Cherry GW. Histomorphologic evaluation of guinea pig skin and soft tissue after controlled tissue expansion. *Plast Reconstr Surg* 1982;70:704–710.
2. Gabbiani G, Ryan GB, Majno G. Presence of modified fibroblasts in granulation tissue and their possible role in wound contraction. *Experientia* 1971;27:549–555.
3. Guber S, Rudolph R. The myofibroblast. *Surg Gynecol Obstet* 1978;146:641–649.
4. Canady JW, Kelly KM, Squier CA, Bardach J. Blood flow in expanded tissue treated with an anticontractile agent. *Ann Otol Rhinol Laryngol* 1991;100:962–965.
5. Madden JW, Morton D, Peacock EE. Contraction of experimental wounds: I, Inhibiting wound contraction by using a topical smooth muscle antagonist. *Surgery* 1974;76:8–15.
6. Ryan GB, Cliff WJ, Gabbiani G, et al. Myofibroblasts in an avascular fibrous tissue. *Lab Invest* 1973;29:197–206.
7. Ryan GB, Cliff WJ, Gabbiani G, et al. Myofibroblasts in human granulation tissue. *Human Pathol* 1974;5:55–67.
8. Pasyk KA, Austad ED, McClatchey KD, Cherry GW. Electron microscopic evaluation of guinea pig skin and soft

tissues expanded with a self-inflating silicone implant. *Plast Reconstr Surg* 1982;70:37–45.
9. Madsen SM, Ribel U. Pharmacokinetics of theophylline and 3-methylxanthine in guinea pigs: I, Single dose administration. *Acta Pharmacol Toxicol* 1981;48:1–6.
10. Matt BH, Squier CA, Kelly KM, Bardach J. Enhancement of expansion of guinea pig skin by local delivery of an anticontractile agent using a new bilumen expander. *Ann Plast Surg* 1990;24:335–341.
11. Lee P, Squier CA, Bardach J. Enhancement of tissue expansion by anticontractile agents. *Plast Reconstr Surg* 1985;76:604–610.
12. Miranda AF, Godman GC, Tanenbaum SW. Action of cytochalasin D on cells of established lines. *J Cell Biol* 1974;62:406–423.
13. Chvapil M, Peacock EE, Carlson EC. et al. Colchicine and wound healing. *J Surg Res* 1980;28:49–56.
14. Netscher DT, Spira M, Peterson R. Adjunctive agents to facilitate tissue expansion. *Ann Plast Surg* 1989;23:412–416.
15. Jacob SW, Herschler R. Pharmacology of DMSO. *Criobiol* 1986;23:14–27.
16. Goldblum OM, Alvarez OM, Mertz PM, Eaglstein WH. Dimethyl sulfoxide (DMSO) does not affect epidermal wound healing. *Proc Soc Exp Biol Med* 1983;172:301–307.
17. Berliner DL, Ruhmann AG. The influence of dimethyl sulfoxide on fibroblastic proliferation. *Ann NY Acad Sci* 1964;141:159–166.
18. Gorog P, Kovacs LB. Effect of dimethyl sulfoxide (DMSO) on various experimental inflammations. *Curr Ther Res* 1968;10:486–491.
19. Weissman G, Sessa G, Bevans V. Effect of DMSO on the stabilization of lysosomes by cortisone and chloroquine in vitro. *Ann NY Acad Sci* 1967;141:326–331.
20. Haigler HJ. Comparison of the analgesic effects of dimethyl sulfoxide and morphine. *Ann NY Acad Sci* 1983;411:19–27.
21. Erk Y, Raskin DJ, Mace M, Spira M. Dimethyl sulfoxide alteration of collagen. *Ann NY Acad Sci* 1983;141:364–368.
22. Gries G, Bublitz G, Lindner J. The effect of dimethyl sulfoxide on the components of connective tissues (clinical and experimental investigations). *Ann NY Acad Sci* 1967;141:630–637.
23. Scherbel AL, McCormack LI, Layle JK. Further observations on the effect of dimethyl sulfoxide in patients with generalized scleroderma (progressive systemic sclerosis). *Ann NY Acad Sci* 1967;141:613–629.
24. Engle MF. Indications and contraindications for the use of DMSO in clinical dermatology. *Ann NY Acad Sci,* 1967;141:638–644.
25. Wong KK, Wang GM, Dreyfuss J, Schreiber EC. Absorption, excretion, and biotransformation of dimethyl sulfoxide in man and miniature pigs after topical application as an 80% gel. *J Invest Dermatol* 1971;56:44–48.
26. Adamson JE, Horton CE, Crawford HH, Ayer S. Studies on action of dimethyl sulfoxide on experimental pedicled flap. *Plast Reconstr Surg* 1967;39:142–146.
27. Arturson G, Khanna NN. The effects of hyperbaric oxygen, dimethyl sulfoxide and complamin on the survival of experimental skin flaps. *Scand J Plast Reconstr Surg* 1970;4:8–10.
28. Haller J, Trachy R, Cummings CW. Effect of dimethyl sulfoxide on island flap perfusion and survival in rats. *Arch Otolaryngol Head Neck Surg* 1987;113:859–863.
29. Koehnlein HE, Lemperle G. Experimental studies on the effect of dimethyl sulfoxide on pedicle flaps. *Surgery* 1970;67:672–677.
30. Liang MD, Dick GO, Narayanan K. Enhancement of tissue expansion using DMSO. *Surg Forum* 1987;38: 593–595.
31. Vinnik CA, Jacob SW. Dimethylsulfoxide (DMSO) for human single-stage intraoperative tissue expansion and circulatory enhancement. *Aesth Plast Surg* 1991;15: 327–337.
32. Coffman JD. Vasodilator drugs in peripheral vascular disease. *New Engl J Med* 1979;300:713–717.
33. Finseth F, Zimmermann J, Liggins D. Prevention of muscle necrosis in an experimental neurovascular island muscle flap by a vasodilator drug: Isoxsuprine. *Plast Reconstr Surg* 1979;63:774–780.
34. Finseth F, Adelberg MG. Prevention of skin flap necrosis by a course of treatment with vasodilator drugs. *Plast Reconstr Surg* 1978;61:738–743.
35. Jonsson CE, Jurell G, Nylen B, Pandeya N. Effect of phentolamine and propranolol on the survival of experimental skin flaps. *Scand J Plast Reconstr Surg* 1975;9:98–100.
36. Voorhees JJ, Duell EA, Kelsey WH, Hayes E. Effects of alpha and beta adrenergic stimulation on cyclic AMP formation and mitosis in epidermis. *Clin Res* 1972;20: 419–423.
37. Sasaki GH, Pang CY. Experimental evidence for involvement of prostaglandins in viability of acute skin flaps: Effects on viability and mode of action. *Plast Reconstr Surg* 1981;67:335–340.
38. Myers MB, Cherry G. Enhancement of survival in devascularized pedicles by the use of phenoxybenzamine. *Plast Reconstr Surg* 1968;41:254–260.
39. Kerrigan CL, Daniel RK. Pharmacologic treatment of the failing skin flap. *Plast Reconstr Surg* 1982;70:541–548.
40. Wexler MR, Kalisman M, Yeschua R, Neuman Z. The effect of phenoxybenzamine, phentolamine and 6-hydroxydopamine on skin flap survival in rats. *J Surg Res* 1975;19:83–87.
41. Olivari N, Olivari B. Beeinflussung der Lappendurchblutung und Nekroserate nach Verschiebeschwenkplastik durch Pentoxifyllin und Dextran 40. *Med Welt* 1975; 26:1973.
42. Takayanagi S, Ogawa Y. Effects of pentoxifylline on flap survival. *Plast Reconstr Surg* 1980;65:763–767.

27

Muscle Vascularized Pedicle Flap (MVP Flap) Using Tissue Expanders: An Experimental Study in Rats

TETSUNORI YOSHIDA, TAKEHIKO OHURA, AND KEN-ICHI HOMMA

In 1982, using a vascular bundle carrying a small amount of muscle on its distal end, Shintomi and Ohura[1] reported a number of successful clinical cases utilizing the muscle vascularized pedicle (MVP) flaps, which are considered to be typical prefabricated flaps. However, this unique flap is not routinely used in clinical practice because the MVP flaps require at least two or three surgical delays, depending upon the size of flaps. To date, there have been no reports concerning the use of prefabricated flap combined with tissue expansion. In the present study, we have used tissue expansion in a rat model to produce MVP flaps without the need for any surgical delay. The goal was to determine if the surviving skin area would have been the same or greater in MVP flaps elevated in expanded tissue or in MVP flaps elevated in nonexpanded tissue.

MATERIALS AND METHODS

A 30-milliliter, 5-centimeter, round tissue expander with a distal port (Koken Corporation, Tokyo, Japan) was selected for use in mature male Wistar rats weighing 400 to 500 grams. Intraperitoneal ketamine HCL (10mg/100gm) was used as the anesthetic agent. The hair on the abdomens and legs was removed with a depilatory cream. A 5×5 cm^2 flap was designed 1 centimeter cephalad to the external orifice of the urethra. The corner of the flap was tattooed and then prepared with 0.5 percent chlorhexidine alcohol. All operations were performed under sterile conditions. The animals were divided into three groups as follows (Figure 27-1):[2]

- Group I: with a tissue expander. (n = 10)
- Group II: with a tissue expander but not expanded. (n = 10)
- Group III: with a thin silastic sheet (Biobrane Dow B. Hickam, Inc., Sugar Land, Texas) (n = 10)

Standard surgical procedures and prepared vascular pedicles were used in the three groups. A 3-centimeter incision was made 1-centimeter cephalad to the external orifice of the urethra to create a 5×5 cm^2 pocket beneath the panniculus carnosus for inserting a tissue expander or a Biobrane sheet. A 2-centimeter incision was made over the saphenous vessels extending from the inguinal region to the lower leg. A composite pedicle of the right saphenous artery, vein, and nerve with an island of 8×10 mm^2 gracilis muscle on its distal end was isolated, ligated distally, and divided (Figure 27-2). The epigastric vessels were preserved to ensure viability of the abdominal skin. The pedicle, continuous proximally with the femoral vessels, was subcutaneously tunneled medially across the epigastric vessels. The pedicle muscle stump was tacked under the center of the overlying panniculus carnosus with 5-0 nylon sutures. Both the muscle stump and its vascular pedicle were sandwiched between the tissue expander or the Biobrane and the cutaneous flap (Figure 27-3). The wounds were closed with 5-0 nylon sutures. In group I, the expander was inflated 1 week later up to a maximum capacity of 30 milliliters with a single injection by percutaneously adding saline through the reservoir. Three weeks after the first operation, a 5×5 cm^2 island flap, outlined by the tattooed points, was

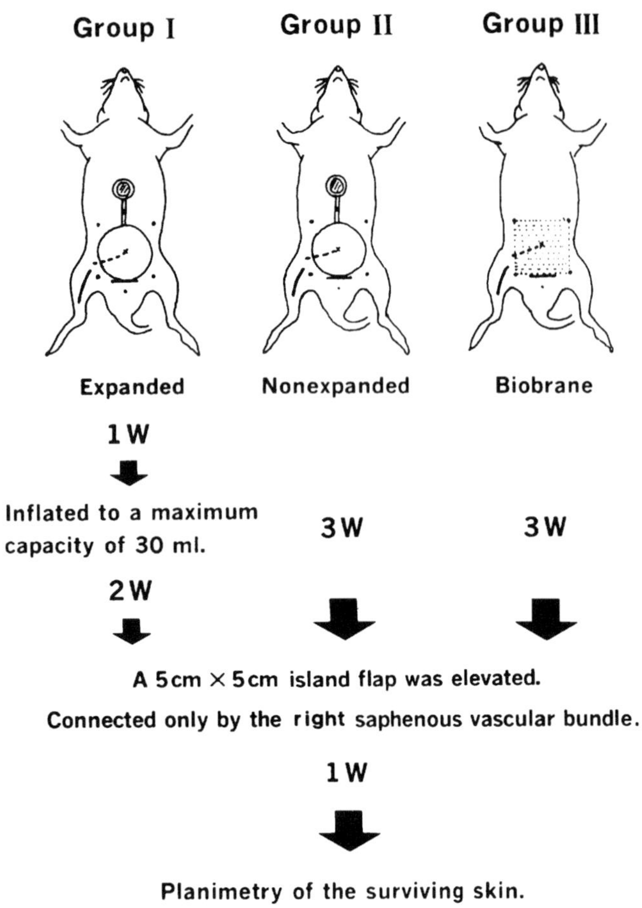

FIGURE 27-1

Schema of the three rat models used in this study. (Reprinted with permission from Homma K, Ohura T, Sugihara T, Yoshida T, Hasegawa T. Prefabricated flaps using tissue expanders: An experimental study in rats. Plast Reconstr Surg *1993; 91:1098–1107.)*

elevated. The flap was resutured into position, connected only by the transplanted vascular bundle. A thin silastic sheet (Biobrane) was used to isolate the flap and prevent vascular regrowth from the underlying muscle of the abdominal wall. The percentage of surviving skin area of the flap was determined by planimetry 7 days later. The surviving area within each flap was calculated as shown in Figure 27-4. A selective microangiography of the right saphenous artery was performed under an operating microscope by inserting a 30-gauge cannula into the right saphenous artery. The vascular system of the flap was washed out with heparinized saline and perfused with a mixture of 30 percent fine barium powder and 5 percent gelatin. This study was performed with six flaps in each group 3 weeks after implantation of the vascular pedicle. It was also performed with flaps exhibiting nearly total survival in groups I and II at the end of the evaluation study.

The results were analyzed using a Wilcoxon test for independent samples, and a statistical significance was accepted at a 5 percent level as shown in Table 27-1.

RESULTS

Flap survival

The percentage of surviving skin area in each group is shown in Table 27-1 and Figures 27-5 and 27-6. The median of group I was larger than that of group II or III. In group I, the mean percentage of surviving skin area was 96.6 ± 3.6 percent, and every flap showed over 91 percent flap survival. This survival area of group I was statistically higher than that of group II (73.2 ± 24.7 percent) ($p < .05$) and group III (59.5 ± 25.1 percent) ($p < .005$). There was no statistically significant difference between groups II and III. However, even in group II there were four cases that showed over 90 percent flap survival. In both groups II and III the surviving area varied from 20% to 30% to as high as 80% to 100% (Figure 27-6). In all experimental groups, the area just overlying the vascular pedicle survived. The results differed beyond the midline where the vascular pedicle was not present. When the surviving area was near 50 percent, breakdown of the distal portion of the skin flap occurred. This open wound induced a contraction of the flap and the surviving area decreased to 20% to 30% (Figure 27-5).

Microangiography

In group I, all six flaps showed well-visualized vascular trees of both the pedicle and the epigastric vessels. They were expanded and elongated compared with nonexpanded and Biobrane groups (Figure 27-7A). Three weeks after implantation of the pedicle, small new vessels were visible from the pedicle muscle stump. The newly formed vessels had joined to form communications to inferior epigastric vessels and increased the vascularity of the flap. Vasodilatation of the pedicle itself was extensive in group I (Figure 27-7B).

In group II, four flaps showed the vascular tree of bilateral epigastric vessels (Figure 27-8A,B), but two flaps demonstrated only the right half of the epigastric vessels (Figure 27-8C, D). Newly formed small vessels from the pedicle were present in every flap, but vasodilatation of the pedicle was not observed.

In group III, three flaps showed bilateral vascular trees (Figure 27-9A, B), but in the other three cases, epigastric vessels were not well visualized (Figure 27-9C, D) and the network of new vessels between the muscle cuff and the flap was poor. An efflux of the perfusate

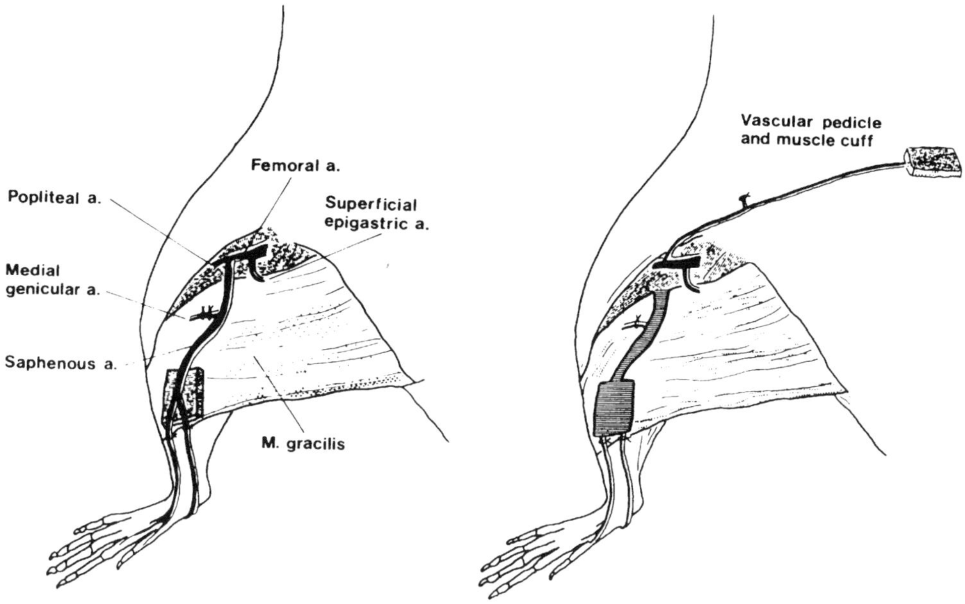

FIGURE 27-2
Anatomy of the rat lower extremity and the elevated vascular pedicle and muscle cuff (MVP). (Reprinted with permission from Homma K, Ohura T, Sugihara T, Yoshida T, Hasegawa T. Prefabricated flaps using tissue expanders: An experimental study in rats. Plast Reconstr Surg *1993;91: 1098–1107.)*

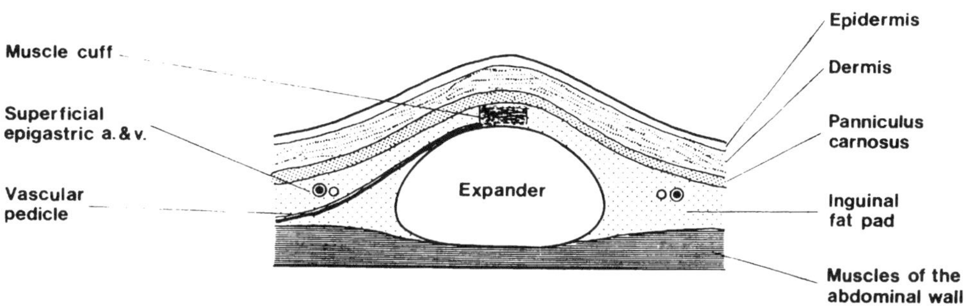

FIGURE 27-3
Cross section of the abdominal region. (Reprinted with permission from Homma K, Ohura T, Sugihara T, Yoshida T, Hasegawa T. Prefabricated flaps using tissue expanders: An experimental study in rats. Plast Reconstr Surg *1993;91:1098–1107.)*

was observed. In these cases, vasodilatation of the pedicle was apparent (Figure 27-9D).

We completed a selective microangiography of the right saphenous artery (MVP artery). After injection, the direction of flow was observed under an operating microscope. Information from both the radiographs and the records of flow was used to produce the schematic representation of the blood flow (Figure 27-10). In expanded MVP flaps, the injectant was soon drained from the accompanying vein (MVP vein) and gradually filled the vascular tree of the epigastric vessels through the newly formed small vessels (Figure 27-10A). In nonexpanded and Biobrane groups, backflow from the MVP vein was seen. But major flow of the injectant passed through the new channels between the MVP and the original vessels and drained into the epigastric veins (Figure 27-10B).

With the increased vascularity of the expanded skin, there seemed to be good filling of the transplanted vascular pedicle (MVP), for reasons that appear unclear.

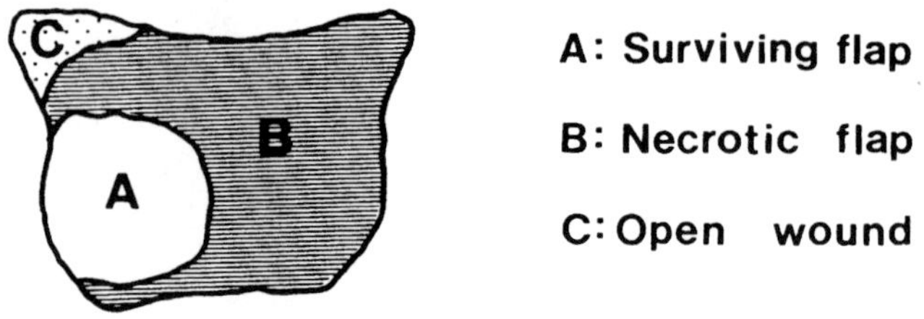

$$\text{Surviving area} = \frac{A}{A + B + C} \times 100\ (\%)$$

FIGURE 27-4
Planimetry of the surviving skin. (Reprinted with permission from Homma K, Ohura T, Sugihara T, Yoshida T, Hasegawa T. Prefabricated flaps using tissue expanders: An experimental study in rats. Plast Reconstr Surg *1993;91:1098–1107.)*

TABLE 27-1
Percentage of the area surviving

	Group I	*Group II*	*Group III*
	91.0%	63.9%	88.4%
	94.5	44.5	23.0
	100	99.3	21.8
	97.6	35.7	71.2
	91.1	92.9	26.6
	94.3	73.7	71.8
	99.4	38.8	55.5
	100	99.6	78.7
	98.0	100	69.3
	100	83.1	88.9
Mean	96.6%* **	73.2%*	59.5%**
S.D.	3.6	24.7	25.1
	(n = 10)	(n = 10)	(n = 10)

*Wilcoxon test $p < 0.05$
**Wilcoxon test $p < 0.005$
(Reprinted with permission from Homma K, Ohura T, Sugihara T, Yoshida T, Hasegawa T. Prefabricated flaps using tissue expanders: An experimental study in rats. *Plast Reconstr Surg* 1993; 91:1098–1107.)

Perhaps increased microcirculation of the expanded MVP flaps provided the venous drainage into the MVP veins. This increase in vascularity played an important role in the augmented flap survival of MVP flaps elevated in the expanded tissue in group I.

At the time of evaluation of the surviving flap, 7 days after flap elevation, microangiography demonstrated different vascular patterns than were observed before flap elevation. The structure of the original epigastric vessels had changed to a pedicle-dominant structure. A newly formed small vascular plexus, centripetal to the pedicle, was present (Figure 27-7C).

DISCUSSION

Previously, Hasegawa et al.[3] conducted experiments in animals to study how the vascular network of the MVP flap was established. The authors used 60 Wistar rats—12 for creating control tubed pedicle flaps and 48 for tubed pedicle MVP flaps. They compared the survival length of both types of flaps, performed microangiograms, India ink injection studies, and histologic examinations. In these studies, they found that the survival lengths of MVP flaps were statistically longer than those of control flaps. The neovascularization of the MVP flap started at 3 days postoperatively and increased its vascularity toward day 14. But even in the MVP flap group, the mean survival length of a 3 cm × 10 cm rectangular flap was only 25 percent because the flap was elevated beyond the vascular pedicle. From these results, we concluded that a surgical delay was essential to obtain a larger MVP flap.

Cherry et al.[4] compared acute, delayed, and nonexpanded skin flaps with expanded skin flaps raised on pig skin. Their expanded flaps had a mean increased survival length of 117 percent over acutely raised control flaps. The authors found an increased vascularity, as well as vasodilatation in the angiograms of the expanded skin. They concluded that the increase in vascularity was undoubtedly important in the survival of the flaps raised in the expanded tissue. The precise mechanism of increased vascularity associated with tissue expansion is still unclear. Cherry et al. suggested that mechanical forces associated with expansion play a role in new vessel growth.[4]

Sasaki and Pang[5], compared acute, delayed, nonexpanded, and expanded skin flaps in a pig model. They demonstrated the relationship between capillary blood flow and skin-flap viability by using a radioactive microsphere technique. The total capillary blood flow was significantly higher in delayed and expanded random-pattern flaps as compared to acute random-pattern skin flaps.

Leighton et al.[6,7] investigated flap viability in expanded cutaneous buttock flaps and myocutaneous latissimus dorsi flaps in a pig model. There was a statistically significant increase in blood flow into expanded axial-pattern cutaneous and myocutaneous flaps compared to the control using an electromagnetic flowmeter. In their histologic study, pronounced neovascularization was seen in the papillary dermis and capsular layers. But there was no significant difference in survival between control and expanded cutaneous and myocutaneous flaps after elevation solely on their axial pedicles. Van Beek and Adson[8] pointed out in their discussion of the latter article that it was difficult to make a conclusion about flap viability in their study because the limits of the latissimus dorsi myocuta-

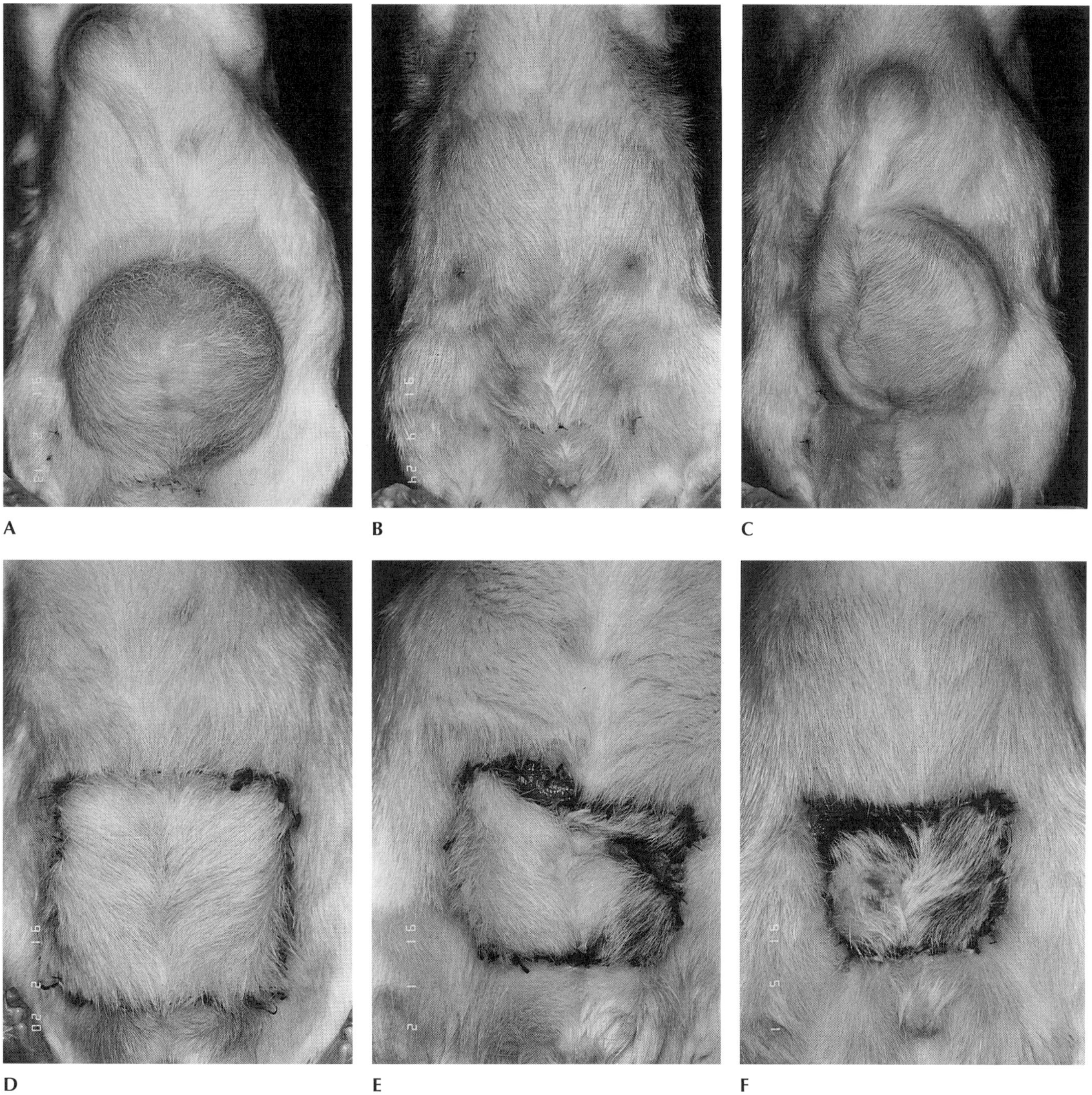

FIGURE 27-5

(A, B, C) Before flap elevation. (D, E, F) Same flaps 1 week after elevation. (Reprinted with permission from Homma K, Ohura T, Sugihara T, Yoshida T, Hasegawa T. Prefabricated flaps using tissue expanders: An experimental study in rats. Plast Reconstr Surg *1993;91: 1098–1107.)*

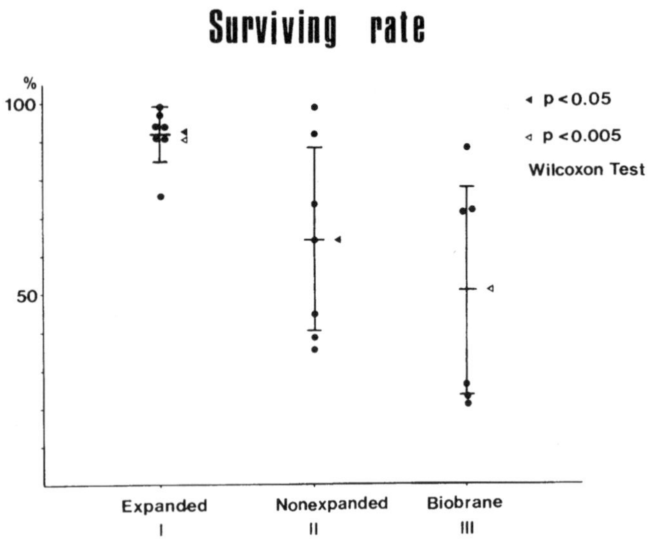

FIGURE 27-6
The percentage of surviving skin area of each group. The survival area of the expanded group (96.6 ± 3.6 percent) was statistically higher than that of the nonexpanded group (73.2 ± 24.7 percent) (p < .05) and the Biobrane group (59.5 ± 25.1 percent) (p < .005). There were no statistically significant differences between groups II and III. (Reprinted with permission from Homma K, Ohura T, Sugihara T, Yoshida T, Hasegawa T. Prefabricated flaps using tissue expanders: An experimental study in rats. Plast Reconstr Surg *1993;91:1098–1107.)*

neous flap in the pig had not been defined, and also the size of the buttock flap chosen by the authors was smaller than that originally described by Daniel and Kerrigan.[9]

There is no experimental study on the expansion of prefabricated flaps. The present experiments, carried out in a rat model, show that MVP flaps elevated in expanded tissue survive in greater areas than MVP flaps elevated in nonexpanded tissue. The variability in the survival area of groups II and III proved the unstable vascularity of MVP flaps in which no axial-pattern vessel had been present. The reasons that nonexpanded and Biobrane groups also had some flaps that survived very well are unclear. However, anatomical variety of epigastric vessels in rats could be a factor. Petry and Wortham[10] investigated the anatomy of the epigastric flap in experimental rats. In their study, the pattern of the medial branches of the superficial epigastric trunk was variable. If there were some branches at the center of the 5 cm × 5 cm flap, spontaneous anastomosis of vessels between transplanted pedicle and epigastric vessels would occur, as Smahel and Jentsch[11] demonstrated. In such cases, we speculate that neovascularization of the flap would be well established.

In most clinical cases, prefabricated flaps are made where axial-pattern vessels are absent. If we want to get a larger flap beyond the vascular pedicle, surgical delay is necessary to get a reliable flap. In our clinical experience with MVP flaps, the most important concern was how to get a good venous drainage into the MVP vein. Although the mechanism for the increased vascularity following expansion has not been determined, the fact that vasodilatation of the vascular pedicle and the epigastric vessels was most apparent in group I would appear to support the theory that good venous drainage into the transplanted vascular pedicle is essential in the prefabricated flaps. By whatever mechanism, the results of this work demonstrate that MVP flaps elevated in expanded skin have a significantly augmented survival area. Using a tissue expander, we are able to produce a reliable MVP flap without any surgical delay, while obtaining the added advantage of primary donor site closure.

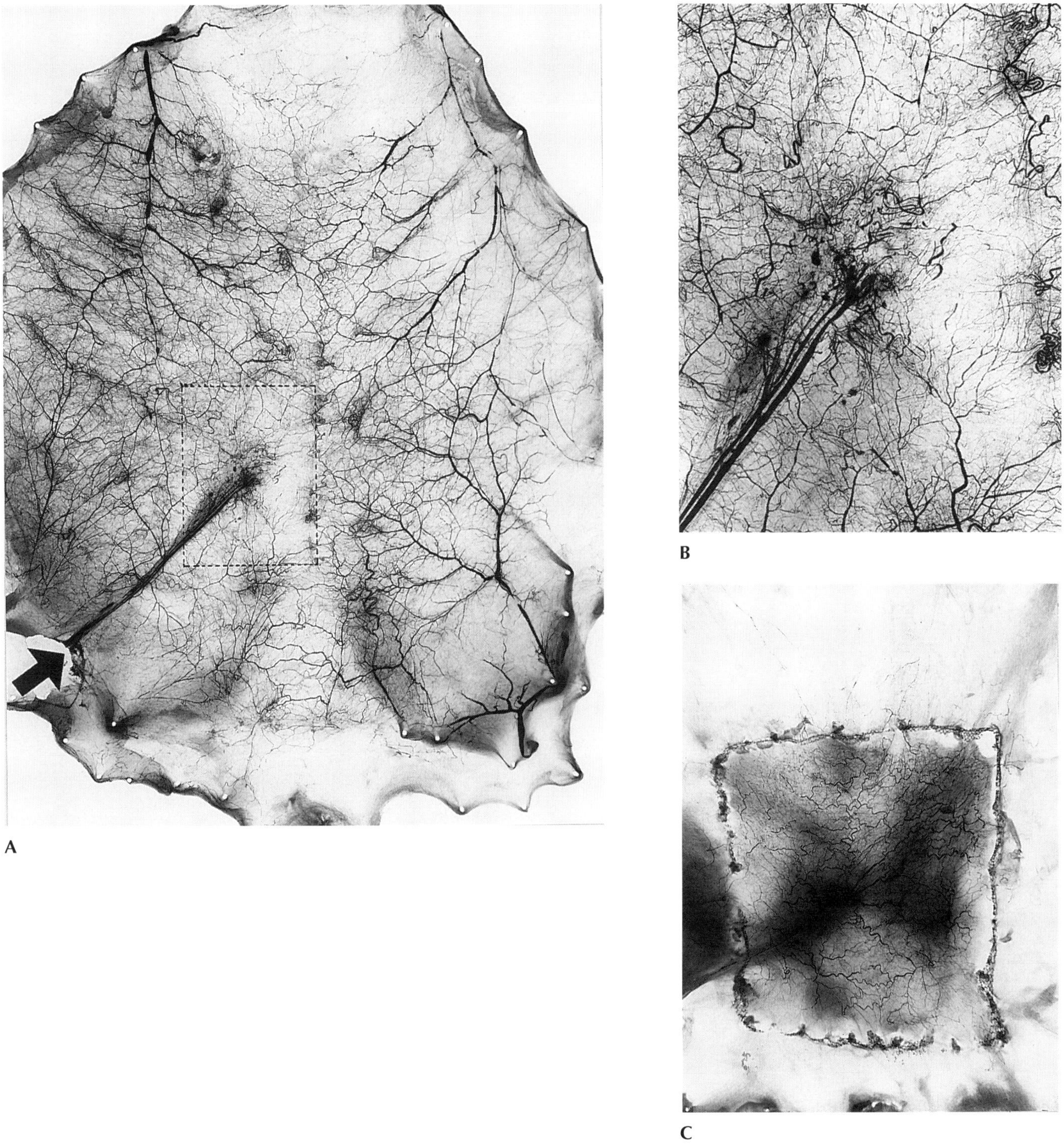

A

B

C

FIGURE 27-7

Microangiograms of group I (expanded) (arrow indicates the cannulated artery). (A) I-3W. Three weeks after implantation of the MVP. The original roentgenogram was used for contact printing. The vascular trees of both the MVP and the bilateral epigastric vessels were well visualized. They were expanded and elongated compared with the nonexpanded and Biobrane groups (see Figures 27-8 and 27-9). (B) I-3W. Direct image amplification of dotted area in (A) where the muscle cuff was transplanted, × 3. The newly formed vessels between the vascular pedicle and the inferior epigastric vessels were observed. Note the greater vascularity of the MVP artery and vein. (C) I-4W. Seven days after flap elevation. A newly formed small vascular plexus, centripetal to the pedicle, was present. (Reprinted with permission from Homma K, Ohura T, Sugihara T, Yoshida T, Hasegawa T. Prefabricated flaps using tissue expanders: An experimental study in rats. Plast Reconstr Surg *1993;91:1098–1107.)*

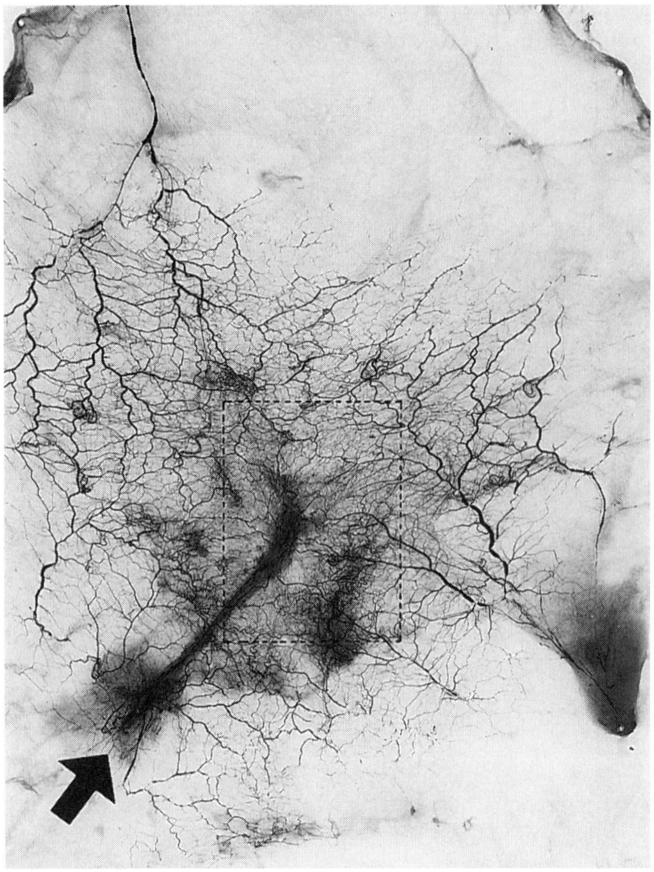

A

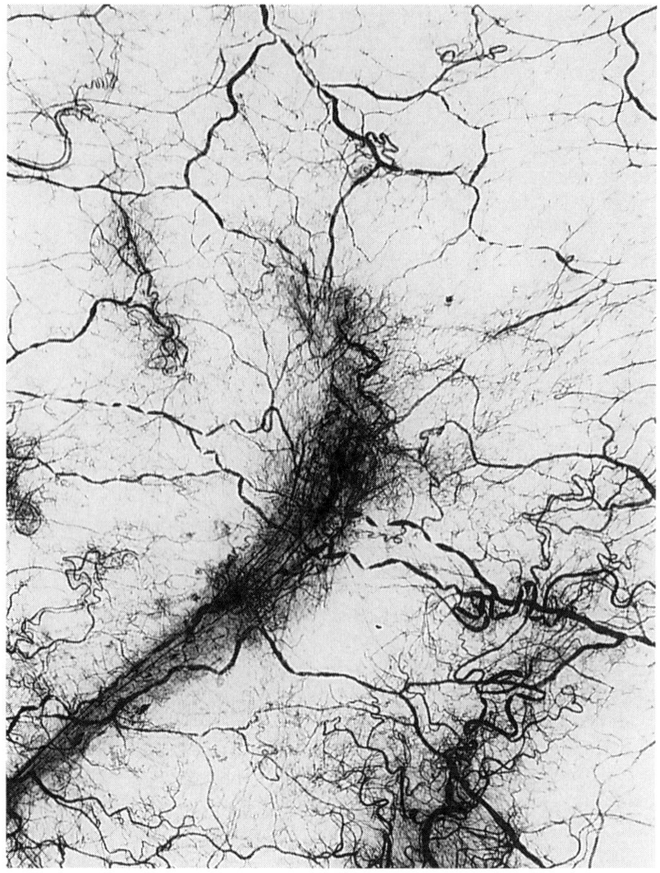

B

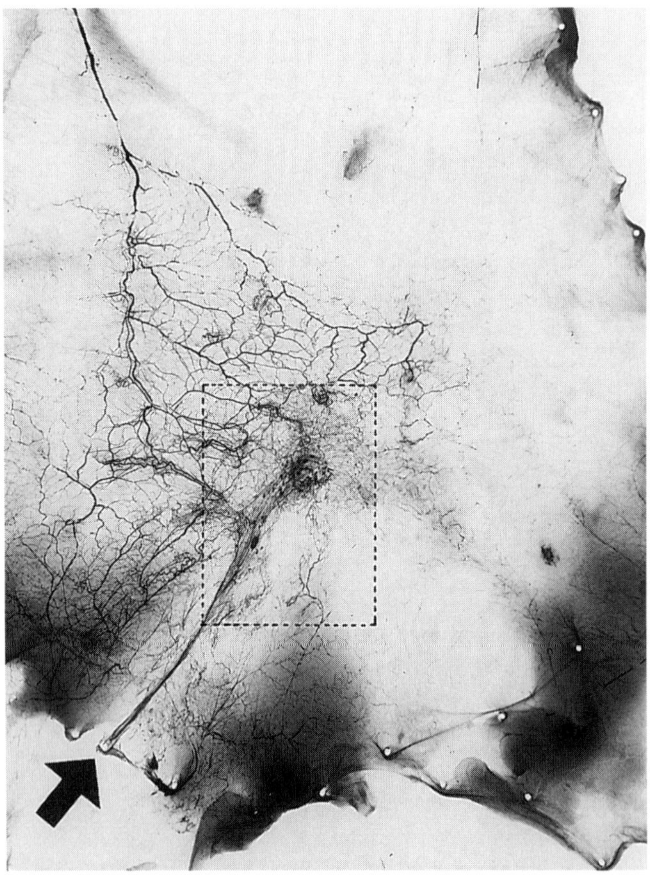

C

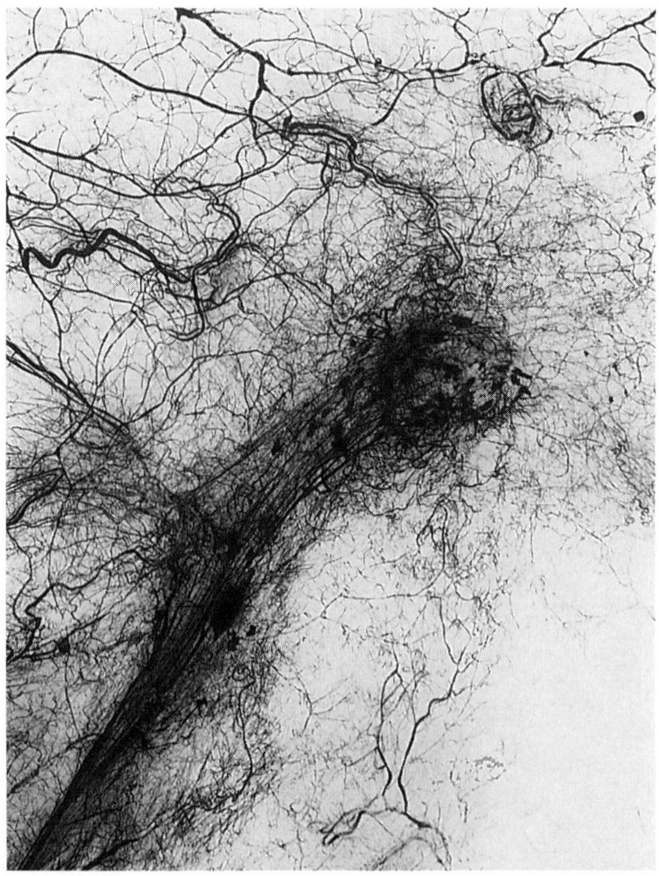

D

FIGURE 27-8

Microangiograms of group II (nonexpanded). (A) IIa-3W. Three weeks after implantation of the MVP. Contact printing. (B) IIa-3W. × 3. The newly formed vessels from the pedicle were present. (C) IIb-3W. Another case from the nonexpanded group. Contact printing. (D) IIb-3W. × 3. In this case, beyond the midline, the left epigastric vessels were not well visualized. In both IIa and IIb, vasodilatation of the MVP artery and vein was not observed. (Reprinted with permission from Homma K, Ohura T, Sugihara T, Yoshida T, Hasegawa T. Prefabricated flaps using tissue expanders: An experimental study in rats. Plast Reconstr Surg *1993;91:1098–1107.)*

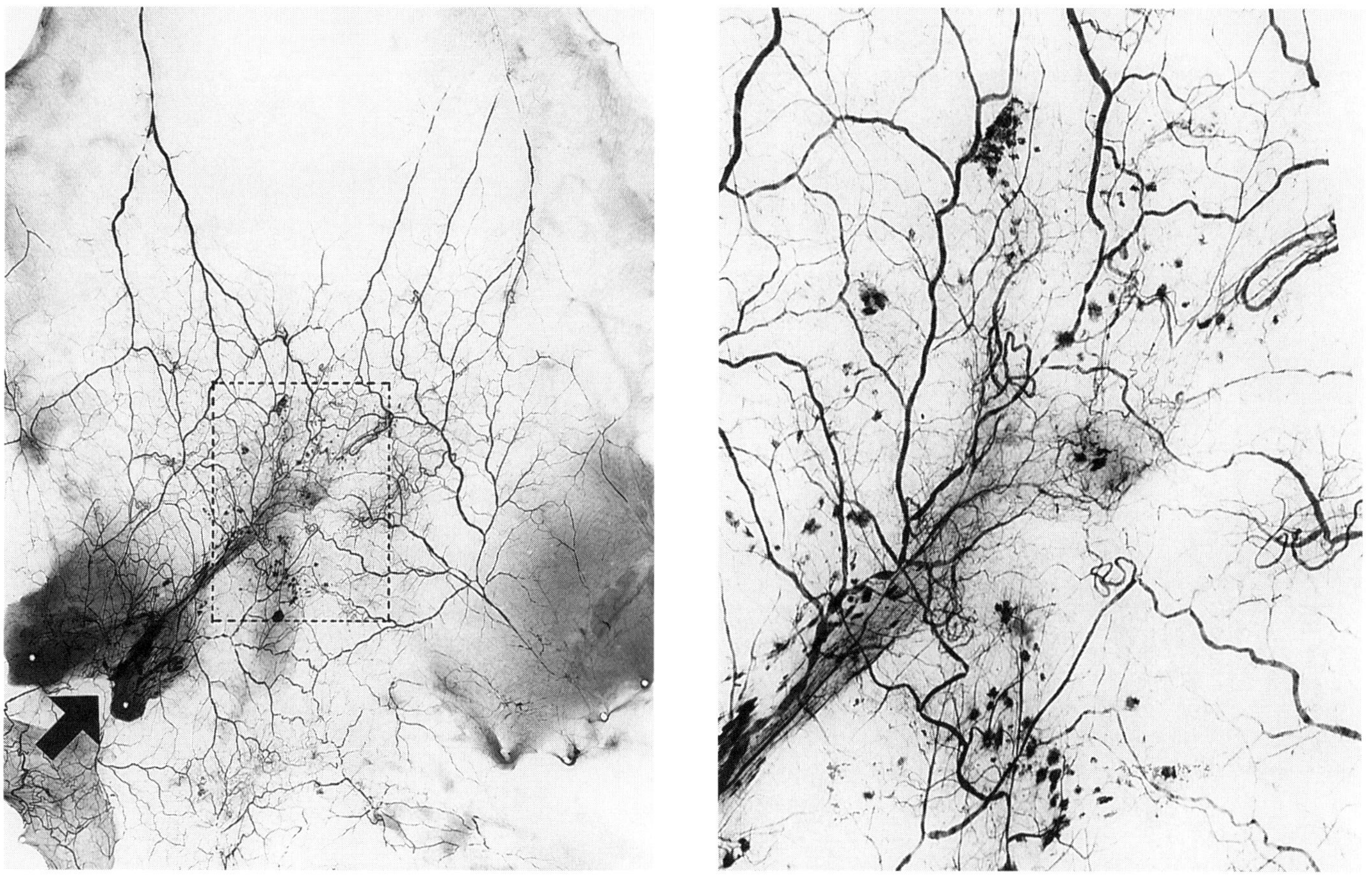

A B

FIGURE 27-9

Microangiograms of group III (Biobrane). (A) IIIa-3W. Three weeks after implantation of the MVP. Contact printing. (B) IIIa-3W. × 3. Vasodilatation of the epigastric veins was visualized, but that of the MVP was not. (C) IIIb-3W. Another case of Biobrane group. (D) IIIb-3W. × 3. The network of the new vessels between the muscle cuff and the flap was poor. An efflux of the perfusate from the muscle cuff was seen and the entire vascular tree of the epigastric vessels was not perfused. (Reprinted with permission from Homma K, Ohura T, Sugihara T, Yoshida T, Hasegawa T. Prefabricated flaps using tissue expanders: An experimental study in rats. Plast Reconstr Surg *1993;91:1098–1107.)*

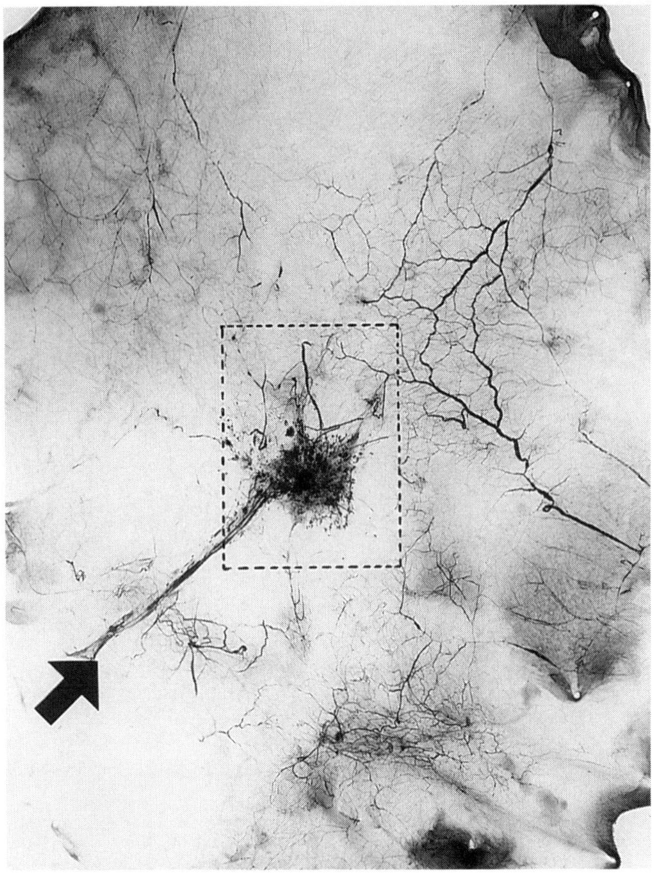

C

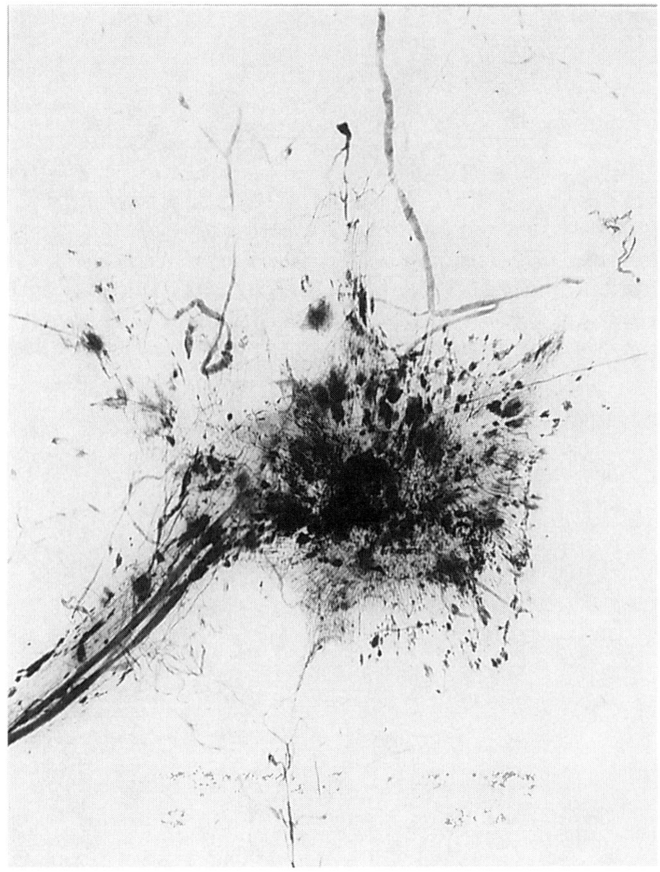

D

FIGURE 27-9 *(Continued)*

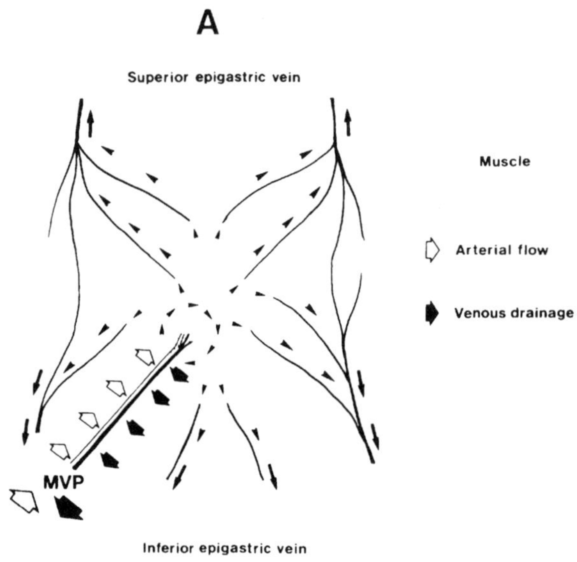

FIGURE 27-10

Schematic representation of the blood flow to correspond with angiograms. (A) The perfusate was selectively injected into the MVP artery under an operating microscope. In expanded MVP flaps, the injectant was soon drained from the MVP vein. It gradually filled the entire vascular tree of the epigastric vessels through the newly formed small vessels. (B) In nonexpanded and Biobrane groups, backflow from the MVP vein was seen. But major flow of the injectant passed through the new channels between MVP and the original vessels and drained into the epigastric veins. (Reprinted with permission from Homma K, Ohura T, Sugihara T, Yoshida T, Hasegawa T. Prefabricated flaps using tissue expanders: An experimental study in rats. Plast Reconstr Surg *1993;91: 1098–1107.)*

REFERENCES

1. Shintomi Y, Ohura T. The use of muscle vascularized pedicle flaps. *Plast Reconstr Surg* 1982;70:725–734.
2. Homma K, Ohura T, Sugihara T, Yoshida T, Hasegawa T. Prefabricated flaps using tissue expanders: An experimental study in rats. *Plast Reconstr Surg* 1993;91:1098–1107.
3. Hasegawa T, Igawa H, Sugihara T, Ohura T. An experimental study of the neovascular network in an MVP flap. *J Jpn Plast Reconstr Surg* 1991;11:609–616.
4. Cherry GW, Austad E, Pasyk K, McClatchey K, Robrich RJ. Increased survival and vascularity of random-pattern skin flaps elevated in controlled, expanded skin. *Plast Reconstr Surg* 1983;72:680–685.
5. Sasaki GH, Pang CY. Pathophysiology of skin flaps raised on expanded pig skin. *Plast Reconstr Surg* 1984; 74:59–65.
6. Leighton WD, Russell RC, Marcus DE, Eriksson E, Suchy H, Zook EG. Experimental pretransfer expansion of free-flap donor sites: I, Flap viability and expansion characteristics. *Plast Reconstr Surg* 1988;82:69–75.
7. Leighton WD, Russell RC, Feller AM, Eriksson E, Mathur A, Zook EG. Experimental pretransfer expansion of free-flap donor sites: II, Physiology, histology, and clinical correlation. *Plast Reconstr Surg* 1988;82:76–84.
8. Van Beek AL, Adson MH. Discussion. Experimental pretransfer expansion of free-flap donor sites: I, Flap viability and expansion characteristics; II, Physiology, histology, and clinical correlation. *Plast Reconstr Surg* 1988;82:85–87.
9. Daniel RK, Kerrigan CL. The omnipotential pig buttock flap. *Plast Reconstr Surg* 1982;70:11–15.
10. Petry JJ, Wortham KA. The anatomy of the epigastric flap in the experimental rat. *Plast Reconstr Surg* 1984;74: 410–413.
11. Smahel J, Jentsch B. Spontaneous anastomosis of vessels approximately 100 μ in diameter: An experimental study. *Br J Plast Surg* 1984;37:236–240.

Bibliography

Abramo AC, Viola JC, Angelo AJ. Intraoperative rapid expansion in cleft palate repair. *Plast Reconstr Surg* 1993; 91:441.

Adamson JG. Nasal reconstruction with the expanded forehead flap. *Plast Reconstr Surg* 1988;81:12.

Adson MH, Anderson RD, Argenta LC. Scalp expansion in the treatment of male pattern baldness. *Plast Reconstr Surg* 1987;79:906.

Agrawal K. Tissue expansion: Its cost factor (letter) (published erratum appears in *Arch Otolaryngol Head Neck Surg* 1993;119:349). *Arch Otolaryngol Head Neck Surg* 1992; 118:1003.

Ai YF. Planning of a local flap of expanded scalp for repair of alopecia cicatrisata. *Chung Hua Wai Ko Tsa Chih* 1989; 27:558.

Amiraslanov IA, Svetukhin AM, Sarkisov DS, Mitish VA, Lipatov KV. Staged tissue expansion in the surgery of suppurative wounds. *Vestn Ross Akad Med Nauk* 1994:41.

Anderson JM. Acceleration of capsule formation around silicone implants by infection in a guinea pig model (discussion). *Plast Reconstr Surg* 1984;73:97.

Anderson RD. Expansion assisted treatment of male pattern baldness. *Clin Plast Surg* 1987;14:477.

Anderson RD, Argenta LC. Tissue expansion for the treatment of alopecia. In Unger WP, Nordström REA (eds.), *Hair transplantation*. New York: Marcel Dekker, 1987, pp. 519–561.

Anger J. The reuse of tissue expanders in developing countries (letter). *Plast Reconstr Surg* 1991;88:1114.

Anger J, Szego T. Use of videoscopy as an aid in the placement of tissue expanders. *Rev Paul Med* 1993;111:363.

Antonyshyn O, Gruss JS, Mackinnon SE, et al. Complications of soft tissue-expansion. *Br J Plast Surg* 1988;41:239.

Apesos J, Perofsky HJ. The expanded forehead flap for nasal reconstruction. *Ann Plast Surg* 1993;30:411.

Apfelberg DB, Maser MR, White DN, Lash H. High- and low-technology solutions for massive cavernous hemangioma of the face: Lasers and leeches to the rescue. *Ann Plast Surg* 1989;23:341.

Argenta LC. Controlled tissue expansion in reconstructive surgery. *Br J Plast Surg* 1984;37:520.

Argenta LC. Reconstruction of the breast by tissue expansion. *Clin Plast Surg* 1984;11:257.

Argenta LC. Tissue expansion: Tissue expansion in head and neck reconstruction. *Adv Plast Reconstr Surg* 1984;1:122.

Argenta LC. Controlled tissue expansion. *Surg Rounds* 1986; 2:65.

Argenta LC, Adson MH, Iacobucci JJ. Tissue expansion revisited. *Adv Plast Reconstr Surg* 1988;4:113.

Argenta LC, Austad ED. Principles and techniques of tissue expansion. In McCarthy JG (ed.), *Plastic surgery* Philadelphia, PA: W.B. Saunders, 1990, pp. 475–507.

Argenta LC, Dingman RO. Total reconstruction of aplasia cutis congenita involving scalp, skull and dura. *Plast Reconstr Surg* 1986;77:650.

Argenta LC, Marks MW, Grabb WC. Selective use of serial expansion in breast reconstruction. *Ann Plast Surg* 1983;11:188.

Argenta LC, Marks MW, Grabb WC. Selective use of serial expansion in breast reconstruction. *Ann Plast Surg* 1985; 12:159.

Argenta LC, Marks MW, Pasyk KA. Advances in tissue expansion. *Clin Plast Surg* 1985;12:159.

Argenta LC, Van der Kolk CA. Tissue expansion in craniofacial surgery. *Clin Plast Surg* 1987;14:143.

Argenta LC, Van der Kolk CA, Friedman RJ, et al. Refinements in reconstruction of congenital breast deformity. *Plast Reconstr Surg* 1985;76:73.

Argenta LC, Watanabe MJ, Grabb WC, et al. Soft tissue expanders in head and neck surgery: A new method of reconstruction. *Plast Surg Forum* 1981;4:55.

Argenta LC, Watanabe MJ, Grabb WC. The use of tissue expansion in head and neck reconstruction. *Ann Plast Surg* 1983;11:31.

Armstrong RW, Berkowitz RL, Bolding F. Infection following breast reconstruction. *Ann Plast Surg* 1989;23:284.

Artz JS, Dinner MI, Foglietti MA, et al. Breast reconstruction utilizing subcutaneous tissue expansion followed by polyurethane-covered silicone implants: A 6-year experience. *Plast Reconstr Surg* 1991;88:635.

Artz JS, Dinner MI, Sampliner J. Breast reconstruction with a subcutaneous tissue expander followed with a polyurethane-covered silicone breast implant. *Ann Plast Surg* 1988;20:517.

Ashall G, Quaba A. A haemorrhagic hazard of tissue expansion. *Plast Reconstr Surg* 1987;79:627.

Atabey A, Barutcu A. Some useful techniques for avoiding complications of tissue expansion. *Plast Reconstr Surg* 1994;94:897.

Atabay K, Cenetoglu S, Aydogdu M, et al. Vulva reconstruction with a tissue expander. *Plast Reconstr Surg* 1987;90:520.

Atar D, Grant AD, Silver L, et al. The use of a tissue expander in club-foot surgery. A case report and review. *J Bone Joint Surg Br* 1990;72:574.

Aubert JP. Etude clinique et expérimentale de l'expansion tissulaire. Thèse Médecine, Marseille, 1987.

Aubert JP, Magalon G. Skin expansion. A new weapon in the therapeutic arsenal of plastic surgeons. *J Chir Paris* 1993;129:107.

Aubert JP, Magalon G. Treatment of skin lesions. Contribution of tissue expansion. *Presse Med* 1993;22:417.

Aubert JP, Paulhe P, Bardot J, et al. L'apport des valves externes dans l'expansion cutanée. *Ann Chir Plast Esthet* 1991;36:218.

Aubert JP, Paulhe P, Magalon G. Crossed leg expanded flaps. *Ann Chir Plast Esthet* 1991;36:447.

Aubert JP, Paulhe P, Magalon G. Forum on tissue expansion. Expansion of the upper limb. *Ann Chir Plast Esthet* 1993; 38:34.

Audry G, Buis J, Vazquez MP, Gruner M. Amputation of penis after circumcision—penoplasty using expandable prosthesis. *Eur J Pediatr Surg* 1994;4:44.

Auletta MJ, Matarasso SL, Glogau RG, Tromovitch TA. Comparison of skin hooks and Foley catheters for immediate tissue expansion. *J Dermatol Surg Oncol* 1993;19:1084.

Austad ED. The Caulking Gun: An aid to inflate tissue-expander prostheses (discussion). *Plast Reconstr Surg* 1983; 72:403.

Austad ED. Repeated skin expansion to resurface a massive thigh wound (discussion). *Plast Reconstr Surg* 1986;77:658.

Austad ED. Complications in tissue expansion. *Clin Plast Surg* 1987;14:431.

Austad ED. The origin of expanded tissue. *Clin Plast Surg* 1987;14:429.

Austad ED. Tissue expansion techniques (editorial). *Arch Dermatol* 1987c;1123:588.

Austad ED. Contraindications and complications in tissue-expansion. *Facial Plast Surg* 1988;5:379.

Austad ED. Evolution of the concept of tissue expansion. *Fac Plast Surg* 1988;5:277.

Austad ED, Pasyk KA. Rapid elongation of arteries and veins in rats with a tissue expander (discussion). *Plast Reconstr Surg* 1987;80:579.

Austad ED, Pasyk KA, McClatchey KD, et al. Histomorphologic evaluation of guinea pig skin and soft tissue after controlled tissue expansion. *Plast Reconstr Surg* 1982;70:704.

Austad ED, Rose GL. A self-inflating tissue expander. *Plast Reconstr Surg* 1982;70:588.

Austad ED, Thomas SB, Pasyk KA. Tissue expansion: "Dividend or loan?" *Plast Reconstr Surg* 1986;78:63.

Azzolini A, Riberti C, Cavalca D. Skin expansion in head and neck reconstructive surgery. *Plast Reconstr Surg* 1992; 90:799.

Babovic S, Angel MF, Im MJ, Ress AM, Manson PN. Effects of tissue expansion on secondary ischemic tolerance in experimental free flaps. *Ann Plast Surg* 1995;34:593.

Bachynsky T, Antonyshyn OM, Ross JB. Dissecting folliculitis of the scalp. A case report of combined treatment using tissue expansion, radical excision, and isotretinoin. *J Dermatol Surg Oncol* 1992;18:877.

Bahat O, Handelsman M. Controlled tissue expansion in reconstructive periodontal surgery. *Int J Periodontics Restorative Dent* 1991;11:32.

Baker SR. Fundamentals of expanded tissue. *Head-Neck* 1991;13:327.

Baker SR, Johnson TM, Nelson BR. Technical aspects of prolonged scalp expansion. *Arch Otolaryngol Head Neck Surg* 1994;120:431.

Baker SR, Swanson NA. Clinical applications of tissue expansion in head and neck surgery. *Laryngoscope* 1990; 100:313.

Baker SR, Swanson NA. Tissue expansion of the head and neck. Indications technique and complications. *Arch Otolaryngol Head Neck Surg* 1990;116:1147.

Baker SR, Swanson NA. Rapid intraoperative tissue expansion in reconstruction of the head and neck. *Arch Otolaryngol Head Neck Surg* 1990;116:1431.

Baker SR, Swanson NA. Reconstruction of midfacial defects following surgical management of skin cancer. The role of tissue expansion. *J Dermatol Surg Oncol* 1994;20:133.

Balakrishnan C. The preexpanded radial free flap (letter comment). *Plast Reconstr Surg* 1991;88:171.

Balakrishnan C. Tissue expansion in the treatment of pressure ulcers (letter). *Plast Reconstr Surg* 1991;88:1108.

Balakrishnan C, Papini R. Breast reconstructions with expander (letter comment). *Plast Reconstr Surg* 1991;88:372.

Bannerot H, Garnier D, Ricbourg B. Forum on tissue expansion. Fast continuous expansion. A 3-year evaluation of its use from a retrospective study of 78 cases. *Ann Chir Plast Esthet* 1993;38:41.

Baron JL, Reynaud JP, Bousquet P, et al. Déhiscences cutanées en cours d'expansion: Utilisation d'un film plastique chirurgical adhésif. *Ann Chir Plast Esthet* 1991;36:167.

Barone FE, Perry L, Keller T, et al. The biomechanical and histopathologic effects of surface texturing with silicone and polyurethane in tissue implantation and expansion. *Plast Reconstr Surg* 1992;90:77.

Barsan RE, Kent JN. Hydroxylapatite reconstruction of alveolar ridge deficiency with an open mucosal flap technique. *Oral Surg Oral Med Oral Pathol* 1985;59:113.

Bartell TH, Mustoe TA. Animal models of human tissue expansion. *Plast Reconstr Surg* 1989;83:681.

Bashir AH. Wound closure by skin traction: An application of tissue expansion. *Br J Plast Surg* 1987;40:582.

Bass CB. Herniated areolar complex. *Ann Plast Surg* 1978;1:40.

Battiston B, Buffoli P, Vigasio A, et al. The effects of lengthening on nerves. *Ital J Orthop Traumatol* 1992;18:79.

Bauer BS. Reconstruction of the microtic ear. *J Pediatr Surg* 1984;19:440.

Bauer BS. The role of tissue expansion in reconstruction of the ear. *Clin Plast Surg* 1990;17:319–325.

Bauer BS, Johnson PG, Lovato G. Applications of soft tissue expansion in children. *Pediatr Dermatol* 1986;3:281.

Bauer BS, Vicari FA. Approach to excisions of congenital giant pigmented naevi in infancy and early childhood. *Plast Reconstr Surg* 1988;82:1012.

Bauer BS, Vicari FA, Richard ME. The role of tissue expansion in pediatric plastic surgery. *Clin Plast Surg* 1990;17:101.

Bauer BS, Vicari FA, Richard ME, Schwed R. Expanded full-thickness skin grafts in children: Case selection planning and management. *Plast Reconstr Surg* 1993;92:59.

Baux S, Mimoun M, Hilligot P. Cutaneous expansion in burns sequelae. *Chirurgie* 1990;116:373.

Baux S, Mimoun M, Kirsch JM, et al. Les prothéses d'expansion cutanée et leur utilisation en dehors de la reconstruction mammaire. A propos de 54 cas. *Ann Chir Plast Esthet* 1988;33:159.

Bax NM, Van-der-Zee DC, Pulltergunne AJ, Rovekamp MH. Treatment of giant omphalocele by enlargement of the abdominal cavity with a tissue expander. *J Pediatr Surg* 1993;28:1181.

Beasley ME. Eighty-four consecutive breast reconstructions using a textured silicone tissue expander (discussion). *Plast Reconstr Surg* 1992;89:1035.

Beauchene JG, Chambers MM, Peterson AE, Scott PG. Biochemical, biomechanical, and physical changes in the skin in an experimental animal model of therapeutic tissue expansion. *J Surg Res* 1989;47:507.

Becker H. Breast reconstruction using an inflatable breast implant with detachable reservoir. *Plast Reconstr Surg* 1984;73:678.

Becker H. The permanent tissue expander. *Clin Plast Surg* 1987;14:519.

Becker H. Breast augmentation using the expander mammary prosthesis. *Plast Reconstr Surg* 1987;79:192.

Becker H. The expandable mammary implant. *Plast Reconstr Surg* 1987;79:631.

Becker H. The expandable mammary implant: An update. *Persp Plast Surg* 1989;3:164.

Becker H, Maraist F. Immediate breast reconstruction after mastectomy using a permanent tissue expander. *South Med J* 1987;80:154.

Belkoff SM, Naylor EC, Walshaw R, Lanigan E, Colony L, Haut RC. Effects of subcutaneous expansion on the mechanical properties of porcine skin. *J Surg Res* 1995;58:117.

Bengtson BP, Ringler SL, George ER, DeHaan MR, Mills KA. Capsular tissue: A new local flap. *Plast Reconstr Surg* 1993;91:1073.

Ben Hur N, Neuman Z. Malignant tumor formation following subcutaneous injection of silicone fluid in white mice. *Isr Med J* 1963;22:15.

Bennett RG, Hirt M. A history of tissue expansion. Concepts, controversies and complications. *J Dermatol Surg Oncol* 1993;19:1066.

Beraka GJ. Tissue expansion for rhytidectomy. *Plast Reconstr Surg* 1990;86:802.

Berg A, Jonsson CE. Tissue expansion to cover amputation stumps. A case report. *Acta Orthop Scand* 1992;63:566.

Berger U, Schumann D, Wollina U, et al. Experimental studies on tissue expansion in reconstructive surgery. *Dtsch Zahn Mund Kieferheilkd Zentralbl* 1992;80:415.

Berman DE, Lettieri J, Herold D, et al. Lidocaine permeability in silicone tissue expanders: An in vitro analysis. *Plast Reconstr Surg* 1989;84:621.

Berrino P. On reconstructing the ptotic breast with an expander. *Plast Reconstr Surg* 1988;82:1101.

Berry FD. A modified tissue expander for socket enlargement in clinical anophthalmos. *Ophthalmol Plast Reconstr Surg* 1991;7:41.

Bertsch S, Csontos K, Schweizter JAL. Effect of mechanical stimulation on cell proliferation in mouse epidermis and on growth regulation by endogenous factors (chalones). *Cell Tissue Kinet* 1976;9:455.

Birnbaum L. Discussion on breast reconstruction using modified tissue expansion. *Plast Reconstr Surg* 1986;77:620.

Birnbaum L, Olsen JA. Breast reconstruction following radical mastectomy using custom designed implants. *Plast Reconstr Surg* 1978;61:355.

Bochlogyros PM, Henser R, Becker R, et al. A modified HA implant material. *J Maxillofac Surg* 1985;13:213.

Bock NJJ, Rose R. Der subperiostale Gewebeexpander als Hilfsmittel bei der Alveolarkammplastik. *Dtsch Z Mund Kieferheilkd Gesichtschir* 1987;11:443.

Bodin Y. Rendement des plasties aprés expansion tissulaire à l'exclusion des reconstructions mammaraires. Thèse Médecine, Toulouse, 1987.

Boezeman JBM, Bauer FW, DeGrood RM. Flowcytometric analysis of the recruitment of G30 cells in human epidermis "in vivo" following tape stripping. *Cell Tissue Kinet* 1987;20:99.

Bolton LL, Chandrasekhar B, Gottlieb ME. Forehead expansion and total nasal reconstruction. *Ann Plast Surg* 1988; 21:210.

Bonomo D. Subperiostal tissue expander for ridge augmentation. *Dent Soc Rev* 1986;34:186.

Boot JC, Naftel AJ, Ramli AR. Bodymap: An image processing system for the measurement of body surface profiles encountered in skin expansion surgery. *Int J Biomed Comput* 1992;31:189.

Borenstein A, Yaffe B, Seidman DS, et al. Tissue expansion in reconstruction of postburn contracture of the first web space of the hand. *Ann Plast Surg* 1991;26:463.

Borges Filho PT, Neves RI, Gemperli R, et al. Soft-tissue expansion in lower extremity reconstruction. *Clin Plast Surg* 1991;18:593.

Brandy DA. A special tissue expander in combination with extensive scalp-lifting. *Aesthet Plast Surg* 1991;15:349.

Brar M. Tissue expansion with rhytidectomy. *Plast Reconstr Surg* 1990;85:996.

Brent B. Reconstruction of traumatic ear deformities. Symposium on deformities of the external ear. *Clin Plast Surg* 1978;5:437.

Bricout N, Le Danvic D, De Roquancourt M, et al. Prothèses et accidents des silicones. *Ann Chir Plast Esthet* 1986; 31:268.

Brobmann GF, Huber J. Effects of different shaped tissue expanders on transluminal pressure, oxygen tension, histopathologic changes, and skin expansion in pigs. *Plast Reconstr Surg* 1985;76:731.

Brody GS. Biological creep (letter). *Plast Reconstr Surg* 1993; 92:1202.

Brophy D, Lobitz WC. Injury and reinjury of the human epidermis. Epidermal basal cell response. *J Invest Dermatol* 1959;32:495.

Brotherston TM, Dhir A, Balakrishnan C, et al. Locating tissue expander injection ports with a metal detector (letter). *Plast Reconstr Surg* 1992;90:146.

Brown IA. A scanning electron microscope study of the effect of uniaxial tension on human skin. *Br J Dermatol* 1973; 89:383.

Brown MD, Johnson TM. Complications of tissue expansion. *J Dermatol Surg Oncol* 1993;19:1120.

Brunette DM. Mechanical stretching increases the number of epithelial cells synthetising DNA in culture. *J Cell Sci* 1984;69:35.

Buebendorf ND, Concannon MJ, Gaines RW, Puckett CL. Skin expansion as preparation for an opening wedge osteotomy of the mid-foot in arthrogryposis. *Mo Med* 1992; 89:671.

Buhrer DP, Huang TT, Yee HW, et al. Treatment of burn alopecia with tissue expanders in children. *Plast Reconstr Surg* 1988;81:512.

Buis J. Tissue expansion. Techniques and indications for children. *Soins Gynecol Obstet Pueric Pediatr* 1994;156:4.

Bullough WS. The control of mitotic activity in adult mammalian tissues. *Biol Res* 1962;37:307.

Bullough WS, Mitrani EN. The significance of vertical mitosis in epidermis. *Br J Dermatol* 1978;99:603.

Burckardt BR. Comparing contracture rates: Probability theory and the unilateral contracture. *Plast Reconstr Surg* 1984;74:528.

Burnett W. Yank melts native. *Natl Geogr* 1945;88:124.

Butow KW, deWitt TW. Bilateral oblique facial cleft-tissue expansion with primary reconstruction. *Tydskr Tandheelkd Ver S-Afr* 1990;45:507.

Byrd HS, Hobar PC. Abdominal wall expansion in congenital defects. *Plast Reconstr Surg* 1989;84:347.

Cairns TS, de Villiers W. Capsular contracture after breast augmentation: A comparison between gel and saline filled prostheses. *S Afr Med J* 1980;57:951.

Caleel RT. Breast reconstruction utilizing the tissue expander. *Am J Cosmetic Surg* 1985;2:19.

Caleffi C, Fasano D, Bocchi A, et al. Expanded tubed flap in scalp reconstruction. A modern option in an old technique. *Plast Reconstr Surg* 1988;82:1075.

Canady JW, Squier CA, Kelly KM, Bardach J. Serial measurement of blood flow in expanded tissue by laser Doppler velocimetry. *Otolaryngol Head Neck Surg* 1990;103:986.

Canady JW, Squier CA, Kelly KM, Bardach J. Blood flow in expanded tissue treated with an anticontractile agent. *Ann Otol Rhinol Laryngol* 1991;100:962.

Cano I, Encinas A, Herrero E, et al. The indications, techniques and results of the use of tissue expanders in pediatric surgery. *Cir Pediatr* 1991;4:173.

Caputo R. Ethiopia: Revolution in an ancient empire. *Natl Geogr* 1983;165:614.

Cariou JL, Hilligot P, Arrouvel C, Banzet P. Periprosthetic membrane island flap with axial pedicle. Experimental study on the effect of time and expanded volume. *Ann Chir Plast Esthet* 1992;37:139.

Carneiro R, Dichiara J. A protocol for tissue expansion in upper extremity reconstruction. *J Hand Surg* 1991;147:51.

Carruthers A. Tissue expansion and Mohs micrographic surgery. *J Dermatol Surg Oncol* 1993;19:1106.

Cattaneo R, Villa A, Catagni M, et al. Traitement des pseudarthroses diaphysaires septiques ou non septiques selon la méthode d'Ilizarov en compression monofocale. *RCO* 1985;71:223.

Cattaneo R, Villa A, Catagni M, et al. Application de la méthode d'Ilizarov dans l'allongement de l'humérus. *RCO* 1986;72:203.

Cawood JJ, Howell RA. A classification of the edentulous jaws. *Int J Oral Maxillofac Surg* 1988;17:232.

Chaibi M, Houssemand N. Removal of the expander. Taking charge and postoperative care. *Soins Gynecol Obstet Pueric Pediatr* 1994;156:22.

Chang B, Tuchler RE, Siebert JW, et al. The effect of tissue expansion on dermal fibroblast contraction. *Ann Plast Surg* 1992;28:315.

Chang TS, Jin YT. Applications of tissue expansion to the treatment of post burn skin contracture and alopecia. *Eur J Plast Surg* 1986;9:7.

Charles AR. The ultrastructure of the epidermis. The mammalian epidermis and its derivatives. *Symp Zool* 1964; 12:39.

Chassagne JF, Brice M, Maxant P, et al. Evolution des idées dans les techniques de réparation du scalp. *Ann Chir Plast Esthet* 1986;31:325.

Cheng NX, Chen YF. Advances in clinical uses of soft tissue expansion. *Chung Hua Cheng Hsing Shao Shang Wai Ko Tsa Chih* 1994;10:40.

Cherry GW, Austad ED, Pasyk KA, et al. Increased survival and vascularity of random pattern skin flaps elevated in controlled expanded skin. *Plast Reconstr Surg* 1983;72:680.

Cherry GW, Austad ED, Pasyk KA, et al. Increased survival and vascularity of skin flaps elevated in controlled expanded skin. *Microvasc Res* 1984;27:390.

Cherskov M. Tissue expansion "Future of plastic surgery, for next 20 years." *JAMA* 1982;247:3039.

Chiarini L, Tanza D, Cantoni E, Ferronato G. Total atrophy of the mandibular alveolar crest. A new therapeutic proposal (the use of tissue expander) and a clinical case report. *Minerva Stomatol* 1992;41:193.

Chippaux C. Mutilations et déformations chroniques dans les races humaines. *Hist Med* 1961;11:43.

Chisholm EM, Marr S, Macfie J, et al. Post mastectomy breast reconstruction using the inflatable tissue-expander. *Br J Surg* 1986;73:817.

Chretien-Marquet B. Rapid intraoperative distension using isotonic saline solution. *Plast Reconstr Surg* 1995;96:158.

Chretien-Marquet B, Benaceur S, Cerceau, et al. Cutaneous expansion using enforced position in the treatment of large skin defects. *Plast Reconstr Surg* 1994;93:337.

Clausen UPF, Lundmo T. Regenerative proliferation of mouse epidermal cells following adhesive tape stripping. *Cell Tissue Kinet* 1976;9:573.

Clauss LC, Biomatériaux et biocompatibilité: Les silicones médicales. Thése Medecine, Paris, 1985.

Cleall JF, Bayne DI, Posen JM, et al. Expansion of the midpalatal suture in the monkey. *Angle Orthodont* 1965;35:23.

Clodius L. Temporary expander. *Plast Reconstr Surg* 1983;71:279.

Codivilla A. On the means of lengthening in the lower limbs the muscle and tissues which are shortened through deformity. *Am J Orth Surg* 1905;2:405.

Coessens B, De-Mey A, Vadoud-Seyedi J, Vandeweyer E. Use of tissue expansion in head and neck. Review of 33 cases. *Acta Chir Belg* 1993;93:8.

Cohen BE, Casso D, Whetstone M. Analysis of risks and aesthetics in a consecutive series of tissue expansion breast reconstructions. *Plast Reconstr Surg* 1992;89:840.

Cohen BH. Navigating through tissue expansion terminology. *J Dermatol Surg Oncol* 1993;19:614.

Cohen BH, Cosmetto AJ. The suture tension adjustment reel. A new device for the management of skin closure. *J Dermatol Surg Oncol* 1992;18:112.

Cohen IK. Antibiotics in the tissue expander to decrease the rate of infection (reply). *Plast Reconstr Surg* 1988;81:138.

Cohen IK, Dunn JD. Future aspects in tissue expansion. *Fac Plast Surg* 1988;5:383.

Cohen IK, Roberts C. Lidocaine to relieve pain with tissue expansion of the breast. *Plast Reconstr Surg* 1987;79:489.

Cohen IK, Turner D. Immediate breast reconstruction with tissue expanders. *Clin Plast Surg* 1987;14:779.

Cohen M, Marschall MA. Tissue expansion. An alternative technique in reconstructive surgery. *Surg Ann* 1990;22:343.

Cohen M, Marschall MA, Schafer ME. Tissue expansion for the reconstruction of burn defects. *J Trauma* 1988;28:158.

Cole RP, Gault DT, Mayou GT, et al. Pain and forehead expansion. *Br J Plast Surg* 1991;44:41.

Cole WG. Tissue expansion in the lower limbs of children and young adults. *J Bone Joint Surg (Br)* 1991;72:578.

Coleman DJ. Use of expanded temporal flaps to resurface the skin grafted forehead. *Br J Plast Surg* 1987;40:171.

Collins SA, Swanson NA. Chronic tissue expansion. *J Dermatol Surg Oncol* 1993;19:1090.

Converse J. Reconstructive Plastic Surgery. Philadelphia: W.B. Saunders, 1977.

Coombs CJ, Mutimer KL. Tissue expansion for the treatment of complete syndactyly of the first web. *J Hand Surg Am* 1994;19:968.

Cooper RL, Brown D. Pretransfer tissue expansion of a scalp free flap for burn alopecia reconstruction in a child: A case report. *J Reconstr Microsurg* 1990;6:339.

Craig GT. Obituary: Radovan, Ch., M.D. 1932–1984. *Plast Reconstr Surg* 1986;77:356.

Cullen KW, Clarke JA, Mac Lean NR. The complications of tissue expansion in the burned scalp. *Burns* 1986;12:273.

Cullen KW, Powell B. Tissue expanders in surgery. *Br J Clin Pract* 1989;43:75.

Cuono CB. Caulking gun for tissue expansion. *Plast Reconstr Surg* 1989;83:923.

Curtis ASG, Seehar GM. The control of cell division by tension or diffusion. *Nature* 1978;274:52.

Dalenius E. Tissue expansion: A sufficient factual basis is missing. *Lakartidningen*. 1995;92:2173.

Daly CH. The biomechanical characteristics of human skin. Glasgow: Stathclyde University, Pergamon Press, 1966.

Da Matta A. Reconstruction of postburn sequelae with expanded flaps. *Burns* 1989;15:407.

Danielson DA. Human skin as an elastic membrane. *J Biochem* 1973;6:539.

D'Anjou B, Allieu Y, Cenac P, et al. Nerve expansion. *Ann Chir Main* 1989;8:336.

Dardour JC. Utilisation d'une prothèse d'expansion dans le traitement de la calvitie masculine. *Ann Chir Plast Esthet* 1988;33:347.

Datubo-Brown DD, Khalid KN, Levick PL. Tissue-expanded visor flap in burn surgery. *Ann Plast Surg* 1994;32:205.

Day TA, Aarstad RF, Hoasjoe DK, Stucker FJ. The use of tissue expanders in the correction of ectropion in an animal model. *Laryngoscope* 1995;105:365.

De Agustin JC, Morris SF, Zuker RM. Tissue expansion in pediatric burn reconstruction. *J Burn Care Rehabil* 1993;14:43.

De Agustin JC, Morris SF, Zuker RM. Tissue expansion in the reconstruction of post-burn lesions in childhood. *Cir Pediatr* 1994;7:174.

de Greff A. Tissue expansion in reconstructive surgery. *South Afr Med J* 1986;69:705.

Dehaan MR, Hammond DC, Mann RJ. Controlled tissue expansion of a groin flap for upper extremity reconstruction. *Plast Reconstr Surg* 1990;86:979.

De Koomen HA. De verhoging van de geresoorbeerde mandibula. Thesis, University of Nijmegen, The Netherlands, 1982.

Denny AD. Expanded midline forehead flap for coverage of non-nasal facial defects. *Ann Plast Surg* 1992;29:576.

Derby LD, Sinow JD, Bowers LD, et al. Quantitative analysis of lidocaine hydrochloride delivery by diffusion across tissue expander membranes. *Plast Reconstr Surg* 1992;89:900.

Devine JW, Howard PS, Nordström REA. Treatment of male pattern baldness. *Adv Plast Reconstr Surg* 1987;13:668.

Dick GD, Brown SA. Breast reconstruction using modified tissue expansion. *Plast Reconstr Surg* 1986;77:613.

Dickson MG. Breast reconstruction by tissue expansion: A previously unreported complication (letter). *Plast Reconstr Surg* 1987;80:474.

Dickson MG, Batchelor AG. A technical aid to tissue-expansion. *Br J Plast Surg* 1987;40:305.

Dickson MG, Sharpe DT. The complications of tissue-expansion in breast reconstruction: A review of 75 cases. *Br J Plast Surg* 1987;40:629.

Dickson MG, Sharpe DT, Dickson WA, et al. Breast reconstruction by tissue expansion. *Ann R Coll Surg Eng* 1987;69:18.

Dickson WA, Sharpe DT, Jackson IT. Experience with an external valve in small volume tissue expanded. *Br J Plast Surg* 1988;41:373.

DiGregorio VR. Two unusual postburn reconstructions using tissue expansion: One exogenous and one endogenous. *Ann Plast Surg* 1993;30:549.

Dingman RO, Argenta LC. The surgical repair of traumatic defects of the scalp. *Clin Plast Surg* 1982;9:131.

Dinner MJ, Dowden RV. The tubular/tuberous breast syndrome. *Ann Plast Surg* 1987;19:414.

Dion V. Putting the expander in place. Taking charge and postoperative monitoring. *Soins Gynecol Obstet Pueric Pediatr* 1994;156:15.

Ditmars DM, Jr. The caulking gun: An aid to inflate tissue-expander prosthesis. *Plast Reconstr Surg* 1983;71:401.

Donelan MB. Complications of controlled tissue expansion in the pediatric burn patient (discussion). *Plast Reconstr Surg* 1988;82:846.

Don Parsa F. Observations with the use of textured tissue expanders in breast reconstruction (letter). *Plast Reconstr Surg* 1991;87:1142.

Dover R. How does skin expansion work? *Plast Reconstr Surg* 1987;79:672.

Dowden RV, Marks KE, Stulberg BN. Expandable breast implant reconstruction of Tikhoff-Linberg shoulder deformity. *Plast Reconstr Surg* 1990;85:112.

Dowden RV. Selection criteria for successful immediate breast reconstruction. *Plast Reconstr Surg* 1991;88:628.

Dowden RV. Breast reconstruction with the inflatable expander. *Plast Reconstr Surg* 1991;88:735.

Dubert T. L'expansion cutanée contrôlée en dehors des indications liées à la mastectomie. A propos de 86 cas suivis à l'hôpital Rotschild. Thèse Médecine, Cochin Paris, 1986.

Ducours JL, Failloux A, Ardanza B, et al. The value of a previously expanded musculo-cutaneous flap of the lower eyelid for the reconstruction of a full thickness loss of tissue substance of the upper eyelid. *Rev Stomatol Chir Maxillofac* 1992;93:40.

Duffy JS, Shuter M. Evaluation of soft-tissue properties under contolled expansion for reconstructive surgical use. *Med Eng Phys* 1994;16:304.

Duits EHA, Molenaar J, van Rappard JHA. The modeling of skin expanders. *Plast Reconstr Surg* 1989;83:362.

Duskova M. Use of Becker's prosthesis in breast reconstruction. *Acta Chir Plast* 1992;34:249.

Edmond JA, Padilla JF 3rd. "Off the Shelf" techniques for continuous tissue expansion. *Ann Plast Surg* 1993;30:552.

Edmond JA, Padilla JF 3rd. Preexpansion galeal scoring. *Plast Reconstr Surg* 1994;93:1087.

Ehlert TK, Thomas JR. Rapid intraoperative tissue expansion for closure of facial defects. *Arch Otolaryngol Head Neck Surg* 1991;117:1043.

El Deeb M. Comparison of three methods of stabilisation of particulate hydroxilapatite for augmentation of the mandibular ridge. *J Oral Maxillofac Surg* 1988;46:758.

Elisevich K, Bite U, Colcleugh R. Microorbitalism: A technique for orbital rim expansion. *Plast Reconstr Surg* 1991; 88:609.

Ellenbogen R, Collini FJ. Tissue expansion for rhytidectomy. *Plast Reconstr Surg* 1990;85:995.

Elliott MP. A musculocutaneous transposition flap mammaplasty for correction of the tuberous breast. *Ann Plast Surg* 1988;20:153.

Elliott MP, Dubrul W. Magna site tissue expander: An innovation for injection site location. *Plast Reconstr Surg* 1988; 81:605.

Endo T, Nakayama Y. Histologic examination of peripheral nerves elongated by tissue expanders. *Br J Plast Surg* 1993; 46:421.

Esposito G, Di Caprio G, Ziccardi P, et al. Tissue expansion in the treatment of pressure ulcers. *Plast Reconstr Surg* 1991; 87:501.

Esposito G, Ziccardi P, Di-Caprio G, Scuderi N. Reconstruction of ischial pressure ulcers by skin expansion. *Scand J Plast Reconstr Surg Hand Surg* 1993;27:133.

Fan ZH. Experimental study of the survival length of random flap influenced by rapid tissue expansion. *Chung Hua Cheng Hsing Shao Shang Wai Ko Tsa Chih* 1992;8:208.

Fan Z, Guan W, Jin Y. Experimental study of changes in biomechanical properties of pig skin after rapid expansion. *Chung Hua Cheng Hsing Shao Shang Wai Ko Tsa Chih* 1994; 10:34.

Fazili M, Noorman van der Dussen MF, Van Waas MAJ. Long term results of augmentation of the atrophic mandible. *Int J Oral Maxillofac Surg* 1986;15:513.

Fenton O. Expanding possibilities. *Br Med J* 1987;295:684.

Fenton O. Complications of soft-tissue expansion (discussion). *Br J Plast Surg* 1988;41:249.

Fernandez-Villoria JM, Abad-Morenilla JM. Tissue expansion for thumb and first web space reconstruction. *J Hand Surg Am* 1994;19:663.

Ferrand JY, Watier E, Faivre J, Pailheret JP. Forum for tissue expansion. Value of cutaneous expansion for the covering of synthetic materials used in cranioplasty apropos of a case. *Ann Chir Plast Esthet* 1993;38:23.

Ferguson DJ. Tissue-expanders complicating intubation (letter). *Anaesthesiology* 1985;40:822.

Flageul G, Vial P, Gholam C, et al. Prothèse d'expansion définitive et derniers raffinements en chirurgie reconstructice mammaire. *Ann Chir Plast Esthet* 1989;34:489.

Foo IT, Coleman DJ, Holmes JD, et al. Delay between expansion and expander/implant exchange in breast reconstruction: A prospective study. *Br J Plast Surg* 1992;45:279.

Foo IT, Ward CM. Salvage of an exposed arterial prosthesis by tissue expansion. *Br J Surg* 1989;76:1320.

Forte V, Middleton WG, Briant TD. Expansion of myocutaneous flaps. *Arch Otolaryngol* 1985;111:371.

Foyatier JL, Delay E, Comparin JP, et al. Forum for tissue expansion. Expansion of the scalp: Surgical techniques and clinical applications. *Ann Chir Plast Esthet* 1993;38:11.

Foyatier JL, Comparin JP, Latarjet J, Delay E, Spitalier P, Masson CL. Forum on tissue expansion. Repair of sequelae of facial burns by cervical cutaneous expansion. *Ann Chir Plast Esthet* 1993;38:27.

Frame JW, Rout PGJ, Browne RM. Ridge augmentation using solid and porous hydroxylapatite particles with and without autogenous bone or plaster. *J Oral Maxillofac Surg* 1987; 45:771.

Francis AJ, Marks R. Skin stretching and epidermopoiesis. *Br J Exper Path* 1977;58:35.

Francis GA, Jayarajah M. Bradycardia associated with insertion and inflation of tissue-expanders under the scalp. *Acta Anaesthesiol Scand* 1991;35:366.

Frechet P. Scalp extension. *J Dermatol Surg Oncol* 1993;19:616.

Frisch CE. *Biomaterials in reconstructive surgery.* Mosby, St. Louis, 1983.

Frodel JL, Jr., Whitaker DC. Primary reconstruction of congenital facial lesion defects with tissue expansion. *J Dermatol Surg Oncol* 1993;19:1110.

Frykberg T, Olsen L. Tissue expansion facilitates operation of large myelomeningoceles. *Z Kinderchir* 1990;45:242.

Fudem GM, Orgel MG. Full-thickness erosion of the skull secondary to tissue-expansion for scalp reconstruction (letter). *Plast Reconstr Surg* 1988;82:368.

Fukuta K, Jackson IT, Noreldin AA, Pieper DR. Efficacy of cycled hyperinflation for rapid tissue expansion. *Plast Reconstr Surg* 1993;91:846.

Fung YCB. Elasticity of soft tissues in simple elongation. *Am J Physiol* 1967;213:1532.

Galli A, Berrino P, Santi PL. A systematic aesthetic approach to primary closure of the donor site following transposition of vertical forehead flaps. *Aesth Plast Surg* 1991; 15:245.

Garnier D. La microvascularisation cutanée après expansion tissulaire. Thèse Médecine, Besançon, 1988.

Garnier D, Bachour K, Coste Y, Ricbourg B. Rapid tissue expansion: Value of the measurement of the transcutaneous oxygen pressure and intraprosthetic pressure. *Ann Chir Plast Esthet* 1991;36:223.

Garnier D, Costes Y, Ricbourg B. La microvascularisation cutanée après expansìon tissulaire: Étude expérimentale chez le rat. Proposition d'une méthode de surveillance de l'expansion par la mesure de la pression transcutanée d'oxygène (TcPO$_2$). *Ann Chir Plast Esthet* 1989;34:531.

Gary-Bobo A, Rodriguez L. Breast reconstruction using skin expansion after radiotherapy: Limitations of the procedure apropos of 10 cases. *Ann Chir Plast Esthet* 1989;34:427.

Gary-Bobo A, Rodriguez L. Mammaplasty on breast implants (or cutaneous expansion in treating distended skin). *Ann Chir Plast Esthet* 1989;35:147.

Gemperli R, Cardin V, Schmid RDP, et al. The use of tissue expanders for induction of facial skeletal growth. *J Craniofac Surg* 1991;2:42.

Gemperli R, Ferreira MC, Manders E, et al. Use of tissue expanders on the scalp. *Rev Hosp Clin Fac Med Sao Paulo* 1991;46:112.

Gemperli R, Tuma P, Jr., Bonamichi GT, Ferreira MC. The correction of cryptotia with tissue expanders. *Rev Hosp Clin Fac Med Sao Paulo* 1992;47:261.

Geter RK, Puckett CL. Salvage of infected expanded scalp without loss of flap length. *Plast Reconstr Surg* 1987;80:720.

Ghestem M, Guilbert F, Dhellemmes-Defoort S, et al. Congenital microphthalmos. Orbital and palpebral expansion. *Rev Stomatol Chir Maxillofac* 1991;92:77.

Gibney J. Tissue expansion: Breast reconstruction. *Adv Plast Reconstr Surg* 1984;1:122.

Gibney J. The long term results of tissue expansion for breast reconstruction. *Clin Plast Surg* 1987;14:509.

Gibney J. Use of a permanent tissue expander for breast reconstruction. *Plast Reconstr Surg* 1989;84:607.

Gibney J. Clinical experience with immediate breast reconstruction using tissue expansion or transverse rectus abdominis musculocutaneous flaps. *Ann Plast Surg* 1991; 27:586.

Gibson T. *Biomechanics in plastic surgery*. Glasgow: University of Stathclyde, Pergamon Press, 1964.

Gibson T. The physical properties of skin. In Converse JM (ed.), *Reconstructive plastic surgery*. Philadelphia: W.B. Saunders, 1977, pp. 70–77.

Gibson T. Reconstruction of the tip of the nose and ala by load cycling of the nasal skin and harnessing of extra skin (discussion). *Plast Reconstr Surg* 1986;70:320.

Gibson T, Kenedi RM. Biomechanical properties of skin. *Surg Clin North Am* 1967;47:279.

Gibson T, Kenedi RM, Craik JK. The mobile micro architecture of dermal collagen: A bioengineering study. *Br J Plast Surg* 1965;52:764.

Gilchrest BA. Aging of skin. *Dermatol Gen Med* 1979;11:146.

Giordano P, Courdi A, Gioanni J, et al. Forum on tissue expansion. Tissue expansion and radiotherapy. Changes in epidermal mitotic index induced by expansion and/or irradiation in the rat. *Ann Chir Plast Esthet* 1993;38:62.

Godfrey PM, Godfrey NV, Romita MC, et al. The teres major muscle flap in breast reconstruction. *Ann Plast Surg* 1990; 25:402.

Goin M, Goin J. Growing pains: The psychological experience of breast reconstruction with tissue expansion. *Ann Plast Surg* 1988;21:217.

Goldstein J, Mase C, Newman MH. Fixed membranous bone graft survival after recipient bed alteration. *Plast Reconstr Surg* 1993;91:589.

Goldstein RD, Schuster SH. Methylene blue, a simple adjunct to aid in soft tissue-expansion. *Plast Reconstr Surg* 1986; 77:452.

Gordon DJ, Millar R, Harley JM. The use of a tissue expansion technique in the reconstruction of the female genitalia in bladder extrophy. *Br J Obstet Gynaecol* 1992;99:771.

Gottlieb LJ, Greenwald DP. Quantitative analysis of lidocaine hydrochloride delivery by diffusion across tissue expander membranes (discussion). *Plast Reconstr Surg* 1992; 89:908.

Gouet O, Boureau M, Pradet G, Iselin F. Expansion of the facial integument in children. *Ann Chir Plast Esthet* 1989; 34:421.

Grabb WC. Breast reconstruction after mastectomy using the temporary expander (discussion). *Plast Reconstr Surg* 1982; 69:207.

Grahame M. A method for measuring human skin elasticity in vivo with observations on the effect of age, sex and pregnancy. *Clin Sci* 1970;39:223.

Greenbaum SS. Intraoperative tissue expansion with the Foley catheter. *J Dermatol Surg Oncol* 1993;19:1079.

Greenbaum SS, Burke LA. Intraoperative tissue expansion: A review. *Dermatol Nurs* 1992;4:117.

Greenbaum SS, Greenbaum CH. Intraoperative tissue expansion using a Foley catheter following excision of a basal cell carcinoma. *J Dermatol Surg Oncol* 1990;16:45.

Grigor'eva TG, Nevzorov VP, Nevzorova OF, et al. Ultrastructure of skin flaps obtained in vivo by graded tissue expansion of different intensity. *Klin Khir* 1990;6:42.

Grosshans E, Samsoen M. Histologie de la peau normale: Histochimie et microscopie électronique. *EMC Dermatol* 1977; 12:4.

Grossman JAI, McGraw JB, McDougal HD, et al. Histopathology of human skin expansion. *Plast Surg Forum* 1984;7:95.

Guero SJ, LaCroix PA. An experimental model of scalp expansion: A method of optimizing flap size and shape. *Plast Reconstr Surg* 1991;87:776.

Guida RA, Cohen JI, Cook TA, et al. Assessment of survival and microscopic changes in porcine skin flaps undergoing immediate intraoperative tissue expansion. *Otolaryngol Head Neck Surg* 1993;109:926.

Guillou-Garnier MF, de-Mortillet S, Ballon G. Forum on tissue expansion. Breast reconstruction by expansion in a young woman with sequelae of thoracic amniotic disease. Apropos of a case. *Ann Chir Plast Esthet* 1993;38:48.

Gullestad HP, Bretteville G, Lundar T, Enger E. Tissue expansion for the treatment of myelomeningocele. Case report. *Scand J Plast Reconstr Surg Hand Surg* 1993;27:149.

Haas AJ. Rapid expansion of the maxillary dental arch and nasal cavity by opening the midpalatal suture. *Angle Orthodont* 1961;31:73.

Haas AJ. Palatal expansion: Just the beginning of dentofacial orthopedics. *Am J Orthodont* 1970;57:219.

Haas AJ. Long-term post-treatment evaluation of rapid palatal expansion. *Angle Orthodont* 1980;50:189.

Hagerty MC, Zubowitz VN. Tissue expansion in the treatment of hypertrophic scars and scar contractures. *South Med J* 1986;79:432.

Hall GD, Van Way CW 3rd. A comparison of nerve grafting and tissue expansion techniques in the rat. *Microsurgery* 1994;15:439.

Hall GD, Van-Way CW, Kung FT, Compton-Allen M. Peripheral nerve elongation with tissue expansion techniques. *J Trauma* 1993;34:401.

Hallock GG. Tissue expansion. *Contemp Surg* 1986;29:34.

Hallock GG. Maximum overinflation of tissue expanders. *Plast Reconstr Surg* 1987;80:567.

Hallock GG. Tissue expansion techniques in burn reconstruction. *Am Plast Surg* 1987;18:274.

Hallock GG. Refinement of the radial forearm flap donor site using skin expansion. *Plast Reconstr Surg* 1988;81:21.

Hallock GG. Puncture threshold prior to leakage from tissue expander reservoirs. *Plast Reconstr Surg* 1988;82:666.

Hallock GG. Color duplex imaging for identifying perforators prior to pretransfer expansion of fasciocutaneous free flaps. *Ann Plast Surg* 1994;32:595.

Hallock GG. Safety of clinical overinflation of tissue expanders. *Plast Reconstr Surg* 1995;96:153.

Hallock GG, Rice DC. Objective monitoring for safe tissue expansion. *Plast Reconstr Surg* 1986;77:416.

Hallock GG, Rice DC. Increased sensitivity in objective monitoring of tissue expansion. *Plast Reconstr Surg* 1993;91:217.

Harbon S, Labbe D, Compere JF, et al. Rhinopoïèse: Essai comparatif entre lambeau scalpant de Converse et lambeau médio-frontal expansé. A propos de 16 cas. *Ann Chir Plast Esthet* 1990;35:351.

Harving S, Ravnsbaek J, Johansen LV. Salvage of exposed implants. *Scand J Plast Reconstr Surg Hand Surg* 1989;23:143.

Hata Y, Hosokawa K, Yano K, et al. Correction of congenital microtia using the tissue expander. *Plast Reconstr Surg* 1989;84:741.

He ZW. Hemodynamics in skin and tissue expansion. *Chung Hua Cheng Hsing Shao Shang Wai Ko Tsa Chih* 1992;30:755.

Hemmer KM, Marsh JL, Picker S. Calvarial erosion after scalp expansion. *Ann Plast Surg* 1987;19:454.

Hems TE, Godfrey A. Cosmetic surgery for Cornelia de Lange syndrome. *Br J Plast Surg* 1990;43:489.

Herndl E, Muhlbauer W. Direct closure of donor defects of the radial flap by preliminary stretching of the skin with a skin expander. *Handchir Mikrochir Plast Chir* 1986;18:289.

Hetter GP. Satisfactions and dissatisfactions of patients with augmentation mammaplasty. *Plast Reconstr Surg* 1979; 64:151.

Highet WB, Holmes W. Traction injuries to the lateral popliteal nerve and traction injuries to peripheral nerves after suture. *Br J Surg* 1943;30:212.

Hinojosa RJ. Breast reconstruction through tissue expansion. *Plast Surg Nurs* 1991;11:52.

Hirshowitz B. The two stage face lift. *Br J Plast Surg* 1978; 31:159.

Hirshowitz B, Jackson IT. An attempt to harness the viscoelastic properties of skin in face lift operations. A preliminary report. *Ann Plast Surg* 1987;18:188.

Hirshowitz B, Kaufman T, Ulman J. Reconstruction of the tip of the nose and ala by load cycling of the nasal skin and harnessing of extra skin. *Plast Reconstr Surg* 1986;77:716.

Hirshowitz B, Lindenbaum E, Har-Shai Y. A skin-stretching device for the harnessing of the viscoelastic properties of the skin. *Plast Reconstr Surg* 1993;92:260.

Hobar PC, Rohrich RJ, Byrd HS. Abdominal-wall reconstruction with expanded musculofascial tissue in a posttraumatic defect. *Plast Reconstr Surg* 1994;94:379.

Hochberg J, Ardenghy MF, Gustafson RA, Murray GF. Repair of thoracoabdominal ectopia cordis with myocutaneous flaps and intraoperative tissue expansion. *Plast Reconstr Surg* 1995;95:148.

Hochman M, Branham G, Thomas JR. Relative effects of intraoperative tissue expansion and undermining on wound closing tensions. *Arch Otolaryngol Head Neck Surg* 1992;118:1185.

Hodkinson DJ. Skin graft as an alternative to Marlex mesh in conjunction with tissue expansion. *Plast Reconstr Surg* 1992;90:933.

Hoffman HT, Baker SR. Nasal reconstruction with the rapidly expanded forehead flap. *Laryngoscop* 1989;99:1096.

Hoffman S. Breast reconstruction by larger implants (letter). *Plast Reconstr Surg* 1986;78:695.

Holbrook KA, Wolff K. The structure and development of skin. In Fitzpatrick TB (ed.), *Dermatology in general medicine* New York: McGraw Hill, 1979, p. 93.

Holmes JD. Capsular contracture after breast reconstruction with tissue expansion. *Br J Plast Surg* 1989;42:591.

Holstrom H. Clinical and pathological features of maxillonasal dysplasia (Binder syndrome). Significance of the prenasal fossa or etiology. *Plast Reconstr Surg* 1986;78:559.

Hori S. Thermographic observation of tissue expansion. *Biomed Thermol* 1989;9:151.

Hu H, Sun G, Zhang H, He W, Song W. Full-load expansion technique and preliminary clinical application. *Plast Reconstr Surg* 1994;93:1459.

Hudson DA, Grobbelaar AO. The use of tissue expansion in children with burns of the head and neck. *Burns* 1995; 21:209.

Iconomou TG, Michelow BJ, Zuker RM. The relative risk of tissue expansion in the pediatric patient with burns. *J Burn Care Rehabil* 1993;14:51.

Iconomou TG, Michelow BJ, Zuker RM. Tissue expansion in the pediatric patient. *Ann Plast Surg* 1993;31:134.

Imber G, Shwagen RG, Gutherie RH, et al. Fibrous capsule formation after subcutaneous implantation of synthetic materials in experimentation animals. *Plast Reconstr Surg* 1974;54:183.

Iwahira Y. An investigation of tissue expansion techniques centering around sub-acute rapid tissue expansion. *Jpn J Plast Reconstr Surg* 1988;31:632.

Iwahira Y, Maruyama Y. Expanded preauricular full-thickness free skin graft. *Plast Reconstr Surg* 1991;87:150.

Iwahira Y, Maruyama Y. Combined tissue expansion: Clinical attempt to decrease pain and shorten placement time. *Plast Reconstr Surg* 1993;91:408.

Iwahira Y, Maruyama Y. Expanded unilateral forehead flap (sail flap) for coverage of opposite forehead defect. *Plast Reconstr Surg* 1993;92:1052.

Jackson IT. Expanders with a stylette to position the tubing (letter). *Plast Reconstr Surg* 1989;83:1082.

Jackson IT. Immediate breast reconstruction with prolonged overexpansion using the Becker permanent expander prosthesis. *Eur J Plast Surg* 1992;15:79.

Jackson IT, Grothaus R. Treatment options in hemifacial atrophy. *Eur J Plast Surg* 1986;9:22.

Jackson IT, Sharpe DT, Polley J, et al. Use of external reservoirs in tissue-expansion. *Plast Reconstr Surg* 1987;80: 266.

Jemec GBE, Jemec B, Jemec BIE, et al. The effect of superficial hydration on the mechanical properties of human skin in vivo: Implications for plastic surgery. *Plast Reconstr Surg* 1990;85:100.

Jiang BG, Shibata M, Wang SH. Experimental study of peripheral nerve expansion with a tissue expander. *Chung Hua Cheng Hsing Shao Shang Wai Ko Tsa Chih* 1994;32:93.

Jin YT. Nasal reconstruction using frontal tissue expansion. *Chung Hua Erh Pi Yen Hou Ko Tsa Chih* 1990;23:284.

Jin YT. Scalp expansion in the treatment of cicatricial alopecia. *Chung Hua Cheng Hsing Shao Shang Wai Ko Tsa Chih* 1990;6:22.

Johnson FP, Coppins DA, Brook IM. Fabrication of custom made tissue spacers for use in tissue augmentation. *Br J Oral Maxillofac Surg* 1991;29:282.

Johnson GR, Han P, Giacopelli JA. Tissue expansion as an alternative to skin grafting for closure of skin deficits. *J Am Podiatr Med Assoc* 1992;82:249.

Johnson N, Batchelor A, Lilford RJ. Experience with tissue expansion vaginoplasty. *Br J Obstet Gynaecol* 1991;98:564.

Johnson PE, Kernahan DA, Bauer RS. Dermal and epidermal response to soft tissue expansion in the pig. *Plast Reconstr Surg* 1988;81:390.

Johnson TM, Brown MD, Sullivan MJ, et al. Immediate intraoperative tissue expansion. *J Am Acad Dermatol* 1990; 22:283.

Johnson TM, Lowe L, Brown MD, et al. Histology and physiology of tissue expansion. *J Dermatol Surg Oncol* 1993; 19:1074.

Joss GS, Zoltie N, Chapman P. Tissue expansion technique and the transposition flap. *Br J Plast Surg* 1990;43:328.

Kabaker SS, Kridel RWH, Kraigman ME, et al. Tissue expansion in the treatment of alopecia. *Arch Otolaryngol* 1986; 112:720.

Kabaker SS, Moss RM, Swenson RW. Expanded pedicle scalp flaps for baldness: Is a delay necessary? *Fac Plast Surg* 1988;5:356.

Kane WJ, McCaffrey TV, Wang TD, Koval TM. The effect of tissue expansion on previously irradiated skin. *Arch Otolaryngol Head Neck Surg* 1992;118:419.

Kane WJ, McCaffrey TV, Wang TD, Koval TM. The effect of tissue expansion on the random flap viability and wound tensile strength of previously irradiated rabbit skin. *Arch Otolaryngol Head Neck Surg* 1993;119:417.

Karacaoglan N, Uysal A. Reconstruction of postburn scar contracture of the neck by expanded skin flaps. *Burns* 1994;20:547.

Kawashima T, Yamada A, Hashimoto M. Problems in the use of tissue expanders for scalp reconstruction. *Jpn J Plast Reconstr Surg* 1988;8:513.

Kawashima T, Yamada A, Ueda K, Asato H, Harii K. Tissue expansion in facial reconstruction. *Plast Reconstr Surg* 1994;94:944.

Keller WG, Aron DN, Rakich PM, Crowe DT, Marks MA. Rapid tissue expansion for the development of rotational skin flaps in the distal portion of the hindlimb of dogs: An experimental study. *Vet Surg* 1994;23:31.

Kenedi RM, Gibson T. Etude expérimentale des tensions de la peau dans le corps humain Systéme de mesure des forces et des résultats. *Rev Fr Mec* 1962;4:121.

Kenedi RM, Gibson T, Daly CH. Bioengineering studies of the human skin. In Kenedi RM (ed.), *Biomechanics and related bioengineering topics*. Oxford: Pergamon Press, 1965, p. 147.

Kenna DM, Robotti EB, Bocchiotti G, Blazek J, Grossman JA. Diffusion of antibiotics across tissue expanders: An in vitro study. *Ann Plast Surg* 1994;32:346.

Kenney JG, DiMercurio S, Angel M. Tissue expanded radial forearm free flap in neck burn contracture. *J Burn Care Rehabil* 1990;11:443.

Kent JN, Quinn JH, Zide MF, et al. Alveolar ridge augmentation using non-resorbable hydroxylapatite with or without autogenous cancellous bone. *J Oral Maxillofac Surg* 1983; 41:629.

Kent JN, Finger IM, Quinn JH, et al. Hydroxylapatite alveolar ridge reconstruction: Clinical experiences, complications and technical modifications. *J Oral Maxillofac Surg* 1986;44:37.

Kerrigan GL, Zelt RG, Daniel RK. Secondary critical ischemia time of experimental skin flaps. *Plast Reconstr Surg* 1984; 74:522.

Kessler A, Orenstein A, Kaplan H, et al. Neck contracture as a rare complication of cervical soft tissue expansion. *Burns* 1989;15:392.

Kessler A, Shovron A, Tsur H. Rapid expansion of the scalp using a soft tissue expander. *Harefuah* 1990;118:625–626.

Khouri RK, Ozbek MR, Hruza GJ, Young VL. Facial reconstruction with prefabricated induced expanded (PIE) supraclavicular skin flaps. *Plast Reconstr Surg* 1995;95: 1007.

Kibblewhite D, Larrabee WF, Jr., Sutton D. The effect of relaxin on tissue expansion. *Arch Otolaryngol Head Neck Surg* 1990;118:153.

Kim KH, Hong C, Futrell JW. Histomorphologic changes in expanded skeletal muscle in rats. *Plast Reconstr Surg* 1993; 92:710.

Kissin RW, Kark AE. Post mastectomy breast reconstruction with an inflatable prosthesis. *Br J Surg* 1984;71:802.

Kivanc O, Acarturk S, Kaya E, et al. Ultrastructural and surface-area changes of acute and traditionally expanded full thickness skin grafts. *Br J Plast Surg* 1992;45:239.

Kiyono M, Matsuo K, Fujiwara T, et al. Repair of scalp defects using a tissue expander and Marlex mesh. *Plast Reconstr Surg* 1992;89:349.

Knight KR, McCann JJ, Van der Kolk CA, et al. The redistribution of collagen in expanded pig skin. *Br J Plast Surg* 1990; 43:565.

Kolbusz RV, Bielinski KB. Basal cell carcinoma: Excision with immediate intraoperative tissue expansion. *Cutis* 1990; 46:419.

Kolbusz RV, Bielinski KB. The combined use of immediate intraoperative tissue expansion and meshing technique. *Arch Dermatol* 1993;129:152.

Kostakoglu N, Kecik A, Ozyilmaz F, et al. Expansion of fascial flaps: Histopathologic changes and clinical benefits. *Plast Reconstr Surg* 1993;91:72.

Kreis RW, Mackie DP, Hermans RR, Vloemans AR. Expansion techniques for skin grafts: Comparison between mesh and Meek island (sandwich-) grafts. *Burns* 1994;20:S39.

Kroll SS. Forehead flap nasal reconstruction with tissue expansion and delayed pedicle separation. *Laryngoscope* 1989; 99:448.

Kroll SS. The early management of flap necrosis in breast reconstruction. *Plast Reconstr Surg* 1991;87:893.

Kroll SS, Baldwin B. A comparison of outcomes using three different methods of breast reconstruction. *Plast Reconstr Surg* 1992;90:455.

Krüger E. Alevolar kammaufbau im Unterkiefer mit hydroxylapatite *Dtsch Z Mund Kiefer Gesichtschir* 1985;9:194.

Kuriloff DB, Goldsher M, Blaugrund SM, et al. Controlled laryngoplasty for vocal cord medialization: A technique using tissue expansion. *Laryngoscope* 1990;100:615.

Kuske RR, Schuster R, Klein E, et al. Radiotherapy and breast reconstruction: Clinical results and dosimetry. *Int J Radiat Oncol Biol Physics* 1989;21:339.

Laitung JKG, Batchelor AG. Successful preexpansion of a free scapular flap. *Ann Plast Surg* 1990;25:205.

Laitung JKG, Brough MD, Orton CI. Scalp expansion flaps. *Br J Plast Surg* 1986;39:542.

Lam AC, Nguyen QH, Tahery DP, Cohen BH, Sasaki GH, Moy RL. Decrease in skin-closing tension intraoperatively with suture tension adjustment reel, balloon expansion, and undermining. *J Dermatol Surg Oncol* 1994;20:368.

Lamont E. Expanders for male pattern baldness. *Plast Reconstr Surg* 1986;77:859.

Landsman L, Mandy SH. Adjuncts to scalp reduction surgery intraoperative tissue expanders and hyaluronidase *J Dermatol Surg Oncol* 1991;17:670.

Langer K. On the anatomy and physiology of the skin. *Br J Plast Surg* 1978;31:93.

Langrana NA, Alexander H, Stranoer I, et al. Effect of mechanical load in wound healing. *Ann Plast Surg* 1983; 10:2001.

Lapin R, Daniel D, Hutchins H, et al. Primary breast reconstruction following mastectomy using a skin expander prosthesis. *Breast* 1980;6:20.

Lapin R, Elliott M, Juri H. The use of an integral tissue expander for primary breast reconstruction. *Aesthet Plast Surg* 1985;9:221.

Lari AA. Tissue expansion. *J R Coll Surg Edinb* 1992;37:149.

La Trenta GS, McCarthy JG. Bone graft survival in expanded skin. *Plast Reconstr Surg* 1988;81:406.

Laurence WT. Huber needles for tissue expander inflation. *Plast Reconstr Surg* 1986;78:541.

Laurence WT. Isolated necrosis of deep tissue as a complication of rapid tissue expansion. *Plast Reconstr Surg* 1989; 84:172.

Lee P, Squier CA, Bardach J. Enhancement of tissue expansion by anticontractile agents. *Plast Reconstr Surg* 1985;76:604.

Leighton WD, Johnson ML, Friedland JA. Use of the temporary soft tissue expander in post traumatic alopecia. *Plast Reconstr Surg* 1986;77:737.

Leighton WD, Russell MC, Marcus DE, et al. Experimental expansion of cutaneous and myocutaneous free flap donor sites: Anatomical physiological and histological changes. *Plast Surg Forum* 1986;9:262.

Leighton WD, Russell MC, Marcus DE, et al. Experimental pretransfer expansion of free-flap donor sites: II, Physiology, histology and clinical correlation. *Plast Reconstr Surg* 1988;82:76.

Leonard AG, Small JO. Tissue expansion in the treatment of alopecia. *Br J Plast Surg* 1986;39:42.

Levine RA. Surgical product failures (letter). *Plast Reconstr Surg* 1991;88:916.

Lew D, Amos EA, Unhold GP. An open procedure for placement of a tissue expander over the atrophic alveolar ridge. *J Oral Maxillofac Surg* 1988;46:161.

Lew D, Amos EL, Shroyer JV. The use of a subperiosteal tissue expander in rib reconstruction of an atrophic mandible. *J Oral Maxillofac Surg* 1988;48:229.

Lew D, Clark R, Shahbazian T. Use of a soft tissue expander in alveolar ridge augmentation: A preliminary report. *J Oral Maxillofac Surg* 1986;44:516.

Lew D, Fuseler JW. The effect of stepwise expansion on the mitotic activity and vascularity of subdermal tissue and induced capsule in the rat. *J Oral Maxillofac Surg* 1991;49:848.

Lew D, Fuseler JW. The effect of pulsed expansion of subfascially placed expanders on the extent and duration of mitosis in the capsule and rat integument. *J Oral Maxillofac Surg* 1993;51:154.

Lew D, Hinkle RM, Collins SF. Use of subperiosteal implants with distal filling ports in the correction of atrophic alveolar ridge. *Int J Oral Maxillofac Surg* 1991;20:15.

Lew D, Shroyer JV, Jr., Unhold GP, et al. The use of tissue expanders in the reconstruction of orofacial defects secondary to congenital rubella: Case report. *J Oral Maxillofac Surg* 1989;47:1202.

Le Winn LR, Ruetschi MS. Tissue expansion: A fascinating frontier of reconstructive surgery. *Postgrad Med* 1986; 79:189.

Liang MD, Briggs P, Heckler FR, et al. Presuturing—A new technique for closing large skin defects: Clinical and experimental studies. *Plast Reconstr Surg* 1988;81:694.

Liang MD, Dick GO, Narayana K. Enhancement of tissue expansion using DMSO. *Am Coll Surg Forum* 1987;38:593.

Liang MD, Narayanan K, Kanal E. Magnetic ports in tissue expanders—A caution for MRI. *Magn Res Imaging* 1989; 7:541.

Liang MD, Narayanan K, Ravilochan K, Roche K. The permeability of tissue expanders to bacteria: An experimental study. *Plast Reconstr Surg* 1993;92:1294.

Liu R. A new method of expanding free skin grafts. *Chung Hua Cheng Hsing Shao Shang Wai Ko Tsa Chih* 1992;30:278.

Livingston DH, Sharma PK, Glantz AI. Tissue expanders for abdominal wall reconstruction following severe trauma: Technical note and case reports. *J Trauma* 1992;32:82.

Lo AMK, Colcleugh RG, Allen L, et al. The role of tissue expanders in an anophthalmic animal model. *Plast Reconstr Surg* 1990;86:399.

Logan SE, Hayden JJ. A control unit for maximal-rate continuous tissue expansion (CTE). *Biomed Sci Instrum* 1989;25:27.

Long J, Ma E, Liu F. Cross transposition of expanded scalp flaps for the treatment of postburn cicatricial alopecia. *Chung Hua Cheng Hsing Shao Shang Wai Ko Tsa Chih* 1994; 10:38.

Longacre JJ. Correction of the hypoplastic breast with special reference to reconstruction of the "nipple type breast" with local dermo-fat pedicle flap. *Plast Reconstr Surg* 1957; 14:431.

Lorber M, Milobsky SA. Stretching of the skin in vivo. A method of influencing cell division and migration in the rat epidermis. *J Invest Dermatol* 1968;51:395.

Lu KH. Expanded forehead flap for reconstruction of the nose. *Chung Hua Cheng Hsing Shao Shang Wai Ko Tsa Chih* 1989;5:182.

Luce EA. Tissue expansion for rhytidectomy (letter). *Plast Reconstr Surg* 1990;86:602.

Lyos AT, Stal S, Stafford G. Topical anesthesia in soft-tissue expansion. *Plast Reconstr Surg* 1995;95:940.

MacCann JJ, Mitchell GM, O'Brien B, et al. Comparative viability of expanded and unexpanded axial pattern skin flaps in pigs. *Br J Plast Surg* 1988;41:294.

MacGuire MF. Studies of the excisional wound: Biomechanical effects of undermining and wound orientation on closing tension and work. *Plast Reconstr Surg* 1980;66:419.

Machida BK, Liu-Shindo M, Sasaki GG, et al. Immediate versus chronic tissue expansion. *Ann Plast Surg* 1991;26:227.

MacKay DR, Saggers GC, Kotwal N, et al. Stretching skin: Undermining is more important than intraoperative expansion. *Plast Reconstr Surg* 1990;86:722.

MacKensie IC. The effects of frictional stimulation on mouse ear epidermis: Cell proliferation. *J Invest Dermatol* 1974; 62:80.

MacKinnon SE, Gruss JS. Soft tissue expansion in upper limb surgery. *J Hand Surg* 1987;12b:73.

MacKinnon SE, Lee Dellon AL. Soft tissue expanders in hand reconstruction. *J Hand Surg* 1987;12b:79.

MacMillan RW, Arias JD, Stayman JW III. Management of radiation necrosis of the chest wall following mastectomy: A new treatment option. *Plast Reconstr Surg* 1986;77:832.

Madazimov MM, Grishkevich VM, Moroz VIU, et al. Treatment of cicatricial alopecia by the method of tissue expansion. *Khirurgiia Mosk* 1990;9:81.

Madazimov MM, Moroz VIU, Grishkevich VM, et al. Elimination of cicatricial deformities, defects of the skin by the method of soft tissue expansion. *Sov Med* 1990;5:97.

Madison JB. Tissue expansion. *Vet Clin North Am Equine Pract* 1989;5:633.

Magalon G. L'expansion tissulaire. *Rev Prat* 1990;40:2466.

Magalon G, Aubert JP. Apport de l'expansion dans le traitement des tumeurs malignes cutanées. *Cancer Comm* 1988;2:413.

Magalon G, Aubert JP, Vasse D. Expansion tissulaire au membre supérieur. In Gilbert A, Masquelet AC, Hentz RV (eds.), *Les lambeaux artériels pédiculés du membre supérieur*. Expansion Scientifique Française, 1990, pp. 171–176.

Mahomed N, McKee N, Solomon P, Lahoda L, Gross AE. Soft-tissue expansion before total knee arthroplasty in arthrodesed joints. A report of two cases. *J Bone Joint Surg Br* 1994; 76:88.

Maladry D, Berard V, Dupuis P, Mitz V, Lemerle JP. Scalp reconstruction after avulsion: Emergency microsurgery and secondary tissue expansion. Apropos of 3 cases caused by farming machines. *Ann Chir Plast Esthet* 1994;39:169.

Maladry D, Brabant B, Berard V, Dupuis P, Mitz V. Secondary expansion of a totally replanted scalp for aesthetic adjustment. *Plast Reconstr Surg* 1994;94:1052.

Malata CM, Williams NW, Sharpe DT. Tissue expansion: An overview. *J Wound Care* 1995;4:37.

Malata CM, Williams NW, Sharpe DT. Tissue expansion: Clinical applications. *J Wound Care* 1995;4:88.

Malcom MP. Primary breast reconstruction with a subpectoral silicone tissue expander. *Ann Plast Surg* 1985;15:404.

Malis DJ, MacMillan JG, Kelley JL, McGath JH. Tissue expansion of the facial nerve in an animal model. *Ear Nose Throat J* 1995;74:261.

Man D. Stretching and tissue expansion for rhytidectomy: An improved approach. *Plast Reconstr Surg* 1989;84:561.

Man D. Stretching and tissue expansion for face lift. *Fac Plast Surg* 1992;8:52.

Manders EK. Nasal reconstruction by expansion in the patient with cancer of the head and neck. *Surg Clin North Am* 1986;66:1.

Manders EK. Reconstruction by expansion in the patient with cancer of the head and neck. *Surg Clin North Am* 1986; 66:201.

Manders EK. Used tissue expanders for developing countries: You can help. *Plast Reconstr Surg* 1987;80:643.

Manders EK. Soft tissue expansion in the lower extremities. *Plast Reconstr Surg* 1988;81:208.

Manders EK, Graham WP. Alopecia reduction by scalp expansion. *J Dermatol Surg Oncol* 1984;10:967.

Manders EK, Graham WP, Schenden MJ, et al. Skin expansion to eliminate large scalp defects. *Ann Plast Surg* 1984;12:305.

Manders EK, MacKay DR, Saggers GC. Tissue expansion for rhytidectomy. *Plast Reconstr Surg* 1990;85:995.

Manders EK, Saggers G, Diaz P, et al. Elongation of peripheral nerve and viscera containing smooth muscle. *Clin Plast Surg* 1987;14:551.

Manders EK, Schenden MJ, Furrey JA. Soft tissue expansion: Concept and complications. *Plast Reconstr Surg* 1984;74:493.

Mandy SH. Intraoperative expander-assisted scalp reduction. *J Dermatol Surg Oncol* 1993;19:1117.

Manger G. Coping with soft-tissue expansion. *Can Nurse* 1990;86:21.

Marchac D, Cophignon J, Achard E, et al. Orbital expansion for anophthalmia and microorbitism. *Plast Reconstr Surg* 1977;59:486.

Marchac D, Laregue M. L'expansion cutanée par implant gonflable. *Ann Dermatol Venerol* 1986;113:329.

Marconi F, Messina P, Pavanello P, De Castro R. Cosmetic reconstruction of the mons veneris and lower abdominal wall by skin expansion as the last stage of the surgical treatment of bladder exstrophy: A report of three cases. *Plast Reconstr Surg* 1993;91:551.

Marcus J, Esterly NB, Bauer BS. Tissue expansion in a patient with extensive nevus comedonicus. *Ann Plast Surg* 1992; 29:362.

Marcus J, Horan DB, Robinson JK. Tissue expansion: Past, present, and future. *J Am Acad Dermatol* 1990;23:813.

Marion RB. Expanders for male pattern baldness. *Plast Reconstr Surg* 1985;76:797.

Marks MW. *Mechanical properties of the skin. Biochemistry and physiology of the skin.* Oxford: Oxford University Press, 1983.

Marks MW. The use of soft tissue expansion in reconstructive surgery. *J La State Med Soc* 1984;136:40.

Marks MW, Argenta LC. Skin expansion in reconstructive surgery. *Fac Plast Surg* 1988;5:311.

Marks MW, Argenta LC, Thornton JW. Burn management: The role of tissue expansion. *Clin Plast Surg* 1987;14:543.

Marks MW, Argenta LC, Thornton JW. Rapid expansion: Experimental and clinical experience. *Clin Plast Surg* 1987; 14:455.

Marks MW, Burney RE, MacKenzie JR, et al. Enhanced capillary blood flow in rapidly expanded random pattern flaps. *J Trauma* 1986;26:913.

Marks MW, MacKenzie JR, Burney RE, et al. Response of random skin flaps to rapid expansion. *J Trauma* 1985;25:947.

Marks MW, Payne PA. Bioengineering and the skin. New York: MTP Press, 1981.

Masser MR. A twin tissue expander used in the elimination of alopecia. *Plast Reconstr Surg* 1988;81:444.

Masser MR. The preexpanded radial free flap. *Plast Reconstr Surg* 1990;86:295.

Masser MR. Tissue-expansion: A reconstructive revolution or a cornucopia of complications? *Br J Plast Surg* 1990;43:344.

Masser MR. The preexpanded radial free flap (reply). *Plast Reconstr Surg* 1991;88:171.

Masson CL, Foyatier JL, Dessapt B, Comparin JP. Rapid tissue expansion. *Ann Chir Plast Esthet* 1989;34:521.

Matarasso SL, Matarasso A, Glogau RG. Permanent tissue expansion. *J Dermatol Surg Oncol* 1993;19:1103.

Matev IB. Thumb reconstruction after amputation at the metacarpophalangeal joint by bone lengthening. *J Bone Joint Surg* 1970;52B:957.

Matt BH, Squier CA, Kelly KM, Bardach J. Enhancement of expansion of Guinea pig skin by local delivery of an anticontractile agent using a new bilumen expander. *Ann Plast Surg* 1990;24:335.

Matthews RN, Missoten FEM. Early tissue expansion to close a traumatic defect of scalp and pericranium. *Br J Plast Surg* 1986;39:417.

Matton GE, Tonnard PL, Monstrey SJ, Van-Landuyt KH. A universal incision for tissue expander insertion. *Br J Plast Surg* 1995;48:172.

Matturri L, Azzolini A, Riberti C, et al. Long-term histopathologic evaluation of human expanded skin. *Plast Reconstr Surg* 1992;90:636.

Matturri L, Campiglio GL, Lavezzi AM, et al. Cell kinetics and DNA content (ploidy) of human skin under expansion. *Eur J Basic Appl Histochem* 1991;35:73.

Maxwell GP. Immediate breast reconstruction utilizing a textured silicone tissue expander. *Plast Surg Forum* 1989; 12:146.

Maxwell GP. Breast reconstruction utilizing subcutaneous tissue expansion followed by polyurethane-covered silicone implants: A 6-year experience (discussion). *Plast Reconstr Surg* 1991;88:640.

Maxwell GP, Falcone PA. Eighty-four consecutive breast reconstructions using a textured silicone tissue expander. *Plast Reconstr Surg* 1992;89:1022.

May JW, Attwood J, Bartlett S. Staged use of soft tissue expansion and lower thoracic advancement flap in breast reconstruction. *Plast Reconstr Surg* 1987;79:272.

Mayou BJ, Gault DT, Crock JG. Tissue expanded free flaps. *Br J Plast Surg* 1992;45:413.

McCafferty LR, Howard PS. Minimal expansion for correction of male pattern baldness. *Ann Plast Surg* 1995;34:379.

McCauley RL, Oliphant JR, Robson MC. Tissue expansion in the correction of burn alopecia: Classification and methods of correction. *Ann Plast Surg* 1990;25:103.

McGuire JJ, Caffee HH. In vivo diffusion of lidocaine through tissue expanders. *Plast Reconstr Surg* 1992;89:675.

McGrath MH, Burkhardt BR. The safety and efficacy of breast implants for augmentation mammaplasty. *Plast Reconstr Surg* 1984;74:550.

McIvor NP, Fong MW, Berger KJ, Freeman JL. Use of tissue expansion in head and neck reconstruction. *J Otolaryngol* 1994;23:46.

McKinney P, Edelson R, Terrasse A, et al. Chest wall deformity following soft tissue-expansion for breast reconstruction. *Plast Reconstr Surg* 1987;80:442.

Mehlisch DR, Taylor TD, Leibold DG, et al. Evaluation of collagen/hydroxylapatite for augmenting deficient alveolar ridges: A preliminary report. *J Oral Maxillofac Surg* 1987; 45:408.

Middleton WG, Ellis DA, Trimas SJ. Expanded pectoralis major myocutaneous flaps in head and neck surgery. *J Otolaryngol* 1995;24:42.

Miller LB, Kanter M, Wolfort F. Treatment of webbed neck in Turner's syndrome with tissue expansion. *Ann Plast Surg* 1990;24:447.

Miller SH. Soft-tissue expansion in the lower extremities (discussion). *Plast Reconstr Surg* 1987;81:218.

Miller SH. In vivo diffusion of lidocaine through tissue expanders (discussion). *Plast Reconstr Surg* 1992;89:678.

Milner RH. The effect of tissue expansion on peripheral nerves. *Br J Plast Surg* 1989;42:414.

Milner RH, Wilkins PR. The recovery of peripheral nerves following tissue expansion. *J Hand Surg Br* 1992;17:78.

Mimoun M, Baux S, Kirsch JM, et al. Apport des prothèses d'expansion cutanée à la réalisation des lambeaux de cuir chevelu. Application au lambeau de Dufourmentel. *Ann Chir Plast Esthet* 1987;32:386.

Min Z, Svensson H, Svedman P. On expander pressure and skin blood flow during tissue expansion in the pig. *Ann Plast Surg* 1988;21:134.

Moelleken BRW, Mathes SJ, Cann CE, et al. Long-term effects of tissue expansion on cranial and skeletal bone development in neonatal miniature swine: Clinical findings and histomorphometric correlates. *Plast Reconstr Surg* 1990;86:825.

Moghari A, Emami A, Sheen R, O'Brien BM. Lower limb reconstruction in children using expanded free flaps. *Br J Plast Surg* 1989;42:649.

Mohammed A, Rahamatalla A, Wynne-Jones CH. Tissue expansion in late repair of tendo Achillis rupture. *J Bone Joint Surg Br* 1995;77:64.

Monti M, Vergine M, Lai G, et al. Direct post-mastectomy reconstruction with skin expansion. *Minerva Chir* 1989; 44:2321.

Moore MH, Trott JA, David DJ. Soft tissue expansion in the management of the rare craniofacial clefts. *Br J Plast Surg* 1992;45:155.

Morgan RF, Edgerton MT. Tissue expansion in reconstructive hand surgery: Case report. *J Hand Surg* 1985;10a:754.

Moroz VIU, Grishkevich VM, Madazimov MM, Ganzha PF. A method of expansion of soft tissues in plastic and reconstruction surgery. *Khirurgiia Mosk* 1989;11:140.

Moroz VI, Grishkevich VM, Vaganova NA. Skin expansion in the reconstructive and plastic surgery of burns. *Vestn Ross Akad Med Nauk* 1994:37.

Morris SF, Pang CY, Mahoney J, et al. Effect of capsulectomy on the hemodynamics and viability of random-pattern skin flaps raised on expanded skin in the pig. *Plast Reconstr Surg* 1989;84:314.

Morselli PG, Marconi F, Pistorale T, Cavina C. Tissue expansion in head and neck regions. *Acta Otorhinolaryngol Ital* 1994;14:575.

Morykwas MJ, Marks MW, Argenta LC. Surface area and tissue volume increases with differential expansion. *Ann Plast Surg* 1992;28:311.

Mottaleb M, Wong RKM, Manders EK, et al. Tissue expansion. In Riley WB (ed.), *Instructional Courses, Vol 1: Plastic Surgery Educational Foundation*. St. Louis: Mosby, 1988, p. 276.

Muenker R. Nasal reconstruction with tissue expansion. *Fac Plast Surg* 1988;5:328.

Muenker R. Various devices available for tissue expansion and clinical experience. *Fac Plast Surg* 1988;5:291.

Murray JM, Cleal JF. Early tissue response to rapid maxillary expansion in the midpalatal suture for the rhesus monkey. *J Dental Res* 1971;50:1654.

Mustoe TA, Bartell TH, Garner WL. Physical biomechanical histologic and biochemical effects of rapid versus conventional tissue expansion. *Plast Reconstr Surg* 1989; 83:687.

Mustoe TA, Gifford GH, Lach E. Rapid tissue expansion in the treatment of myelomeningocele. *Ann Plast Surg* 1988;21:70.

Mutaf M, Celebioglu S, Sensoz O. Growth potential of autogenous cartilage grafts in expanded skin: An experimental study in rabbits. *Ann Plast Surg* 1994;32:126.

Mutaf M, Ustuner ET, Celebioglu S, Kocer U, Sensoz O. Tissue expansion-assisted prefabrication of the forehead flap for nasal reconstruction. *Ann Plast Surg* 1995;34:478.

Mutimer KL, Mulliken JB. Correction of cryptotia using tissue expansion. *Plast Reconstr Surg* 1988;81:601.

Myers B, Donovan W. An evaluation of eight methods of using fluorescein to predict the viability of skin flaps in the pig. *Plast Reconstr Surg* 1984;75:245.

Nahai F, McCain L. A method for the exteriorization of tissue expander tubing. *Plast Reconstr Surg* 1988;82:723.

Namba K, Kaji H. On the use of a free full-thickness skin graft stretched by tissue expander. *Jpn J Plast Reconstr Surg* 1988;31:601.

Nataf J. Scalp expansion for male pattern baldness (letter). *Plast Reconstr Surg* 1988;81:999.

Neale HW, High RM, Billmire DA, et al. Complications of controlled tissue expansion in the pediatric burn patient. *Plast Reconstr Surg* 1988;82:840.

Neale HW, Kurtzman L. Postburn breast reconstruction: Tissue expansion prior to contracture release (discussion). *Plast Reconstr Surg* 1992;90:672.

Neale HW, Kurtzman LC, Goh KB, et al. Tissue expanders in the lower face and anterior neck in pediatric burn patients: limitations and pitfalls. *Plast Reconstr Surg* 1993;91:624.

Neligan PC, Peters WJ. Advances in burn scar reconstruction: The use of tissue expansion. *Ann Plast Surg* 1989;22:203.

Nelson CV, Miller SF, Eppinger M, Finley RK. Micrografts:II, Evaluation of 25:1, 50:1, and 100:1 expansion skin grafts in the porcine model. *J Burn Care Rehabil* 1995;16:31.

Netscher DT, Spira M, Peterson R. Adjunctive agents to facilitate rapid tissue expansion. *Ann Plast Surg* 1989;23:412.

Neumann CG. The expansion of an area of skin by progressive distension of a subcutaneous balloon. *Plast Reconstr Surg* 1957;19:124.

Neves RI, Kahler SH, Banducci DR, Manders EK. Tissue expansion of sensate skin for pressure sores. *Ann Plast Surg* 1992;29:433.

Nicoletis C, Gholam D. Frontal stage in cervicofacial lift: The "helmet crest" technique associated with rapid expansion. *Ann Chir Plast Esthet* 1992;37:27.

Niranjan NS. Webbing of the neck: Correction by tissue expansion. *Plast Reconstr Surg* 1989;84:985.

Noone RB, Frazier TG, Hayward CZ, et al. Patient acceptance of immediate reconstruction following mastectomy. *Plast Reconstr Surg* 1982;67:632.

Nordström REA. Scalp kinetics in multiple excision for correction of male pattern baldness. *J Dermatol Surg Oncol* 1984;10:991.

Nordström REA. Stretch back in scalp reductions for male pattern baldness. *Plast Reconstr Surg* 1984;73:422.

Nordström REA. Antibiotics in the tissue expander to decrease the rate of infection. *Plast Reconstr Surg* 1988;81:137.

Nordström REA. Tissue expansion and flaps for surgical correction of male pattern baldness. *Br J Plast Surg* 1988;41:154.

Nordström REA. Tissue expansion and flaps for surgical correction of male pattern baldness. *Fac Plast Surg* 1988;5:347.

Nordström REA. Continuous versus conventional tissue expansion (letter). *Plast Reconstr Surg* 1991;88:1108.

Nordström REA. Vevsekspansjon. *Tidsskr Norske Laegeforen* 1992;112:1832.

Nordström REA, Devine JW. Scalp stretching with a tissue expander for closure of scalp defects. *Plast Reconstr Surg* 1985;75:578.

Nordström REA, Pietila JP, Rintala AC. Our clinical experience of tissue expansion. *Fac Plast Surg* 1988;5:317.

Nordström REA, Pietila JP, Voutilainen PE, et al. Tissue expander injection dome leakage. *Plast Reconstr Surg* 1988; 81:26.

Nordström REA, Salo AP, Rintala AC. Auricle reconstruction with the help of tissue expansion. *Fac Plast Surg* 1988;5:338.

Nosaki M. Soft tissue expansion: Procedures and complications. *Jpn J Plast Reconstr Surg* 1988;31:641.

O'Connel HB, Chalfamian T. Use of transillumination in tissue expansion. *Plast Reconstr Surg* 1989;83:574.

Ogawa Y, Kasai K, Doi H, et al. The preoperative use of extratissue expander for syndactyly. *Ann Plast Surg* 1989;23:552.

Ohana J. Utilisation du dispositif d'expansion tissulaire de Radovan en chirurgie reconstructive. *Ann Chir Plast Esthet* 1986;31:86.

Olenius M, Dalsgaard CJ, Wickman M. Mitotic activity in expanded human skin. *Plast Reconstr Surg* 1993;91:213.

Olenius M, Forslind B, Johansson O. Morphometric evaluation of collagen fibril dimensions in expanded human breast skin. *Int J Biol Macromol* 1991;13:162.

Olenius M, Jurell G. Breast reconstruction using tissue expansion. *Scand J Plast Reconstr Surg Hand Surg* 1992;26:83.

Olenius M, Wickman M, Malm M, Jurell G. Skin thickness in expanded human breast skin. *Plast Reconstr Surg* 1994; 93:1428.

Olenius M, Johansson O. Capsular contracture after breast reconstruction with the tissue expansion technique. *Scand J Plast Reconstr Surg Hand Surg* 1995;29:15.

Olshansky K. Tissue expansion in conjunction with rhytidectomy. *Plast Reconstr Surg* 1990;85:828.

Oneal RM, Rohrich RJ, Izenberg PH. Skin expansion as an adjunct to reconstruction of the external ear. *Br J Plast Surg* 1984;37:517.

Ordman LJ, Gillman T. Studies in healing of cutaneous wounds. *Arch Surg* 1966;93:857.

Ozgur FF, Kocabalkan O, Gursu KG. Tissue expansion in median facial cleft reconstruction: A case report. *Int J Oral Maxillofac Surg* 1994;23:137.

Ozgur F, Ozcan G. Scrotal construction by expansion of labia majora in biological female transsexuals. *Ann Plast Surg* 1995;34:451.

Ozun G. A dirigible balloon in surgery: A tissue expansion. *Bull Acad Natl Chir Dent* 1987;33:63.

Ozun G. Traitement chirurgical de la calvitie aprés expansion du cuir chevelu. *Concours Med* 1988;12:3390.

Ozun G. Forum on tissue expansion. Conclusion. How to master expansion. *Ann Chir Plast Esthet* 1993;38:75.

Paletta CE, Bass J, Shehadi SI. Outer table skull erosion causing rupture of scalp expander. *Ann Plast Surg* 1989;23:538.

Paletta CE, Campbell E, Shehadi SI. Tissue expanders in children. *J Pediatr Surg* 1991;26:22.

Palmer JH. With reference to the caulking gun syringe (letter). *Plast Reconstr Surg* 1989;83:1075.

Panje WR, Dobleman TJ. Facial reconstruction for radiation-induced skin cancer. *Arch Otolaryngol Head Neck Surg* 1990;116:470.

Pasyk KA, Argenta LC, Austad ED. Histopathology of human expanded tissue. *Clin Plast Surg* 1987;14:435.

Pasyk KA, Argenta LC, Hassett C. Quantitative analysis of the thickness of human skin and subcutaneous tissue following controlled expansion with a silicone implant. *Plast Reconstr Surg* 1988;81:516.

Pasyk KA, Austad ED, Cherry GW. Intracellular collagen fibers in the capsule around silicone expanders in guinea pigs. *J Surg Res* 1984;36:125.

Pasyk KA, Austad ED, McClatchey KD, et al. Electron microscopic evaluation of guinea pig skin and soft tissues "expanded" with self-inflating silicone implant. *Plast Reconstr Surg* 1982;70:37.

Patel PK. Estimating the tissue expander volume: A poor man's recipe. *Plast Reconstr Surg* 1986;78:426.

Patterson TJS, Chir M. The survival of skin flaps in the pig. *Br J Plast Surg* 1968;21:113.

Paul MD. Primary breast reconstruction with a subpectoral silicone tissue expander. *Ann Plast Surg* 1985;15:404.

Paulhe P, Aubert JP, Magalon G. Forum on tissue expansion. Are tissue expansion and radiotherapy compatible? Apropos of a series of 50 consecutive breast reconstructions. *Ann Chir Plast Esthet* 1993;38:54.

Payement G, Cariou JL, Arcila M, Arrouvel C, Banzet P. Arterial phenomenon of autonomization of vascularized island flaps reviewed by expansion of the vascular pedicle. An experimental study in the rat. *Ann Chir Plast Esthet* 1994; 39:779.

Pearl RM. Pathophysiology of skin flaps raised on expanded pig skin. *Plast Reconstr Surg* 1984;74:66.

Penneys NS, Rademaker B, Jackson IT, et al. Loss of factor XIIIa in pig dermis during tissue expansion. *J Dermatol Sci* 1991;2:62.

Pennisi VR. Making a definitive inframammary fold under a reconstructed breast. *Plast Reconstr Surg* 1977;60:523.

Penoff J. Skin expansion—a sword that "stretches" two ways: Scalp expansion and bone erosion. *J Craniofac Surg* 1990; 1:103.

Persing JA, Babler WJ, Nagorsky MJ, et al. Skull expansion in experimental craniosynostosis. *Plast Reconstr Surg* 1986; 78:594.

Persing JA, Gampper TJ, Morgan E, Wolcott P. Experimental expansion of the maxillary sinus. *J Craniofac Surg* 1994;5:11.

Persing JA, Morgan EP, Cronin AJ, et al. Skull base expansion: Craniofacial effects. *Plast Reconstr Surg* 1991;87:1028.

Persoff MM. Achieving symmetry in the tissue-expanded breast reconstruction. *Aesthet Plast Surg* 1991;15:133.

Persoff MM. Expansion-augmentation of the breast. *Plast Reconstr Surg* 1993;91:393.

Persson BM, Broome A. Lengthening a short femoral amputation stump. A case of tissue expander and endoprosthesis. *Acta Orthop Scand* 1994;65:99.

Pestalardo CM, Cordero A Jr., Ansorena JM, Bestue M, Martinho A. Acne keloidalis nuchae. Tissue expansion treatment. *Dermatol Surg* 1995;21:723.

Pham H. The use of an open splint in ridge augmentation with hydroxylapatite. *J Oral Maxillofac Surg* 1986;44:80.

Pierce HE. Possible use of the Radovan tissue expander in hair replacement surgery. *J Dermatol Surg Oncol* 1985; 11:413.

Pietila JP. Tissue expansion and skin circulation. Simultaneous monitoring by laser Doppler flowmetry and transcutaneous oximetry. *Scand J Plast Reconstr Surg Hand Surg* 1990;24:135.

Pietila JP, Nordström REA, Numminen MK, et al. Suitability of the laser doppler flowmetry for measuring the effect of the tissue expander pressure on skin circulation. *Scand J Plast Reconstr Surg Hand Surg* 1988;22:135.

Pietila JP, Nordström REA, Virkkunen PJ, et al. Accelerated tissue-expansion with the "overfilling" technique. *Plast Reconstr Surg* 1988;81:204.

Pitanguy I. Tissue expansion and multiple partial resection. *Rev Bras Cirurg* 1987;87:41.

Propper RH. A technique for controlled placement of hydroxylapatite over atrophic mandibular ridges. *J Oral Maxillofac Surg* 1985;45:469.

Quaba A. Reconstruction of a post-traumatic ear defect using tissue expansion: 30 years after Neumann. *Plast Reconstr Surg* 1988;82:521.

Quayle AA, Marouf H, Holland I. Alveolar ridge augmentation using a new design of inflatable tissue expander: Surgical technique and preliminary results. *Br J Oral Maxillofac Surg* 1990;28:375.

Radovan C. Reconstruction of the breast after radical mastectomy using a temporary expander. *Plast Surg Forum* 1978; 1:41.

Radovan C. Development of adjacent flaps using a temporary expander. *Plast Surg Forum* 1979;2:62.

Radovan C. Advantages and complications of breast reconstruction using a temporary expander. *Plast Surg Forum* 1980;3:63.

Radovan C. Breast reconstruction after mastectomy using a temporary expander. *Plast Reconstr Surg* 1982;69:195.

Radovan C. A self-inflating tissue-expander (discussion). *Plast Reconstr Surg* 1982;70:594.

Radovan C. Tissue expansion in soft tissue reconstruction. *Plast Reconstr Surg* 1984;74:482.

Ramasastry SS, Weinstein LW, Zerbe A, et al. Regression of local and distant tumor growth by tissue expansion: An experimental study of mammary carcinoma in rats. *Plast Reconstr Surg* 1991;87:1.

Ramaswamy CN. Tissue expander for developing countries. *Plast Reconstr Surg* 1988;82:553.

Ramirez OM, Orlando JC. Chest-wall deformity during tissue expansion in breast reconstruction. *Plast Surg Forum* 1986;9:148.

Ramon Y, Moscona R. Tissue expanders a new reconstructive aid. *Harefuah* 1990;118:658.

Ramstad T, Bretteville G. Nasal stenosis after operations on the nose: Expansion and subsequent maintenance of the nasal airway. Case report. *Scand J Plast Reconstr Surg Hand Surg* 1994;28:235.

Reali UM, Chiarugi C, De Siena GM, Giannotti V. Sonographic evaluation of dermis and subcutaneous tissue during and after skin expansion. *Plast Reconstr Surg* 1994; 93:1050.

Rees TD. *Aesthetic plastic surgery*. Philadelphia: W.B. Saunders, 1980.

Rees TD, Aston S. The tuberous breast. *Clin Plast Surg* 1973; 3:339.

Rees TD, Guy CL, Coburn RJ. The use of inflatable breast implants. *Plast Reconstr Surg* 1973;52:609.

Rees RS, Nanney LB, Fleming P, et al. Tissue expansion: Its role in traumatic below knee amputations. *Plast Reconstr Surg* 1986;77:133.

Regnault P, Baker TJ, Gleason MC, et al. Clinical trial and evaluation of proposed new inflatable mammary prosthesis. *Plast Reconstr Surg* 1972;50:220.

Reifen E, Freeman JL. The use of tissue expansion in nasal reconstruction. *J Otolaryngol* 1991;20:5.

Remmler D, McCoy FJ, O'Neil D, et al. Osseous expansion of the cranial vault. *Plast Reconstr Surg* 1992;89:787.

Reynolds J. Tissue expansion. *Br J Nurs* 1992;1:134.

Riberti C, Azzolini C, Parodi PC. Scrotal reconstruction by cutaneous expansion. *Arch Esp Urol* 1994;47:453.

Ricbourg B. Utilisation de colle de fibrine après mise en place d'une protèsse d'expansion. *Ann Chir Plast Esthet* 1990; 35:317.

Ricbourg B, Le-Quang C. Forum on tissue expansion in practice: Plastic surgery in full expansion. *Ann Chir Plast Esthet* 1993;38:85.

Richard L. L'expansion tissularie en chirurgie réparatrice. Thèse Médecine, Lyon, 1987.

Richard ME. Preparation of the child for the tissue expansion: A developmental approach. *Plast Surg Nurs* 1990;10: 157–160.

Richard-Kadio M, Dick R, Keli E, et al. Skin expansion in plastic surgery. *Med Trop Mars* 1990;50:293.

Ridge MD, Wright V. The directional effect of skin. A bioengineering study of skin with particular reference to Langer's lines. *J Invest Dermatol* 1966;46:341.

Rigdon RH. Plasties and carcinogenesis. *South Med J* 1974; 67:459.

Roaf LD. Giraffe-necked women. A myth exploded. *J Bone Joint Surg* 1961;43b:114.

Roberts AHN. The correction of scalp hair defects by tissue expansion. *Ann R Coll Surg Eng* 1986;68:89.

Robin TW, Yuan MD. The use of tissue expansion in lower extremity wounds in paraplegic patients. *Plast Reconstr Surg* 1989;83:892.

Robinson JK. Tissue expansion. A new way to achieve an old goal in dermatologic surgery (editorial). *J Dermatol Surg Oncol* 1993;19:1063.

Robotti E, Grossman JAI, Verna G, et al. Lembo frontale espanso per la ricostruzione del naso. *Riv Ital Chir Plast* 1990;22:347.

Roeder JA, White SK. Tissue expansion for the treatment of keloids. *Plast Surg Nurs* 1990;10:114.

Roenigk RK, Wheeland RG. Tissue expansion in cicatricial alopecia. *Arch Dermatol* 1987;123:641.

Rokitskii MR, Akhunzianov AA, Naneishvili GT. Instruments for measuring the degree of tissue tension. *Klin Khir* 1991;59:241.

Romo T, Goldberg J. Versatile use of skin expanders in facial plastic surgery. *Arch Otolaryngol Head Neck Surg* 1992; 118:333.

Romo T 3rd, Jablonski RD, Shapiro AL, McCormick SA. Long-term nasal mucosal tissue expansion use in repair of large nasoseptal perforations. *Arch Otolaryngol Head Neck Surg* 1995;121:327.

Rosen PB, Jabs AD, Kister SJ, et al. Clinical experience with immediate breast reconstruction using tissue expansion or transverse rectus abdominis musculocutaneous flaps. *Ann Plast Surg* 1990;25:249.

Rosen HM. Facial skeletal expansion: Treatment strategies and rationale. *Plast Reconstr Surg* 1992;89:798.

Roth AG, Dado DV. Venous air embolism during removal of tissue expander in a supine child. *Anaesthesiology* 1988; 68:457.

Rotondo M, Parlato C, Iaccarino C, Scrocca A. Tissue expansion in neurosurgical reconstructive technique: Case report. *Surg Neurol* 1995;43:201.

Rowell AR, Godfrey AM, Richards MA. The thinned latissimus dorsi free flap: A case report. *Br J Plast Surg* 1986; 39:210.

Rubin G, Lachman CJ, Scheinhaus A. Controlled tissue expansion for correcting amputation stump scar. *Orthop Rev* 1987;16:331.

Ruiz-Razura A, Branfman GS, Cohen BE. Tissue expanders in microvascular surgery: Acute intraoperative arterial elongation *Surg Forum* 1989;610:72.

Russel FE, Simmers MH, Hirst AE. Tumors associated with embedded polymers. *J Natl Cancer Res* 1959;23:305.

Russell IS, Collins JP, Holmes AD, et al. The use of tissue expansion for immediate breast reconstruction after mastectomy. *Med J Aust* 1990;152:632.

Russell RC, Khouri RK, Upton J, Jones TR, Bush K, Lantieri LA. The expanded scapular flap. *Plast Reconstr Surg* 1995;96:884.

Ryan JJ. Discussion about "stage use of soft tissue expansion and lower thoracic advancement flap in breast reconstruction." *Plast Reconstr Surg* 1979;79:276.

Ryan JJ. A lower thoracic advancement flap in breast reconstruction after mastectomy. *Plast Reconstr Surg* 1982;70:153.

Sadove AM, Eppley BL. Instrument technique in scalp dissection for tissue expander placement. *Aesthet Plast Surg* 1993;17:257.

Samayoa V, Valiente E, Serra-Renom JM. Estudio experimental de los cambios histopatologicos y de la formacion capsular durante expansion tisular progresiva. *Rev Senol Patol Mamm* 1988;1:3.

Samayoa V, Valiente E, Serra-Renom JM. Estudio experimental de la formacion capsular después de la sobre expansion y la posterior colocacion de protesis mamaria de menor tamano. *Rev Senol Patol Mamm* 1988;1:9.

Saouma S, Bricout N, Servant JM, et al. Technique de l'espansion tissulaire. *J Chir* 1989;126:34.

Sasaki GH. Tissue expansion: Extremity soft tissue reconstruction. *Adv Plast Reconstr Surg* 1984;1:117.

Sasaki GH. Refinements of tissue expansion. In Jurkiewicz MJ, Krizek TJ, Mathes SJ, Ariyan S (eds.), *Plastic surgery: Principles and practice.* St. Louis: Mosby Co., 1986, p. 325.

Sasaki GH. Scalp repair by tissue expansion. In Brent B (ed.), *The artistry of reconstructive surgery.* St. Louis: Mosby Co., 1987, p. 459.

Sasaki GH. Intraoperative sustained limited expansion (ISLE) as an immediate reconstructive technique. *Clin Plast Surg* 1987;14:563.

Sasaki GH. Intraoperative expansion as an immediate reconstructive technique. *Fac Plast Surg* 1988;5:362.

Sasaki GH. Tissue expansion. In Jurkiewicz JM, Krizek TJ, Mathes SJ, Ariyan S (eds.), *Plastic surgery. Principles and practice.* St. Louis: Mosby Co., 1990, p. 286.

Sasaki GH. Tissue expansion in reconstruction of acquired auricular defects. *Clin Plast Surg* 1990;17:317.

Sasaki GH, Pang CY. Pathophysiology of skin flaps raised on expanded pig skin. *Plast Reconstr Surg* 1984;74:59.

Satoh K, Shigehara T. Clinical trial of a prefabricated secondary hypogastric flap pedicled on the deep inferior epigastric vessel with or without a tissue expander in three patients. *Plast Reconstr Surg* 1995;96:905.

Saxby PJ. Inelastic tape to treat problems of tissue-expansion (letter). *Br J Plast Surg* 1988;41:666.

Saxby PJ. Survival of island flaps after tissue expansion: A pig model. *Plast Reconstr Surg* 1988;81:30.

Scheepers JH, Quaba AA. Tissue expansion in the treatment of tubular breast deformity. *Br J Plast Surg* 1992;45:529.

Schmidt SC, Logan SE, Hayden JM, et al. Continuous versus conventional tissue expansion: Experimental verification of a new technique. *Plast Reconstr Surg* 1991;87:10.

Schneider MS, Borkow JE, et al. The tensiometric properties of expanded guinea pig skin. Discussion. Austad, ED. *Plast Reconstr Surg* 1988;81:398.

Schneider MS, Wyatt DB, Konvolinka CW, et al. Comparison of rapid versus slow tissue expansion on skin-flap viability. *Plast Reconstr Surg* 1993;92:1126.

Schubert W, Guzman-Stein G, Cunningham BL. A new method for placement of tissue expanders. *Ann Plast Surg* 1988;20:191.

Schubert W, Shons AR. Intraoperative tissue expansion for rhytidectomy. *Plast Reconstr Surg* 1990;86:603.

Schwartz HC, Relle RJ. Extraoral placement of a subperiosteal tissue expander for reconstruction with hydroxylapatite of the severely atrophic mandibular alveolar ridge. *J Oral Maxillofac Surg* 1990;48:157.

Seckel BR, Hyland WT. Soft tissue expander for delayed and immediate breast reconstruction. *Surg Clin North Am* 1985;65:383.

Sellers DS, Miller SH, Demuth RJ, et al. Repeated skin expansion to resurface a massive thigh wound. *Plast Reconstr Surg* 1986;77:654.

Sellers DS, Miller SH, Demuth RJ. Tissue expansion as an adjunctive technique for the management of difficult wounds. *Am J Surg* 1986;151:603.

Sellers DS, Miller SH, Demuth RJ, et al. Use of ultrasound to locate a lost tissue expander injection port. *Plast Reconstr Surg* 1986;78:250.

Sengezer M, Sadove RC. Scrotal construction by expansion of labia majora in biological female transsexuals. *Ann Plast Surg* 1993;31:372.

Serra JM, Mesa F, Paloma V, Ballesteros A. Use of calf prosthesis and tissue expansion in aesthetic reconstruction of the leg. *Plast Reconstr Surg* 1992;89:684.

Serra JM, Muiraggui A, Recalde M. Reconstruction mamaria postmastectomia con expansores tissulares. *Rev Med Univ Navarra* 1985;29:84.

Serra JM, Samayoa V, Valiente E. Reconstruction mamaria y correccion de asimetrias mamarias mediante la protesis-expansora de Becker. *Rev Senol Patol Mamm* 1988;1:27.

Serra JM, Samayoa V, Valiente E, et al. Reconstruction mamaria postmastectomia mediante expansion tisular. *Rev Senol Patol Mamm* 1988;1:31.

Serra JM, Sanz J, Ballesteros A, et al. Surgical treatment for congenital absence of the vagina using tissue expansion. *Surg Gynecol Obstet* 1993;177:158.

Serra JM, Valiente E, Samayoa V. Expansion tisular en reconstruccion mamaria después de una mastectomia subcutanea antigua. *Rev Senol Patol Mamm* 1988;1:23.

Serup J, Northeved A. Skin elasticity in localized scleroderma introduction of a biaxial in vivo method for measurement of tensile distensibility. *J Dermatol* 1985;12:52.

Seyfer AE. Management of the chest wall deformity in male patients with Poland's syndrome. *Plast Reconstr Surg* 1991;87:679.

Seyfer AE, Icochea R, Graeber GM. Natural history and long-term results of chest wall reconstruction in 33 patients. *Ann Plast Surg* 1988;23:776.

Shafir R, Weiss J, Herman O. Use of the expander for estimating required volume for breast reconstruction. *Plast Reconstr Surg* 1989;83:397.

Sharpe DT, Burd RM. Tissue expansion in perspective. *Ann R Coll Surg Engl* 1989;71:175.

Sharpe DT, Palmer JH. Tissue expander tube exteriorization using a stylet. *Plast Reconstr Surg* 1989;84:363.

Shehadi SI, Donovan P, Steigman CK, Vogler CA, Vogler GA. Implantation and expansion of split-thickness skin grafts: A new source of prefabricated pedicle flaps and grafts. *Plast Reconstr Surg* 1994;93:1449.

Shenaq SM. Pretransfer expansion of a sensate lateral arm free flap. *Ann Plast Surg* 1987;19:558.

Shimotakahara SG, Larrabee WF, Jr. The effect of tissue expansion on previously irradiated skin (comment). *Arch Otolaryngol Head Neck Surg* 1992;118:534.

Shively RE. Skin expander volume estimator. *Plast Reconstr Surg* 1986;77:482.

Shively RE. Surface-area increase in tissue expansion. *Plast Reconstr Surg* 1988;82:838.

Shively RE, Bermant MA, Bucholz RD. Separation of craniopagous twins utilizing tissue expanders. *Plast Reconstr Surg* 1985;76:765.

Shuter ML, Malata CM, Duffy JS, Sharpe DT. Determination of relative contributions of "mechanical" and "biological" creep in tissue expansion using in vivo pressure monitoring: A preliminary report. *Med Eng Phys* 1994;16:24.

Siegel R. Silicone implant under a skin graft for breast reconstruction after radical mastectomy. *Plast Reconstr Surg* 1983;72:720.

Siegert R, Weerda H. Tissue expansion: I, Technical and physiologic principles. *HNO* 1994;42:124.

Siegert R, Weerda H. Skin expansion: II, Clinical use and complications. *HNO* 1994;42:182.

Siegert R, Weerda H, Hoffmann S, Mohadjer C. Clinical and experimental evaluation of intermittent intraoperative short-term expansion. *Plast Reconstr Surg* 1993;92:248.

Siegert R, Weerda H, Loffler JA, Baretton G. Epidermal proliferation rate after skin expansion in the dog model. *Laryngorhinootologie* 1994;73:206.

Silver L, Grant AD, Atar D, Lehman WB. Use of tissue expansion in clubfoot surgery. *Foot-Ankle* 1993;14:117.

Sinow JD, Cunningham BL. Postmastectomy brachial plexus injury exacerbated by tissue expansion. *Ann Plast Surg* 1991;27:368.

Sinow JD, Cunningham BL. Intraluminal lidocaine for analgesia after tissue expansion: A double-blind prospective trial in breast reconstruction. *Ann Plast Surg* 1992; 28:320.

Sinow JD, Halvorsen RA, Jr., Matts JP, et al. Chest-wall deformity after tissue expansion for breast reconstruction. *Plast Reconstr Surg* 1991;88:998.

Sinow JD, Halvorsen RA, Schubert W, et al. Chest-wall deformity with tissue expansion: A prospective longitudinal study. *Plast Surg Forum* 1989;12:174.

Skoulis TG, Lovice D, von-Fricken K, Terzis JK. Nerve expansion. The optimal answer for the short nerve gap. Behavioral analysis. *Clin Orthop* 1995;314:84.

Slator RC, Wilson GR, Sharpe DT. Postburn breast reconstruction: Tissue expansion prior to contracture release. *Plast Reconstr Surg* 1992;90:668.

Slavin SA. Improving the latissimus dorsi myocutaneous flap with tissue expansion. *Plast Reconstr Surg* 1994;93:811.

Slavin SA, Colen SR. Sixty consecutive breast reconstructions with the inflatable expander: A critical appraisal. *Plast Reconstr Surg* 1990;86:910.

Slavin SA, Colen SR. Analysis of risks and aesthetics in a consecutive series of tissue expansion breast reconstructions (discussion). *Plast Reconstr Surg* 1992;89:844.

Snyder CC. Discussion about "separation of craniopagous twins utilizing tissue expanders" by R.E. Shively. *Plast Reconstr Surg* 1985;76:773.

Snyderman RK, Guthrie RH. Reconstruction of the female breast following radical mastectomy. *Plast Reconstr Surg* 1971;47:565.

Spence RJ. Experience with novel uses of tissue expanders in burn reconstruction of the face and neck. *Ann Plast Surg* 1992;28:453.

Spilker G, Oeking G, Biemer E, et al. Möglichkeiten der Defektdeckung am Schädeldach nach Traumen oder ausgedehnten Tumorresektionen. *Handchirurgie* 1985;17: 247.

Spitz L, Capps SN, Kiely EM. Xiphoomphaloischiopagus tripus conjoined twins: Successful separation following abdominal wall expansion. *J Pediatr Surg* 1991;26:26.

Spodnick GJ, Pavletic MM, Clark GN, et al. Controlled tissue expansion in the distal extremities of dogs. *Vet Surg* 1993;22:436.

Squier CA. The stretching of mouse skin in vivo: Effect on epidermal proliferation and thickness. *J Invest Dermatol* 1980; 74:68.

Stark GB. Total reconstruction of scrotum by tissue expansion (letter). *Plast Reconstr Surg* 1991;87:192.

Stark GB. Histomorphologic findings in human expanded tissue. *Handchir Mikrochir Plast Chir* 1992;24:50.

Stark GB, Dorer A, Walgenbach KJ, et al. The creation of a small bowel pouch by tissue expansion—an experimental study in pigs. *Langenbecks Arch Chir* 1990;375:145.

Stark GB, Jaeger K. Failed tissue expansion in a patient with osteogenesis imperfecta. *Ann Plast Surg* 1989;22:156.

Stark HL. Directional variations in the extensibility of human skin. *Br J Plast Surg* 1977;30:105.

Stark HL, Hong C, Narayanan H. The use of tissue expander to elongate axial blood vessels. *Plast Surg Forum* 1986;9:151.

Stark HL, Hong C, Futrell JW. Rapid elongation of arteries and veins in rats with a tissue expander. *Plast Reconstr Surg* 1987;80:570.

Steenfos H, Tarnow P, Blomqvist G. Skin expansion. Long term follow up of complications and costs of care. *Scand J Plast Reconstr Surg Hand Surg* 1993;27:137.

Steinau HU, Askenazy L, Biemer E, Gradinger J. Bilateral arm amputation caused by high voltage burns in childhood: Stump reconstruction by tissue expansion and fibula transfer. *Chirurgie* 1991;62:426.

Stevens LA, MacGrath MH, Druss RG. The psychological impact of immediate breast reconstruction for women with early breast cancer. *Plast Reconstr Surg* 1984;73:19.

Still EF, Goodman RC. Total reconstruction of a two-compartment scrotum by tissue expansion. *Plast Reconstr Surg* 1990;85:805.

Stuffer M, Papp C. Reconstruction of large scar areas with controlled tissue expansion combined with fasciocutaneous flaps. *Burns* 1991;17:166.

Swanson NA, Argenta LC. Tissue expansion. *Adv Dermatol* 1988;3:243.

Tagliacozzi G. *De curtorum chirurgia per insitionem.* Venice, Italy: B. Bindonus, 1597.

Tanino K, Miyasaka M. Reconstruction of microtia using tissue expander. *Clin Plast Surg* 1990;17:339.

Tebbetts JB. Transaxillary subpectoral augmentation mammaplasty: Long-term follow-up and refinements. *Plast Reconstr Surg* 1984;74:636.

Teimourian B, Adham NM. Surgical correction of the tuberous breast. *Ann Plast Surg* 1983;10:190.

Ten Cate AR, Freeman E, Dickinson JB. Sutural development: Structure and its response to rapid expansion. *Am J Orthodont* 1977;71:622.

Thaw LC, Stout Thaw M. Trans-Africa Safari. *Natl Geogr* 1938;74:327.

Thompson AM, Chetty U. Appraisal of breast reconstruction with the tissue expansion technique. *J R Coll Surg Edinb* 1990;35:4.

Thomson HG. The fate of the pseudo sheath pocket around silicone implants. *Plast Reconstr Surg* 1973;51:667.

Thomson SJ, Tanner NS, Mercer DM, Web neck deformity anatomical consideration and option in surgical management. *Br J Plast Surg* 1990;43:94.

Thornton JW, Marks MW, Adson MH. Burn scar management: The role of tissue expander. *Plast Surg Forum* 1986; 9:151.

Thornton JW, Marks MW, Izenberg PH, et al. Expanded myocutaneous flaps: Their clinical use. *Clin Plast Surg* 1987; 14:529.

Timmenga EJ. Skin expansion. Thesis, University of Amsterdam, The Netherlands, 1992.

Timmenga EJ, Andreassen TT, Houthoff HJ, Klopper PJ. The effect of mechanical stress on healing skin wounds: An experimental study in rabbits using tissue expansion. *Br J Plast Surg* 1991;44:514.

Timmenga EJ, Das PK. Histomorphological observations on dermal repair in expanded rabbit skin: A preliminary report. *Br J Plast Surg* 1992;45:503.

Timmenga EJ, Schoorl R, Bos JD, et al. An improved model for tissue expansion and flap research in the rabbit. *Br J Plast Surg* 1989;42:301.

Timmenga EJ, Schoorl R, Klopper PJ. Biomechanical and histomorphological changes in expanded rabbit skin. *Br J Plast Surg* 1990;43:101.

Tominaga K, Matsuo T, Kuga Y, Mizuno A. An animal model for subperiosteal tissue expansion. *J Oral Maxillofac Surg* 1993;51:1244.

Tominaga K, Matsuo T, Kuga Y, Mizuno A. Subperiosteal tissue expansion for mandibular augmentation with hydroxylapatite particles: An experimental study. *J Oral Maxillofac Surg* 1994;52:945.

Tonnard PL, Monstrey SJ, Van-Landuyt KH, Vermeersch HK, Matton G. Scar-length reduction by ring-shaped expansion. *Ann Plast Surg* 1994;33:647.

Toranto IR. Two-stage correction of tuberous breasts. *Plast Reconstr Surg* 1981;67:642.

Toth BA, Glafkides MC, Wandel A. The role of tissue expansion in the treatment of atypical facial clefting. *Plast Reconstr Surg* 1990;86:119.

Trott JA. A preliminary report on a strategy for treatment of male pattern baldness: Bilateral vertical flaps plus tissue expansion. *Br J Plast Surg* 1993;46:611.

Tse DT, McCafferty LR. Controlled tissue expansion in periocular reconstructive surgery. *Ophthalmology* 1993;100:260.

Tsur H, Winkler E, Kessler A. Oblique facial cleft with anaophthalmia in mentally normal child. *Ann Plast Surg* 1991;26:449.

Tucker SM, Sapp N, Collin R. Orbital expansion of the congenitally anophthalmic socket. *Br J Ophthalmol* 1995;79:667.

Unger WP, Nordström REA. *Hair Transplantation*, New York: Marcel Dekker, 1988.

Valauri FA, Buncke HJ, Alpert BS, et al. Microvascular transplantation of expanded free scalp flaps between identical twins. *Plast Reconstr Surg* 1990;85:432.

Van Damme PA, Heidbüchel KLWM, Kuijpers-Jagtman AM, et al. Cranio-maxillo-facial tissue expansion experimentally based or clinically empiric? A review of the literature. *J Craniomaxillofac Surg* 1992;20:61.

Van Damme PA. Soft-tissue expansion in cleft palate surgery. *Plast Reconstr Surg* 1994;93:1307.

Van Damme PA, Freihofer HP, Van't-Hof MA, Kuijpers-Jagtman AM, Maltha JC, Spijkers JM. Radiologic analysis of the effects of subperiosteal palatal soft-tissue expansion in growing cats. *Int J Oral Maxillofac Surg* 1994;23:393.

Van Damme PH. Surgical treatment of cicatricial alopecia of the hairy scalp using tissue expansion (letter). *Ned Tijdschr Geneeskd* 1992;136:1034.

Van der Kolk CA, McCann JJ, Knight KR, et al. Some further characteristics of expanded tissue. *Clin Plast Surg* 1987; 14:447.

Van der Kolk CA, McCann JJ, Mitchell GM, et al. Changes in area and thickness of expanded and unexpanded axial pattern skin flaps in pigs. *Br J Plast Surg* 1988;41:294.

Van der Wey LP, Gabreels-Festen AA, Merks MH, Polder TW, Stegeman DF, Spauwen PH, Gabreels FJ. Peripheral nerve elongation by laser Doppler flowmetry controlled expansion: Morphological aspects. *Acta Neuropathol Berl* 1995;89:166.

Van Heerden JA, Jackson IT, Martin JK, Jr., Fisher J. Surgical technique and pitfalls of breast reconstruction immediately after mastectomy for carcinoma: Initial experience. *Mayo Clinic Proc* 1987;62:185.

Van Rappard JHA. Controlled tissue expansion in reconstructive surgery. Thesis, University of Groningen, The Netherlands, 1988.

Van Rappard JHA. Capsule formation in controlled tissue-expansion. In van Rappard JHA (ed.), *Controlled tissue-expansion in reconstructive surgery* Nijmegen: SSN Press, 1988, pp. 51–88.

Van Rappard JHA, Bauer FW, Grubben MJAL, et al. Epidermopoiesis in controlled tissue-expansion. In Van Rappard JHA (ed.), *Controlled tissue-expansion in reconstructive surgery* Nijmegen: SSN Press, 1988, pp. 41–49.

Van Rappard JHA, Jerusalem C, Sonneveld GJ,Borghouts JMHM. Histologic changes in soft tissue due to tissue expansion. *Fac Plast Surg* 1988;5:280.

Van Rappard JHA, Molenaar J, van Doorn K, et al. Surface area increase in tissue expansion. Discussion: Shively, R. *Plast Reconstr Surg* 1988;82:833.

Van Rappard JHA, Sonneveld GJ, Borghouts JMHM. Geometric planning and the shape of the expander. *Fac Plast Surg* 1988;5:287.

Van Veen R, Pulles HJ. The surgical treatment of cicatrix alopecia of the hairy scalp using tissue expansion. *Ned Tijdschr Geneeskd* 1992;136:627.

Van Wey LP, Polder TW, Hoogbergen MM, Spauwen PH. A model for monitoring nerve blood flow during expansion by laser Doppler flowmetry in the rabbit. *J Neurol Sci* 1993;117:79.

Vecchione TR. A method of recontouring the domed nipple. *Plast Reconstr Surg* 1976;57:30.

Vergnes P, Taieb A, Maleville J, et al. Repeated skin expansion for excision of congenital giant nevi in infancy and childhood. *Plast Reconstr Surg* 1993;91:450.

Vergote T, Arnaud E. Forum on tissue expansion. Neovascularization of skin flap from expanded anatomical arteriovenous pedicle. An initial experimental study. *Ann Chir Plast Esthet* 1993;38:69.

Vergote T, Revol M, Servant JM, Banzet P. Microsurgical expanded musculocutaneous flaps of the latissimus dorsi. *Ann Chir Plast Esthet* 1994;38:323.

Versaci AD. Refinements in reconstruction of congenital breast deformities (discussion). *Plast Reconstr Surg* 1985; 76:81.

Versaci AD. A method of reconstructing a pendulous breast utilizing the tissue expander. *Plast Surg Forum* 1986;9:14.

Versaci AD. A method of reconstructing a pendulous breast utilizing the tissue expander. *Plast Reconstr Surg* 1987; 80:387.

Versaci AD. Reconstruction of a pendulous breast utilizing a tissue expander. *Clin Plast Surg* 1987;14:499.

Versaci AD. Reconstruction of a pendulous breast utilizing a shaped expander. *Ann Plast Surg* 1989;23:469.

Versaci AD. Treatment of tuberous breasts utilizing tissue expansion. *Aesthet Plast Surg* 1991;15:307.

Versaci AD, Balkovich ME. Tissue expansion: History advances. *Plast Reconstr Surg* 1984;1:95.

Versaci AD, Balkovich ME, Goldstein SA. Breast reconstruction by tissue expansion for congenital and burn deformities. *Ann Plast Surg* 1986;16:20.

Victor WH, Hurwitz M, Gruss JL. The development of a new tissue expander for use in ophthalmic plastic surgery. *Ophtal Surg* 1986;17:661.

Victor WH, Hurwitz JJ. Cicatricial ectropion following blepharoplasty: Treatment by tissue expansion. *Can J Ophthalmol* 1984;19:317.

Vinnik CA, Jacob SW. Dimethylsulfoxide (DMSO) for human single-stage intraoperative tissue expansion and circulatory enhancement. *Aesthet Plast Surg* 1991;15:327.

Vistnes LM. Discussion about "tissue expansion in soft tissue reconstruction" by C. Radovan. *Plast Reconstr Surg* 1984; 74:491.

Von Heimburg D, Lemperle G, Dippe B, Kruger S. Free transplantation of fat autografts expanded by tissue expanders in rats. *Br J Plast Surg* 1994;47:470.

Von Smitten K, Sundell B. The impact of adjuvant radiotherapy and cytotoxic chemotherapy on the outcome of immediate breast reconstruction by tissue expansion after mastectomy for breast cancer. *Eur J Surg Oncol* 1992; 18:119.

Von Szalay L. Reconstruction of the burned face using skin expanders. *HNO* 1985;33:413.

Von Szalay L. Scar correction by tissue expander. *Handchir Mikrochir Plast Chir* 1985;17:61.

Wang CZ, Niu XT, Cheng DM. Experimental study and clinical use of rapid expansion of skin. *Chung Hua Cheng Hsing Shao Shang Wai Ko Tsa Chih* 1994;10:30.

Wang G, Jiang C, Ruan K. Study on clinical application of expanded skin tubes. *Chung Hua Cheng Hsing Shao Shang Wai Ko Tsa Chih* 1994;10:26.

Wang SC, Chen FL, Li JY. Soft tissue expansion in the treatment of scar alopecia in preschool children. *Chin Med J Engl* 1991;104:164.

Ward J, Cohen IK, Knaysi GA, et al. Immediate breast reconstruction with tissue expansion. *Plast Reconstr Surg* 1987; 80:559.

Webster HR, Marshall DR. Nasal augmentation in chondrodysplasia punctata using tissue expansion. *Br J Plast Surg* 1991;44:384.

Wee SS, Logan SE, Mustoe TA. Continuous versus intraoperative expansion in the pig model. *Plast Reconstr Surg* 1992; 90:808.

Weeks GS. Into the heart of Africa. *Natl Geogr* 1956;110:257.

Weiner SR, Paulus HE. Chronic arthropathy occurring after augmentation mammaplasty. *Plast Reconstr Surg* 1986; 77:185.

Weinsten GD, Boucek RJ. Collagen and elastin of human dermis. *J Invest Dermatol* 1960;35:227.

Weinzberg N, Dowden RV, Stulberg BN. The use of tissue expansion to allow reconstruction of the knee. A case report. *JBJS* 1987;69:1238.

Whitney TM, Ramasastry S, Futrell JW. Combined tissue expansion and free tissue transfer for reconstruction of the hand in epidermolysis bullosa-associated malignancy. *Ann Plast Surg* 1993;31:552.

Whitton JT. New values for epidermal thickness and their importance. *Health Physiol* 1973;24:436.

Whitton JT, Everall JO. The thickness of the epidermis. *Br J Dermatol* 1973;89:467.

Wickman M. Comparison between rapid and slow tissue expansion in breast reconstruction. *Plast Reconstr Surg* 1993; 91:663.

Wickman M. Rapid versus slow tissue expansion for breast reconstruction: A three-year follow-up. *Plast Reconstr Surg* 1995;95:712.

Wickman M, Hedén P, Jurell G. Circulatory and metabolic changes in expanded pig skin flaps. *Plast Reconstr Surg* 1991;88:650.

Wickman M, Johansson O, Forslind B. Dimensions of capsular collagen fibrils: Image analysis of rapid compared with slow tissue expansion for breast reconstruction. *Scand J Plast Surg Hand Surg* 1992;26:281.

Wickman M, Olenius M, Malm M, et al. Alterations in skin properties during rapid and slow tissue expansion for breast reconstruction. *Plast Reconstr Surg* 1992;90:945.

Wiese KG. Osmotically induced tissue expansion with hydrogels: A new dimension in tissue expansion? A preliminary report. *J Craniomaxillofac Surg* 1993;21:309.

Wieslander JB. Repeated tissue expansion in reconstruction of a huge combined scalp-forehead avulsion injury. *Ann Plast Surg* 1988;20:347.

Wieslander JB. Tissue expansion in functional and aesthetic reconstruction of the trunk and extremities. Case report. *Scand J Plast Reconstr Surg Hand Surg* 1991;25:285.

Wieslander JB. Tissue expansion in the head and neck. A 6-year review. *Scand J Plast Reconstr Surg Hand Surg* 1991;25:47.

Wieslander JB. Tissue expansion a revolutionary technique in the reconstructive surgery. *Lakartidningen* 1991;88:2224.

Wieslander JB. Preservation of bone length in traumatic amputations. Tissue expansion may prevent reamputation. *Lakartidningen* 1995;92:1215.

Wilk A, Rodier-Bruant C, Benyacoub N, Herman D. Tissue expansion in mammary reconstruction and asymmetry. Apropos of 24 prostheses. *Ann Chir Plast Esthet* 1994;39:221.

Wilke CP, Woods JE. Use of tissue expanders in reconstruction after excision of multiple large silicone granulomas. *Ann Plast Surg* 1993;30:367.

Williams JE, Hoffman S. Mammaplasty for tubular breasts. *Aesthet Plast Surg* 1981;5:51.

Wilmshurst AD, Sharpe DT. Immediate placement of tissue expanders in the management of large excisional defects on the face. *Br J Plast Surg* 1990;43:150.

Wiss K. Lasers, tissue expansion and sun protection in pediatric dermatology. *Cutis* 1990;45:331.

Wittkampf ARM. Augmentation of the maxillary alveolar ridge with hydroxylapatite and fibrin glue. *J Oral Maxillofac Surg* 1988;46:1019.

Wittkampf ARM. Fibrin glue as a cement for HA-granules. *J Craniomaxillofac Surg* 1989;17:179.

Wittkampf ARM. Short-term experience with the subperiosteal tissue expander in reconstruction of the mandibular alveolar rigde. *J Oral Maxillofac Surg* 1989;47:469.

Wittkampf ARM. Placement of tissue expanders on the mandibular ridge (letter comment). *J Oral Maxillofac Surg* 1990;48:1352.

Wood FM, McMahon SB. The response of the peripheral nerve field to controlled soft tissue expansion. *Br J Plast Surg* 1989;42:682.

Wood RJ, Adson MH, VanBeek AL, et al. Controlled expansion of peripheral nerves: Comparison of nerve grafting and nerve expansion/repair for canine sciatic nerve defects. *J Trauma* 1991;31:686.

Woods JE, Mangan MA. Breast reconstruction with tissue expanders: Obtaining an optimal result. *Ann Plast Surg* 1992; 28:390.

Worthen E. Surgical treatment of electrical burns of the scalp and skull past and present. *Clin Plast Surg* 1982;9:161.

Worton EW, Seifert LN, Sherwood R. Late leakage of inflatable silicone breast prostheses. *Plast Reconstr Surg* 1980;65:302.

Wyllie F, Gowar JP, Levick PL. Use of tissue expanders after burns and other injuries. *Burns* 1986;12:277.

Xiao NK. Uses of fibronectin on eschar-excised wound to promote expansion of skin grafts. *Chung Hua Cheng Hsing Shao Shang Wai Ko Tsa Chih* 1992;8:130.

Yang CC. Pitfalls in percutaneous filling of the tissue expander. *Plast Reconstr Surg* 1986;77:339.

Yuan RTW. The use of tissue expansion in lower extremity wounds in paraplegic patients. *Plast Reconstr Surg* 1989; 83:892.

Zellweger G, Kunzi W. Tissue expanders in reconstruction of burn sequelae. *Ann Plast Surg* 1991;26:380.

Zha YK. Complication of soft tissue expander dilation and its prevention. *Chung Hua Cheng Hsing Shao Shang Wai Ko Tsa Chih* 1990;6:113.

Zha YK. An evaluation of problems arising from the use of soft tissue expander. *Chung Hua Cheng Hsing Shao Shang Wai Ko Tsa Chih* 1993;9:259.

Zhong DC, Lu KH, Ai YF. Total nose reconstruction by expanded forehead flap. *Chung Hua Cheng Hsing Shao Shang Wai Ko Tsa Chih* 1994;10:178.

Zhou XT. Effects of tissue expansion on the ratio of length to width of random-pattern skin flaps. *Chung Hua Cheng Hsing Shao Shang Wai Ko Tsa Chih* 1989;27:418.

Zide BM, Karp NS. Maximizing gain from rectangular tissue expanders. *Plast Reconstr Surg* 1992;90:500.

Zoltie N, Chapman P, Joss G. Tissue expansion: A unit review of nonscalp nonbreast expansion. *Br J Plast Surg* 1990; 43:325.

Index